COMPUTATIONAL PHYLOGENETICS
An Introduction to Designing Methods for Phylogeny Estimation

A comprehensive account of both basic and advanced material in phylogeny estimation, focusing on computational and statistical issues. No background in biology or computer science is assumed, and there is minimal use of mathematical formulas, meaning that students from many disciplines, including biology, computer science, statistics, and applied mathematics, will find the text accessible.

The mathematical and statistical foundations of phylogeny estimation are presented rigorously, following which more advanced material is covered. This includes substantial chapters on multi-locus phylogeny estimation, supertree methods, Markov models of sequence evolution, multiple sequence alignment techniques, and designing methods for large-scale phylogeny estimation. The author provides key analytical techniques to prove theoretical properties about methods, as well as addressing performance in practice for methods for estimating trees. Research problems requiring novel computational methods are also presented, so that graduate students and researchers from varying disciplines will be able to enter the broad and exciting field of computational phylogenetics.

TANDY WARNOW is a Founder Professor of Engineering at the University of Illinois at Urbana-Champaign. Her awards include the National Science Foundation Young Investigator Award (1994), the David and Lucile Packard Foundation Award in Science and Engineering (1996), a Radcliffe Institute for Advanced Study Fellowship (2003), and a John Simon Guggenheim Memorial Foundation Fellowship (2011). She was elected a Fellow of the Association for Computing Machinery (ACM) in 2006, and of the International Society for Computational Biology (ISCB) in 2017.

COMPUTATIONAL PHYLOGENETICS

An Introduction to Designing Methods for Phylogeny Estimation

TANDY WARNOW
University of Illinois, Urbana-Champaign

CAMBRIDGE
UNIVERSITY PRESS

University Printing House, Cambridge CB2 8BS, United Kingdom

One Liberty Plaza, 20th Floor, New York, NY 10006, USA

477 Williamstown Road, Port Melbourne, VIC 3207, Australia

4843/24, 2nd Floor, Ansari Road, Daryaganj, Delhi – 110002, India

79 Anson Road, #06–04/06, Singapore 079906

Cambridge University Press is part of the University of Cambridge.

It furthers the Universitys mission by disseminating knowledge in the pursuit of education, learning, and research at the highest international levels of excellence.

www.cambridge.org
Information on this title: www.cambridge.org/9781107184718
DOI: 10.1017/9781316882313

© Cambridge University Press 2018

This publication is in copyright. Subject to statutory exception and to the provisions of relevant collective licensing agreements, no reproduction of any part may take place without the written permission of Cambridge University Press.

First published 2018

Printed in the United Kingdom by TJ International Ltd. Padstow Cornwall

A catalogue record for this publication is available from the British Library.

ISBN 978-1-107-18471-8 Hardback

Cambridge University Press has no responsibility for the persistence or accuracy of URLs for external or third-party internet websites referred to in this publication and does not guarantee that any content on such websites is, or will remain, accurate or appropriate.

Dedicated to Eugene Leighton (Gene) Lawler (1933–1994), my doctoral advisor, whose enthusiasm and generosity inspired me throughout graduate school, and who introduced me to the field of computational phylogenetics.

Contents

Preface	*page* xiii
Glossary	xvii
Notation	xviii

PART I   BASIC TECHNIQUES — 1

1 Brief Introduction to Phylogenetic Estimation — 3
- 1.1   The Cavender–Farris–Neyman Model — 4
- 1.2   An Analogy: Determining Whether a Coin is Biased Toward Heads or Tails — 6
- 1.3   Estimating the Cavender–Farris–Neyman Tree — 7
- 1.4   Some Comments about the CFN Model — 16
- 1.5   Phylogeny Estimation Methods Used in Practice — 16
- 1.6   Measuring Error Rates on Simulated Datasets — 18
- 1.7   Getting Branch Support — 20
- 1.8   Using Simulations to Understand Methods — 20
- 1.9   Genome-Scale Evolution — 23
- 1.10 Designing Methods for Improved Accuracy and Scalability — 24
- 1.11 Summary — 24
- 1.12 Review Questions — 26
- 1.13 Homework Problems — 27

2 Trees — 29
- 2.1   Introduction — 29
- 2.2   Rooted Trees — 29
- 2.3   Unrooted Trees — 35
- 2.4   Constructing the Strict Consensus Tree — 41
- 2.5   Quantifying Error in Estimated Trees — 41
- 2.6   The Number of Binary Trees on n Leaves — 43
- 2.7   Rogue Taxa — 43
- 2.8   Difficulties in Rooting Trees — 44

	2.9	Homeomorphic Subtrees	45
	2.10	Some Special Trees	45
	2.11	Further Reading	46
	2.12	Review Questions	47
	2.13	Homework Problems	47
3	**Constructing Trees from True Subtrees**		**51**
	3.1	Introduction	51
	3.2	Tree Compatibility	51
	3.3	The Algorithm of Aho, Sagiv, Szymanski, and Ullman: Constructing Rooted Trees from Rooted Triples	52
	3.4	Constructing Unrooted Binary Trees from Quartet Subtrees	53
	3.5	Testing Compatibility of a Set of Trees	56
	3.6	Further Reading	57
	3.7	Review Questions	58
	3.8	Homework Problems	58
4	**Constructing Trees from Qualitative Characters**		**61**
	4.1	Introduction	61
	4.2	Terminology	62
	4.3	Tree Construction Based on Maximum Parsimony	63
	4.4	Constructing Trees from Compatible Characters	69
	4.5	Tree Construction Based on Maximum Compatibility	72
	4.6	Treatment of Missing Data	75
	4.7	Informative and Uninformative Characters	75
	4.8	Further Reading	77
	4.9	Review Questions	78
	4.10	Homework Problems	78
5	**Distance-based Tree Estimation Methods**		**83**
	5.1	Introduction	83
	5.2	UPGMA	84
	5.3	Additive Matrices	86
	5.4	Estimating Four-Leaf Trees: The Four Point Method	87
	5.5	Quartet-based Methods	89
	5.6	Neighbor Joining	91
	5.7	Distance-based Methods as Functions	92
	5.8	Optimization Problems	94
	5.9	Minimum Evolution	95
	5.10	The Safety Radius	96
	5.11	Comparing Methods	99
	5.12	Further Reading	100
	5.13	Review Questions	103
	5.14	Homework Problems	104

6	**Consensus and Agreement Trees**		109
	6.1	Introduction	109
	6.2	Consensus Trees	109
	6.3	Agreement Subtrees	116
	6.4	Clustering Sets of Trees	117
	6.5	Further Reading	117
	6.6	Review Questions	118
	6.7	Homework Problems	118
7	**Supertrees**		121
	7.1	Introduction	121
	7.2	Compatibility Supertrees	123
	7.3	Asymmetric Median Supertrees	123
	7.4	Robinson–Foulds Supertrees	124
	7.5	Matrix Representation with Parsimony	126
	7.6	Matrix Representation with Likelihood	128
	7.7	Quartet-based Supertrees	128
	7.8	The Strict Consensus Merger	132
	7.9	SuperFine: A Meta-Method to Improve Supertree Methods	135
	7.10	Further Reading	139
	7.11	Review Questions	142
	7.12	Homework Problems	142

	PART II MOLECULAR PHYLOGENETICS		143
8	**Statistical Gene Tree Estimation Methods**		145
	8.1	Introduction to Statistical Estimation in Phylogenetics	145
	8.2	Models of Site Evolution	146
	8.3	Model Selection	151
	8.4	Distance-based Estimation	152
	8.5	Calculating the Probability of a Set of Sequences on a Model Tree	154
	8.6	Maximum Likelihood	157
	8.7	Bayesian Phylogenetics	159
	8.8	Statistical Properties of Maximum Parsimony and Maximum Compatibility	161
	8.9	The Impact of Taxon Sampling on Phylogenetic Estimation	164
	8.10	Estimating Branch Support	165
	8.11	Beyond Statistical Consistency: Sample Complexity	167
	8.12	Absolute Fast Converging Methods	167
	8.13	Heterotachy and the No Common Mechanism Model	170
	8.14	Further Reading	172
	8.15	Review Questions	173

	8.16	Homework Problems	174
9	**Multiple Sequence Alignment**		**178**
	9.1	Introduction	178
	9.2	Evolutionary History and Sequence Alignment	180
	9.3	Computing Differences Between Two Multiple Sequence Alignments	180
	9.4	Edit Distances and How to Compute Them	184
	9.5	Optimization Problems for Multiple Sequence Alignment	190
	9.6	Sequence Profiles	194
	9.7	Profile Hidden Markov Models	198
	9.8	Reference-based Alignments	204
	9.9	Template-based Methods	205
	9.10	Seed Alignment Methods	206
	9.11	Aligning Alignments	207
	9.12	Progressive Alignment	209
	9.13	Consistency	212
	9.14	Weighted Homology Pair Methods	213
	9.15	Divide-and-Conquer Methods	214
	9.16	Co-estimation of Alignments and Trees	215
	9.17	Ensembles of HMMs	220
	9.18	Consensus Alignments	224
	9.19	Discussion	226
	9.20	Further Reading	227
	9.21	Review Questions	231
	9.22	Homework Problems	231
10	**Phylogenomics: Constructing Species Phylogenies from Multi-Locus Data**		**234**
	10.1	Introduction	234
	10.2	The Multi-Species Coalescent Model (MSC)	235
	10.3	Using Standard Phylogeny Estimation Methods in the Presence of ILS	238
	10.4	Probabilities of Gene Trees under the MSC	239
	10.5	Coalescent-based Methods for Species Tree Estimation	241
	10.6	Improving Scalability of Coalescent-based Methods	253
	10.7	Species Tree Estimation under Duplication and Loss Models	254
	10.8	Constructing Trees in the Presence of Horizontal Gene Transfer	259
	10.9	Phylogenetic Networks	260
	10.10	Further Reading	268
	10.11	Review Questions	272
	10.12	Homework Problems	272
11	**Designing Methods for Large-Scale Phylogeny Estimation**		**274**
	11.1	Introduction	274
	11.2	Standard Approaches	274

11.3	Introduction to Disk-Covering Methods (DCMs)	279
11.4	DCMs that Use Distance Matrices	282
11.5	Tree-based DCMs	285
11.6	Recursive Decompositions of Triangulated Graphs	288
11.7	Creating Multiple Trees	288
11.8	DACTAL: A General Purpose DCM	289
11.9	Triangulated Graphs	293
11.10	Further Reading	296
11.11	Review Questions	297
11.12	Homework Problems	298
Appendix A	**Primer on Biological Data and Evolution**	299
Appendix B	**Algorithm Design and Analysis**	304
Appendix C	**Guidelines for Writing Papers About Computational Methods**	327
Appendix D	**Projects**	331
References		339
Index		376

Preface

Overview

The evolutionary history of a set of genes, species, or individuals provides a context in which biological questions can be addressed. For this reason, phylogeny estimation is a fundamental step in many biological studies, with many applications throughout biology, such as protein structure and function prediction, analyses of microbiomes, inference of human migrations, etc. In fact, there is a famous saying by Dobzhansky that "Nothing in biology makes sense except in the light of evolution" (Dobzhansky, 1973).

Because phylogenies represent what has happened in the past, they cannot be directly observed but rather must be estimated. Consequently, increasingly sophisticated statistical models of sequence evolution have been developed, and are now used to estimate phylogenetic trees. Indeed, over the last few decades, hundreds of software packages and novel algorithms have been developed for phylogeny estimation, and this influx of computational approaches into phylogenetic estimation has transformed systematics. The availability of sophisticated computational methods, fast computers and high-performance computing (HPC) platforms, and large sequence datasets enabled through DNA sequencing technologies, has led to the expectation that highly accurate large-scale phylogeny estimation, potentially answering open questions about how life evolved on earth, should be achievable.

Yet large-scale phylogeny estimation turns out to be much more difficult than expected, for multiple reasons. First, all the best methods are computationally intensive, and standard techniques do not scale well to large datasets; for example, maximum likelihood phylogeny estimation is NP-hard, so exact solutions cannot be found efficiently (unless P = NP), and Bayesian MCMC methods can take a long time to reach stationarity. While massive parallelism can ameliorate these challenges to some extent, it doesn't really address the basic challenge inherent in searching an exponential search space. However, another issue is that the statistical models of sequence evolution that properly address genomic data are substantially more complex than the ones that model individual loci, and methods to estimate genome-scale phylogenies are (relatively speaking) in their infancy compared to methods for single gene phylogenies. Finally, there is a substantial gap between performance as suggested by mathematical theory (which is used to establish guarantees about methods under statistical models of evolution) and how well the methods actually perform on data – even on data generated under the same statistical models! Indeed, this gap is one of the most interesting things about doing research in computational phylogenetics, because it means that the most impactful research in the area must draw on mathematical theory (especially probability theory and graph theory) as well as on observations from data.

The main goal of this text is to enable researchers (typically graduate students in computer science, applied mathematics, or statistics) to be able to contribute new methods for phylogeny estimation, and in particular to develop methods that are capable of providing improved accuracy for large heterogeneous datasets that are characteristic of the types of inputs that are increasingly of interest in practice. The secondary goal is to enable biologists to understand the methods and their statistical guarantees under these models of evolution, so that they can select appropriate methods for their datasets, and select appropriate datasets given the available methods.

Some of the material in the textbook is fairly mathematical, and presumes undergraduate coursework in discrete mathematics and algorithm design and analysis. However, no background in biology is assumed, and the assumed statistics background is relatively lightweight. While some students without the expected background in computer science may find it difficult to understand some of the proofs, my goal has been to enable all students to understand the theoretical guarantees for phylogeny estimation methods and the statistical models on which they are based, so that they can adequately critique the scientific literature, and also choose methods and datasets that are best able to address the scientific questions they wish to answer.

Outline of the Textbook

Part I provides the "Discrete Mathematics for Phylogenetics" foundations for the textbook; the concepts and mathematics introduced in this part are the building blocks for algorithm design in phylogenetics, especially for developing methods that can scale to large datasets; understanding these concepts makes it possible to understand theoretical guarantees of methods under statistical models of evolution. Chapter 1 introduces the major themes involved in computational phylogenetics, addressing both theory (e.g., statistical consistency under a statistical model of evolution) and performance on both simulated and biological data. This chapter uses the Cavender–Farris–Neyman model of binary sequence evolution since understanding issues in analyzing data generated by this very simple model is helpful to understanding statistical estimation under the commonly used models of molecular sequence evolution. Chapter 2 introduces trees as graph-theoretic objects, and presents different representations of trees that will be useful for method development. Chapters 3, 4, and 5 present different types of methods for phylogenetic tree estimation (based on combining subtrees, using character data, or using distances, respectively). Chapter 6 presents methods for analyzing sets of trees, each on the same set of taxa, and for computing consensus trees and agreement subtrees; it also discusses how these methods are used to estimate support for different phylogenetic hypotheses. Chapter 7 examines the topic of supertree estimation, where the input is a set of trees on overlapping sets of taxa and the objective is a tree on the full set of taxa. Supertree methods are of interest in their own right and also because they are key algorithmic ingredients in divide-and-conquer methods, a topic we return to in Chapter 11.

Part II of the textbook is concerned with molecular phylogenetics. Chapter 8 presents commonly used statistical models of molecular sequence evolution and statistical methods for phylogeny estimation under these models. However, standard sequence evolution models do not include events such as insertions, deletions, and duplications, which can change the sequence length. These are very common processes, so biological sequences are usually of different lengths and must first be put into a *multiple sequence alignment* before they can be analyzed using phylogeny estimation methods; the subject of how to compute a multiple sequence alignment is covered in Chapter 9. Constructing a species tree or even a phylogenetic network from different gene trees in the presence of gene tree heterogeneity due to incomplete lineage sorting, gene duplication and loss, horizontal gene transfer, or hybridization is a fascinating research area that we present in Chapter 10. We end with Chapter 11, which addresses method development for estimating trees on large datasets. Large-scale phylogeny estimation is increasingly important since nearly all good approaches to phylogeny and multiple sequence alignment estimation are computationally intensive (either heuristics for NP-hard optimization problems or Bayesian methods), and many large datasets are being assembled that cannot be accurately analyzed using existing methods.

Each chapter ends with a set of review questions and homework problems. The review questions are easy to answer and do not require any significant problem solving or calculation. The homework problems are largely pen and paper problems that reinforce the mathematical content of the text.

The textbook comes with four appendices. Appendix A provides an introduction to biological evolution and data; the textbook can be read without it, but the reader who wishes to analyze biological data will benefit from this material. Appendix B provides an introduction to algorithm design and analysis; this material is not necessary for students with undergraduate computer science backgrounds, but may be a helpful introduction for students without this background. Appendix C provides some guidelines about how to write papers that introduce new methods or evaluate existing methods. Appendix D provides computational projects ranging from short term (i.e., a few days) to research projects that could lead to publications. In fact, several of the final projects for my Computational Phylogenetics courses have grown into journal publications (e.g., Bayzid et al. (2014); Zimmermann et al. (2014); Davidson et al. (2015); Chou et al. (2015); Nute and Warnow (2016)).

I wish to thank my editor, David Tranah, for his detailed and insightful comments on the many earlier versions of the text. I also wish to thank my students, colleagues, and family members who gave helpful criticism, including Edward Braun, Sarah Christensen, Steve Evans, Dan Gusfield, Joseph Herman, Ally Kaminsky, Laura Kubatko, Ari Löytynoja, Siavash Mirarab, Erin Molloy, Luay Nakhleh, Mike Nute, David Posada, Bhanu Renukuntla, Ehsan Saleh, Erfan Sayyari, Kimmen Sjölander, Travis Wheeler, and Mark Wilkinson. The several anonymous reviewers also gave very useful comments.

The images of the Monterey Cypress tree on the front and back covers are in honor of the CIPRES project (www.phylo.org), an NSF-funded project for phylogenetic research that I co-led with Bernard Moret from 2003–2010. Many of the algorithmic advances discussed in the text came out of research supported by CIPRES.

Glossary

afc: Absolute fast-converging methods

ASSU: The algorithm by Aho, Sagiv, Szymanski, and Ullman for determining if a set of rooted triplet trees is compatible, and constructing the compatibility tree if it exists

centroid edge: An edge in a tree T whose deletion defines a decomposition of the leafset into two parts that is as close to balanced as possible

c-gene: A region within a set of genomes that is recombination-free

GTR: General Time Reversible model

GTR-GAMMA: The GTR model with gamma-distributed rates across sites

HMM: Hidden Markov model

homologous: Two sequences are homologous if they have descended from a common ancestor

homoplasy: Evolution with back-mutation or parallel evolution

indels: Insertions and deletions

JC69: Jukes–Cantor model

K2P: Kimura 2-parameter model

MRCA: Most recent common ancestor

MSC: Multi-species coalescent model

NP: The class of decision (i.e., yes/no) problems for which the yes-instances can be verified in polynomial time

NP-hard: A problem that is at least as difficult as the hardest problems in the class NP

NP-complete: A decision problem that is NP-hard and also in the class NP

polytomy: A node in an unrooted tree of degree greater than three, or a node in a rooted tree with more than two children

Notation

- λ: The empty string
- $ab|cd$: Quartet tree on leafset a,b,c,d with one internal edge separating a,b from c,d
- $(a,(b,c))$: Rooted tree on three leaves a,b,c in which b and c are siblings
- $Clades(T)$: The set of clades of a rooted tree T, where a clade is the set of leaves below some internal node in T
- $C(T)$: The set of bipartitions on the leafset induced by edge deletions in a tree T
- $C_I(T)$: The set of non-trivial bipartitions on the leafset induced by deletions of internal edges in a tree T
- $\mathscr{L}(T)$: The set of leaves of a tree T
- $L_\infty(\mathbf{M},\mathbf{M}')$: For matrices $\mathbf{M}$ and $\mathbf{M}'$ with the same dimensions, this is $max_{ij}|M_{ij} - M'_{ij}|$
- $\mathbf{M}[i,j]$: For matrix $\mathbf{M}$, this is the entry in row i and column j. This is also denoted by M_{ij}.
- $MP(T,M)$: The maximum parsimony score of a tree T given the character matrix $\mathbf{M}$
- $Q(T)$: The set of homeomorphic unrooted quartet trees induced by T on its leafset
- $Q_r(T)$: The set of unrooted fully resolved (i.e., binary) quartet trees in $Q(T)$
- $|S|$: The number of elements in the set S
- $S \setminus S'$: The set $\{x : x \in S \text{ and } x \notin S'\}$ (i.e., the elements of S that are not in S')
- $T|X$: The subtree of T induced on leafset X, with nodes of degree two suppressed
- T_u: The unrooted tree obtained by suppressing the root for T

Part I
Basic Techniques

1
Brief Introduction to Phylogenetic Estimation

The construction of evolutionary trees is one of the major steps in many biological research studies. For example, evolutionary trees of different species tell us how the species evolved from a common ancestor, and perhaps also shed insights into the morphology of their common ancestors, the speed at which the different lineages evolved, how they and their common ancestors adapted to different environmental conditions, etc.

Because the true evolutionary histories cannot be known and can only be estimated, evolutionary trees are *hypotheses* about what has happened in the past. The tree in Figure 1.1 presents a hypothesis about the evolutionary history of a group of mammals and one bird. According to this tree, cats and gray seals are more closely related to each other than either is to blue whales, and cows are more closely related to blue whales than they are to rats or opossums (or anything else in the figure). Not surprisingly, chickens (since they are birds) are the outgroup, since the others are all mammals. The tree shown is the result of a statistical analysis of molecular sequences taken from the genomes of these species. Hence, the estimation of evolutionary trees, also known as phylogenetic reconstruction, is a computational problem that involves statistical inference. Furthermore, because it is a statistical inference problem, the accuracy of the tree depends on the model assumptions, the method used to analyze the data, and the data themselves. In other words, the estimation

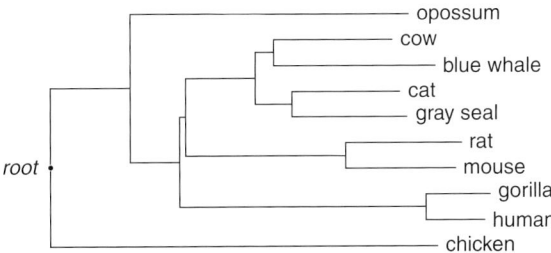

Figure 1.1 (Figure 3.5 from Huson et al. (2010)) A hypothesis of the evolutionary tree of various animals. The tree is rooted on the left, and has been estimated using molecular sequence data; branch lengths are proportional to evolutionary distances, and not necessarily to time.

of phylogenies is complicated and difficult. How this phylogeny construction is done, and how to do it *better*, is the point of this text.

Most modern phylogenetic estimation uses molecular sequence data (typically DNA sequences, which can be considered strings over the nucleotide alphabet $\{A, C, T, G\}$), and computes a tree from the set. While there are many ways to compute trees from a set of sequences, understanding when the methods are likely to be accurate requires having some kind of model for how the sequences relate to each other, and more specifically how they *evolved from a common ancestor*.

DNA sequences evolve under fairly complicated processes. At the simplest level, these sequences evolve down trees under processes in which single nucleotides are substituted by other single nucleotides. Many stochastic models have been developed to describe the evolution of sequences down trees under these substitution-only models, and most phylogenetic estimation is based on these models.

However, sequence evolution is more complicated than this. For example, many sequences for the same gene have different lengths, with the changes in length due to processes such as insertions and deletions (jointly called **indels**) of nucleotides, and in some cases duplications or rearrangements of regions within the sequences. Multiple sequence alignments are used to put the sequences into a matrix form so that each column has nucleotides that have a common ancestor, which are then used in phylogeny estimation. Stochastic models to describe sequence evolution have been developed that extend the simpler substitution-only models to include indel events, and are sometimes used to co-estimate alignments and trees.

Genome-scale evolution is even more complicated, since different parts of the genome can evolve under more complicated processes than the models that govern individual portions of the genomes. In particular, due to processes such as incomplete lineage sorting, gene duplication and loss, horizontal gene transfer, hybridization, and recombination, different parts of the genome can evolve down different trees (Maddison, 1997). Again, stochastic models have been developed to describe genome-scale evolution, and are used to estimate genome-scale phylogenies in the presence of one or more of these processes.

Thus, phylogeny estimation is addressed largely through statistical inference under an assumed stochastic model of evolution. While biologically realistic models are fairly complicated, the basic techniques that are used can be described even in the context of very simple models. In this chapter, we introduce the key concepts, issues, and techniques in phylogeny estimation in the context of a very simple binary model of sequence evolution.

1.1 The Cavender–Farris–Neyman Model

The **Cavender–Farris–Neyman** (CFN) model describes how a trait (which can either be present or absent) evolves down a tree (Neyman, 1971; Farris, 1973; Cavender, 1978). Hence, a CFN model has a rooted binary tree T (i.e., a tree in which every node is either a leaf or has two children) with numerical parameters that describe the evolutionary process of a trait. Under the CFN model, the probability of absence (0) or presence (1) is the same at the root, but the state can change on the edges (also called branches) of the tree. Thus, we associate a parameter $p(e)$ to every edge e in the tree, where $p(e)$ denotes the probability that the endpoints of the edge e have different states. In other words, $p(e)$ is the probability of changing state (from 1 to 0, or vice versa). For reasons that we will explain later, we require that $0 < p(e) < 0.5$.

Under the CFN model, a trait (which is also called a "character") evolves down the tree under this random process, and hence attains a state at every node in the tree, and in particular at the leaves of the tree. You could write a computer program for a CFN model tree that would generate 0s and 1s at the leaves of the tree; thus, CFN is a *generative model*. Each time you ran the program you would get another pattern of 0s and 1s at the leaves of the tree. Thus, if you repeated the process ten times, each time independently generating a new trait down the tree, you would produce sequences of length ten at the leaves of the tree.

The task of phylogenetic estimation is generally the inference of the tree from the sequences we see at the leaves of the tree. To do this, we assume that we know something about the sequence evolution model that generated the data. For example, we might assume (whether rightly or wrongly) that the sequences we see were generated by some unknown CFN model tree. Then, we would use what we know about CFN models to estimate the tree. Thus, we can also treat the CFN model as a tool for inference; i.e., CFN can be an *inferential model*. However, suppose we were lucky and the sequences we observe were, in fact, generated by some CFN model tree. Would we be able to reconstruct the tree T from the sequences?

To do this inference, we would assume that each of the sites (i.e., positions) in the sequences we observe had evolved down the same tree, and that each of them had evolved identically and independently (*i.i.d.*). It should be obvious that the ability to infer the tree correctly requires having enough data – i.e., long enough sequences – since otherwise we just can't distinguish between trees. For example, if we have 100 sequences, each of length one, there just isn't enough information to select the true tree with any level of confidence. Therefore, we ask "If sequence length were not an issue, so that we had sequences that were extremely long, would we have enough information in the input sequences to construct the tree exactly with high probability?" We can also formulate this more precisely, as "Suppose M is a method for constructing trees from binary sequences, (T, Θ) is a CFN model tree, and $\varepsilon > 0$. Is there a constant $k > 0$ such that the probability that M returns T given sequences of length at least k is at least $1 - \varepsilon$?"

A positive answer to this question would imply that for *any* level of confidence that is desired, there is some sequence length so that the method M would be accurate with that desired probability given sequences of at least that length. A positive answer also indicates that M has this property for all CFN model trees, and not just for some. A method M for which the answer is *Yes* is said to be **statistically consistent** under the CFN model. Thus, what we are actually asking is: *Are there any statistically consistent methods for the CFN model?*

1.2 An Analogy: Determining Whether a Coin is Biased Toward Heads or Tails

Let's consider a related but obviously simpler question. Suppose you have a coin that is biased either toward heads or toward tails, but you don't know which. Can you run an experiment to figure out which type of coin you have?

After a little thought, the answer may seem obvious – toss the coin many times, and see whether heads comes up more often than tails. If it does, say the coin is biased toward heads; otherwise, say it is biased toward tails. The probability that you guess correctly will approach 1 as you increase the number of coin tosses. We express this statement by saying that this method is a *statistically consistent* technique for determining which way the coin is biased (toward heads or toward tails). However, the probability of being correct will clearly depend on the number of coin tosses, so you may need to toss it many times before the probability that you answer correctly will be high.

Now suppose you don't get to toss the coin yourself, but are instead shown a sequence of coin tosses of some length that is chosen by someone else. Now, you can still guess whether the coin is biased toward heads or tails, but the probability of being correct may be small if the coin is not tossed enough times. Note that for this problem – deciding whether the coin is biased toward heads or tails – you will either be 100 percent correct or 100 percent wrong. The reason you can be 100 percent correct is that there are only a finite number of choices. Note also that the probability of being 100 percent correct can be high, but will never actually be equal to 1; in other words, for any finite number of times you can toss the coin, you will always have some probability of error. Also, the probability of error will depend on how many coin tosses you have and the probability of heads for that coin!

The problem changes in interesting ways if you want to estimate the *actual probability* of a head for that coin, instead of just whether it is biased toward heads or toward tails. However, it's pretty straightforward what you should do – toss the coin as many times as you can, and report the fraction of the coin tosses that come up heads. Note that in this problem your estimations of the probability of a head will generally have error. (For example, if the probability of a head is irrational, then this technique can *never* be completely correct.) Despite this, your estimate *will converge* to the true probability of a head as the number of coin tosses increases. This is expressed by saying that the method is statistically consistent for estimating the probability of a head.

The problem of constructing a CFN tree is very similar to the problem of determining whether a coin is biased toward heads or tails. There are only a finite number of different trees on n distinctly labeled leaves, and you are asked to select from among these. Then, if you have a sequence of samples of a random process, you are trying to use the samples to select the tree from that finite set; this is very much like deciding between the two types of biased coins. As we will show, it is possible to *correctly* construct the CFN tree with arbitrarily high probability, given sufficiently long sequences generated on the tree. The problem of constructing the substitution probabilities on the edges of the CFN tree is similar to the problem of determining the actual probability of a head, in that these are real-valued parameters, and so some error will always be expected.

However, if good methods are used, then as the sequence lengths increase the error in the estimated substitution probabilities will decrease, and the estimates will converge to the true values.

While estimating the numeric parameters of a CFN model tree is important for many tasks, we'll focus here on the challenge of estimating the tree T, rather than the numeric parameters. We describe some techniques for estimating this tree from binary sequences, and discuss whether they can estimate the tree correctly with high probability, given sufficiently long sequences.

1.3 Estimating the Cavender–Farris–Neyman Tree

Recall that the CFN model tree is a pair (T, θ) where T is the rooted binary tree with leaves labelled $s_1, s_2, \ldots, s_n$ and θ provides the values of $p(e)$ for every edge $e \in E(T)$. The CFN model tree describes the evolution of a sequence down the tree, where every site (i.e., position) within the sequence evolves down the model tree identically and independently. Thus the substitution probabilities $p(e)$ on each edge describe the evolutionary process operating on each site in the sequence.

However, this stochastic process can also be described differently, and in a way that is helpful for understanding why some methods can have good statistical properties for estimating CFN model trees. Under the CFN model, the number of substitutions on an edge is modeled by a Poisson random variable $N(e)$ with expected value $\lambda(e)$. Thus, if $N(e) = 0$, then there is no substitution on the edge, while if $N(e) = 1$, then there is a substitution on the edge. Furthermore, if $N(e)$ is even then the endpoints of the edge have the same state, while if $N(e)$ is odd then the endpoints of the edge have different states; hence, $p(e)$ is the probability that $N(e)$ is odd, since we only observe a change on the edge if there is an odd number of substitutions.

Using $\lambda(e)$ (the expected value of $N(e)$) instead of $p(e)$ turns out to be very useful in developing methods for phylogeny estimation. Note that $0 < \lambda(e)$ for all e since $p(e) > 0$. Using the properties of Poisson random variables, it can be shown that

$$\lambda(e) = -\frac{1}{2} ln(1 - 2p(e)).$$

Note that as $p(e) \to 0.5$, $\lambda(e) \to \infty$.

1.3.1 Estimating the CFN Tree When Evolution is Clocklike

An assumption that is sometimes made is that sequence evolution is *clocklike* (also referred to as obeying the **strict molecular clock**), which means that the expected number of changes is proportional to time. If we assume that the leaves represent extant (i.e., living) species, then under the assumption of a strict molecular clock, the total expected number of changes from the root to any leaf is the same. Under the assumption of a strict molecular

Inferring Clocklike Evolution

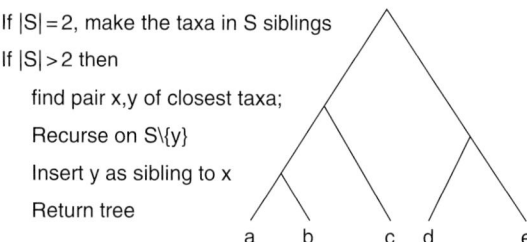

If |S|=2, make the taxa in S siblings
If |S|>2 then
    find pair x,y of closest taxa;
    Recurse on S\{y}
    Insert y as sibling to x
    Return tree

Figure 1.2 Constructing trees when evolution is clocklike. We show a cartoon of a model tree, with branch lengths drawn proportional to the expected number of changes. When evolution is clocklike, as it is for this cartoon model, simple techniques such as the one described in the figure will reconstruct the model tree with probability that converges to 1 as the sequence length increases.

clock, the matrix of expected distances between the leaves in the tree has properties that make it "ultrametric":

Definition 1.1 An **ultrametric matrix** is an $n \times n$ matrix M corresponding to distances between the leaves in a rooted edge-weighted tree T (with non-negative edge weights) where the sum of the edge weights in the path from the root to any leaf of T does not depend on the selected leaf.

Constructing trees from ultrametric matrices is much easier than the general problem of constructing trees from distance matrices that are not ultrametric. However, the assumption of clocklike evolution may not hold on a given dataset, and is generally not considered realistic (Li and Tanimura, 1987). Furthermore, the ability to reconstruct the tree using a particular technique may depend on whether the evolutionary process is in fact clocklike.

Even though clocklike evolution is generally unlikely, there are some conditions where evolution is close to clocklike. So, let's assume we have a clocklike evolutionary process operating on a CFN tree (T, θ), and so the total number of expected changes from the root to any leaf is the same. We consider a very simple case where the tree T has three leaves, a, b, and c. To reconstruct the tree T we need to be able to infer which pair of leaves are siblings, from the sequences we observe at a, b, and c. How should we do this?

One very natural approach to estimating the tree would be to select as siblings the two sequences that are the most similar to each other from the three possible pairs. Because the sequence evolution model is clocklike, this technique will correctly construct rooted three-leaf trees with high probability. Furthermore, the method can even be extended to work on trees with more than three leaves, using recursion. For example, consider the model tree given in Figure 1.2, where the branch lengths indicate the expected number of substitutions on the branch. Note that this model tree is *ultrametric*. Thus, under this model, the sequences at leaves a and b will be the most similar to each other of all the possible pairs of sequences at the leaves of the tree. Hence, to estimate this tree, we would

1.3 Estimating the Cavender–Farris–Neyman Tree

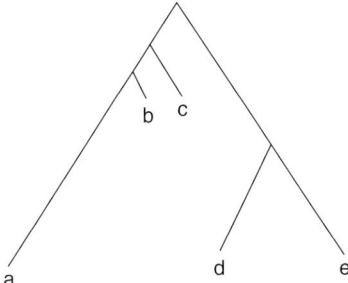

Figure 1.3 Constructing evolutionary trees when evolution is not clocklike. We show a cartoon of a model tree on five leaves, with branch lengths drawn proportionally to the expected number of changes of a random site (i.e., position in the sequence alignment). Note that leaves b and c are not siblings, but have the smallest evolutionary distance (measured in terms of expected number of changes). Hence, methods that make the most similar sequences siblings will likely fail to reconstruct the model tree, and the probability of failure will increase to 1 as the number of sites (i.e., sequence length) increases.

first compare all pairs of sequences to find which pair is the most similar, and we'd select a and b as this pair. We'd then correctly infer that species a and b are siblings. We could then remove one of these two sequences (say, a), and reconstruct the tree on what remains. Finally, we would add a into the tree we construct on b,c,d,e, by making it a sibling to b.

It is easy to see that a tree computed using this approach, which is a variant of the UPGMA (Sokal and Michener, 1958) method (Unweighted Pair Group Method with Arithmetic Mean), will converge to the true tree as the sequence length increases. That is, it is possible to make mistakes in the construction of the tree – but the probability of making a mistake decreases as the sequence length increases.

However, what if the evolutionary process isn't clocklike? Suppose, for example, that we have a three-leaf CFN model tree with leaves a,b, and c, in which a and b are siblings. Suppose, however, that the substitution probabilities on the edges leading from the root to b and c are extremely small, while the substitution probability on the single edge incident with a is very large. Then, applying the technique described above would return the tree with b and c siblings – i.e., the wrong tree. In other words, this simplified version of UPGMA would converge to a tree other than the true tree as the sequence length increases. This is clearly an undesirable property of a phylogeny estimation method, and is referred to by saying the method is **positively misleading**. An example of a model tree where UPGMA and its variants would not construct the correct tree – even as the sequence length increases – is given in Figure 1.3; the probability of selecting b and c as the first sibling pair would increase to 1 as the sequence length increases, and so UPGMA and its variants would return the wrong tree.

Clearly, when there is no clock, then sequence evolution can result in datasets for which the inference problem seems to be much harder. Furthermore, for the CFN and other sequence evolution models, if we drop the assumption of the molecular clock, the correct inference of rooted three-leaf trees with high probability is not possible. In fact, the best that can be hoped for is the correct estimation of the *unrooted* version of the model tree with high probability.

1.3.2 Estimating the Unrooted CFN Tree when Evolution is Not Clocklike

We now discuss how to estimate the underlying unrooted CFN tree from sequences, without assuming clocklike evolution. We will begin with an idealized situation, in which we have something we will call "CFN model distances," and show how we can construct the tree from these distances. Afterwards, we will show how to construct the tree from estimated distances rather than from model distances.

CFN model distances: Let (T, θ) be a CFN model tree on leaves $s_1, s_2, \ldots, s_n$, so that T is the rooted binary tree and θ gives all the edge parameters $\lambda(e)$. Let $\lambda_{i,j}$ denote the expected number of changes for a random site on the path $P_{i,j}$ between leaves s_i and s_j in the CFN model tree T; it follows that

$$\lambda_{i,j} = \sum_{e \in P_{i,j}} \lambda(e).$$

The matrix λ is referred to as the **CFN model distance** matrix.

Note that by definition, λ is the matrix of path distances in a tree, where the path distance between two leaves is the sum of the branch lengths and all branch lengths are positive. Matrices that have this property have special mathematical properties, and in particular are examples of **additive** matrices.

Definition 1.2 An $n \times n$ matrix M is **additive** if there is a tree T with leaves labeled $1, 2, \ldots n$ and non-negative lengths (or weights) on the edges, so that the path distance between i and j in T is equal to $M[i,j]$. An example of an additive matrix is given in Figure 1.4.

In other words, additive matrices correspond to edge-weighted trees in which all edge weights are non-negative; therefore, distance matrices arising from CFN model trees are

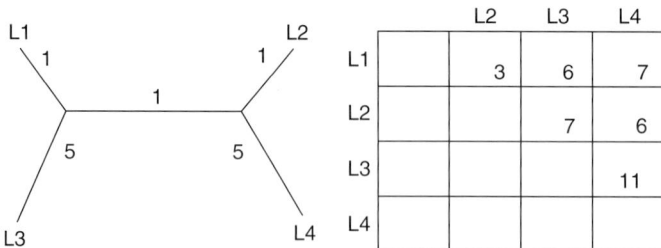

Figure 1.4 Additive matrix and its edge-weighted tree.

necessarily additive. Furthermore, CFN model trees have strictly positive branch lengths, and this property additionally constrains the additive matrices corresponding to CFN model trees and makes the inference of CFN model trees particularly easy to do. Techniques to compute trees from additive distance matrices (and even from noisy versions of additive distance matrices) are presented in Chapter 5, and briefly summarized here in the context of CFN distance matrices.

Let's consider the case where the CFN tree T has $n \geq 4$ leaves, and that s_1, s_2, s_3, and s_4 are four of its leaves. Without loss of generality, assume the tree T has one or more internal edges that separate s_1 and s_2 from s_3 and s_4. We describe this by saying that T **induces the quartet tree** $s_1 s_2 | s_3 s_4$. We first show how to compute the quartet tree on these leaves induced by T, and then we will show how to use all these quartet trees to construct T.

Suppose we have the values of $\lambda(e)$ for every edge in T and hence also the additive matrix λ of path distances in the tree. Consider the three following pairwise sums:

- $\lambda_{1,2} + \lambda_{3,4}$
- $\lambda_{1,3} + \lambda_{2,4}$
- $\lambda_{1,4} + \lambda_{2,3}$

Since the weights of the edges are all positive, the smallest of these three pairwise sums has to be $\lambda_{1,2} + \lambda_{3,4}$, since it covers all the edges of T connecting these four leaves *except* for the ones on the path P separating s_1, s_2 from s_3, s_4. Furthermore, the two larger of the three pairwise sums have to be identical, since they cover the same set of edges (every edge in T connecting the four leaves is covered either once or twice, with only the edges in P covered twice). Letting $w(P)$ denote the total weight of the edges in the path P, and assuming that T induces the quartet tree $s_1 s_2 | s_3 s_4$,

$$\lambda_{1,2} + \lambda_{3,4} + 2w(P) = \lambda_{1,3} + \lambda_{2,4} = \lambda_{1,4} + \lambda_{2,3}.$$

Since $\lambda(e) > 0$ for every edge e, $w(P) > 0$. Hence, $\lambda_{1,2} + \lambda_{3,4}$ is strictly smaller than the other two pairwise sums.

The **Four Point Condition** is the statement that the two largest values of the three pairwise sums are the same. Hence, the Four Point Condition holds for any additive matrix, which allows branch lengths to be zero (as long as they are never negative). The additional property that the smallest of the three pairwise sums is strictly smaller than the other two is not part of the Four Point Condition, and can fail to hold on additive matrices. It is worth noting that a matrix is additive if and only if it satisfies the Four Point Condition on every four indices.

Now, if we are given a 4×4 additive matrix $\mathbf{D}$ that corresponds to a tree T with positive branch weights, then we can easily compute T from $\mathbf{D}$: We calculate the three pairwise sums, we determine which of the three pairwise sums is the smallest, and use that one to define the split for the four leaves into two sets of two leaves each. We refer to this method as the **Four Point Method**.

Given an $n \times n$ additive matrix $\mathbf{M}$ with $n \geq 5$ associated to a binary tree T with positive branch lengths, we can construct T using a two-step technique that we now describe. In

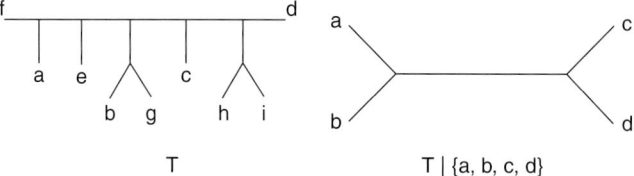

Figure 1.5 Tree T and its homeomorphic subtree on a,b,c,d (i.e., quartet tree).

Step 1, we compute a quartet tree on every four leaves by applying the Four Point Method to each 4×4 submatrix of **M**. In Step 2, we assemble the quartet trees into a tree on the full set of leaves. Step 1 is straightforward, so we only need to describe the technique we use in Step 2, which is called the "All Quartets Method."

The All Quartets Method: Suppose we are given a set Q of fully resolved (i.e., binary) quartet trees, with one tree on every four leaves, and we want to construct a tree T that agrees with all the quartet trees in Q. To see what is meant by "agrees with all the quartets," consider Figure 1.5, which shows an unrooted tree T on leafset $\{a,b,c,d,e,f,g,h,i\}$ and the quartet it defines on $\{a,b,c,d\}$.

The reason this is the quartet tree is that T contains an edge that separates leaves a and b from leaves c and d. We can apply the same reasoning to any unrooted tree and subset of four leaves. In addition, we can even apply this analysis to rooted trees, by first considering them as unrooted trees. For example, Figure 1.3 shows a rooted tree with five leaves: if we ignore the location of the root, we get an unrooted tree with five leaves that we will refer to as T. Now, consider the quartet tree on $\{b,c,d,e\}$ defined by T. With a little care, you should be able to see that T has an edge that separates b,c from d,e; hence, T induces the quartet tree $bc|de$. T also induces the quartet tree $ab|de$, because T has at least one edge (in this case, two) that separates a,b from d,e. We will refer to the set of quartet trees for T constructed in this way by $Q(T)$, and note that

$$Q(T) = \{ab|cd, ab|ce, ab|de, ac|de, bc|de\}.$$

Now, suppose we were given $Q(T)$; could we infer T?

The All Quartets Method constructs a binary tree T given its set $Q(T)$ of quartet trees, and has a simple recursive design. We say that a pair of leaves x,y in T that have a common neighbor are **siblings**, thus generalizing the concept of siblinghood to the unrooted tree case. Note that every unrooted binary tree with at least four leaves has at least two disjoint sibling pairs. To construct the unrooted tree T from $Q = Q(T)$, we will use a recursive approach.

First, if $|Q| = 1$, we just return the tree in Q. Otherwise, we search for a pair x,y of leaves that is always together in any quartet tree in Q that contains both x and y. (In other words, for all a,b, the quartet tree in Q on $\{x,y,a,b\}$ is $xy|ab$.) Note that any pair of leaves that are siblings in T will satisfy this property. Furthermore, because Q is the set of quartet trees

1.3 Estimating the Cavender–Farris–Neyman Tree

for T, we are guaranteed to find at least two such sibling pairs, and we only need one. Once we find such a sibling pair x, y, we remove one of the leaves in the pair, say x. Removing x corresponds to removing all quartet trees in Q that contain leaf x, and so reduces the size of Q. This reduced set Q' of quartet trees corresponds to all the quartet trees that T defines on $S \setminus \{x\}$. We then recurse on Q', obtaining the unrooted tree T' realizing Q'. To finish the computation, we add x into T' by making it a sibling of y. Hence, we can prove the following:

Theorem 1.3 *Let* **D** *be an additive matrix corresponding to a binary tree* T *with positive edge weights. Then the two-step algorithm described above (where we construct a set of quartet trees using the Four Point Method and then apply the All Quartets Method to the set of quartet trees) returns* T *and runs in polynomial time.*

The proof of correctness uses induction (since the algorithm is recursive) and is straightforward. The running time is also easily verified to be polynomial. Now that we know that the two-step process has good theoretical properties, we ask how we can use this approach when we have binary sequences that have evolved down some unknown CFN tree. To use this approach, we first need to estimate the CFN model distances from sequences, and then possibly modify the algorithm so that we can obtain good results on estimated distances rather than on additive distances.

Estimating CFN distances: Let $p(i,j)$ denote the probability that the leaves s_i and s_j have different states for a random site; the expected number of changes on the path between i and j is therefore

$$\lambda_{i,j} = -\frac{1}{2} \ln(1 - 2p(i,j)).$$

If we knew all the $p(i,j)$ exactly, we could compute all the $\lambda_{i,j}$ exactly, and hence we would have an additive matrix for the tree T; this means we could reconstruct the model tree and its branch lengths perfectly. (The reconstruction of the model tree topology is straightforward, as we just showed; the calculation of the branch lengths involves some more effort, but is not too difficult.)

Unfortunately, unless we are told the model tree, we cannot know any $p(i,j)$ exactly. Nevertheless, we can *estimate* these values from the data we observe, in a natural way. That is, given sequences s_i and s_j of length k that evolve down the tree T, we can estimate $p(i,j)$. For example, a natural technique would be to use the *fraction* of the number of positions in which s_i and s_j have different states. To put this precisely, letting $H(i,j)$ be the **Hamming distance** between s_i and s_j (i.e., the number of positions in which they are different), then since k is the sequence length, $\frac{H(i,j)}{k}$ is the fraction in which the two sequences are different. Furthermore, as $k \to \infty$, then by the law of large numbers $\frac{H(i,j)}{k} \to p(i,j)$. Hence, we can estimate $\lambda_{i,j}$, the CFN model distance (also known as true evolutionary distance) between sequences s_i and s_j, using the following formula:

$$\hat{\lambda}_{i,j} = -\frac{1}{2} \ln(1 - 2\frac{H(i,j)}{k}).$$

Also, as $k \to \infty$, $\hat{\lambda}_{i,j} \to \lambda_{i,j}$. We call this the **Cavender–Farris–Neyman distance correction**, and the distances that we compute using this distance correction are the **Cavender–Farris–Neyman distances**. The distance matrix computed using the CFN distance correction is an estimate of the model CFN distance matrix, and converges to the model distance matrix as the sequence length increases.

To say that $\hat{\lambda}_{i,j}$ converges to $\lambda_{i,j}$ for all i,j as the sequence length increases means that for any $\varepsilon > 0$ and $\delta > 0$, there is a sequence length K so that the distance matrix $\hat{\lambda}$ will satisfy $|\hat{\lambda}_{ij} - \lambda_{ij}| < \delta$ for all i,j with probability at least $1 - \varepsilon$, given sequence length at least K.

CFN distance matrices may *not* satisfy the triangle inequality, which states that $d_{ik} \leq d_{ij} + d_{jk}$ for all i,j,k, and hence are not properly speaking "distance matrices." However, these estimated distance matrices are symmetric (i.e., $d_{ij} = d_{ji}$) and zero on the diagonal, and so are referred to as **dissimilarity matrices**.

The Naive Quartet Method: We now show how to estimate the unrooted topology of a CFN model tree from a matrix of estimated CFN distances. We can *almost* use exactly the same technique as we described before when the input was an additive matrix, but we need to make two changes to allow for the chance of failure. First, to use the Four Point Method on estimated 4×4 distance matrices, we compute the three pairwise sums, and if the minimum is unique then we return the quartet tree corresponding to that smallest pairwise sum; otherwise, we return *Fail*. This modification is necessary since the input matrix of estimated distances may not uniquely determine the quartet tree. Second, the input Q to the All Quartets Method may not be equal to $Q(T)$ for any tree T; therefore, we have to modify the All Quartets Method so that it can recognize when this happens. The modification is also straightforward: If we fail to find a sibling pair of leaves during some recursive call, we return *Fail*. Also, even if we do construct a tree T, we verify that $Q = Q(T)$, and return *Fail* if $Q \neq Q(T)$.

We call this method the **Naive Quartet Method** because of its simplistic approach to tree estimation. It was originally proposed in Erdös et al. (1999a), where it was called the "Naive Method." We summarize this as follows:

Step 1: Apply the Four Point Method to every four leaves; if any four-leaf subset fails to return a tree, return *Fail*, and exit.

Step 2: Use the All Quartets Method to construct a tree that agrees with all the quartet trees computed in Step 1, if it exists, and otherwise return *Fail*.

This Naive Quartet Method has desirable properties, as we now show:

Theorem 1.4 *Let* **d** *be a* n × n *dissimilarity matrix and* **D** *be a* n × n *additive matrix defined by a binary tree* T *with positive edge weights. Suppose that*

$$\max_{ij} |d_{ij} - D_{ij}| < f/2,$$

1.3 Estimating the Cavender–Farris–Neyman Tree

where f *is the smallest weight of any internal edge in* T. *Then, the Naive Quartet Method applied to* **d** *will return* T*, and runs in polynomial time.*

This theorem and its proof appears in Chapter 5, but the essence of the proof is as follows. When applying the Four Point Method to any 4×4 submatrix of the additive matrix **D**, the gap between the smallest of the three pairwise sums and the other two pairwise sums is at least $2f$. If the entries in the matrix **d** are less than $f/2$ from the entries in matrix **D**, then the smallest of the three pairwise sums given by **d** will have the same 2:2 split as the smallest of the three pairwise sums given by **D**. Hence, the application of the Four Point Method to any 4×4 submatrix of **d** will return the same quartet tree as when it is applied to **D**. Hence, Step 1 will return $Q(T)$. Then, the All Quartets Method will construct T, given $Q(T)$.

We summarize this argument, which shows that the Naive Quartet Method is **statistically consistent** under the CFN model (i.e., that it will reconstruct the true tree with probability increasing to 1 as the sequence length k increases):

- We showed that the Naive Quartet Method will reconstruct the unrooted tree topology T of the model CFN tree given an additive matrix defining the model CFN tree.
- We showed that the matrix of estimated CFN distances converges to an additive matrix for the CFN model tree topology as the sequence length goes to infinity.
- We stated (although we did not provide the proof) that whenever the estimated distances are all within $f/2$ of an additive matrix for the model tree T (where f is the length of the shortest internal edge in the model tree), then the Naive Quartet Method will return the unrooted tree topology T.
- Hence, as the sequence length increases, the tree returned by the Naive Quartet Method will be the unrooted topology of the CFN model tree with probability increasing to 1.

The Naive Quartet Method is just one such statistically consistent method, and many other methods have been developed to construct CFN trees from sequence data that have the same basic theoretical guarantee of converging to the true tree as the sequence length increases.

Although the Naive Quartet Method is statistically consistent under the CFN model, when the input dissimilarity matrix **d** is not sufficiently close to additive, this two-step process can fail to return anything! For example, the Four Point Method can fail to determine a unique quartet tree (if the smallest of the three pairwise sums is not unique), and the whole process can fail in that case. Or, even if the Four Point Method returns a unique tree for every set of four leaves, the set of quartet trees may not be compatible, and so the second step can fail to construct a tree on the full dataset. Thus, the two-step process will only succeed in returning a tree under fairly restricted conditions. For this reason, even though this two-step process for constructing trees has nice theoretical guarantees, it is not used in practice. This is why Erdös et al. (1999a) used the word *Naive* in the name of this method – to suggest that the method is really a mathematical construct rather than a practical tool.

1.4 Some Comments about the CFN Model

In the CFN model, we constrain the substitution probabilities $p(e)$ to be strictly between 0 and 0.5. If we were to allow $p(e) = 0$ on some edge e, then there will be no change on e, and hence it would be impossible to reconstruct the edge with probability converging to 1. At the other extreme, if we were to allow $p(e) = 0.5$ on some edge e, then the two sequences at the endpoints of e will look random with respect to each other, and correctly connecting the two halves of the tree with high probability would be impossible. This is why $p(e)$ is constrained to be strictly between 0 and 0.5.

In the CFN model, the sites evolve down the same model tree. In other words, given any two sites and any edge e in the tree, the expected numbers of changes of the two sites on that edge e are the same, so that all the sites have the *same rate of change*. This assumption is typically relaxed so that each site draws its rate independently from a distribution of rates-across-sites. The meaning of "rates-across-sites" is that each rate gives a multiple for the expected number of changes. Thus, if site i draws rate 2 and site j draws rate 1, then site i has twice as many expected changes as site j on every edge of the tree. Typically, the distribution of rates across sites is modeled using the gamma distribution, but some other distributions (such as gamma plus invariable) are also sometimes used. Note that although the sites can have different rates, they draw their rates independently, and hence all sites evolve under the "same process." This is called the *i.i.d.* assumption. Finally, given a particular gamma distribution, the entire stochastic model of evolution is fully described by the model tree topology T, the branch lengths, and the gamma distribution.

Biological data typically are not binary sequences, and instead are typically molecular sequences, either of nucleotides (which are over a four-letter alphabet) or amino acids (which are over a 20-letter alphabet). Statistical models of nucleotide and amino acid sequence evolution (discussed in Chapter 8) have also been developed, and methods for estimating trees under these more complex multi-state models have been developed to estimate trees under these models. Despite the increased complexity of the models and methods, for most of these models the theoretical framework and analysis for these more sophisticated methods are basically the same as that which we've described under the CFN model. Thus, even under more biologically realistic models it is possible to reconstruct the unrooted topology of the true tree with high probability, given sufficiently long sequences generated on the tree.

1.5 Phylogeny Estimation Methods Used in Practice

There are many phylogeny estimation methods that have been developed, some of which are statistically consistent under the standard statistical models of sequence evolution. One of the methods that has been used to construct trees is the UPGMA method alluded to earlier; UPGMA is an agglomerative clustering method that computes a distance between every pair of sequences, then selects the closest pair of sequences to be siblings, updates the matrix, and repeats the process until a tree is computed for the full dataset. Yet, as we have noted, UPGMA can fail to be statistically consistent under some model conditions.

Maximum parsimony is another approach that has been used to construct many trees. The objective is a tree T in which the input sequences are placed at the leaves of T and additional sequences are placed at the internal nodes of T so that the total treelength, defined to be the total number of changes over the entire tree, is minimized. Another way of defining maximum parsimony is that it is the Hamming Distance Steiner Tree Problem: The input is a set of sequences, and the output is a tree connecting these sequences (which are at the leaves) with other sequences (i.e., the Steiner points) at the internal nodes, that minimizes the total of the Hamming distances on the edges of the tree. Since the Hamming distance between two sequences of the same length is the number of positions in which they are different, the total of the Hamming distances on the edges of the tree is the same as its treelength.

Finding the best tree under the maximum parsimony criterion is an NP-hard problem (Foulds and Graham, 1982), and hence heuristics, typically based on a combination of hill-climbing and randomization to get out of local optima, are used to find good, though not provably globally optimal, solutions. Maximum parsimony heuristics have improved over the years, but can still be computationally very intensive on large datasets. However, suppose we could solve maximum parsimony exactly (i.e., find global optima); would maximum parsimony then be statistically consistent under the CFN model, or other models?

Unfortunately, maximum parsimony has been proven to be statistically inconsistent under the CFN model and also under standard DNA sequence evolution models, and may even converge to the wrong tree as the sequence length increases (Felsenstein, 1978) (i.e., it can even be *positively misleading*, just like UPGMA). Although UPGMA and maximum parsimony are both statistically inconsistent under standard DNA sequence evolution models, other methods have been developed that are statistically consistent under these models, and are commonly used in practice. Examples of these methods include polynomial time distance-based methods such as neighbor joining (Saitou and Nei, 1987) and FastME (Lefort et al., 2015). The Naive Quartet Method is statistically consistent under the CFN model, and also under standard nucleotide sequence evolution models, and its statistical consistency is extremely easy to prove. The Naive Quartet Method is also polynomial time, and so is a polynomial time statistically consistent method for estimating trees under standard sequence evolution models.

Maximum likelihood is another method that is statistically consistent under standard sequence evolution models (Neyman, 1971; Felsenstein, 1981). To understand maximum likelihood, we describe its use for estimating CFN trees. First, given a CFN model tree (T, θ) and a set S of binary sequences of the same length, we can compute the probability that S was generated by the model tree (we will prove this in Chapter 8). Thus, given S, we can search for the model tree (i.e., the tree topology and the substitution probabilities on the edges) that has the largest probability of generating S. Finding the maximum likelihood tree is NP-hard (Roch, 2006), but if solved exactly it is a statistically consistent estimator of the tree.

Bayesian estimation of phylogenetic trees using Markov Chain Monte Carlo (MCMC) (described in Section 8.7) has many theoretical advantages over maximum likelihood estimation and other approaches (Huelsenbeck et al., 2001) and is also able to produce a statistically consistent estimation of the true tree under standard DNA sequence evolution models (Steel, 2013); however, Bayesian MCMC methods needs to run for a long time to have good accuracy. Thus, maximum likelihood and Bayesian MCMC estimation of phylogenetic trees are generally much slower than distance-based estimation methods.

Based on this, one could presume that methods like neighbor joining and the Naive Quartet Method would dominate in practice, since they are polynomial time and statistically consistent, while other methods are statistically inconsistent (e.g., maximum parsimony and UPGMA) or computationally intensive (e.g., maximum likelihood, Bayesian methods, and maximum parsimony), or sometimes both.

1.6 Measuring Error Rates on Simulated Datasets

Phylogeny estimation methods are evaluated for accuracy, primarily with respect to the tree topology (as an unrooted tree), using both simulated and biological datasets. Because the true evolutionary history of a biological dataset can rarely be known with confidence, most performance studies are based on simulated datasets. In a simulation study, a model tree is created, and then sequences are evolved down the tree. These sequences are then used to compute a tree, and the computed tree is compared to the model tree. Under the simplest evolutionary models, the sequences evolve just with substitutions, so that individual letters (i.e., nucleotides or amino acids) within the sequences can change during the evolutionary process, but the length of the sequence does not change. More complex models include other processes, such as insertions and deletions (jointly called "indels"), so that the sequences change in length over time. If the sequence evolution process includes insertions and deletions, then a multiple sequence alignment is typically first computed before the tree is estimated. See Figure 1.6 for a graphical description of how a simulation study is performed.

Because the true tree and true alignment are rarely known on any biological dataset, simulation studies are the norm for evaluating phylogeny estimation methods, and are also frequently used to evaluate multiple sequence alignment methods. In a simulation study, a model tree is created, often using a simulation process where a tree is produced under a mathematical model for speciation (e.g., a birth–death process), and then sequences are evolved down the tree under a model that describes the evolutionary process. Often, these models will assume a substitution-only process (such as the CFN model for binary sequences that we discussed earlier, but also under models such as the Jukes–Cantor (Jukes and Cantor, 1997) and Generalised Time Reversible (Tavaré, 1986) models, which model DNA sequence evolution). When alignment estimation is also of interest, then other models are used in which sequences evolve with insertions, deletions, and sometimes other events.

Thus, in one run of the simulation procedure, a set of sequences is generated for which we know the entire evolutionary history relating the sequences, and hence we know the true alignment. Once the sequences are generated, an alignment can be estimated from the unaligned sequences, and a tree can be estimated on the estimated alignment. The estimated alignment and estimated tree can be compared to the true (model) tree and true alignment, and the error can be quantified. By varying the model parameters, the robustness of the estimation methods to different conditions can be explored, and methods can be compared for accuracy.

There are many ways to quantify error in phylogeny estimation, but the most common one measures the distance between two trees in terms of their bipartitions, which we now define. If you take an edge (but not its endpoints) out of a tree, it separates the tree into two subtrees, and so defines a bipartition on the leaves of the tree. Each edge in a tree thus defines a bipartition on the leafset. The bipartitions that are present in the model tree but not in the estimated tree are called **false negatives** (FNs), and the bipartitions that are present in the estimated tree but not in the model tree are referred to as **false positives** (FPs). Since the bipartitions are defined by edges, we sometimes refer to these as FP and FN edges (or as FN and FP branches). The edges that are incident with leaves define the trivial bipartitions, and are present in any tree that has the same leafset; the internal edges (which are the ones that are not incident with leaves) define the non-trivial bipartitions. The **Robinson–Foulds** (RF) distance (also called the bipartition distance) between two trees is the number of non-trivial bipartitions that are present in one or the other tree but not in both trees.

Each of these ways of quantifying error in an estimated tree can be normalized to produce a proportion between 0 and 1 (equivalently, a percentage between 0 and 100). For example, the FN error rate would be the percentage of the non-trivial model tree bipartitions that are not present in the estimated tree, and the FP error rate would be the percentage of the non-trivial bipartitions in the estimated tree that are not present in the model tree. Finally, the **Robinson–Foulds error rate** is the RF distance divided by $2n - 6$, where n is the number of leaves in the model tree; note that $2n - 6$ is the maximum possible RF distance between two trees on the same set of n leaves.

Figure 1.7 provides an example of this comparison; note that the model tree (called the true tree in the figure) is rooted, but the inferred tree is unrooted. To compute the tree error, we unroot the true tree, and treat it only as an unrooted tree. Since both trees are binary (i.e., each non-leaf node has degree three), there are only two internal edges. Each of the two trees have the non-trivial bipartition separating S_1, S_2 from S_3, S_4, S_5, but each tree also has a bipartition that is not in the other tree. Hence, the RF distance between the two trees is 2, out of a maximum possible of 4, and so the RF error rate is 50 percent. Note also that there is one true positive edge and one false positive edge in the inferred tree, so that the inferred tree has FN and FP rates of 50 percent.

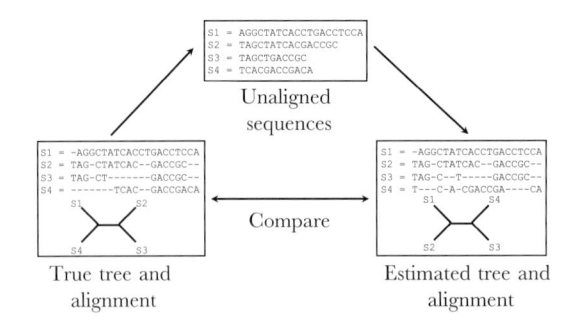

Figure 1.6 A simulation study protocol. Sequences are evolved down a model tree under a process that includes insertions and deletions; hence, the true alignment and true tree are known. An alignment and tree are estimated on the generated sequences, and then compared to the true alignment and true tree.

1.7 Getting Branch Support

The methods we have described output trees, and most of them output a single tree (i.e., a "point estimate"). The only real exceptions to this are the Bayesian MCMC methods, which output a distribution on trees, and maximum parsimony analyses, which can return the set of all the best trees found during the heuristic search. However, a single tree (or even a collection of trees) is not generally sufficient; typically the biological analysis also needs to have a sense of the statistical support for each edge in the tree.

The estimation of support values of edges in a phylogenetic tree is often performed using non-parametric bootstrapping, where "bootstrap replicate" datasets are created by sampling sites with replacement from the input sequence alignment, and then trees are computed on these bootstrap replicate datasets. The proportion of these trees that have a particular edge (as defined by its bipartition on the leafset) is used as the statistical support for the edge. Bayesian methods output a distribution on trees, and can use the trees in the distribution to compute the support on each edge.

1.8 Using Simulations to Understand Methods

Our discussion has introduced the basic theoretical framework for phylogeny estimation, including statistical models of sequence evolution and some simple methods for estimating trees under these models. We have also noted that many methods, including maximum likelihood, Bayesian MCMC, neighbor joining, and the Naive Quartet Method, are statistically consistent methods for estimating the true unrooted tree under standard stochastic models of evolution. In contrast, we showed that UPGMA and maximum parsimony are *not* statistically consistent under the same stochastic models, and are even positively misleading

1.8 Using Simulations to Understand Methods

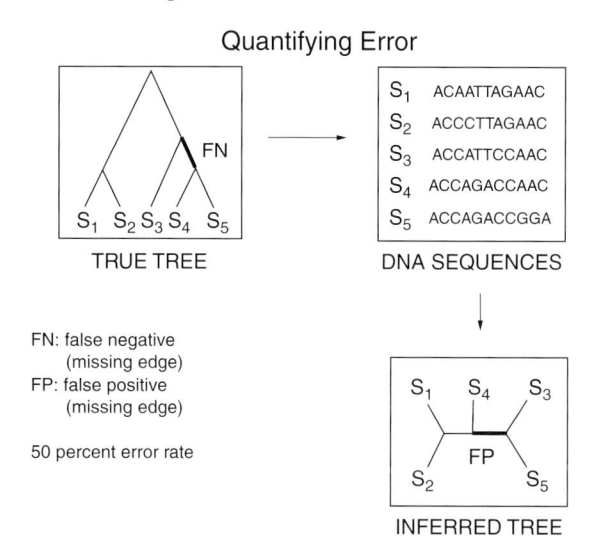

Quantifying Error

Figure 1.7 How tree estimation error is calculated in a simulation study. In a simulation study, the true tree is known, and so can be used to measure error in an estimated tree. Although the true tree is rooted and the inferred tree is unrooted, the error calculation is based on the (non-trivial) bipartitions induced by the internal edges, and so the true tree is interpreted as an unrooted tree. Some of the edges in the two trees are labeled, but others are not. The edges that are not labeled induce bipartitions that are in both trees; all other edges define bipartitions for only one of the two trees. False positive (FP) edges are those that are in the estimated tree but not the model tree, while false negative (FN) edges are those that are in the model tree but not the estimated tree. In this example, one of the two internal edges in the inferred tree is a false positive, and the other is a true positive; hence the false positive rate is 50 percent. Similarly, although the true tree is rooted, when we treat it as an unrooted tree, one of its internal edges is a true positive and the other is a false negative; hence the false negative rate is 50 percent. The number of false positive plus false negative edges, divided by $2n - 6$ (where n is the number of leaves in each tree) is the Robinson–Foulds (RF) error rate. When both trees are binary, the FN, FP, and RF rates are identical.

in that they will produce the wrong tree with probability increasing to 1 as the sequence length increases under some conditions. On the face of it, this would seem to suggest that UPGMA and maximum parsimony are both inferior to the Naive Quartet Method and neighbor joining. Thus, perhaps maximum parsimony should never be used instead of the Naive Quartet Method or neighbor joining.

Yet the Naive Quartet Method will only return the true tree if every quartet tree is computed exactly correctly. As many have observed, some quartet trees can be very difficult to compute, even given sequences that have thousands of sites (Huelsenbeck and Hillis, 1993; Hillis et al., 1994). Furthermore, as the number of sequences in the dataset increases, the probability of correctly reconstructing every quartet tree would decrease. Hence, the Naive Quartet Method would seem to be a rather poor choice of method for phylogeny estimation for any large dataset, even though it is statistically consistent and runs in polynomial time.

Indeed, the Naive Quartet Method may not even be useful on most moderate-sized datasets. In comparison, UPGMA, neighbor joining, and maximum parsimony always return a tree, and so will not have this kind of dramatic failure that the Naive Quartet Method has.

What does the theory suggest about the relative performance between neighbor joining and maximum parsimony? Or, put differently, since neighbor joining is polynomial time and statistically consistent whereas maximum parsimony is neither, does this mean that neighbor joining should be more accurate than maximum parsimony? The answer, perhaps surprisingly, is *no*: there are model conditions and sequence lengths where trees computed using maximum parsimony heuristics are substantially more accurate than trees computed using neighbor joining.

As an example of this phenomenon, see Figure 1.8, which shows some of the results from Nakhleh et al. (2002) comparing a heuristic for maximum parsimony, neighbor joining, and a variant of neighbor joining called Weighbor (Bruno et al., 2000). The results shown here are for simulated data that evolve down a Kimura 2-parameter (K2P) model tree (see Figure 8.1) with 400 leaves, under varying rates of evolution from low (diameter = 0.2) to high (diameter = 2.0), where the diameter indicates the expected number of changes for a random site on the longest leaf-to-leaf path. The y-axis shows the RF error rate (i.e., the normalized RF distance), so the maximum possible is 1.0.

The K2P model is a DNA sequence evolution model, and neighbor joining and Weighbor are known to be statistically consistent under this model. However, maximum parsimony is not statistically consistent under the K2P model. Hence, the relative performance of neighbor joining and maximum parsimony on these data is striking, since neighbor joining is less accurate than maximum parsimony under all the tested conditions.

This figure also shows other trends that are very interesting. First, Weighbor is very similar to neighbor joining for low diameters, but for high diameters Weighbor is clearly more accurate. The difference between Weighbor and neighbor joining is most noticeable for the highest diameter condition. In fact, Weighbor is designed explicitly to deal better with high rates of evolution, and it does this by considering the statistical model of evolution in a more nuanced way (in particular, by noting that large distances have high variance).

The third interesting observation is the bell-shaped curve for the three methods: at the lowest diameter, the errors start off somewhat high, then decrease as the diameter increases, and then increase again. Bell-shaped curves are quite common, and the explanation is interesting. At the lowest evolutionary rates, there may not be sufficient amounts of change to reconstruct one or more of the edges in the tree. At the highest evolutionary rates, sequences can seem nearly random, and it can be difficult to distinguish signal from noise. Indeed, there tends to be a "sweet spot" at which the rate of evolution is sufficient to reconstruct the tree with high accuracy, but not so high that the noise overcomes the signal.

This study also shows that not all methods respond the same to increases in evolutionary rates; as seen here, neighbor joining in particular has a stronger negative reaction to high evolutionary rates than Weighbor, which seems to not be negatively impacted at all; this

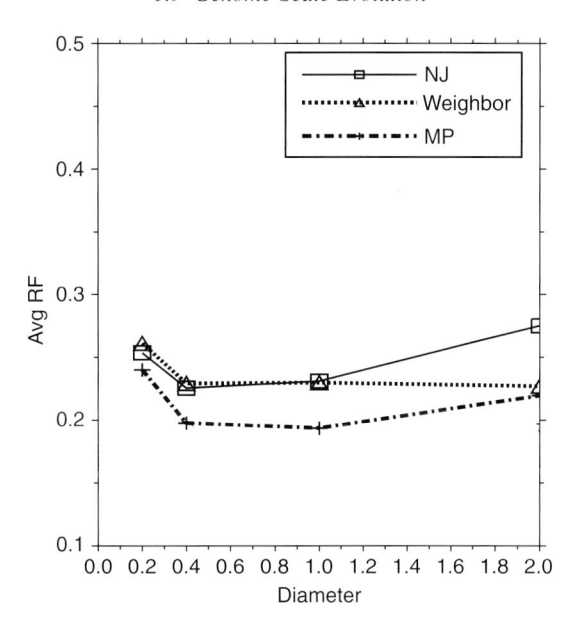

Figure 1.8 (Adapted from Nakhleh et al., 2002) Tree error of three phylogeny estimation methods on simulated datasets with 400 sequences, as a function of the evolutionary diameter (expected number of changes of a random site across the longest path in the tree). The three methods are neighbor joining (NJ), weighted neighbor joining (Weighbor), and maximum parsimony (MP). The sequence datasets were evolved under Kimura 2-parameter (K2P) model trees (Kimura, 1980) with gamma distributed rates across sites, and distances between sequences were computed under the K2P model. The study shows that increasing the evolutionary diameter tends to increase the estimation error, that NJ is the least accurate method on these data, and that MP is the most accurate. The data also suggest that at higher evolutionary diameters, Weighbor might be more accurate than MP.

difference is due to how Weighbor treats large distances in its distance matrix. Maximum parsimony is certainly impacted, but perhaps not as substantially as neighbor joining.

1.9 Genome-Scale Evolution

As we have described it, DNA sequences evolve down a tree under a process that includes substitutions of single nucleotides and insertions and deletions of DNA strings. Yet evolution is even more complex. For example, different biological processes such as horizontal gene transfer, hybridization, gene duplication and loss, and incomplete lineage sorting (Maddison, 1997) can cause different genomic regions to have different evolutionary histories, and make the inference of how a set of species evolved quite challenging. For example, horizontal gene transfer and hybridization require non-tree graphical models to represent the evolutionary history, and so methods that by design can only return trees are unable to correctly reconstruct these evolutionary scenarios. On the other hand, incomplete lineage sorting and gene duplication can also create tremendous heterogeneity, but a species tree is still an appropriate model. With the increased availability of sequence data, phylogenies based on multiple genes sampled from across the genomes of many species are becoming increasingly commonplace (e.g., Jarvis et al., 2014), and it is clear that new methods are needed to estimate species trees (or networks) that take biological causes for gene tree heterogeneity into account. All this makes the design of methods for species phylogeny estimation considerably complex.

1.10 Designing Methods for Improved Accuracy and Scalability

One of the themes in this textbook is the use of algorithmic strategies to improve the accuracy or scalability of phylogeny estimation methods or multiple sequence alignment methods. Examples of such strategies include divide-and-conquer, whereby the sequence dataset is divided into overlapping subsets, trees are computed on the subsets, and a supertree is obtained by combining subset trees. Divide-and-conquer has also been applied to great success in multiple sequence alignment estimation. Iteration is another powerful algorithmic technique, both for alignment estimation and tree estimation. For example, in the context of tree estimation, each iteration uses the tree computed in the previous iteration to decompose the sequence dataset into subsets, constructs trees on subsets using a preferred phylogeny estimation method, and then combines the trees into a tree on the full dataset.

Examples of divide-and-conquer methods include SATé (Liu et al., 2009a, 2012b) and PASTA (Mirarab et al., 2015a), two methods for co-estimating multiple sequence alignments and trees; DACTAL (Nelesen et al., 2012), a method for estimating a tree without estimating an alignment; and DCM1 (Nakhleh et al., 2001a; Warnow et al., 2001), a method that is designed to improve the statistical properties of distance-based methods such as neighbor joining. Similarly, Bayzid et al. (2014) used a modification of DACTAL to improve the scalability and accuracy of MP-EST (Liu et al., 2010), a statistical method for estimating the species tree from trees computed for different parts of the genome, a topic covered in Chapter 10.

In other words, phylogeny estimation methods can be built using other phylogeny estimation methods, with the goal of improving accuracy and/or speed. However, we are also interested in establishing theoretical guarantees under stochastic models of evolution. Therefore, a proper understanding of the graph theory involved in the divide-and-conquer strategies, and the stochastic models of evolution operating on the sequence data that are used to construct the phylogenetic trees (or phylogenetic networks, as the case may be), is also important. The rest of this text provides these foundations.

1.11 Summary

We began with a discussion of some basic (and fairly simple) methods for phylogeny estimation – UPGMA, maximum parsimony, neighbor joining, and the Naive Quartet

1.11 Summary

Method – and how they perform under some simple statistical models of sequence evolution. We observed that these methods have very different theoretical guarantees, and that neighbor joining and the Naive Quartet Method are both statistically consistent under standard sequence evolution models, while UPGMA and maximum parsimony are not. Yet, we also observed that maximum parsimony solved heuristically can be more accurate than neighbor joining, and that the Naive Quartet Method may be unlikely to return any tree at all for large datasets, until the sequence lengths are very large (perhaps unrealistically large). Hence, knowing that a method is statistically consistent and polynomial time does not mean that it is superior on data to another method that may not be statistically consistent.

Later chapters will return to this issue, but under increasingly complex and realistic models of evolution. For example, in Chapter 8 we will discuss the standard sequence evolution models that are used in biological systematics, and the statistical methods that are used to analyze data under these models. Since these models assume sequences evolve only under substitutions, Chapter 9 addresses phylogeny estimation and multiple sequence alignment when sequences evolve also with insertions and deletions. Chapter 10 discusses species tree estimation under genome-scale evolution models in which the different parts of the genome evolve down different trees due to various evolutionary processes. Finally, Chapter 11 describes algorithmic techniques to scale computationally intensive tree estimation methods to large datasets. In each of these chapters, we will explore the theoretical guarantees of methods as well as their performance (in terms of accuracy) on data. In many cases, the theoretical guarantees established for methods provide insight into the conditions in which they will or will not work well, but in some cases there is a gap between theory and practice.

Note that this gap does not imply that the theory is wrong, but only that it does not predict performance very well. In other words, statistical consistency is a statement about asymptotic performance, and so addresses performance given unbounded amounts of data, and theoretical guarantees about asymptotic performance do not have any direct relevance to performance on finite data.

Predicting performance on finite datasets is a fabulously interesting theoretical question, but very little has been established about this. For example, there are some upper bounds that have been established for the sequence lengths that suffice for some methods to return the true tree with high probability under simple sequence evolution models, and some lower bounds as well. But even here, the theory does not provide reliable insights into the relative performance of methods on datasets – even when those datasets are generated under the same models as the theory assumes!

Simply put, it is very difficult to predict the performance of a phylogeny estimation method based just on theory. In other words, the performance of phylogenetic estimation methods is a good example of a more general phenomenon where *in theory, there is no difference between theory and practice, but in practice there is*.[1] The gap between theory and

---

[1] The source of this quote is unknown; it may be Yogi Berra, Jan van de Snepscheut, Walter Savitch, or perhaps others.

practice is one of the major themes in this text, and is one of the reasons that phylogenetic method development and evaluation is such an interesting research area.

This chapter has used simulations to complement the theoretical understanding of methods under stochastic models of evolution. Performing simulation studies is a fundamental part of research in phylogenetics, and is helpful for understanding the performance of existing methods, and hence for designing new methods with improved performance. However, real datasets are also essential, and provide important insight into the difference between how biological datasets evolve and the models of evolution that are used to describe the evolutionary process. The challenge in using biological datasets to evaluate accuracy is that the true evolutionary history is typically at best only partially known, and so differences in trees computed on biological datasets are difficult to evaluate (Iantomo et al., 2013; Morrison et al., 2015). Even so, the best understanding of algorithms and how well they can estimate evolutionary histories depends on using both types of datasets. Appendix C provides further discussion about how to evaluate methods well, including issues such as *how to simulate your data*, *how to vary parameters*, *how to select benchmarks*, and *how to report results*. The challenge to the algorithm developer is to develop methods that have outstanding performance on data and that also have the desirable theoretical guarantees of being statistically consistent and not requiring excessive amounts of data to return the true tree with high probability. Developing the theoretical framework to design methods with strong guarantees, the empirical framework to evaluate methods on data and determine the conditions in which they perform well or poorly, and algorithm design strategies (including divide-and-conquer) that can enable highly accurate methods to scale to large datasets, are the goals of the remaining chapters of this text.

1.12 Review Questions

1. Consider the Cavender–Farris–Neyman (CFN) model. What are the parameters of a CFN model tree? What do these parameters mean?
2. What is meant by the CFN model tree topology?
3. What does it mean to say that a method is statistically consistent for estimating the CFN model tree topology?
4. What is the CFN distance correction? Why is it used?
5. What is the triangle inequality?
6. What are Hamming distances?
7. For a given set S of binary sequences, each of the same length, will the matrix of pairwise Hamming distances satisfy the triangle inequality? Will the matrix of pairwise CFN distances satisfy the triangle inequality?
8. What is the definition of a dissimilarity matrix?
9. What is the definition of an additive matrix?
10. Is a square matrix in which all diagonal entries are 0 and all off-diagonal entries are 1 additive? What about ultrametric?
11. What is the Four Point Condition?

12. What is the Four Point Method? If you were given a 4×4 dissimilarity matrix, would you know how to use the Four Point Method to construct a tree on the matrix?
13. Recall the Naive Quartet Method. What is the input, and how does the Naive Quartet Method operate on the input?
14. Given a model tree and an estimated tree, each on the same set of five leaves, what is the maximum possible number of false positive edges?

1.13 Homework Problems

1. Compute the CFN distance matrix between all pairs of sequences in the set
   - $s_1 = 0011010111$
   - $s_2 = 0011000111$
   - $s_3 = 0011111111$
   - $s_4 = 0011111110$

   Apply the Four Point Method to this dataset. What tree do you get?
2. Suppose e is an edge in a CFN model tree, and $p(e) = 0.1$. What is $\lambda(e)$?
3. Recall the definition of $\lambda(e)$ and $p(e)$ for the CFN model. Write $p(e)$ as a function of $\lambda(e)$.
4. Suppose you have a tree T rooted at leaf R, and R has two children, X and Y, and each of these nodes has two children that are leaves. Hence, T has four leaves: A and B, which are below X, and C and D, which are below Y. Draw T.
5. Suppose you are given a binary tree T on n leaves $s_1, s_2, \ldots, s_n$, with positive branch lengths. Present a polynomial time algorithm to compute the set $Q(T)$ of quartet trees. (See if you can do this in $O(n^4)$ time.)
6. Make up a CFN model tree in which the branch lengths on the edges are all different. Now compute the matrix of the 4×4 distance matrix you get using the branch lengths you wrote down. (Hence your matrix should have values for $\lambda_{A,B}, \lambda_{A,C}, \lambda_{A,D}, \lambda_{B,C}, \lambda_{B,D}$, and $\lambda_{C,D}$.)
   - What is the largest distance in the matrix?
   - What is the smallest distance in the matrix?
7. Consider a rooted tree T where R is the root, the children of R are X and Y, the children of X are A and B, and the children of Y are C and D. Consider the CFN model tree with this rooted topology, where $p(R,X) = p(R,Y) = p(Y,C) = p(Y,D) = 0.1$, and $p(X,A) = p(X,B) = 0.4$
   a. Compute the values for $\lambda(e)$ for every edge e, and draw the CFN tree with these branch lengths.
   b. Compute the CFN distance of the root to every leaf. Is this distance the same for every leaf, or does it depend on the leaf?
   c. Write down the matrix M of leaf-to-leaf CFN distances for this tree.
   d. What is the longest leaf-to-leaf path in this tree?

e. What is the smallest positive value in M?
f. Are the two leaves with this smallest distance siblings in the tree?
g. Write down the three pairwise sums. Which one is the smallest?
h. Is the matrix additive?

8. Consider the same rooted tree T as for the previous problem, but with $p(R,X) = p(R,Y) = p(Y,C) = p(X,A) = 0.1$, and $p(Y,D) = p(X,B) = 0.4$.

    a. Compute the values for $\lambda(e)$ for every edge e, and draw the CFN tree with these branch lengths.
    b. Compute the CFN distance of the root to every leaf. Is this distance the same for every leaf, or does it depend on the leaf?
    c. Compute the matrix M of leaf-to-leaf CFN distances.
    d. What is the longest leaf-to-leaf path in this tree?
    e. What is the smallest positive value in the matrix M? Are the two leaves with this smallest distance siblings in the tree?
    f. Write down the three pairwise sums. Which one is the smallest?
    g. Is the matrix additive?

9. Consider how $\lambda(e)$ is defined by $p(e)$.

    a. Compute $lim_{p(e) \to 0} \lambda(e)$.
    b. Compute $lim_{p(e) \to 0.5} \lambda(e)$.
    c. Graph $\lambda(e)$ as a function of $p(e)$, noting that $0 < p(e) < 0.5$

10. Let A, B, and C be three binary sequences, each of length k, and consider the values for $\hat{\lambda}_{A,B}, \hat{\lambda}_{A,C}$, and $\hat{\lambda}_{B,C}$. Prove or disprove: for all A, B, C, $\hat{\lambda}_{A,B} + \hat{\lambda}_{B,C} \geq \hat{\lambda}_{A,C}$ (i.e., that the triangle inequality holds for estimated CFN distances).

11. Give an example of a 4×4 normalized Hamming distance matrix H so that the Four Point Method applied to H yields a tree T that is different from the tree obtained by using the Four Point Method applied to CFN distances computed for H.

12. Consider the 4×4 matrix with 0 on the diagonal and all off-diagonal entries equal to 4. Prove that the matrix is ultrametric by drawing a rooted tree with edge weights that realizes this matrix, and show that all root-to-leaf paths have the same length.

13. Suppose you have two unrooted trees T_1 and T_2, where T_1 is binary (i.e., all non-leaf nodes have degree three) but T_2 may have nodes with degree greater than three. If you treat T_1 as the true tree and T_2 as the estimated tree, is it necessarily the case that the Robinson–Foulds (RF) error rate is the average of the FN (false negative) error rate and the FP (false positive) error rate? If so, prove it; otherwise give a counterexample.

2
Trees

2.1 Introduction

Mathematically, a tree is a graph (i.e., a pair $G = (V, E)$, where V is the vertex set and E is the edge set) that is connected and acyclic; equivalently, a tree is a graph so that for every pair of vertices v, w in the graph there is a unique path between v and w. When the tree is rooted, we denote its root by $r(T)$. Also, we may denote the set of vertices of T by $V(T)$, the edges by $E(T)$, and the leaves of T by $\mathscr{L}(T)$.

Trees, and especially rooted trees, are used to represent evolutionary histories, and are called "phylogenies," "phylogenetic trees," or "evolutionary trees." Most statistical models of evolution assume that the model tree is a rooted binary tree, so that every node that is not a leaf has exactly two children. However, estimated trees will in general be unrooted and may not be binary. This chapter discusses both rooted and unrooted trees, but we will begin with rooted trees before moving to unrooted trees.

2.2 Rooted Trees

We begin with some basic definitions.

Definition 2.1 In a rooted tree T, we can orient the edges in the direction of the root $r = r(T)$, so that all vertices other than r have outdegree one. Thus, for all nodes $v \neq r$, there is a unique vertex w such that $v \rightarrow w$ is an arc (directed edge) in the tree; w is called the **parent** of v, and v is called the **child** of w. Two or more vertices sharing the same parent are **siblings**. A vertex without any children is called a **leaf** or a **tip**; all other nodes are called **internal nodes**. A vertex with more than two children is a **polytomy**. A rooted tree that has no polytomies is said to be **binary**, **bifurcating**, or **fully resolved**, and is also called a **binary tree**. If a tree is not binary, it is said to be **multifurcating**. The edges of the tree may be referred to as **branches** or even as **inter-nodes**.

If a and b are vertices and there is a directed path from b to a, then a is said to be an **ancestor** of b and conversely b is a **descendent** of a. The most recent common ancestor (**MRCA**) of a set X of leaves in a rooted tree T is the node v that is a common ancestor of all nodes in X, and that is further from the root of T than all other common ancestors.

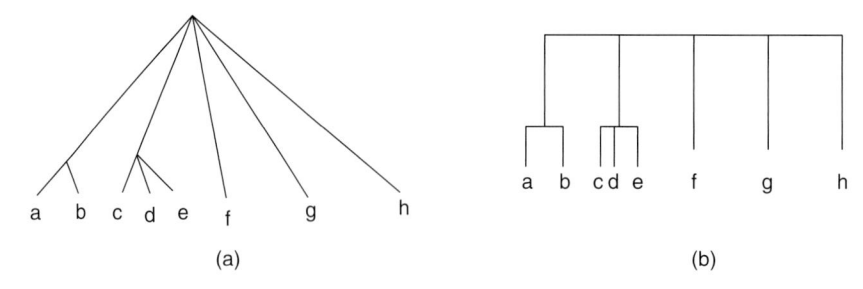

Figure 2.1 Two ways of drawing the same tree

In a phylogenetic tree, the leaves represent the taxa of interest (generally extant species, but sometimes different individuals from the same species) and the internal nodes represent the ancestors of the taxa at the leaves.

A rooted tree can be drawn with the root r on top, on the bottom, on the left, or on the right. However, we'll draw trees with the roots at the top. Graphical representations of trees sometimes include branch lengths, to help suggest relative rates of change and/or actual amounts of elapsed time. The **topology** of the tree is the graphical model without branch lengths.

There are multiple ways to draw the topology of a rooted phylogenetic tree. For example, Figure 2.1 shows two representations of the same evolutionary history. One of these (on the left) is standard in computer science, and the other (on the right) is often found within biological systematics. Note that the root has five children in the tree on the left. Hence, when interpreting a tree in the form given on the right, you should remember that the horizontal lines do not correspond to edges.

2.2.1 Newick Notation for Rooted Trees

Newick notation is the standard way that trees, both rooted and unrooted, are represented in phylogenetic software. The Newick notation for a rooted binary tree with subtrees A and B is given by (A', B'), where A' is the Newick notation for the subtree A and B' is the Newick notation for the subtree B. Since we don't care about the left-to-right ordering of subtrees, it follows that (A', B') is the same thing as (B', A'). Also, the Newick notation for a leaf is the leaf itself.

Some examples of Newick notation should make this process clear. The Newick notation $((a,b),(c,d))$ represents the rooted tree with four leaves, a,b,c,d, with a and b siblings on the left side of the root, and c and d siblings on the right side of the root. The same tree could have been written $((c,d),(a,b))$, or $((b,a),(d,c))$, etc. Thus, the graphical representation is somewhat flexible – swapping sibling nodes (whether leaves or internal vertices in the rooted tree) doesn't change the tree "topology." In fact, there are exactly eight different Newick representations for the rooted tree given in Figure 2.2:

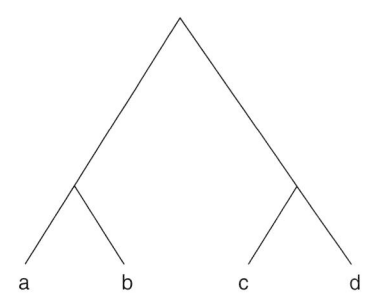

Figure 2.2 Tree $((a,b),(c,d))$

- $((a,b),(c,d))$
- $((b,a),(c,d))$
- $((a,b),(d,c))$
- $((b,a),(d,c))$
- $((c,d),(a,b))$
- $((c,d),(b,a))$
- $((d,c),(a,b))$
- $((d,c),(b,a))$

Similarly, the following Newick strings all refer to the rooted tree in Figure 2.3:

- $((d,e),(c,(a,b)))$
- $((e,d),((a,b),c))$
- $((e,d),(c,(a,b)))$

The second fundamental task is to be able to recognize when two rooted trees are the same when you don't consider branch lengths. For example, the three trees given in Figure 2.3 are different drawings of the same basic tree.

Sometimes the rooted tree you want to represent is not binary. To represent these trees using Newick notation is quite simple. For example, a rooted star tree (i.e., tree without any internal edges) on six leaves a,b,c,d,e,f, is represented by (a,b,c,d,e,f). Similarly, imagine a tree with three children u,v,w off the root, where u has children u_1 and u_2, v has children v_1 and v_2, and w has children w_1 and w_2. A Newick string for the tree is $((u_1,u_2),(v_1,v_2),(w_1,w_2))$. Thus, we can also use Newick strings to represent non-binary trees.

2.2.2 The Clade Representation of a Rooted Tree

Definition 2.2 Let T be a rooted tree in which the leaves are bijectively labeled by a set S; hence, every element of S appears as the label of exactly one leaf. Thus, $\mathscr{L}(T) = S$ and

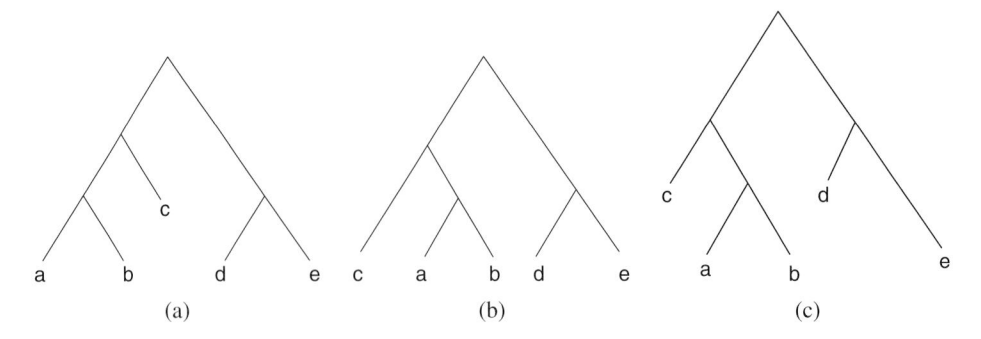

Figure 2.3 Three drawings of the same rooted tree, given by $(((a,b),c),(d,e))$. The trees are considered identical in terms of their topologies (rotations preserve the topology), and branch length does not matter.

is called the **leafset** of T. T_v denotes the subtree of T below node v. The **clades** of T are the subsets of $\mathscr{L}(T)$ that are equal to $\mathscr{L}(T_v)$ for some vertex v.

We now show how to use the clades of a tree to compare it to other trees.

Definition 2.3 Let T be a rooted tree on leafset S. We define the set **Clades(T)** = $\{\mathscr{L}(T_v) : v \in V(T)\}$. Thus, Clades(T) has all the singleton sets (each containing one leaf), a set containing all of S (defined by the root of T), and a clade associated to each vertex of T. The clades that always appear in every possible tree with leafset S are called the **trivial clades**, and all other clades are called the **non-trivial clades**. Thus, all the singleton clades and the set S are trivial clades.

Example 2.4 Consider the rooted tree $T = ((a,b),(c,(d,e)))$. The trivial clades are $\{a\}$, $\{b\}$, $\{c\}$, $\{d\}$, $\{e\}$, and $\{a,b,c,d,e\}$; these appear in every possible tree on the leafset of T. The non-trivial clades are $\{a,b\}$, $\{c,d,e\}$, and $\{d,e\}$. Hence,

$$Clades(T) = \{\{a\},\{b\},\{c\},\{d\},\{e\},\{a,b\},\{d,e\},\{c,d,e\},\{a,b,c,d,e\}\}.$$

Testing if two rooted trees are identical: Determining if two rooted leaf-labeled trees are the same (with all leaves labeled distinctly) can be difficult if they are drawn differently; but this is easy if you examine the clades! Thus, to determine if two trees T and T' are the same, you can write down the set of clades for the two trees, and see if the sets are identical. If $Clades(T) = Clades(T')$, then $T = T'$; otherwise, $T \neq T'$. For example, if you examine the rooted trees in Figure 2.3, you'll see that they all have the same clade sets. Thus, they are all identical.

2.2.3 Using Hasse Diagrams to Construct Rooted Trees from Sets of Clades

To compute a tree from its set of clades, we will draw a Hasse Diagram based on the binary relation R on the set $Clades(T)$, where $\langle A, B \rangle \in R$ if and only if $A \subseteq B$ (see Section B.1.3). It is not hard to see that R is a partial order, and so the set of clades of a tree, under this relation, is a partially ordered set.

To construct the Hasse Diagram for this partially ordered set, we make a graph with vertex set $Clades(T)$ and a directed edge from a node x to a node y if $x \subset y$. Since containment is transitive, if $x \subset y$ and $y \subset z$, then $x \subset z$. Hence, if we have directed edges from x to y, and from y to z, then we know that $x \subset z$, and so can remove the directed edge from x to z without loss of information. This is the basis of the Hasse Diagram: you take the graphical representation of a partially ordered set, and you remove directed edges that are implied by transitivity. Equivalently, for a given subset x, you find the smallest subsets y such that $x \subset y$, and you put a directed edge from x to y.

As we will see, the Hasse Diagram formed for a set $Clades(T)$ is the tree T itself. You can run the algorithm on an arbitrary set of subsets of a taxon set S, but the output may or may not be a tree.

Example 2.5 Consider
$$A = \{\{a\}, \{a,b,c,d\}, \{a,d,e,f\}, \{a,b,c,d,e,f\}\}.$$
On this input, there are four sets, and so the Hasse Diagram will have four vertices. Let v_1 denote the set $\{a\}$, v_2 denote the set $\{a,b,c,d\}$, v_3 denote the set $\{a,d,e,f\}$, and v_4 denote the set $\{a,b,c,d,e,f\}$. Then, in the Hasse Diagram, we will have the following directed edges: $v_1 \to v_2$, $v_1 \to v_3$, $v_2 \to v_4$, and $v_3 \to v_4$. This is not a tree, since it has a cycle (even though this is only a cycle when considering the graph as an undirected graph).

Theorem 2.6 *Let* T *be a rooted tree in which every internal node has at least two children. Then the Hasse Diagram constructed for* Clades(T) *is identical to* T.

Proof We prove this by strong induction on the number n of leaves in T. For $n = 1$, then T consists of a single node (since every node has at least two children). When we construct the Hasse Diagram for T, we obtain a single node, which is the same as T.

The inductive hypothesis is that the statement is true for all positive n up to $N - 1$, for some arbitrary positive integer N. We now consider a tree T with N leaves for which every internal node has at least two children. Since the root of T has at least two children, we denote the subtrees of the root as $t_1, t_2, \ldots, t_k$ (with $k \geq 2$). Note that $Clades(T) = Clades(t_1) \cup Clades(t_2) \cup \ldots \cup Clades(t_k) \cup \mathscr{L}(T)$. The set of vertices for the Hasse Diagram on T contains one vertex for $\mathscr{L}(T)$ and then each of the vertices for the Hasse Diagrams on the trees t_i, $i = 1, 2, \ldots, k$. Also, every directed edge in the Hasse Diagram on T is either a directed edge in the Hasse Diagram on some t_i, or is the directed edge from $\mathscr{L}(t_i)$ to $\mathscr{L}(T)$. By the inductive hypothesis, the Hasse Diagram defined on $Clades(t_i)$ is isomorphic to t_i for $i = 1, 2$, and hence the Hasse Diagram defined on $Clades(T)$ is isomorphic to T. □

2.2.4 Compatible Sets of Clades

Up to now, we have assumed we were given the set $Clades(T)$ for a binary tree T, and we wanted to construct the tree T from that set. Here we consider a related question: Given a set $\mathscr{X}$ of subsets of a set S of taxa, is there a tree T so that $\mathscr{X} \subseteq Clades(T)$?

Definition 2.7 A set $\mathscr{X}$ of sets is said to be **compatible** if and only if there is a rooted tree T with each leaf in T given a different label, so that $\mathscr{X} \subseteq Clades(T)$.

To answer this, see what happens when you construct the Hasse Diagram for the set $\mathscr{X}$.

Example 2.8 We begin with a simple example, $\mathscr{X}_1 = \{\{a,b,c\},\{d,e,f\},\{a,b\}\}$. Note that $\mathscr{X}_1$ contains three subsets and the set S contains six elements. Thus, $\mathscr{X}_1$ does not contain the singleton sets, nor the full set of leaves, and so it is not possible for $\mathscr{X}_1$ to be equal to the set of clades of any tree; and as we observe, the Hasse Diagram we construct is not connected and so is not a tree. Therefore, we add all the trivial clades (the singletons and the full set of leaves) to $\mathscr{X}_1$ and obtain $\mathscr{X}_1'$. We then compute the Hasse Diagram on this set. The result is a tree T, with Newick string $(((a,b),c),(d,e,f))$. This is not a binary tree, but it *is* a tree, and $\mathscr{X}_1 \subset Clades(T)$. However, there are other trees, such as T' denoted by $(((a,b),c),(d,(e,f)))$, that also satisfy $\mathscr{X}_1 \subset Clades(T')$. Note that T can be derived from T' by contracting an edge in T' (i.e., removing an edge and merging the endpoints into a single vertex).

Example 2.9 Consider $\mathscr{X}_2 = \{\{a,b\},\{b,e\},\{c,d\}\}$. Note that the set S contains five elements, but these singleton sets do not appear in $\mathscr{X}_2$. Similarly, the full set S is not in $\mathscr{X}_2$. Hence, $\mathscr{X}_2$ is not the set of clades of any tree. We add all the trivial clades to $\mathscr{X}_2$ to obtain $\mathscr{X}_2'$, and construct the Hasse Diagram for $\mathscr{X}_2'$. Note that $\{b\} \subset \{a,b\}$ and $\{b\} \subset \{b,e\}$. Hence, the Hasse Diagram for $\mathscr{X}_2'$ has a node with outdegree two – which is inconsistent with X_2' being the subset of $Clades(T)$ for any tree T.

2.2.5 Hasse Diagram Algorithm

These two examples suggest an algorithm that you can use to determine if a set of clades is compatible. We call the algorithm the Hasse Diagram algorithm, since it operates by computing a Hasse Diagram, and then checking to see if the Hasse Diagram is a tree.

The input will be a set $\mathscr{X}$ of subsets of a taxon set. The output will either be a rooted tree T for which $\mathscr{X} \subseteq Clades(T)$, establishing that $\mathscr{X}$ is compatible, or *Fail*.

- Step 1: Compute $S = \cup_{X \in \mathscr{X}} X$ (i.e., all the elements that appear in any set in $\mathscr{X}$). Let $S = \{s_1, s_2, \ldots, s_n\}$. Define $\mathscr{X}' = \mathscr{X} \cup S \cup \{s_1\} \cup \{s_2\} \cup \ldots \cup \{s_n\}$; in other words, $\mathscr{X}'$ is the set of subsets of S obtained by adding the full set S and all the singletons to $\mathscr{X}$.
- Step 2: Construct the Hasse Diagram for $\mathscr{X}'$.
- Step 3: If the Hasse Diagram is a tree T, then return T; otherwise return *Fail*.

If the only interest is in determining whether the set of clades is compatible, the following well-known theorem, perhaps originally from Estabrook et al. (1975), is useful:

Lemma 2.10 *A set $\mathcal{X}$ of subsets is compatible if and only if for any two elements X_1 and X_2 in $\mathcal{X}$, either X_1 and X_2 are disjoint or one contains the other.*

Proof If a set $\mathcal{X}$ of subsets is compatible, then there is a rooted tree T on leafset S, in which every leaf has a different label, so that each element in $\mathcal{X}$ is the set of leaves below some vertex in T. Let X_1 and X_2 be two elements in $\mathcal{X}$, and let x_1 be the vertex of T associated to X_1 and x_2 be the vertex associated to X_2. If x_1 is an ancestor of x_2, then X_1 contains X_2, and similarly if x_2 is an ancestor of x_1 then X_2 contains X_1. Otherwise neither is an ancestor of the other, and the two sets are disjoint. For the reverse direction, note that when all pairs of elements in set $\mathcal{X}$ satisfy this property, then the Hasse Diagram will be a tree T so that $\mathcal{X} = Clades(T)$. □

The following corollary is very useful in algorithm design, and its proof is left to the reader.

Corollary 2.11 *A set $\mathcal{X}$ of subsets of S is compatible if and only if every pair of elements in $\mathcal{X}$ is compatible.*

2.3 Unrooted Trees

According to its mathematical definition, a tree is just a connected acyclic graph; hence, unrooted trees are actually the norm in mathematics. The use of rooted trees is dominant in computer science, however, and also makes sense in phylogenetics. However, the output of phylogenetic software is generally just an unrooted tree. One reason that the output is generally unrooted is that most methods for computing trees are based on time-reversible models of sequence evolution, which means that the root of a tree is not identifiable. However, in practice biologists often use outgroups (i.e., taxa that are clearly not as closely related to the rest of the input taxa as the remaining taxa are to each other) to root trees. Unfortunately, outgroup rooting is also challenging (for reasons that we will discuss later). Hence, typically phylogenetic trees are unrooted trees, and being able to switch between thinking about rooted trees and their unrooted versions is an important skill.

Definition 2.12 The **unrooted version** of a rooted tree T is denoted by T_u, and is formed by suppressing the root and then treating the tree as unrooted. Thus, if the root of T has two children, then these two children are made adjacent to each other, and if the root has more than two children then the graphical model is unchanged.

Consider, for example, the rooted tree T given in Figure 2.4. To turn this into its unrooted version T_u, we would ignore the location of the root, and we'd obtain the unrooted tree given in Figure 2.5. Thus, each rooted tree has a unique unrooted version. However, if we were given the unrooted tree in Figure 2.5, there would be multiple ways of producing rooted

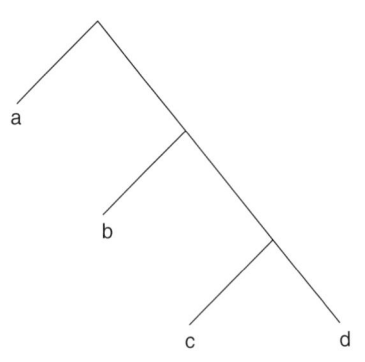

Figure 2.4 Rooted tree $(a,(b,(c,d)))$.

versions. For example, Figures 2.4 and 2.6 are both rooted versions of the same unrooted tree. In fact, you can generate rooted trees from an unrooted tree by picking up the tree at any edge, or at any node. You can even pick up the tree at one of its leaves, but then the tree is rooted at one of its own taxa – which we generally don't do (in that case, we'd root it at the edge leading to that leaf instead, thus keeping the leafset the same).

Definition 2.13 Every node in an unrooted tree is either a **leaf** (in which case it has degree one) or an **internal node**. Two or more leaves with a common neighbor are **siblings**.

The unrooted tree in Figure 2.5 has four leaves, a,b,c,d. We consider a and b to be siblings because they share a common neighbor, and similarly c and d are siblings.

2.3.1 Newick Notation for Unrooted Trees

Notice that the unrooted tree shown in Figure 2.5 was presented with its Newick notation. In fact, to obtain a Newick string for an unrooted tree T, just look at a rooted version of T and write down its Newick string. However, each rooted tree has multiple ways of representing it in Newick format, and every unrooted tree can be rooted in multiple ways. Thus, every unrooted tree will have several Newick representations, each of which is completely valid.

2.3.2 The Bipartitions of an Unrooted Tree

To determine if two unrooted trees are the same, we do something similar to what we did to determine if two rooted trees are the same. The bipartitions of an unrooted tree are formed by taking each edge in turn, and writing down the two sets of leaves that would be formed by deleting that edge. We use $\pi(e)$ to denote the bipartition defined by edge e, with $\pi(e) = A|B$, where A is one half of the bipartition and B is the other. Note that when the edge e is incident to a leaf, then $\pi(e)$ splits the set of leaves into one set with a

2.3 Unrooted Trees

Figure 2.5 Unrooted tree $((a,b),(c,d))$

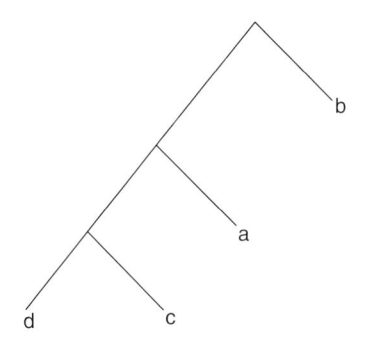

Figure 2.6 Rooted tree $(b,(a,(c,d)))$.

single leaf, and the other set with the remaining leaves. These bipartitions are present in all trees with any given leafset, and hence are called **trivial bipartitions**. Hence, we will focus the discussion just on the **non-trivial bipartitions**. We summarize this discussion with the following definition:

Definition 2.14 Given an unrooted tree T with no nodes of degree two, the **bipartition encoding** of T, denoted by $C(T) = \{\pi(e) : e \in E(T)\}$, is the set of bipartitions defined by each edge in T, where $\pi(e)$ is the bipartition on the leafset of T produced by removing the edge e (but not its endpoints) from T. We also refer to $C(T)$ as the character encoding or split encoding of T; see Figure 2.7 for an example of a tree T and its split encoding $C(T)$. If we restrict this set to the bipartitions formed by the internal edges of the tree T, we obtain $C_I(T)$.

2.3.3 Representing Unresolved Unrooted Trees

Sometimes the unrooted tree we wish to represent is not fully resolved, which means it has nodes of degree greater than three. How do we represent such a tree? For example, consider the tree that has one internal node and four leaves, a,b,c,d. We can represent this

(a) Unrooted tree T

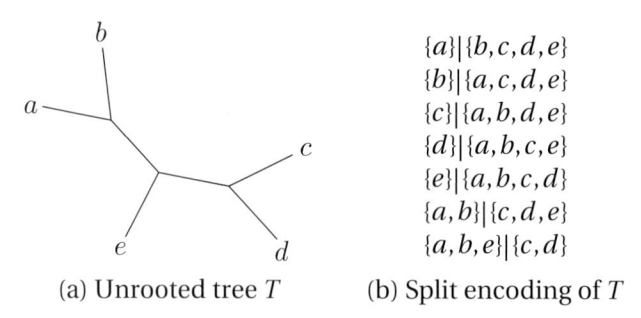

$$\{a\}|\{b,c,d,e\}$$
$$\{b\}|\{a,c,d,e\}$$
$$\{c\}|\{a,b,d,e\}$$
$$\{d\}|\{a,b,c,e\}$$
$$\{e\}|\{a,b,c,d\}$$
$$\{a,b\}|\{c,d,e\}$$
$$\{a,b,e\}|\{c,d\}$$

(b) Split encoding of T

Figure 2.7 (Figure 5.2 in Huson et al., 2010) We show an unrooted tree T and its set of bipartitions, $C(T)$, referred to here as the split encoding of T. The bottom two splits are the non-trivial splits, and come from the internal edges in T; the top five splits are derived from the edges that are incident with leaves, and are the trivial splits.

simply by (a,b,c,d). Note also that representing it by $(a,(b,c,d))$ yields the same *unrooted* tree. Similarly, what about a tree that has six leaves, a,b,c,d,e,f, and one internal edge that separates a,b, from c,d,e,f? We can represent this unrooted tree by $(a,b,(c,d,e,f))$, or any of the alternatives that also yield one single bipartition separating a,b from the remaining leaves.

Sometimes, if the tree has only a single bipartition, we will simplify our representation by just giving the bipartition; i.e., we represent the tree with one edge separating a,b from c,d,e,f by $\{a,b\}|\{c,d,e,f\}$, or more simply by $ab|cdef$. In other words, the representations for trees that appear in the mathematical literature are quite flexible. (Of course, representations of trees in software must be done very precisely, using the requirements for the software ... but that is another matter.)

2.3.4 Comparing Unrooted Trees Using Their Bipartitions

It is easy to see that we can write down the set of bipartitions of any given unrooted tree, and that two unrooted trees are identical if they have the same set of bipartitions. An important additional concept is **refinement**, which we now describe.

Definition 2.15 Given two trees T and T' on the same set of leaves (and each leaf given a different label), tree T is said to **refine** T' if T' can be obtained from T by contracting a set of edges in T. We also express this by saying T **is a refinement of** T' and T' **is a contraction of** T. In fact, T refines T' if and only if $C(T') \subseteq C(T)$.

Note that each tree refines itself and is also a contraction of itself (since we can choose to contract an empty set of edges).

Definition 2.16 An unrooted tree T is **fully resolved** if there is no tree $T' \neq T$ that refines T. Equivalently, T is fully resolved if all the nodes in T have degree one or three.

2.3 Unrooted Trees

An unrooted tree that is fully resolved is also called a **binary tree**. (Note, however, that we also referred to rooted binary trees, so that "binary tree" has a slightly different meaning for rooted and unrooted trees.)

2.3.5 Constructing Unrooted Trees from Their Bipartitions

Sometimes we are given a set A of bipartitions, and we are asked whether these bipartitions could co-exist within a tree (i.e., whether there exists a tree T so that $A \subseteq C(T)$). When this is true, the set of bipartitions is said to be *compatible*, and otherwise the set is said to be *incompatible*.

Definition 2.17 A set A of bipartitions on the set S is **compatible** if there exists an unrooted tree T in which every leaf has a distinct label from a set S, so that $A \subseteq C(T)$.

To construct a tree from a compatible set A of bipartitions, we will use A to construct a set C of clades that will be compatible if and only if the set A is compatible. We will then run the Hasse Diagram Algorithm from Section 2.2.5 on the set C. If C is a compatible set of clades, this will return a rooted tree T realizing the set C. Then, to construct the unrooted tree for A, we will return T_u, the unrooted version of T (see Definition 2.12). T_u is then the **canonical tree** for the set A.

To complete this description, we only need to say how we compute the set C of clades given A. First, we add all of the missing trivial bipartitions (the ones of the form $x|S \setminus \{x\}$) to A. Then, pick any leaf (call it "r") in the set to function as a root. This has the result of turning the unrooted tree into a rooted tree, and therefore turns the bipartitions into clades! In other words, for each bipartition $A|B$, we write down the subset that does not contain r, and denote it as a clade. We also include S (the full set of taxa) and $\{x\}$ for each $x \in S$ (the singleton sets). This is set C of clades we obtain from the set A of bipartitions.

Example 2.18 We will determine if the set A of bipartitions given by

$$A = \{(123|456789), (12345|6789), (12|3456789), (89|1234567)\}$$

is compatible, and if so we will construct its canonical tree. First, we decide to root the tree at leaf 1. We look at each bipartition, and select the half of the bipartition that does not contain 1. Thus, we obtain the following set of clades:

$$\{\{4,5,6,7,8,9\}, \{6,7,8,9\}, \{3,4,5,6,7,8,9\}, \{8,9\}\}$$

We then add the full set S and all the singleton sets, and construct a Hasse Diagram for this set of sets. Note that every non-root node in the Hasse Diagram for this set of sets has outdegree 1, and hence defines a rooted tree given by $(1,(2,(3,(4,5,((6,7),(8,9))))))$. Although we treat 1 as a root in order to form clades, this technique produces a tree in which 1 is a leaf and not the root. We then unroot this tree to obtain the tree given in Figure 2.8.

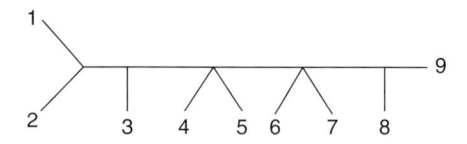

Figure 2.8 Unrooted tree on $\{1\ldots9\}$, obtained by running the Hasse Diagram algorithm on the set $A = \{(123|456789), (12345|6789), (12|3456789), (89|1234567)\}$; see Example 2.18.

2.3.6 Testing Compatibility of a Set of Bipartitions

What we have described is an algorithm that will construct a tree from a set of compatible bipartitions. With a simple modification, we can use the algorithm to test if a set of bipartitions is compatible: When we construct the Hasse Diagram, we check that it creates a tree. If it does, then we return "Compatible" and otherwise we return "Not Compatible." It is easy to verify that this method returns the correct answer when the set is compatible. What about when the set is not compatible? We demonstrate this with an example.

Example 2.19 Suppose the set of bipartitions has two bipartitions $ab|cd$ and $ac|bd$. We root the bipartitions at leaf a, and obtain the non-trivial clades $\{c,d\}, \{b,d\}, \{b,c,d\}$. We add $\{a,b,c,d\}$ and the singleton sets. When we compute the Hasse Diagram, we note that the graph has a cycle (as an undirected graph) on the vertices for clades $\{d\}, \{c,d\}, \{b,d\}$, and $\{b,c,d\}$. Hence, the Hasse Diagram is not a tree, and the algorithm returns "Not Compatible."

Pairwise compatibility ensures setwise compatibility: Just as we saw with testing compatibility for clades, it turns out that bipartition compatibility has a simple characterization, and pairwise compatibility ensures setwise compatibility.

Theorem 2.20 *A set* A *of bipartitions on a set* S *is compatible if and only if every pair of bipartitions is compatible. Furthermore, a pair of bipartitions* $X_1|X_2$ *and* $Y_1|Y_2$ *of bipartitions is compatible if and only if at least one of the four pairwise intersections* $X_i \cap Y_j$ *is empty.*

Proof We begin by proving that a pair of bipartitions is compatible if and only if at least one of the four pairwise intersections is empty. It is easy to see that a pair of bipartitions is compatible if and only if the clades produced (for any way of selecting the root) are compatible. So let's assume that we set some arbitrary element s of S to be the root, and that $s \in X_1 \cap Y_1$. Therefore, X and Y are compatible as bipartitions if and only if X_2 and Y_2 are compatible as clades. Therefore, X and Y are compatible as bipartitions if and only if one of the following statements holds:

- $X_2 \subseteq Y_2$
- $Y_2 \subseteq X_2$
- $X_2 \cap Y_2 = \emptyset$

If the first condition holds, then $X_2 \cap Y_1 = \emptyset$, and at least one of the four pairwise intersections is empty. Similarly, if the second condition holds, then $Y_2 \cap X_1 = \emptyset$, and at least one of the four pairwise intersections is empty. If the third condition holds, then directly at least one of the four pairwise intersections is empty. Thus, if X and Y are compatible as bipartitions, then at least one of the four pairwise intersections is empty.

For the converse, suppose that X and Y are bipartitions on S, and at least one of the four pairwise intersections is empty; we will show that X and Y are compatible as bipartitions. First, not all four pairwise intersections can be empty. For example, suppose $X_1 \cap Y_1 \neq \emptyset$ and let $s \in X_1 \cap Y_1$. To show that X and Y are compatible as bipartitions it will suffice to show that X_2 and Y_2 are compatible as clades. Since $X_1 \cap Y_1 \neq \emptyset$, the pair that produced the empty intersection must be one of the other pairs: i.e., one of the following must be true: $X_1 \cap Y_2 = \emptyset, X_2 \cap Y_2 = \emptyset$, or $X_2 \cap Y_1 = \emptyset$. If $X_1 \cap Y_2 = \emptyset$, then $Y_2 \subseteq X_2$, and X_2 and Y_2 are compatible clades; thus, X and Y are compatible bipartitions. If $X_2 \cap Y_1 = \emptyset$, then a similar analysis shows that $X_2 \subseteq Y_2$, and so X and Y are compatible bipartitions. Finally, if $X_2 \cap Y_2 = \emptyset$, then directly X_2 and Y_2 are compatible clades, and so X and Y are compatible bipartitions.

Now that we have established that two bipartitions are compatible if and only if at least one of the four pairwise intersections is empty, we show that a set of bipartitions is compatible if and only if every pair of bipartitions is compatible. So let $s \in S$ be selected arbitrarily as the root, and consider all the clades (halves of bipartitions) that do not contain s. This set of subsets of S is compatible if and only if every pair of subsets is compatible, by Lemma 2.10. Hence, the theorem is proven. □

Hence, if all we want to do is determine if a set of bipartitions is compatible but we don't need to construct the tree demonstrating the compatibility, we can just check that every pair of bipartitions is compatible. If all pairs are compatible, then we return "Compatible," and otherwise we return "Not Compatible."

2.4 Constructing the Strict Consensus Tree

A common event in phylogeny estimation is that a set of trees is computed for a given dataset, and we will wish to compute a consensus of these trees.

Definition 2.21 The **strict consensus tree** of a set $\mathscr{T} = \{T_1, T_2, \ldots, T_k\}$ of trees is the most resolved common contraction of the trees in $\mathscr{T}$. Hence, T is the strict consensus of $\mathscr{T}$ if and only if every tree $T_i \in \mathscr{T}$ refines T, and every other tree satisfying this property is a refinement of T (see Definition 2.15). Furthermore, the strict consensus tree T satisfies $C(T) = \cap_i C(T_i)$.

2.5 Quantifying Error in Estimated Trees

Since the true tree is unknown, determining how close an estimated tree is to the true tree is typically difficult. However, for the purposes of this section, we will presume that the true tree is known (perhaps because we have performed a simulation), so that we can compare estimated trees to the true tree. Furthermore, since estimated trees are generally unrooted, we will compare the unrooted estimated tree to the unrooted version of the true tree.

Let us presume that the tree T_0 on leafset S is the true tree, and that another tree T is an estimated tree for the same leafset. There are several techniques that have been used to quantify errors in T with respect to T_0, of which the dominant ones are these:

False negatives (FN): The **false negatives** are those edges in T_0 inducing bipartitions that do not appear in $C(T)$; these are also called "missing branches." The **false negative rate** is the fraction of the total number of non-trivial bipartitions that are missing, or $\frac{|C(T_0) \setminus C(T)|}{|C(T_0)|}$.

False positives (FP): The **false positives** in a tree T with respect to the tree T_0 are those edges in T that induce bipartitions that do not appear in $C(T_0)$. The **false positive rate** is the fraction of the total number of non-trivial bipartitions in T that are false positives, or $\frac{|C(T) \setminus C(T_0)|}{|C(T)|}$.

Robinson–Foulds (RF): The most typically used error metric is the sum of the number of FPs and FNs, and is called the **Robinson–Foulds distance** or the **bipartition distance**. The RF rate is obtained by dividing the number of FNs and FPs by $2n - 6$ where n is the number of leaves in each tree.

When both the estimated and true trees are binary, then FN and FP rates are equal, and these also equal the RF distance. The main advantage in splitting the error rate into two parts (FN and FP) is that many estimated trees are not binary. In this case, when the true tree is presumed to be binary, the FP error rate will be less than the FN error rate. Note also that the reverse can happen – the FN error rate could be smaller than the FP error rate – when the true tree is not binary. Also note that the RF error rate is not necessarily equal to the average of the FN and FP error rates. Finally, the RF rate of a **star tree** (i.e., a tree with no internal edges, see Definition 2.25) with respect to a binary true tree is 50 percent, which is the same as the RF rate for a completely resolved tree that has half of its edges correct. Using the RF rate has been criticized because of this phenomenon, since it tends to favor unresolved trees (Rannala et al., 1998). Therefore, when estimated trees may not be binary, it makes sense to report both FN and FP rates instead of just the RF rate.

Observation 2.22 Let T be the true tree, and T_1 and T_2 be two estimated trees for the same leafset. If T_1 is a refinement of T_2, then the number of FNs of T_1 will be less than or equal to that of T_2, and the number of FPs of T_1 will be at least that of T_2.

This observation will turn out to be important in understanding the relationship between the error rates of consensus trees (described in Chapter 6) and how they compare to the error rates of the trees on which they are based.

2.6 The Number of Binary Trees on n Leaves

Since we are interested in estimating phylogenetic trees, knowing the number of possible leaf-labeled binary trees (rooted or unrooted), where each leaf has a different label drawn from $\{1, 2, \ldots, n\}$, is relevant to understanding the computational challenges in exploring "treespace."

We first consider the unrooted case. For $n = 1, 2$, or 3, the answer is 1: there is only one unrooted binary tree when $n \leq 3$. However, when $n = 4$, there are three possible trees. Furthermore, it is easy to see this algorithmically: to construct a tree on $n = 4$ leaves, $s_1, s_2, s_3,$ and s_4, take a tree T on $n = 3$ leaves, and then add the remaining leaf by subdividing an edge in the tree T, and making s_4 adjacent to this newly introduced node. Thus, the number of possible trees on $n = 4$ leaves is equal to the number of edges in T. Since T has three leaves, it has exactly three edges (you can see this by drawing it). Hence, there are three unrooted binary trees on four leaves.

Things become a bit more difficult for larger values of n, but the same algorithmic analysis applies. Take a tree T on $n-1$ leaves, pick an edge in T and subdivide it, and make s_n adjacent to the newly created node. The number of unrooted binary trees on n leaves is therefore equal to the product of the number t_{n-1} of unrooted binary trees on $n-1$ leaves and the number e_{n-1} of edges in an unrooted binary tree on $n-1$ leaves. It is not hard to see that $e_{n-1} = 2(n-1) - 3 = 2n - 5$. Hence, t_n satisfies $t_n = t_{n-1}(2n-5)$, and $t_3 = 1$. Thus, for $n \geq 3$, $t_n = (2n-5)!! = (2n-5)(2n-7)\ldots 3$.

Now, we examine the number of different rooted binary trees on n leaves. To compute this, note that every rooted binary tree T on n leaves defines an unrooted binary tree T_u (obtained by ignoring the root of T), and that every unrooted binary tree T_u corresponds to $2n-3$ rooted binary trees formed by rooting the tree T_u on one of its edges. Hence, the number of rooted binary trees on n leaves is $(2n-3)!! = (2n-3)(2n-5)\ldots 3$.

2.7 Rogue Taxa

Sometimes a phylogenetic analysis of a collection of taxa will produce many trees with nearly identical scores (for whatever optimization problem is used) that differ primarily in terms of where a particular taxon is placed. Such a taxon is called a "rogue taxon" in the biological literature (Sanderson and Shaffer, 2002). Because the inclusion of rogue taxa in a phylogenetic analysis can increase the error of the phylogenetic analysis, they are often removed from the dataset before the final tree is reported.

Causes for rogue taxa vary, but a common cause is having a distantly related outgroup taxon in the dataset (the subject of the next section). The sequences for such taxa can be extremely different from all other sequences in the dataset, so that there is close to no similarity beyond what two random sequences would have to each other. In the extreme case of using a random sequence, the taxon with the random sequence could fit equally well into any location of the tree, and hence its location cannot be inferred with any reliability. When a phylogenetic analysis explores multiple optimal or near-optimal trees for the dataset, this will mean that the profile (set of trees) computed for the dataset will include trees that differ substantially in the placement of the rogue taxon. The strict consensus tree of any such set of trees will be highly unresolved, and may even be a star tree (see Definition 2.25).

2.8 Difficulties in Rooting Trees

Although evolutionary trees are rooted, estimations of evolutionary trees are almost always unrooted, for a variety of reasons. In particular, unless the evolutionary process obeys the strict molecular clock, so that the expected number of changes is proportional to the time elapsed since a common ancestor, rooting trees requires additional information. The typical technique is to use an "outgroup" (a taxon which is not as closely related to the remaining taxa as they are to each other). The outgroup taxon is added to the set of taxa and an unrooted tree is estimated on the enlarged set. This unrooted tree is then rooted by "picking up" the unrooted tree at the outgroup. See Figure 2.9, where we added a fly to a group of mammals. If you root the tree on the edge incident with fly, you obtain the rooted tree $(fly, (cow, (chimp, human)))$, showing that chimp and human have a more recent common ancestor than cow has to either human or chimp.

The problem with this technique is subtle: While it is generally easy to pick outgroups, the less closely related they are to the remaining taxa, the less accurately they are placed in the tree. That is, very distantly related taxa tend to fit equally well into many places in the tree, and thus produce incorrect rootings. See Figure 2.10, where the outgroup attaches to two different places within the tree on the remaining taxa.

Figure 2.9 Tree on some mammals with fly as the outgroup.

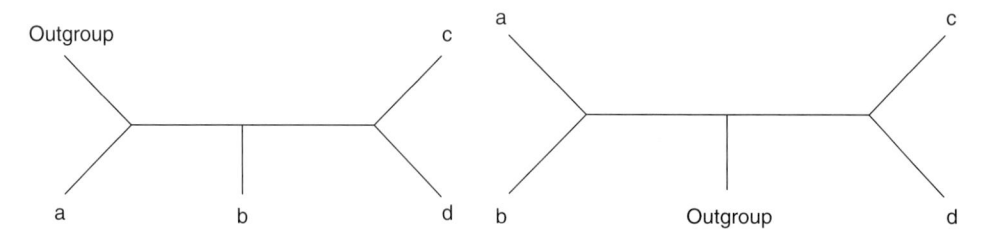

Figure 2.10 Two unrooted trees that differ only in the placement of the outgroup. If these trees were rooted at the outgroup, they would produce different rooted trees on the ingroup taxa a, b, c, d.

Furthermore, it is often difficult to distinguish between an outgroup taxon that is closely related to the ingroup taxa, and a taxon that is, in fact, a member of the same group that branched off early in the group's history. When this happens, even if the unrooted tree is correct, the rooted version of the unrooted tree will be wrong. For this reason, even the use of outgroups is somewhat difficult.

2.9 Homeomorphic Subtrees

Many of the approaches to constructing large trees operate by computing smaller trees and then combining these smaller trees. To understand these techniques, we need to understand the notion of "homeomorphic subtree," which we now define.

Suppose we are given a set X of taxa and an unrooted tree T with leafset S, and suppose $X \subset S$. Now, consider the tree that is obtained by removing from T every node and edge that is not on a path between two leaves in X; this produces a tree that will have at least one node (and possibly many nodes) of degree two. Now, we modify T further so that it has no nodes of degree two, as follows. If T has any nodes of degree two, then it has at least one path $P = v_1, v_2, \ldots, v_t$ such that all of its internal nodes (i.e., v_i for $i = 2, 3, \ldots, t-1$) have degree two, and its endpoints have degree greater than two. Find any such path P, and replace P by the single edge (v_1, v_t). Repeat this process until there are no degree-two nodes. We will denote this final tree by $T|X$, and refer to the final tree produced by this process as the **homeomorphic subtree of** T **induced by** X (or sometimes more simply just as "induced subtree").

When the tree T is rooted, the process for computing the homeomorphic subtree is nearly the same, except that the path P can never have the root as an internal node, and the root (which may well have degree two) is allowed to be an endpoint of P. Hence, the result of suppressing nodes of degree two maintains the location of the root.

Homeomorphic subtrees are often used when you have a tree T (rooted or unrooted) but you are only interested in what the tree says about a particular subset of its leafset. For example, T could have leafset $\{a, b, c, d, e, f, g, h, i\}$, but you are only interested in the evolutionary history of $a, b, c,$ and d. To understand what T tells you about just their evolutionary history, you should construct the homeomorphic subtree induced by T on $\{a, b, c, d\}$. An example of this is shown in Figure 1.5.

Homeomorphic subtrees are also relevant when we try to construct large trees using divide-and-conquer, so that smaller trees are computed and then combined into larger trees. In that case, the hope is that the smaller trees that have been computed are all "correct" in the sense that they would all be homeomorphic subtrees of a common larger tree.

2.10 Some Special Trees

Some types of trees are used frequently as examples to illustrate different properties of algorithms.

Definition 2.23 The **caterpillar** tree on n leaves $s_1, s_2, \ldots, s_n$ is $(s_1, (s_2, (s_3, (s_4, \ldots))))$. The caterpillar tree is also referred to as the **comb** tree, and the tree is described as being **pectinate** or **ladder-like**. Note that the maximum distance (in terms of the number of edges) between any two leaves in a caterpillar tree on n leaves is $n-1$, and that the maximum pairwise distance in any other tree on n leaves is smaller.

Definition 2.24 The **completely balanced** tree on 2^n leaves is a rooted tree, where the root is the parent of the roots of two completely balanced trees on 2^{n-1} leaves. Note that in the completely balanced tree on 2^n leaves, the maximum distance between any two leaves is only $2n$, and that any other binary tree on 2^n leaves will have a larger maximum pairwise distance.

Definition 2.25 The **star tree** on n leaves has one internal node that is adjacent to all its leaves. Thus, the star tree has no internal edges.

2.11 Further Reading

In this chapter we presented three ways (false positives, false negatives, and Robinson–Foulds) to quantify error rates in phylogeny estimation methods. While these techniques (and especially the RF error metric) remain the most commonly used, they have some limitations, and other methods have been proposed in order to address these limitations.

One of the limitations of these criteria is that they are vulnerable to rogue taxa. For example, consider the following pair of caterpillar trees:

$$T = (s_1, (s_2, (s_3, (\ldots (s_{n-1}, s_n) \ldots)$$
$$T' = ((s_n, s_1), (s_2, (s_3, (\ldots (s_{n-2}, s_{n-1}) \ldots)$$

Thus, T' is obtained by taking s_n and moving it to the other end of T. By design, T and T' share no common bipartitions, and hence have the largest possible FP, FN, and RF distances. Methods for quantifying distances between trees that are less affected by rogue taxa include the quartet distance (i.e., the number of quartets of taxa that the two trees resolve differently), the size of the maximum agreement subtree (i.e., the largest number of taxa on which the two trees have the same homeomorphic subtree see Chapter 6), and a metric proposed by Lin et al. (2012a).

Another weakness of these three metrics is that they do not take branch length into account; this was rectified by Kuhner and Felsenstein (1994), who presented the "branch score," which is the sum of the squares of the differences in branch lengths, where an edge has branch length 0 if it does not appear in the tree. The branch score metric can also be used to take branch support (see Section 1.7) into account, by treating a missing branch as having no support and hence zero branch length, and a branch that is present as having the length equal to its support (however that support is calculated). When one of the trees is the model tree, then support values would be 1 for the edges that are present and otherwise 0.

2.12 Review Questions

1. Is the Newick representation of a rooted tree unique, or can there be multiple Newick representations of any given tree? What about unrooted trees?
2. Let T be a rooted binary tree on ten leaves. How many clades does T have, including the singleton clades and the full set of taxa?
3. Let T_u be an unrooted binary tree on ten leaves; how many bipartitions does T_u have?
4. What is the running time to compute a rooted binary tree from its set of clades?
5. What is the largest possible Robinson–Foulds distance between two unrooted binary trees on the same set of n leaves?
6. What is the number of rooted binary trees on ten leaves?
7. What is the number of unrooted binary trees on ten leaves?
8. What is the definition of the strict consensus tree of a set of trees?

2.13 Homework Problems

1. Draw the rooted tree that is given by $(f,((a,b),(c,(d,e))))$.
2. Draw a rooted tree and give its Newick format representation.
3. Draw the rooted tree given by $(1,(2,(3,(4,(5,6)))))$, and write down the set of clades of that tree.
4. Draw the same rooted tree using the different styles as described in the text.
5. For the rooted tree T given by $(a,((b,c),(d,(e,f))))$,
    - write down at least three other Newick representations;
    - write down the set of clades, and indicate which of the clades is non-trivial.
6. Compute the Hasse Diagram on the partially ordered sets given by the following sets of clades, and then draw the rooted tree for each set.
    - $\{\{a,b\},\{a,b,c\},\{a,b,c,d\},\{e,f\},\{e,f,g\}\}$
    - $\{\{a,b,c\},\{a,b,c,d\},\{e,f\},\{e,f,g\}\}$

    Which one of these trees is *not binary*?
7. Draw all rooted binary trees on leafset $\{a,b,c,d\}$. (Note that trees that can be obtained by swapping siblings are the same.)
8. Draw all rooted trees (not necessarily binary) on leafset $\{a,b,c,d\}$.
9. Give a polynomial time algorithm to determine if two Newick strings represent the same rooted tree. For example,
    - your algorithm should return "YES" on $(a,(b,c))$ and $((c,b),a)$; and
    - should return "NO" on $(a,(b,c))$ and $(b,(a,c))$.
10. Draw the rooted and unrooted versions of the unrooted tree given by the following Newick string: $((a,b),(c,(d,e))$.
11. Draw all the rooted versions of the unrooted tree $(x,(y,(z,w)))$, and give their Newick formats.

12. Draw the unrooted version of the trees given below, and write down the set $C(T)$ of each tree T below. Are the two trees the same as unrooted trees?
    1. $(a,(b,(c,((d,e),(f,g)))))$
    2. $(((a,b),c),((d,e),(f,g)))$

13. Consider the two unrooted trees given below by their bipartition encodings. Draw them. Do you see how one tree can be derived from the other by contracting a single edge? Which one refines the other?
    - T_1 is given by $C(T_1) = \{(ab|cdef),(abcd|ef)\}$.
    - T_2 is given by $C(T_2) = \{(ab|cdef)\}$.

14. Draw two unrooted trees, so that neither can be derived from the other by contracting a set of edges.

15. Draw three different unrooted trees, T_1, T_2, and T_3, on no more than eight leaves, so that T_1 is a contraction of T_2, and T_2 is a contraction of T_3 (identically, T_3 is a refinement of T_2, and T_2 is a refinement of T_1). Write down the bipartition encodings of each tree.

16. Apply the technique for computing unrooted trees from compatible bipartitions to the input given below, using leaf 3 as the root. After you are done, do it again but use a different leaf as the root. Compare the rooted trees you obtained using the different leaves as roots: are they different? Unroot the trees, and compare the two unrooted trees. Are they the same?
    Input: $\{(123|456789),(12345|6789),(12|3456789),(89|1234567)\}$

17. Compute the unrooted trees compatible with the following sets of bipartitions (use the algorithm that operates on clades, using the specified roots):
    - $\{(ab|cdef),(abc|def),(abcd|ef)\}$, with root "b." Then do this again using root c. Are the unrooted trees you get different or the same?
    - $\{(ab|cdef),(abc|def),(abcd|ef)\}$, with root "d."
    - $\{(abcdef|ghij),(abc|defghij),(abcdefg|hij)\}$, using any root you wish.

18. Give a polynomial time algorithm to determine if the unrooted trees defined by two Newick strings are the same.
    - Your algorithm should return "YES" for $(a,(b,(c,d)))$ and $(c,(d,(b,a)))$; and
    - should return "NO" for $(a,(b,(c,d)))$ and $((b,d),(a,c))$.

19. Consider the unrooted tree given by $(1,((2,3),(4,(8,9)),(5,(6,7))))$. Root this tree at leaf 5, draw this rooted tree, and write the Newick string for the rooted tree you obtain.

20. Draw two binary unrooted trees on leafset $\{a,b,c,d,e,f\}$ that induce the same homeomorphic subtree on $\{a,b,c,d,e\}$ but have no non-trivial bipartitions in common.

21. Suppose T_0 is the true tree and T is the estimated tree. Which of the following statements are not possible, under the assumption that both T_0 and T are unrooted trees on ten leaves, and that T_0 is binary and T may not be binary? If you think the statement is impossible, explain why. Else, give an example where it is true.
    - There are five false negatives and three false positives.

- There are three false negatives and five false positives.
- There are three false negatives and three false positives.
- There are eight false negatives and two false positives.
- There are eight false negatives and eight false positives.
- There are seven false negatives and one false positive.
- There are one false negative and seven false positives.

22. Answer the same questions as for the previous problem, but do not assume now that the true tree T_0 is binary, but do require that T is binary.

23. Let T_0 be the unrooted true tree (and hence, binary). Let T_1 and T_2 be estimated unrooted trees on the same leafset as T_0, where T_1 is a star tree (see Definition 2.25) and T_2 is fully resolved (binary).

    1. What is the Robinson–Foulds (RF) rate of T_1 with respect to the true tree?
    2. For what trees T_2 will T_1 have a better RF rate than T_2?
    3. What do you think of using the RF rate as a way of comparing trees? What alternatives would you give?

24. Let T_0 be the unrooted tree given by splits $\{123|456, 12|3456, 1234|56\}$, and let T_1 be an estimated tree. Suppose T_1 is missing split $123|456$, but has a single false positive $124|356$. Draw T_1.

25. Give an algorithm for the following problem:

    Input: unrooted tree T_0 and two sets of bipartitions, C_1 and C_2, where $C_1 \subseteq C(T_0)$ and $C_2 \cap C(T_0) = \emptyset$.
    Output: tree T_1 (if it exists) such that T_1 has false negative set C_1 and false positive set C_2, when T_0 is treated as the true tree. (Equivalently, $C(T_1) = [C(T_0) - C_1] \cup C_2$.)

26. Let T be a caterpillar tree on n leaves (i.e., $T = (s_1, (s_2, (s_3, \ldots, (s_{n-1}, s_n)) \ldots))))$ (see Definition 2.23). Now let $\mathscr{T}$ be the set of trees on $n+1$ leaves formed by adding a new taxon, s_{n+1}, into T in all the possible ways. What is the expected RF distance between two trees picked at random from $\mathscr{T}$?

27. Prove using induction that the number of edges in an unrooted binary tree on $n \geq 2$ distinctly labeled leaves is $2n - 3$.

28. Consider the set $\mathscr{T}_n$ of unrooted binary trees on leafset $S = \{s_1, s_2, \ldots, s_n\}$. If you pick a tree uniformly at random from $\mathscr{T}_n$, what is the probability that s_1 and s_2 are siblings in $\mathscr{T}$?

29. Consider a caterpillar tree T on a set S of n taxa. Suppose there is a very rogue taxon, x, which can be placed into T in any possible position. Consider the set $\mathscr{T}$ that contains all the trees formed by adding x into T.

    1. What is $|\mathscr{T}|$?
    2. What is the strict consensus of all the trees in $\mathscr{T}$? (Give its bipartition set.)

30. For each of the given unrooted trees, draw the homeomorphic subtree induced on $\{a, b, c, d\}$.
    - T_1 has Newick format $(b, (a, (f, (c, (g, (d, e))))))$.

- T_2 has Newick format $(f,(a,(c,(g,(d,(b,e))))))$.

31. a. Give two unrooted trees on $\{a,b,c,d,e,f,g\}$ that induce the same homeomorphic subtree on $\{a,b,c,d\}$ but which are different trees.
    b. Give two unrooted trees on $\{a,b,c,d,e,f,g\}$ that are identical on $\{a,b,c,d\}$ and different on $\{d,e,f,g\}$.
    c. Give two rooted trees on $\{a,b,c,d,e\}$ which are identical on $\{a,b,c\}$ but different on $\{d,e,f\}$.
    d. Let T and T' be two different binary trees on the same leafset S. Suppose T^* is a binary tree on leafset S with $C(T^*) \subset C(T) \cup C(T')$. Must T^* be one of T_1 or T_2? If so, prove it, and otherwise provide a counterexample.
    e. Let T and T' be two different trees (not necessarily binary) on the same leafset S. Suppose T^* is a binary tree on leafset S with $C(T^*) \subset C(T) \cup C(T')$. Must T^* be a refinement of T_1 and T_2? If so, prove it, and otherwise provide a counterexample.

32. Suppose that the reference tree T_0 is not fully resolved, and we want to compute the tree error for an estimated tree T on the same leafset that *is* fully resolved. What can you say about the number of false negatives and false positives? Can they be the same? If not, which must be larger? Why?

3
Constructing Trees from True Subtrees

3.1 Introduction

In many approaches to constructing phylogenetic trees, the input is a set $\mathscr{T}$ of trees, each leaf-labeled by a subset of a set S, and the objective is to construct a tree T on S from $\mathscr{T}$. In this chapter, we discuss methods for constructing trees when the set $\mathscr{T}$ is compatible – which in essence means that it is possible to construct a tree T that agrees with every tree in $\mathscr{T}$. In general, this will only happen when every tree in $\mathscr{T}$ is the true species tree on its leafset, and so we refer to this problem as "Constructing trees from true subtrees." This is a simplification of the more general problem where the input trees may have some estimation error, which is the more realistic case (most estimated species trees have some error); this version of the problem is addressed in Chapter 7. Another variant of this problem is where the input trees are estimated (and perhaps correct) gene trees. Since gene trees can differ from the species tree due to biological processes such as incomplete lineage sorting (ILS) and gene duplication and loss, the estimation of a species tree from a set of gene trees presents its own challenges, which are discussed in Chapter 10.

3.2 Tree Compatibility

One of the key concepts in this chapter is tree compatibility, both for rooted and unrooted trees. In Chapter 2, we defined the concept of compatibility in the context of clades and bipartitions; as we will see, these definitions extend naturally to rooted and unrooted trees, respectively.

3.2.1 Unrooted Tree Compatibility

We begin with unrooted trees. Recall that if we are given a tree T on leafset S and a proper subset X of S, we can compute the homeomorphic tree $T|X$ (see Section 2.9). Now suppose we have two trees, t and T, where t has leafset X and T has leafset S, with $X \subseteq S$. Then we will say that the larger tree T is **compatible** with the smaller tree t if $T|X$ is a refinement of t. Note that we do not require that $t = T|X$; what this means is that t may be unresolved, which can allow $C(T|X)$ to have additional bipartitions beyond those present in $C(t)$. Note also that the term "compatibility" is not symmetric, even when the two trees have the same leafset. For example, the unrooted tree $(1,(2,(3,4)))$ is compatible with the unrooted star tree $(1,2,3,4)$, but not vice versa.

Definition 3.1 Let $\mathscr{T}$ be a set of unrooted trees and let S be the set of taxa that appear at the leaves of trees in $\mathscr{T}$. We will say that $\mathscr{T}$ is **compatible** if and only if there is a tree with leafset S that is compatible with every tree in $\mathscr{T}$. When $\mathscr{T}$ is compatible, then any minimally resolved supertree that is compatible with every tree in $\mathscr{T}$ is called a **compatibility supertree** for the set $\mathscr{X}$. Furthermore, any supertree that is compatible with every tree in $\mathscr{T}$ but that is not minimal is a refinement of some compatibility supertree, and so is called a **refined compatibility supertree**.

Recall that the Robinson–Foulds (RF) distance between two trees is the bipartition distance defined by the edges in the trees.

Theorem 3.2 *Let $\mathscr{T} = \{t_1, t_2, \ldots, t_k\}$ be a set of compatible fully resolved trees with S_i the leafset of t_i. Then $\sum_{i=1}^{k} \mathrm{RF}(T|S_i, t_i) = 0$, where T is a compatibility supertree for $\mathscr{T}$.*

Proof Note that two binary trees on the same leafset that are compatible must be identical. Hence, since every tree $t_i \in \mathscr{T}$ is fully resolved, $t_i = T|S_i$ for $i = 1, 2, \ldots, k$. The rest follows. □

3.2.2 Rooted Tree Compatibility

Just as we said that a set $\mathscr{T}$ of unrooted trees is compatible if there is a compatibility supertree (i.e., a tree on the full set of taxa that is compatible with every tree in $\mathscr{T}$), the same statement can be made for rooted trees. However, to make this precise we need to say what we mean for two rooted trees t and T to be compatible, when the leafset X of t is a subset of the leafset S of T. The only difference between rooted and unrooted trees is that to determine if t and T are compatible, we examine the homeomorphic *rooted* subtree induced in T by X.

3.3 The Algorithm of Aho, Sagiv, Szymanski, and Ullman: Constructing Rooted Trees from Rooted Triples

We now present an algorithm for constructing a rooted tree from its set of "rooted triples," where by "rooted triple" we mean a rooted three-leaf tree. We will also sometimes refer to rooted triples as "triplets," "triplet trees," or "rooted triplet trees." We indicate the rooted triple on a, b, c in which a and b are more closely related by $((a,b),c)$, by $ab|c$, or by any of the equivalent representations. We describe this algorithm for the more general case, where the input is a set of rooted triplets (maybe not one containing a tree on all possible sets of three leaves), and we wish to know if the set is compatible, and construct a tree agreeing with the input if it exists.

Algorithm for determining if a set of rooted triples is compatible. Suppose we are given a set X of rooted triples, and we wish to know if X is compatible, which means that there is a tree T on which all the rooted triples in X agree. Furthermore, when the set X is compatible, we wish to return some tree T on which all the rooted triples agree (i.e., a refined compatibility supertree).

The first algorithm for this problem was developed by Aho et al. (1978), and is widely known in phylogenetics. For the sake of brevity, we will often refer to this algorithm as the ASSU algorithm, after the four authors Aho, Sagiv, Szymanski, and Ullman.

The input to the ASSU algorithm is a pair $(S, Trip)$, where S is a set of taxa and *Trip* is a set of rooted three-leaf trees on S, with at most one tree for any three leaves; furthermore, we assume that every tree in *Trip* is fully resolved (i.e., of the form $((a,b),c)$).

- Group the set S of taxa into disjoint sets, as follows. We make a graph where the vertices are the elements of the set S and we add an edge (a,b) for every $((a,b),c) \in Trip$.
- If the graph is connected, then return *Fail* and exit – no tree is possible. Otherwise, let $C_1, C_2, \ldots, C_k$ ($k \geq 2$) be the connected components of the graph. For each connected component C_i,
  – Let $Trip_i$ be the set of triplets in *Trip* that have all their leaves in C_i.
  – Recurse on $(C_i, Trip_i)$; if this returns *Fail*, then also return *Fail*. Otherwise, let t_i be the output of the recursion.

Make the roots of the trees $t_1, t_2, \ldots, t_k$ all children of a common root, and return the constructed rooted tree.

This surprisingly simple algorithm is provably correct, and runs in polynomial time. A simple proof by induction on the number of leaves establishes correctness (Aho et al., 1978).

3.4 Constructing Unrooted Binary Trees from Quartet Subtrees

3.4.1 Notation

Definition 3.3 The binary quartet tree on a,b,c,d that splits a,b from c,d can be represented by $ab|cd$, $(ab|cd)$, (ab,cd), $(a,b|c,d)$, or $((a,b),(c,d))$. We represent the star tree on a,b,c,d by (a,b,c,d) (see Definition 2.25). Every set X of four leaves defines a quartet tree $T|X$. Therefore, given an unrooted tree T on n distinctly labeled leaves, we denote by **Q(T)** the set of quartet trees induced by T on its leafset, and by $\mathbf{Q}_r(\mathbf{T})$ the set of all fully resolved (i.e., binary) quartet trees in $Q(T)$. Hence, if T is a binary tree, then $Q_r(T) = Q(T)$.

3.4.2 The All Quartets Method

The Quartet Tree Compatibility problem is as follows:

- Input: Set Q of quartet trees.

- Output: Tree T such that $Q \subseteq Q(T)$ (if such a tree T exists) or *Fail*.

The Quartet Tree Compatibility problem is NP-complete, even when all the trees in Q are binary (Steel, 1992), but some special cases of the problem can be solved in polynomial time, as we will show. In particular, the case where all the trees in Q are binary and Q has at least one tree on every four leaves in a set S can be solved in polynomial time, using the All Quartets Method, which we now describe.

The *All Quartets Method* takes as input a set Q of quartet trees (each of them fully resolved), with one tree for every four leaves. The output is either a tree T such that $Q(T) = Q$ or else *Fail*.

We call this algorithm the *All Quartets Method*, because it assumes that the input is the set of all quartet trees for an unknown tree T; indeed, while the method can be applied to proper subsets of $Q(T)$, there are no guarantees that the method will correctly answer whether the input quartet trees are compatible under this more general condition. However, when the input contains a tree on every four leaves, then we can prove that the All Quartets Method is correct.

All Quartets Method:

- Step 1: If $|S| = 4$, then return the tree in Q. Else, find a pair of taxa s_i, s_j that are always grouped together in any quartet that includes both s_i and s_j. If no such pair exists, **return** "Incompatible input" and **exit**. Otherwise, remove s_i.
- Step 2: Recursively compute a tree T' on $S - \{s_i\}$.
- Step 3: Return the tree created by inserting s_i next to s_j in T'.

Example 3.4 Consider the unrooted tree $T = (1,(2,(3,(4,5))))$ and its set of quartet trees $Q(T) = \{12|34, 12|35, 12|45, 13|45, 23|45\}$. Note that taxa 1 and 2 are always grouped together in all the quartets that contain them both, but so also are 4 and 5. On the other hand, no other pair of taxa are always grouped together. If we remove taxon 1, we are left with the single quartet on $2, 3, 4$, and 5. The unrooted quartet tree on that set is $23|45$. We reintroduce the leaf for 1 as sibling to 2, and obtain the unrooted tree given by $(1,(2,(3,(4,5))))$.

3.4.3 Inferring Quartet Trees from Other Quartet Trees

Recall that the All Quartets Method will construct T given $Q(T)$, the set of homeomorphic quartet trees of T. This suggests a method for constructing a tree, in which unrooted trees are estimated on four leaves at a time, and then combined into a tree on the full dataset using the All Quartets Method. This approach will produce the true tree, but only if all of the estimated trees are correct – even one single error will make the entire approach fail.

Since some of the estimated quartet trees might be incorrect, we may wish to try to compute a tree from a proper subset of its quartet trees – ones that look like they may be correctly computed. Here we present one attempt at solving this problem, in which we use just a subset of the possible quartet trees and try to infer the remaining quartet trees. If we succeed, then we can apply the All Quartets Method to the final set of quartet trees, and construct the tree T.

Suppose Q is a set of quartet trees. We will show two rules for inferring new quartet trees from pairs of quartet trees, so that whenever all the trees in Q are true, then the added trees *must also be true*. In other words, we assume that the input set Q satisfies $Q \subseteq Q(T)$ for some (unknown) tree T, and we ensure that any added quartet trees are also in $Q(T)$. In the rules below, we will consider $ab|cd$ to be the same quartet tree as $ba|cd$, $ba|dc$, and $ab|dc$.

- Rule 1: If $ab|cd$ and $ac|de$ are in $Q \subseteq Q(T)$, then $ab|ce, ab|de$, and $bc|de$ are also in $Q(T)$. Hence, if any of these three quartet trees are missing from Q, we can add them to Q.
- Rule 2: If $ab|cd$ and $ab|ce$ are in $Q \subseteq Q(T)$, then $ab|de$ must be in $Q(T)$. Hence, if $ab|de$ is missing from Q, we can add $ab|de$ to Q.

It is easy to see that these rules are valid, and so if the input set Q contains only correct quartet trees (meaning true quartet trees for some unknown tree T), then the quartet trees that are added are also correct quartet trees for that unknown tree T. These two rules are *dyadic* (also called "second order") rules, in that they are based on combining two quartet trees to infer additional quartet trees.

The dyadic closure of a set Q of quartet trees is computed by repeatedly applying Rules 1 and 2 to pairs of quartet trees until no additional quartet trees can be inferred. The final set of quartet trees is the **dyadic closure** of Q, and is denoted by $cl_2(\mathbf{Q})$. The discussion above clearly shows the following:

Theorem 3.5 *Suppose* $Q \subseteq Q(T)$ *for some tree* T. *Then* $cl_2(Q) \subseteq Q(T)$.

Now suppose we were lucky, and $cl_2(Q) = Q(T)$; then we can construct T using the All Quartets Method. However, if Q is too small a set, then $cl_2(Q)$ may not be equal to $Q(T)$. In the next section, we investigate how big Q has to be, in order for $cl_2(Q) = Q(T)$.

3.4.4 Constructing a Tree from a Subset of its Quartet Trees

We begin by defining the "short quartet trees" of an edge-weighted tree T. As we will see, if Q contains all the short quartet trees of a tree T and no incorrect quartet trees, then $cl_2(Q) = Q(T)$.

Definition 3.6 Let T be a binary tree and $w : E(T) \to R^+$ be the positive edge weighting of T. The deletion of an internal edge $e \in E(T)$ (and its endpoints) creates four subtrees, A, B, C, and D. Let a, b, c, d be four leaves nearest to e from these four subtrees, with $a \in A, b \in B, c \in C$ and $d \in D$. The definition of "nearest" is based on the path length, and takes the edge weights into account. Then a, b, c, d is a **short quartet** around e, and the quartet tree on a, b, c, d defined by T is called a **short quartet tree**. Since there can be more than one nearest leaf in a given subtree to the edge e, there can be more than one short quartet around e. The set of all short quartets over all internal edges of T is called the set of **short quartets of** T and is denoted $Q_{short}(T)$, while the set of short quartet trees over all internal edges of T is called the set of **short quartet trees of** T and is denoted $Q^*_{short}(T)$.

Example 3.7 Consider the caterpillar tree $(1,(2,(3,\ldots,(99,100)\ldots)$. There are 97 internal edges of the tree, each of which contributes at least one short quartet. A careful inspection of this tree shows that the set $Q_{short}(T)$ has 99 quartets. For example, $Q_{short}(T)$ includes $\{1,2,3,4\}, \{2,3,4,5\}, \{1,3,4,5\}$, and $\{3,4,5,6\}$. Now consider the set $Q^*_{short}(T)$ of trees on the short quartets of T. A little examination will show that T is the only tree on the same leafset that can contain all the short quartet trees. For example, Rule 1, applied to $12|34$ and $23|45$, produces $13|45, 12|45$, and $12|35$. In other words, applying Rule 1 produced the five quartet trees on four-leaf subsets of $1,2,3,4,5$. Indeed, if we compute $cl_2(Q^*_{short}(T))$, we will obtain $Q(T)$.

In other words, the following theorem can be proven:

Theorem 3.8 *(From Erdös et al. (1997)) If $Q^*_{short}(T) \subseteq Q \subseteq Q(T)$ for some binary unrooted tree* T, *then* $cl_2(Q) = Q(T)$. *In other words, under the assumption that* T *is the true tree, then if* Q *has no incorrect quartet trees and also contains all the short quartet trees of* T, *then the dyadic closure of* Q *will include the true tree on every four leaves in* T, *and nothing beyond that.*

Corollary 3.9 *(From Erdös et al. (1997)) Let* T *be a binary unrooted tree and let* $Q^*_{short}(T)$ *be the set of short quartet trees for* T *for some edge-weighting of* T. *If* T′ *is a tree on the same leafset as* T *and* $Q^*_{short}(T) \subseteq Q(T')$, *then* T′ = T.

Proof Assume that T and T' are on the same leafset, and that $Q^*_{short}(T) \subseteq Q(T')$. We begin by noting that Theorem 3.8 implies that $cl_2(Q^*_{short}(T)) = Q(T)$. Now, since $Q^*_{short}(T) \subseteq Q(T')$, by Theorem 3.5, $cl_2(Q^*_{short}(T)) \subseteq Q(T')$. Hence, $Q(T) \subseteq Q(T')$. Since T and T' are on the same leafset, $Q(T)$ and $Q(T')$ have the same cardinality. Hence, it must follow that $Q(T) = Q(T')$, and so $T = T'$. □

What these results establish is that if we were lucky enough to find such a set Q of quartet trees (one that has no incorrect quartet trees, and yet contains all the short quartet trees), we could use the dyadic closure to infer $Q(T)$, and then construct the tree T using the All Quartets Method. See Section 8.12 for the Dyadic Closure Method (Erdös et al., 1999a), which builds on the theory we have outlined here.

3.5 Testing Compatibility of a Set of Trees

Recall that a compatibility supertree for a set $\mathscr{T}$ of trees is a minimally resolved tree that is compatible with every tree in $\mathscr{T}$ (Definition 3.1).

Example 3.10 Consider the set X of unrooted trees $X = \{(ab|cde), (bc|def), (cd, eg)\}$. The caterpillar tree $(a, (b, (c, (d, (e, (f, g))))))$ (see Definition 2.23) is a refined compatibility supertree for X, and so X is compatible.

Theorem 3.11 *The* Unrooted Tree Compatibility *problem – determining if a set* X *of unrooted trees, each leaf-labeled by elements from* S*, is compatible – is NP-complete, even if all the trees are binary (fully resolved).*

Proof This result follows from the fact that quartet compatibility is NP-complete (Steel, 1992). □

As noted before, some special cases of Unrooted Tree Compatibility can be solved in polynomial time. For example, we already know that the All Quartets Method can construct a tree T from its set $Q(T)$ of quartet trees, and furthermore that the All Quartets Method can be used to determine if a set X of quartet trees is compatible when X contains a tree on every four taxa. Hence, if X is a set of unrooted trees and every four taxa are in at least one tree in X, then we can determine if X is compatible in a straightforward, if brute-force, way: We replace every tree t in X by its set $Q(t)$, and thus make X into a set of quartet trees that contains a tree on every four taxa. We can then apply the All Quartets Method to the set of quartet trees we have created to determine if the quartet trees are compatible.

Similarly, suppose that we are given a set X of rooted leaf-labeled trees and we want to know if there is a rooted tree T that induces each of the trees in X as homeomorphic subtrees. To answer this question, we can *encode* each of the rooted leaf-labeled trees in X by its set of rooted triplet trees, and then run the ASSU algorithm (see Section 3.3) on the resultant set of rooted triplet trees. If the output is a rooted tree that induces all the rooted triplet trees, then it follows that the set X is compatible. The other possible outcome is that the algorithm fails to return a tree (because during at least one of the recursive calls, the graph has a single connected component); in that case, the rooted triplet trees are not compatible, and hence the set X is incompatible. In other words, it is easy to see that testing a set of rooted binary trees for compatibility is a polynomial time problem. We summarize this as follows:

Theorem 3.12 *The* Rooted Binary Tree Compatibility Problem *– determining if a set* X *of rooted binary trees, each leaf-labeled by elements from* S*, is compatible – can be solved in polynomial time.*

The extension of this problem to rooted trees that can have polytomies is also solvable in polynomial time, but the proof of this is left to the reader.

3.6 Further Reading

In this chapter we described several quartet-based methods for tree estimation; in each of these cases, we assumed that the set of quartet trees is computed using some technique, and the objective is to construct a tree that is consistent with the quartet trees. Yet, since quartet trees are not always perfectly estimated, these tree estimation methods can fail to construct any tree. In later chapters, we will return to quartet-based methods for tree estimation, and address this inference problem when the input trees are presumed to have some error.

3.7 Review Questions

1. What is a rooted triple?
2. For each problem below, state whether it is solvable in polynomial time, NP-hard, or of unknown computational complexity:
   - Determining if a set of rooted triples is compatible.
   - Determining if a set of unrooted quartet trees is compatible.
   - Determining if a set of rooted leaf-labeled trees is compatible.
   - Determining if a set of unrooted leaf-labeled trees is compatible.
3. If T is an unrooted leaf-labeled tree, what does $Q(T)$ refer to?
4. What does $ab|cd$ refer to?
5. What is the *All Quartets Method*? Does it run in polynomial time?
6. Suppose you are given a set Q of unrooted quartet trees that contains a tree for *some but not all* of the different sets of four species taken from a species set S. Can you use the All Quartets Method to test for compatibility of the set Q?
7. Suppose you are given a set R of rooted triplet trees that contains a rooted tree for *some but not all* of the different sets of three species taken from a species set S. Can you use the ASSU method to test for compatibility of the set R?
8. What is the difference between a refined compatibility supertree and a compatibility supertree?

3.8 Homework Problems

1. Make up a rooted tree on six leaves, and write down all its rooted triples. Then make up another rooted tree on the same six leaves, and write down all its rooted triples. How many rooted triples do your trees disagree on?
2. Make up two rooted trees on at least five leaves that differ in exactly one rooted triple.
3. Consider the rooted caterpillar tree given by $(1,(2,(3,(4,5))))$.
   a. Write down the set of rooted triples for the tree.
   b. Apply the ASSU algorithm to this set of rooted triples. What do you find?
4. By design, if the ASSU algorithm returns a tree on input set X of rooted triplets, then the tree it returns will be compatible with all the input triplet trees in X. But is it guaranteed to produce a compatibility supertree, or might the output be a refined compatibility supertree?
5. Is it possible to have a compatible set X of rooted triplets for which some pair of leaves i,j is not separated in any rooted triplet in which they both appear, but where i and j are not siblings in *any* tree that is compatible with the set of rooted triplets? If so, provide the example, and otherwise prove it is impossible.

3.8 Homework Problems

6. Suppose we modify the ASSU algorithm as follows. We compute the equivalence relation, and if there is more than two equivalence classes, $C_1, C_2, \ldots, C_k$ (with $k > 2$) we make *two* subproblems, C_1 and $C_2 \cup C_3 \cup \ldots \cup C_k$. Otherwise, we don't change the algorithm. Does this also solve rooted triplet compatibility? (Prove or disprove.)
7. Prove that the ASSU algorithm correctly solves the problem of determining if a set *Trip* of rooted, fully resolved, three-leaf trees is compatible. (Hint: use induction.)
8. Consider input sets $\mathscr{T}$ of rooted trees, each on a subset of taxon set S, and suppose some of them have polytomies. We will consider the polytomies to be *soft*, meaning that we do not consider them to imply any constraint on the tree on the full set S. We would like to find a tree T on the full taxon set that is compatible with every tree in $\mathscr{T}$, meaning that it will either agree with the trees in $\mathscr{T}$ or refine the trees when restricted to the same leafset.

    a. Show how to use the ASSU algorithm so that it solves this problem.
    b. Prove your algorithm correct.

9. Consider input sets *Trip* of rooted triplet trees, each on a subset of taxon set S, and suppose some of them are polytomies (i.e., of the form (a,b,c)). Suppose we consider the triplet tree (a,b,c) to be a *hard polytomy*, meaning that it imposes a constraint on the tree T on S to induce the rooted tree (a,b,c) on $\{a,b,c\}$, rather than just be compatible with it. (A soft polytomy is a polytomy that does not imply such a constraint.) In other words, we would like to find a tree on the full taxon set that induces each triplet tree in *Trip*.

    a. Modify the ASSU algorithm so that it solves this problem.
    b. Prove your algorithm correct.

10. Suppose that we allow triplet trees to represent hard polytomies. For example, we would use (a,b,c) to indicate that the compatibility supertree (if it exists) induces the unresolved tree (a,b,c). Suppose that ASSU ignores these triplet trees. Give an example of an input set *Trip* of triplet trees that is allowed to have these hard polytomies, and show that the ASSU algorithm will not correctly solve the compatibility problem on *Trip*. Thus, either *Trip* should be compatible but the ASSU algorithm should say it is not, or vice versa.
11. In the text, we stated that the ASSU algorithm is polynomial time. Provide a running time analysis, where the input is a set of k rooted triplet trees drawing their leaves from set S of n taxa.
12. Make up two different unrooted trees on the same set of five leaves, but try to make them disagree on as few unrooted quartet trees as possible. How many do they disagree on?
13. Construct a tree on leafset $\{a,b,c,d,e,f\}$ that induces each of the following quartet trees:

    - $(ab|cd)$
    - $(ab|ce)$

- $(ac|de)$
- $(bc|de)$
- $(ab|de)$
- $(ab|cf)$
- $(ab|df)$
- $(ab|ef)$
- $(ac|df)$
- $(ac|ef)$
- $(ad|ef)$

14. Recall that the All Quartets Method is designed to solve the Quartet Compatibility problem when the input is a set of fully resolved (i.e., binary) trees, with exactly one tree on every set of four taxa. Prove that the All Quartets Method is correct for such inputs. (Hint: use induction.)

15. Consider the case where the unrooted tree T is not binary, and so can have a node of degree greater than three. Give an example of such a tree T, so that when the All Quartets Method is applied to $Q(T)$ it fails to recover the tree T.

16. Modify the All Quartets Method so that it will correctly handle inputs Q that contain quartet trees with hard polytomies.

17. Suppose we have a set X of unrooted binary trees, and we encode each tree $T \in X$ by its set $Q(T)$ of quartet trees. Prove or disprove: The set X is a compatible set of unrooted trees if and only if $\bigcup_{T \in X} Q(T)$ is a compatible set of quartet trees.

18. Suppose we have a set X of rooted binary trees, and we encode each tree $T \in X$ by its set $R(T)$ of rooted triplet trees. Prove or disprove: The set X is a compatible set of rooted trees if and only if $\bigcup_{T \in X} R(T)$ is a compatible set of rooted triplet trees.

19. Consider the Split Constrained Quartet Support problem. How would you define the input set X of allowed bipartitions so that the solution to the problem gave an optimal tree over all possible binary trees on the taxon set?

20. Suppose you have a collection $\mathscr{T}$ of unrooted trees, not necessarily binary, all with exactly the same leafset $\{1,2,3,\ldots,n\}$. Suppose that the set $\mathscr{T}$ is compatible. Express the maximum size of $\mathscr{T}$ as a function of n.

21. Suppose you have a collection $\mathscr{T}$ of unrooted binary trees, each of them different, all with exactly the same leafset $\{1,2,3,\ldots,n\}$. Suppose that the set $\mathscr{T}$ is compatible. Express the maximum size of $\mathscr{T}$ as a function of n.

22. Consider the following three unrooted trees:
    - $T_1 = (1,(3,(5,(6,7))))$
    - $T_2 = (1,(2,((4,8),(3,7))))$
    - $T_3 = (2,((4,(3,5)),1))$

    Answer the following questions:

    a. Are these unrooted trees compatible? Justify your answer.
    b. Root all three trees at leaf 1, and draw the rooted versions of these trees. Are these rooted trees compatible? Justify your answer.

4
Constructing Trees from Qualitative Characters

4.1 Introduction

In essence, there is really one primary type of data used to construct trees – **characters**. An example of a character is the nucleotide (A, C, T, or G) that appears in a particular location within a gene, the number of legs (any positive integer), or whether the organism has hair (a Boolean variable). In each of these cases, a character divides the input set into disjoint subsets so that the elements of each subset (which could be molecular sequences or biological species) are equivalent with respect to that character. We will refer to the different sets defined by the character as the **character states**. In other words, a character is an *equivalence relation* on the set S of taxa, and the different equivalence classes are the different character states. However, characters can also be described by functions from the taxon set S to the set of character states.

The number of states the character can take is an important aspect for modeling purposes. Many characters are based on sites within a molecular sequence alignment, and thus have a maximum number of possible states (four for DNA or RNA, and 20 for amino acids). Morphological features can be multi-state (and perhaps even have an unbounded number of possible states), but many morphological features are just based on the presence or absence of a given characteristic, and hence are explicitly binary.

Most stochastic models assume that the taxa are related by a tree, and that a character evolves down the tree under a process that involves substitutions. However, not all evolution is treelike; for example, horizontal gene transfer (HGT) is common in some organisms (e.g., bacteria) and hybridization (whereby two species come together to make a new species) is also frequent for some organisms (e.g., plants).

Character-based methods are the basis of nearly all phylogenetic estimation methods, since characters form the way the input is nearly always described. If the evolutionary process operating on the characters can be modeled adequately, then phylogeny estimation – whether of a tree or of a phylogenetic network – can be performed using statistical methods, such as maximum likelihood or Bayesian MCMC. However, when no statistical model is available (or the statistical models are unreliable), then simple methods that are not based on explicit models of evolution can be used to estimate trees. In this chapter, we will discuss the estimation of trees from character data using two such simple methods – maximum parsimony and maximum compatibility. In later chapters, we will discuss models of character evolution, and statistical methods for estimating trees under these models.

4.2 Terminology

Suppose we have n taxa, $s_1, s_2, \ldots, s_n$ described by k characters, $c_1, c_2, \ldots, c_k$. We will let Σ denote the state space for these characters; for example, Σ could be $\{A, C, T, G\}$. When the taxa are described by strings, then Σ is sometimes referred to as the alphabet.

The input is typically provided in an $n \times k$ matrix **M**, with the taxa occupying rows and different characters occupying the columns. In this case, the entry M_{ij} is the state of the taxon s_i for character c_j. We can also represent this input by just giving the k-tuple representation for each taxon.

When the substitution process produces a state that already appears anywhere else in the tree, this is said to be **homoplastic** evolution (or, more simply, **homoplasy**). **Back-mutation** (reversal to a previous state) and **parallel evolution** (appearance of a state in two separate lineages) are the two types of homoplasy. When all substitutions create new states that do not appear anywhere in the tree, the evolution is said to be **homoplasy-free**. Furthermore, when the tree fits the character data so that no character evolves with any homoplasy, then the tree is called a **perfect phylogeny**.

While homoplasy-free evolution may not be the rule, individual characters can evolve without homoplasy. Therefore, we will say that a character is **compatible on a tree** if it evolves on the tree without homoplasy. We will also say that a set of characters is **compatible** if there is a tree on which all of the characters evolve without homoplasy, and so are compatible on the tree.

In some cases (e.g., some morphological features in biology), the characters are given with a known tree structure relating the character states. For example, for some morphological features, represented by either presence (marked as 1) or absence (marked as 0) of the feature, it is known that the ancestral state is 0 and that the derived state is 1. Some multi-state characters may also have a clear tree structure, where the ancestral state is known, and the progression between the states is also known. We refer to these as **directed characters**, and **directed binary characters** for those binary characters for which the **ancestral state** and **derived state** are known. Homoplasy-free directed characters provide substantial information about the phylogeny, provided that the assumptions (that the evolution is homoplasy-free and the process between the states is known) are valid.

In this chapter we discuss two approaches that have been used to construct trees from undirected characters: maximum parsimony and maximum compatibility. The **parsimony score** (also called the **length**) of a character on a tree is the number of times the character has to change state on the tree. Finding a tree T that has the smallest total parsimony score for an input character matrix is called the **maximum parsimony problem**. The **maximum compatibility problem** is similar, but seeks the tree on which the maximum number of characters are compatible. The relationship between parsimony score and compatibility (homoplasy-free evolution) is that a character is homoplasy-free on a tree T if and only if it changes $r - 1$ times, where r is the number of states of the character at the leaves of the tree. However, parsimony and compatibility treat incompatibility quite differently: a character that is incompatible on a tree only reduces the compatibility score of the tree by 1, but it increases the parsimony score by the total number of extra character state changes it requires on the tree, a number that varies with the character and the tree.

We begin with maximum parsimony, which is much more commonly used in biological phylogenetics than maximum compatibility. However, maximum compatibility, and the related problem of determining if a perfect phylogeny exists and computing it when it does, is relevant to the analysis of phylogenies within species; for example, using SNP (single nucleotide polymorphism) data.

4.3 Tree Construction Based on Maximum Parsimony

We begin this discussion by making a precise statement of what is meant by the number of state changes of a character c on a tree T. We are given a tree T and character state assignments to the leaves of T. Our objective is to find character state assignments to the internal nodes of T so that the number of edges that have different states at the edge's endpoints is as small as possible. This is the "fixed tree parsimony problem." It is also called the "small parsimony problem" to distinguish it from the "large parsimony problem," which seeks to find the best tree and its internal node labels achieving the best possible parsimony score.

Since we are only counting the total number of changes, the parsimony score of a tree is unchanged when the tree is rooted differently. Thus, we will simplify the discussion by assuming T is not rooted. The fixed tree parsimony problem can be formally stated as follows:

Parsimony Score of a Fixed Tree

- Input: Unrooted binary tree T with leafset S, and character $c : S \to \{1, 2, \ldots, r\}$.
- Output: Assignment of character states to the internal nodes of T so as to minimize the number of edges $e = (u, v)$ where $c(u) \neq c(v)$.

As we will see, a simple dynamic programming due to Fitch (1971) will find the parsimony score and an optimal labeling of the internal nodes. The **Fitch algorithm** is explicitly for binary trees; the extension to arbitrary trees was later developed by Hartigan (1973).

4.3.1 The Fitch Algorithm for Fixed-Tree Maximum Parsimony

The input to the small parsimony problem is a tree T whose leaves have assigned states for a character c. The algorithm operates by rooting the tree (arbitrarily) on an edge, and then has two stages. In the first stage it computes the parsimony score for the character on the tree, and in the second stage it computes an assignment of character states to the internal nodes to achieve that score. Thus, the dynamic programming algorithm gives the optimal labeling as well as the maximum parsimony score.

We summarize this description, assuming that the input tree T is already rooted on an edge, so that all nodes other than the leaves have exactly two children. We also assume we have a single character c that has assignments of states to all the leaves in T, and that $c(v)$ is the state of leaf v. We let *root* denote the root of T.

- Initialization:
  - For every leaf x, set $A(x) = \{c(x)\}$.
  - Initialize *score* to be 0.
- Stage 1: Perform a post-order traversal (i.e., starting at the nodes v which have only leaves as children, and processing a vertex only after its children are processed):
  - If v has children w and w', and if $A(w) \cap A(w') \neq \emptyset$, then set $A(v) = A(w) \cap A(w')$. Else, set $A(v) = A(w) \cup A(w')$, and increment *score* by 1.
- Stage 2: Perform a pre-order traversal (i.e., beginning at *root* and processing a vertex only after its parent is processed).
  - Pick an arbitrary state in $A(root)$ to be the state at *root*.
  - For every other internal node v pick a state as follows:
    ○ If the parent of v has been assigned a state that is in $A(v)$, then set the state for v to the same state as was assigned to its parent. Otherwise, pick an arbitrary element in $A(v)$ to be its state.

At the end of Stage 1 you will have computed the parsimony score, and at the end of Stage 2 you will have assigned states to each node in the tree. During the upward phase (Stage 1) the sets $A(v)$ are determined for every node v, and in the downward phase (Stage 2) a state is selected from $A(v)$ for node v. Thus, any node v for which $|A(v)| = 1$ has only one possible assignment in an optimal labeling of that node. The converse, however, is not necessarily true. As we will see, there are cases where $|A(v)| > 1$ and yet every optimal labeling of T has v assigned to the same specific state.

Example 4.1 Suppose that T is the binary tree $(a,(b,(c,d)))$, and assume that character α is defined by $\alpha(d) = 0$ and $\alpha(s) = 1$ for all $s \neq d$. Root the tree on the edge leading to a (so that you get the rooted tree corresponding to the Newick string given above). We will name the internal nodes of this rooted tree with *root* (for the introduced root), p for the parent of c and d, and q for the MRCA of b, c, d.

Stage 1: The variable *score* is initialized to 0. When you run the dynamic programming algorithm, the leaves are processed first, and we obtain $A(a) = A(b) = A(c) = \{1\}$ and $A(d) = \{0\}$. The first internal node that is visited is p, the parent of leaves c and d. Since $A(c) \cap A(d) = \emptyset$, we set $A(p) = A(c) \cup A(d)$, and obtain $A(p) = \{0, 1\}$; we also increment *score* so that it is now 1. We then compute $A(q)$. The children of q are p and b. Since $A(b) = \{1\}$ and $A(p) = \{0, 1\}$, $A(p) \cap A(b) \neq \emptyset$. Hence we set $A(q) = A(p) \cap A(b) = \{1\}$.

Finally we reach *root*, the parent of a and q. Since $A(a) \cap A(q) \neq \emptyset$, we set $A(root) = A(a) \cap A(q) = \{1\}$. This completes Stage 1. Note that at this point, $score = 1$.

Stage 2: We now assign states to the nodes of the tree, beginning at the root. Since $|A(root)| = 1$, we set $\alpha(root) = 1$, the unique element in $A(root)$. We then visit the children of *root*, which are a and q. Since a is a leaf, its character state is already assigned. To assign a state to node q, we note that $A(q) = \{1\}$, and so we set $\alpha(q) = 1$. We then visit the children of q, which are b and p. Since b is a leaf, we do not need to process it. We note that $A(p) = \{0, 1\}$. To assign a state to node p, we then check if p's parent node (q) has been assigned a state in $A(p)$; since $\alpha(q) = 1$, we set $\alpha(p) = 1$. We then visit the children of p, but they are both leaves, and so we are done with Stage 2.

At this point all the nodes of the tree have been assigned states, and we can calculate the parsimony score. Note that there is only one edge on which there is a change, and it is (p, d). Hence, the parsimony score of the character α on the tree is 1, matching the value of *score* computed in Stage 1.

This is a very simple illustration of the Fitch algorithm, but also presents some properties of the algorithm that are applicable in general. For example, the parsimony score (indicated by the variable *score*) of any character on any tree is the number of nodes v for which $A(v)$ is computed by taking the union of $A(w)$ and $A(w')$, where w and w' are the children of v. That is, otherwise we would have $A(v) = A(w) \cap A(w')$, and the character state assignment obtained during the downward pass (Stage 2) would pick the same state for all three nodes, v, w, and w'. Thus, the parsimony score can be computed at the end of Stage 1, without needing to do Stage 2.

To compute assignments of states to the internal nodes, we have to run Stage 2. One observation is that any node v for which $|A(v)| = 1$ can be assigned a state without checking its parent's state. In this particular example, the assignment of states to every node – even at the root – was constrained, and we had no choices. This was even true for p, even though $|A(p)| = 2$: In the downward pass there was only one option for how to set $\alpha(p)$. Hence, the optimal character states at internal nodes can only be determined *after* both stages are complete – they are not determined during the first pass.

Theorem 4.2 *For all unrooted binary trees* T *and characters* c *defined at the leaves of* t, *the parsimony score computed by the Fitch dynamic programming is correct, and the assignment of states to the internal nodes of* T *achieves the reported parsimony score. Furthermore, the algorithm runs in* O(nr) *time, where there are* n *leaves in* T, *and* c *is an r-state character.*

The proof of the correctness of the algorithm follows from the discussion above. Computing $A(v)$ for a node v only depends on being able to compute the intersection and union of sets, each of which has cardinality at most r (where r is the maximum number of states of any character on T).

Corollary 4.3 *The (unweighted) parsimony score of a set of* k *characters on a binary tree* T *can be computed in* O(nkr) *time, where there are* n *leaves and* k *characters, with each* *character having at most* r *states. Hence in particular the parsimony score of a multiple sequence alignment of length* k *over* Σ *on a fixed binary tree* T *can be computed in* O(nkr) *time where* $|\Sigma| = r$.

Proof Follows from Theorem 4.2 and the fact that the parsimony score of each of the *k* characters can be computed separately. □

4.3.2 The Sankoff Algorithm for Fixed Tree Weighted Maximum Parsimony

The parsimony problem is somewhat more complicated if the substitution costs depend on the particular substitution. For example, there are two different types of nucleotides – purines (which are A and G) and pyrimidines (which are C and T). Substitutions that change a purine into a different purine (e.g., changes from A to G) or a pyrimidine into a different pyrimidine (e.g., C to T) are called **transitions**, while substitutions that change a purine into a pyrimidine (e.g., G to C) or vice versa are called **transversions**. Transitions are considered more likely than transversions. Therefore, one variant of maximum parsimony would treat these two types of substitutions differently, so that transitions would have lower costs than transversions, but any two transitions or any two transversions would have the same cost.

The substitution cost matrix is represented by a symmetric $r \times r$ matrix **M** where $M[x,y]$ is the cost of substituting x by y, and r is the number of character states. Clearly $M[x,x]$ should be 0. If all entries off the diagonal are the same, then this is identical to unweighted maximum parsimony, and the previous algorithm works. But what if the entries off the diagonal are different?

As we will see, this problem can be solved using a straightforward dynamic programming algorithm, presented in Sankoff (1975) (but see also Sankoff and Rousseau (1975)). The application of this technique to solving weighted parsimony on a fixed tree is called the **Sankoff algorithm** (Felsenstein, 2004). We present it here, along with a proof of its correctness, because it illustrates the power of dynamic programming algorithms when working with trees. (Also, understanding why the Sankoff algorithm is correct can be helpful in understanding why the Fitch algorithm is correct.)

Let t be an unrooted binary tree with leaves labeled by sequences of length k, all drawn from Σ^k, where Σ is the alphabet. We root t on an edge, thus producing a rooted binary tree T, in which only the leaves are labeled by sequences. We consider a single character (site) at a time, which we will refer to as c; hence $c(v)$ is the state of leaf v for character c. We define the following variables:

- For every vertex v in the rooted tree T, and for every letter x in Σ, we define $Cost(v,x)$ to be the *minimum parsimony cost of the subtree* T_v *over all possible labels at the internal nodes of* T_v, *given that we label* v *by* x.

The Sankoff algorithm, just like the Fitch algorithm, has an initialization step and then two stages. In Stage 1, the algorithm performs a post-order traversal (i.e., starting at the leaves and moving upwards), in which it computes $Cost(v,x)$ for all internal nodes v and all letters $x \in \Sigma$. In Stage 2, it uses backtracking to assign states to the internal nodes in order to achieve the best possible score, which is $min\{Cost(root,x) : x \in \Sigma\}$ where $root$ is the root of the tree. The Sankoff algorithm is thus a bit more complicated than the Fitch algorithm.

The initialization step is simple: for v a leaf in T, we set $Cost(v,x) = 0$ if $c(v) = x$, and otherwise we set $Cost(v,x) = \infty$.

Stage 1 then does the post-order traversal to set $Cost(v,x)$ for every internal node v and every letter x. Let v be a node with children w and w', and assume that we have already computed $Cost(w,y)$ and $Cost(w',y)$ for all letters y. Then:

$$Cost(v,x) = min\{Cost(w,y) + M[x,y] : y \in \Sigma\} + min\{Cost(w',y) + M[x,y] : y \in \Sigma\}$$

When you set $Cost(v,x)$ for a node v and letter x, record the ordered pair of states y, y' such that $Cost(v,x) = Cost(w,y) + M[x,y] + Cost(w',y') + M[x,y']$ where w is the left child of v and w' is the right child of v, and store it in $ChildrenStates(v,x)$. We will use these variables to assign states to the internal nodes during Stage 2. Finally, we set the parsimony score for the tree to be $score = min\{Cost(root,x) : x \in \Sigma\}$.

In Stage 2, we use the calculations from Stage 1 to assign labels (i.e., set the states for c) at every node in the tree. To assign a label at the root, let x_0 satisfy $score = Cost(root, x_0)$, and set $c(root) = x_0$. To label another node w, we use backtracking. When we visit w, we will have already visited its parent v. Thus, v will already have its state set; without loss of generality assume $c(v) = x$. Let $(y, y') = ChildrenStates(v,x)$. If w is the left child of v, then set $c(w) = y$, and otherwise set $c(w) = y'$. The state assignments to the leaves are from the input. Hence, the states at all the internal nodes can be made so as to achieve the parsimony score that Stage 1 reports. It is easy to see that the algorithm uses polynomial time.

The Sankoff algorithm has the following structure. The input is a binary tree T rooted at $root$ and a single character c that has assigned states at the leaves of T.

- Initialization:
  - For all the leaves x, let $c(x)$ denote the state at the leaf x for character c.
- Stage 1: Perform a post-order traversal (just as with the Fitch algorithm):
  - When visiting a node v, note its left child w and right child w'.
  - Set $Cost(v,x) = min\{Cost(w,y) + M[x,y] : y \in \Sigma\} + min\{Cost(w',y) + M[x,y] : y \in \Sigma\}$.
  - Set $ChildrenStates(v,x) = (y, y')$ where y, y' are the states for nodes w, w', respectively, achieving the minimum value for $Cost(v,x)$ in $ChildrenStates(v,x)$. (We will use these for backtracking later.)
  - Let $score = min\{Cost(root,x) : x \in \Sigma\}$.
- Stage 2: Perform a pre-order traversal (just as with the Fitch algorithm):
  - Let $x_0 \in \Sigma$ such that $Cost(root, x_0) = score$, and set $c(root) = x_0$.

- To set the state at a node w whose parent v for which we have already defined $c(v)$, let $(y,y') = ChildrenStates(v, c(v))$. Set the states at w and w' by $c(w) = y$ for w the left child of v and $c(w) = y'$ for w' the right child of v.

We conclude with the following theorem.

Theorem 4.4 *For any input binary tree* T, *character state assignment to the leaves of* T, *and symmetric substitution cost matrix* **M**, *the Sankoff algorithm correctly computes the weighted maximum parsimony score and the optimal assignment of states to the internal nodes of* T, *and does so in* $O(nkr^2)$ *time, where* T *has* n *leaves, and there are* k *characters with states drawn from* $\Sigma = \{1, 2, \ldots, r\}$.

Proof We begin by proving the algorithm correctly computes the parsimony score of the tree for a single character c, and in particular that it sets $Cost(v,x)$ so that it equals the best achievable parsimony score of the rooted tree T_v given that $c(v) = x$. If this holds, then the proof of correctness for the parsimony score for a set of characters follows. In addition, once we establish that it calculates $Cost(v,x)$ correctly for every vertex v and state x, it follows directly that the assignment of states to internal nodes is optimal.

Recall that the algorithm computes $Cost(v,x)$ in a post-order traversal of the tree, so that when we try to set $Cost(v,x)$ we will have already computed $Cost(w,y)$ for all descent nodes w of v and for all states y. The base cases are when v is a leaf in the rooted tree, and are obviously correct. Now assume v is an internal node with two children w and w'. Inductively we will assume that for all $y \in \Sigma$, $Cost(w,y)$ is equal to the best possible parsimony score of T_w given that we assign state y to w, and that the same is true for w'. Let the optimal assignment of states to internal nodes for T_v given that $c(v) = x$, be achieved with $c(w) = a$ and $c(w') = b$. Hence, by the inductive hypothesis, the best possible score achievable for T_v if we label v by x is $Cost(w,a) + Cost(w',b) + M[x,a] + M[x,b]$. Note that this is the same as $min\{Cost(w,y) + M[x,y] : y \in \Sigma\} + min\{Cost(w',y) + M[x,y] : y \in \Sigma\}$, which is how we set $Cost(v,x)$! Therefore, by induction, $Cost(v,x)$ is set so that it equals the best achievable parsimony score for T_v given that we set $c(v) = x$. Finally, the parsimony score of the tree T is $min\{Cost(root,x) : x \in \Sigma\}$, where *root* is the root of the tree.

The running time analysis is simple. The initialization is $O(nr)$ time. In Stage 1, we need to compute $Cost(v,x)$ for every vertex v and every state x. Provided we calculate these in the correct order (from the bottom up), each one can be computed in $O(r)$ time using the previously computed values. Hence Stage 1 takes $O(r^2n)$ time. Stage 2 takes $O(1)$ time per vertex, and so adds $O(n)$ time to the calculation. Hence the total time is $O(r^2n)$ to calculate the parsimony score for a single character. Since the characters can be computed independently, the total time is $O(r^2nk)$. □

4.3.3 Finding Maximum Parsimony Trees

What about finding the best tree, rather than computing the score of a given tree? This is the maximum parsimony problem, also known as the "large parsimony problem."

Maximum Parsimony Problem

Input: Matrix **A** with n rows and k columns, where A_{ij} denotes the state of taxon s_i for character c_j. (Typically **A** represents an input sequence alignment, so each column is a site in the alignment.)

Output: Tree T on leafset $\{s_1, s_2, \ldots, s_n\}$ with the smallest total number of changes for character set $\{c_1, c_2, \ldots, c_k\}$.

Unfortunately, while the small maximum parsimony problem is polynomial time, the large maximum parsimony problem is NP-hard (Foulds and Graham, 1982). Furthermore, exhaustive search or branch-and-bound solutions are limited to small datasets. Fortunately, effective search heuristics exist that enable reasonable analyses on large datasets (with hundreds or even thousands of taxa). Of course, these heuristics are not *guaranteed* to solve the optimization problem exactly, but many seem to produce trees that are possibly close in score and topology to the optimal solution, when run long enough. However, large datasets may be beyond reach, even in terms of near-optima, using current software.

Finding good maximum parsimony trees is of substantial interest in the biology community, since many phylogenies are computed using software optimized for this problem. The challenges involved in developing better methods are discussed in Chapter 11.

4.4 Constructing Trees from Compatible Characters

4.4.1 Constructing Trees from Compatible Binary Characters

Suppose we have a set of binary characters that we assume have evolved without any homoplasy on some unknown tree T, and we wish to infer T from the data. Under this assumption, we will show we can infer an unrooted tree on which all the characters are compatible. Consider one such character c, and the tree T on which it has evolved. Because c is homoplasy-free and has only two states, there must be an edge in T that separates the leaves with one state from the leaves with the other state. In other words, $C(T)$ must have a bipartition π that produces the same split as c. Therefore, given a set of homoplasy-free binary characters, the problem of constructing an unrooted tree consistent with the characters is equivalent to constructing an unrooted tree from a set of bipartitions. This is a problem we studied in Chapter 2, and which can be solved in polynomial time. We describe this algorithm here, for the sake of completeness.

Algorithm to Compute an Unrooted Tree from Compatible Binary Characters

Input: Set $\mathscr{C}$ of compatible binary characters.
Output: Tree T on which the characters in $\mathscr{C}$ are compatible.

- Let S denote the leafset of T. Select one taxon to be the root, and let its state for each character be the ancestral state of that character.

- For every binary character c in the input, denote the set of taxa with the derived state by A_c.
- Let S denote the leafset of T. Compute the Hasse Diagram for the set $\{A_c : c \in \mathscr{C}\} \cup S \cup \{S\}$. This will produce a rooted tree T with leafset S, and which contains each A_c as a clade.
- Return the unrooted version of T.

Note that this algorithm produces a tree that may not be fully resolved. However, any other tree that is consistent with the set of binary characters will be a refinement of this tree. Hence this tree is a *common contraction* of all trees that are consistent with the character set. This is a strong statement, and it allows us to explore (and succinctly characterize) the solution space.

Note that the algorithm we described could be modified to test whether the set of characters is compatible, by returning *Fail* if the Hasse Diagram is not a tree. Also, the algorithm cannot be used when the state of some character for some taxa is unknown. Note that the equivalence between binary characters and bipartitions yields the following:

Theorem 4.5 *A set $\mathscr{C}$ of binary characters is compatible if and only if every pair of binary characters is compatible.*

Example 4.6 Suppose that the input is given by

- $A = (1,0,0,0,1)$
- $B = (1,0,0,0,0)$
- $C = (1,0,0,1,0)$
- $D = (0,0,0,0,0)$
- $E = (0,1,0,0,0)$
- $F = (0,1,1,0,0)$

In this case, there are two non-trivial characters (defined by the first and second positions), but the third through fifth positions define trivial characters. When we apply this algorithm, we pick one taxon as the root. Since the choice of root doesn't matter, we'll pick A as the root. The clades under this rooting are $\{D,E,F\}$ (for the first character) and $\{E,F\}$ (for the second character). We then add the clade $\{B,C,D,E,F\}$ (i.e., everything but the root taxon) and all the singletons. When we apply the algorithm for constructing trees from clades to this set, we get $(A,(B,C,(D,(E,F))))$. When we unroot this tree, we note that it has a node of degree four, and so is not fully resolved.

Now, suppose we have a data matrix in which we have some missing entries, and we would like to know if it is possible to assign values of 0 or 1 to the missing entries in the character matrix so that the resultant data matrix has a perfect phylogeny. Furthermore, if we can assign the values to make the characters compatible, we would also want to demonstrate this by providing a perfect phylogeny for the matrix.

Example 4.7 In the following input, "?" means that the state is unknown.

4.4 Constructing Trees from Compatible Characters

- $A = (0,0,0)$
- $B = (0,1,1)$
- $C = (1,?,1)$
- $D = (1,0,?)$
- $E = (?,0,0)$

We would like to know whether it is possible to set the various missing entries so that the result is a set of compatible characters (i.e., a set of five binary sequences that have a tree on which all the characters are compatible). The answer for this input is *yes*, as we can use the following assignments of states to the missing values:

- $A = (0,0,0)$
- $B = (0,1,1)$
- $C = (1,0,1)$
- $D = (1,0,1)$
- $E = (0,0,0)$

We know this is compatible, because the tree given by $(A,(E,(B,(C,D))))$ is compatible with these characters (i.e., it is a perfect phylogeny).

Example 4.8 In contrast, there is no way to set the values for the missing entries (indicated by ?) in the following matrix to 0 or 1, in order to produce a tree on which all the characters are compatible:

- $A = (0,0,?)$
- $B = (0,1,0)$
- $C = (1,0,0)$
- $D = (1,?,1)$
- $E = (?,1,1)$

Figuring out that these characters are incompatible, no matter how you set the missing data, is not that trivial. But as there are only three missing values, you can try all $2^3 = 8$ possibilities. A more elegant analysis that does not require examining all available settings is possible, and left to the reader.

More generally, however, answering whether an input with missing data admits a perfect phylogeny is NP-complete, even for the case where there are only two states (Bodlaender et al., 1992; Steel, 1992).

4.4.2 Constructing Trees from Compatible Multi-state Characters

Related to the problem of handling missing data is the question of whether a set of multi-state characters are compatible (i.e., that they evolve down some common tree without any homoplasy). Unfortunately, while determining if a perfect phylogeny exists for binary characters is solvable in linear time (Gusfield, 1991), determining if a perfect phylogeny exists for multi-state characters is NP-complete (Bodlaender et al., 1992; Steel, 1992). Furthermore, while the compatibility of two multi-state characters can be determined in polynomial time (Warnow, 1993), pairwise compatibility no longer ensures setwise compatibility, even for three-state characters (Warnow, 1993). Algorithms to determine whether a set of multi-state characters is compatible, and to compute their perfect phylogenies when they exist, have been developed (Dress and Steel, 1992; Agarwala and Fernández-Baca, 1994, 1996; McMorris et al., 1994; Kannan and Warnow, 1997), but these are either limited to special cases, or have running times that grow exponentially in some parameter (e.g., the maximum number of states per character, the number of characters, or the number of taxa) in the input.

Although perfect phylogeny is an idealized concept, there are conditions in which nearly perfect phylogenies are expected to be achievable. Examples of such conditions include population genetics datasets, where the evolutionary distances between individuals is short enough that very few mutations will have occurred in any site, or whole genomes described by large-scale chromosomal rearrangements. For those cases, methods that can compute perfect phylogenies on subsets of the character datasets can be useful.

4.5 Tree Construction Based on Maximum Compatibility

The maximum compatibility problem seeks the tree on which the largest number of characters are compatible. As with maximum parsimony, there are two variants of the problem: the "small" maximum compatibility problem, which is about computing the compatibility score of a fixed tree, and the "large" maximum compatibility problem, which is about finding the best tree (i.e., the tree with the largest compatibility score). As we will see, the small maximum compatibility problem is solvable using a small variation on the algorithm for the small maximum parsimony problem. Like maximum parsimony, the maximum compatibility problem is also NP-hard.

4.5.1 Algorithm for the Fixed Tree Maximum Compatibility Problem

The problem we address here is to compute the compatibility score of a set of characters on a fixed tree, i.e.:

Input: Matrix **A** with n rows and k columns (so that A_{ij} is the state of taxon s_i for character c_j), and a tree T with leaves labeled by the different species, $s_1, s_2, \ldots, s_n$.
Output: The number of characters that are compatible on T.

To address this question, we can take each character c in turn, and see whether we can label the nodes of the tree with character states for c so that c has no homoplasy (back-mutation or parallel evolution). Equivalently, we wish to set the states of the character for the internal nodes of the tree in such a way that for each state of the character, the nodes of the tree that exhibit that state are connected.

4.5 Tree Construction Based on Maximum Compatibility

We begin with a basic observation that allows us to use the algorithm for computing the maximum parsimony score of a character to determine if it is compatible:

Observation 4.9 A character c is compatible on tree T if and only if its parsimony score on T is $r-1$, where r is the number of states exhibited by c at the leaves of T.

Proof If the character is compatible on T, then it must be able to be extended to the internal nodes so that there is no homoplasy. Without loss of generality, suppose that c attains states $1, 2, \ldots, r$ at the leaves, and that the state at the root is 1. Every substitution has to produce a new state, and so there must be exactly $r-1$ edges on which there are substitutions. Hence, c must have parsimony length $r-1$. Conversely, if c has parsimony length $r-1$, then the assignment of states to the internal nodes must be homoplasy free, since c has r states at the leaves. □

It is therefore easy to establish the following corollary:

Corollary 4.10 *Determining if a character* c *is compatible on a tree* T *can be solved in* $O(nr)$ *time using the algorithm for computing the parsimony score on a tree, where* r *is the number of states exhibited by* c *at the leaves of* T *and* n *is the number of leaves in* T. *Hence, the compatibility score of a set of* k *characters on a tree* T *can be computed in* $O(nkr)$ *time.*

On the other hand, testing the compatibility of a character on a tree T can be determined by eye, if the tree is small enough. For a given internal node v in the tree, if v lies on a path between two leaves having the same state x, we assign state x to node v. If this assignment doesn't have any conflicts – that is, as long as we don't try to assign two different states to the same node – then the character can evolve without any homoplasy on the tree, and otherwise, homoplasy-free evolution is not possible on this tree.

Theorem 4.11 *Computing the compatibility score of a tree* T *on* n *leaves with respect to* k *characters can be computed in* $O(nk)$ *time.*

The proof of this statement is left to the reader.

4.5.2 Finding Maximum Compatibility Trees

The maximum compatibility problem attempts to find the tree on which the largest number of characters are compatible.

Maximum Compatibility

Input: Matrix **A** with n rows and k columns (so that A_{ij} is the state of taxon s_i for character c_j).
Output: Tree T on the leafset $S = \{s_1, s_2, \ldots, s_n\}$ on which the number of characters in $C = \{c_1, c_2, \ldots, c_k\}$ that are compatible is maximized.

Finding the tree with the largest compatibility score is an NP-hard problem, even for the special case where all the characters are binary (Day and Sankoff, 1986). We sketch the proof for binary characters.

Theorem 4.12 *Maximum compatibility is NP-hard, when the input is a set of binary characters.*

Proof The decision version of this problem has two inputs: the data matrix defining a set S of taxa in terms of a set $\mathscr{C}$ of binary characters, and an integer B. We will provide a Karp reduction from this decision version of maximum compatibility of binary characters to maximum clique, which is NP-hard, and so establish that the decision problem is NP-hard. Hence, the optimization problem will also be NP-hard.

Let $\mathscr{C}$ be a set of binary characters defined on a set S of taxa, and let integer B be given. We will create a graph $G(\mathscr{C})$ so that $\mathscr{C}$ has a subset of B compatible characters if and only if $G(\mathscr{C})$ has a clique of size B.

The vertices in $G(\mathscr{C})$ will be the characters $c \in \mathscr{C}$, and the edges will be those pairs $(v_c, v_{c'})$ such that c and c' are pairwise compatible. It is easy to see that if $\mathscr{C}$ has a set of B compatible characters, then the corresponding vertices in $G(\mathscr{C})$ are a clique of size B. If $G(\mathscr{C})$ has a clique of size B, then the corresponding characters in $\mathscr{C}$ are pairwise compatible. Since pairwise compatibility of binary characters ensures setwise compatibility, it follows that $\mathscr{C}$ has a set of B compatible characters. Hence, binary character maximum compatibility is NP-hard. □

Note that this was a linear reduction, so that binary character maximum compatibility has the same approximability results as maximum clique! Unfortunately, maximum clique is one of the hardest problems to approximate, as shown in Zuckerman (2006).

Example 4.13 Consider the following set of taxa, represented by three characters.

- $A = (0,0,0)$
- $B = (0,0,3)$
- $C = (1,1,0)$
- $D = (1,1,1)$
- $E = (2,1,0)$
- $F = (2,2,4)$

This set of characters is compatible, since $(A, (B, (E, (F, (C, D)))))$ is a perfect phylogeny for the input.

Example 4.14 The next example is more interesting, because there is no perfect phylogeny:

- $A = (0,1,0)$
- $B = (0,0,0)$
- $C = (1,0,0)$
- $D = (1,1,1)$

Note that the third character is compatible on every tree, but the first two characters are incompatible with each other. Therefore, any tree can have at most one of these first two characters compatible with it, for a total of two compatible characters. One of those trees is given by $((A,B),(C,D))$, and the other is $((A,D),(B,C))$. The third possible unrooted binary tree on these taxa is $((A,C),(B,D))$, but it is incompatible with both of these characters.

4.6 Treatment of Missing Data

Input matrices to maximum parsimony and maximum compatibility often contain missing data, indicated by symbols such as "?". (As we will see later, multiple sequence alignments also contain letters other than nucleotides or amino acids, and the dashes in the alignments are also often treated as missing data.) In a maximum parsimony or maximum compatibility analysis, these missing data entries are typically replaced by character states so as to obtain the best possible score; the output, however, will include the symbols representing the missing data.

An alternative treatment of missing data replaces all the missing entries with the same new state, and then seeks a tree that optimizes the criterion. This approach can produce a different result.

Finally, in some analyses, all columns that are considered to have too much missing data are removed before the tree is computed. The decision about how much missing data is "too much" also varies.

4.7 Informative and Uninformative Characters

Both maximum parsimony and maximum compatibility are computationally intensive if solved exactly. Thus, finding ways to speed up methods to try to solve these problems can be helpful. Here, we consider the question of whether we can eliminate some of the characters from the input, without changing the solution space. In other words, given input matrix **A** (where $A_{i,j}$ is the state of the taxon s_i for the jth character), you would like to know whether removing some specific character (say character **x**) has any impact on the tree that is returned. For example, suppose we wish to solve maximum parsimony. Since removing a character amounts to removing one column in the matrix, this would be the same as saying "If we define matrix **A-x** to be the matrix obtained by taking **A** and removing column **x**, when is it *guaranteed* that the set of optimal parsimony trees for **A-x** are optimal parsimony trees for **A**?" If we can prove that it is safe to delete the column **x**, then removing it could reduce the running time of searching for optimal maximum parsimony trees (and similarly for maximum compatibility). A character that has *no impact* on tree estimation using maximum parsimony methods is called "parsimony uninformative." We formally define this as follows:

Definition 4.15 Let x be a character defined on set S of species. Then x is **parsimony uninformative** if and only if for all matrices **A** for S the set of optimal parsimony trees on **A** is identical to the set of optimal parsimony trees on **A+x**, where **A+x** denotes the matrix obtained by adding column **x** to **A**.

As a consequence, the set of optimal parsimony trees will not change by removing a parsimony uninformative site from any alignment. All other characters are called "parsimony informative." Removing parsimony uninformative characters can result in a speed-up in the search for optimal trees (especially if there are many such characters). Equally importantly, thinking about which characters are parsimony informative or not will help you understand the different impact of different characters on phylogeny estimation using maximum parsimony.

The same property can be asked about *any* phylogeny estimation method, obviously, and so we can ask whether a character is "compatibility informative." It is not hard to see the following:

Lemma 4.16 *A character is parsimony informative and compatibility informative if and only if it has at least two "big states," where a state is "big" if it has at least two taxa in it.*

Proof Suppose a character c exhibits r states on the taxon set S. We will begin by showing that c is parsimony and compatibility uninformative if it does not have two or more big states. There are two possible cases for this: c has exactly one big state (and so all other states are singletons), or it has only singleton states. In either case, let i be one of the largest states for c (thus, i is the unique big state, or i is a singleton state). Given a tree T to be scored, we put i as the state for every internal node in T. Note that c is compatible on the tree T using this labeling for the internal nodes, and that the tree has parsimony score $r-1$. Since $r-1$ is the best possible parsimony score for any tree with r states appearing at the leaves and the tree T was arbitrary, this means that *all* trees have the same parsimony and compatibility score for this character. Therefore, removing c from the set of characters will not change the relative ordering of trees with respect to either parsimony or compatibility. This proves one direction.

For the other direction, let c be a character with at least two big states. Hence, there is a quartet of taxa u,v,x,y for which $c(u) = c(v) \neq c(x) = c(y)$. It is enough to prove this for the case where $C = \{c\}$, so that C only has this one character. If we remove c from C we obtain the empty character set, for which all trees are equally good with respect to both parsimony and compatibility. Hence, all we need to show is that there are at least two trees that have different parsimony and compatibility scores with respect to c. Let $\{1,2,\ldots,r\}$ be the states exhibited by c on the leafset. Consider the tree with a center node a adjacent to r nodes $v_1, v_2, \ldots, v_r$, and where the leaves for the taxa with state i are adjacent to v_i. This tree, although not binary, is compatible with c, and so has compatibility score 1 and parsimony score $r-1$. Furthermore, we can define a second tree T' formed by beginning with quartet tree $ux|vy$, and then attaching all the other leaves arbitrarily to this quartet tree. Since $ux|vy$ is incompatible with c, it follows that c will not be compatible on the tree T', and so its parsimony score will be greater than $r-1$ and its compatibility score will be 0. Thus, c distinguishes T and T' with respect to both parsimony and compatibility. □

Other methods can be used to estimate trees, of course, and so the definition of what constitutes "informative" has to be based on the method.

Example 4.17 Consider the following set of four DNA strings, u, v, w, x:

- u = ACAAAAAG
- v = ACTTTTCG
- w = TTTTTTTG
- x = TTACTGGG

Numbering the eight sites from left to right, we can see that the first, second, and third sites are parsimony informative (each has two big states), but the remaining sites are not parsimony informative. That is, each of the remaining sites either has only one big state, or has no big states. Hence, in a maximum parsimony analysis, only the first three sites will impact the scores, and hence only these three sites need to be considered in solving maximum parsimony. (And of course the same is true for maximum compatibility.)

We now try to solve maximum parsimony on this dataset. There are three possible trees on u, v, w, x, given by $T_1 = uv|wx, T_2 = ux|vw$, and $T_3 = uw|vx$. The first site has cost 1 on T_1 and cost 2 on the other two trees. The second site also has cost 1 on T_1 and cost 2 on the other two trees. The third site has cost 1 on T_2 and cost 2 on the other two trees. Hence, the least cost is obtained on T_1, and so T_1 is the single best maximum parsimony tree for this dataset. Note that if we were to try to solve maximum compatibility on the dataset, the same analysis would work, and so T_1 would be the single best maximum compatibility tree for the dataset.

4.8 Further Reading

The question of whether a perfect phylogeny exists for a character matrix defining a set of taxa by a set of qualitative characters, and constructing the perfect phylogeny when it exists, is an approach to phylogeny estimation that was introduced in LeQuesne (1969), and then elaborated on in a series of papers by various authors. This problem, which is known as the Character Compatibility problem or the Perfect Phylogeny problem, is equivalent to a graph-theoretic problem, which asks whether it is possible to add edges to a given vertex-colored graph G so that it becomes triangulated, without adding edges between vertices of the same color. These two formulations of the problem have led to a collection of algorithms, some combinatorial and others graph-theoretic, that collectively address the fixed-parameter variants of the two problems (i.e., algorithms when the number of states per character is bounded or when the number of characters is bounded). For surveys on this problem, see Warnow (1993) and Fernández-Baca (2000).

4.9 Review Questions

1. What is a perfect phylogeny?
2. Define homoplasy and give an example of a biological characteristic that clearly evolved homoplastically.
3. Give some examples of evolution that are not treelike.
4. What is a character? What is a binary character?
5. Suppose you have a binary character matrix, so the rows represent species and the columns represent characters, and every entry of the matrix is either a 0 or a 1. What does it mean to say that the character set is compatible? How computationally difficult is it to determine if the character set is compatible?
6. Suppose you have a multi-state character matrix, so the rows represent species and the columns represent characters, and the entries of the matrix are any integers. What does it mean to say that the character set is compatible? How computationally difficult is it to determine if the character set is compatible?
7. How computationally difficult is it to test whether a character is compatible on a tree?
8. State the maximum compatibility and maximum parsimony problems. How computationally difficult is it to solve each problem (i.e., are these problems in P, NP-hard, or of unknown computational complexity)?
9. Suppose you are given a tree T and there are DNA sequences (each of length k) at the leaves of the tree. What would you do to calculate the parsimony score of these sequences on the tree?
10. Define parsimony informative, and give an example of a binary character in a character matrix that is not parsimony informative.
11. Let $S = \{s_1, s_2, \ldots, s_n\}$ be a set of binary sequences of length k and let T be a binary tree on the same leafset. Which of the following is the correct running time of the dynamic programming algorithm for computing the parsimony score of T with this set of sequences at the leaves?

    - $\Theta(nk)$
    - $\Theta(2^n k)$
    - $\Theta(2^k n)$
    - $\Theta(n^2 k)$

4.10 Homework Problems

1. Suppose we are given the following input of four taxa described by six-tuples (i.e., six characters), where each character is binary. We let 0 denote the ancestral state and 1 denote the derived state. Construct the rooted tree that is consistent with these characters evolving without homoplasy.

    - $a = (1,1,0,0,1,0)$
    - $b = (1,0,1,0,1,0)$

4.10 Homework Problems

- $c = (0,0,0,1,0,0)$
- $d = (0,0,0,0,1,1)$

2. Take the data matrix from the previous problem, and add in the root sequence r given by $r = (0,0,0,0,0,0)$. Thus, you now have a matrix with five taxa, a,b,c,d,r, defined by six characters (one character for each position in the six-tuple). Divide this matrix into two pieces: the first three characters and the last three characters. Construct the minimally resolved unrooted tree that is compatible with each submatrix. How are these trees different? Are they fully resolved, or do they have polytomies? Compare them to the compatibility tree you obtained on the full matrix. Now, treat the tree on the full matrix as the "true tree," and compute the false negative and false positive rates for these two trees. What do you find?

3. Construct an unrooted tree that is consistent with the following input of four taxa described by four binary characters, under the assumption that all characters evolve without homoplasy. (You may not assume that any particular state is ancestral on any character.)

- $a = (0,0,1,1)$
- $b = (1,0,0,1)$
- $c = (1,1,0,1)$
- $d = (1,0,1,0)$

4. For the tree T given by $((a,(b,(c,(d,(e,f))))))$, determine for each of the characters (columns in the following tuple representation) whether it could have evolved on the tree T without any homoplasy:

- $a = (0,0,0,0,1)$
- $b = (0,1,1,0,0)$
- $c = (1,0,0,1,1)$
- $d = (1,2,0,1,0)$
- $e = (2,0,2,0,1)$
- $f = (2,3,2,0,1)$

5. For the following input, show how to set the entries given with "?" so as to produce a compatible matrix:

- $A = (0,1,0,?)$
- $B = (0,1,1,0)$
- $C = (0,0,1,0)$
- $D = (1,0,1,1)$
- $E = (1,0,?,1)$

Explain how you derived your solution.

6. In Example 4.8, we said that there was no way to set the values for the missing entries in the input of five sequences A,B,C,D, and E given below, in order to produce a tree on which all the characters are compatible. Prove this assertion without examining each of the eight ways to set the missing values.

- $A = (0,0,?)$
- $B = (0,1,0)$
- $C = (1,0,0)$
- $D = (1,?,1)$
- $E = (?,1,1)$

7. Suppose T and T' are two trees on the same leafset, and T' refines T.
    - Prove or disprove: If character c is compatible on T then it is compatible on T'.
    - Prove or disprove: If character c is compatible on T' then it is compatible on T.

8. The maximum parsimony problem asks us to find a tree that has the best maximum parsimony score with respect to a matrix A. Consider the following variant, binary tree maximum parsimony: Given a matrix A, find a *binary tree* that optimizes maximum parsimony.
    a. Is it possible for a solution to the binary tree maximum parsimony problem to not be optimal for the standard maximum parsimony problem?
    b. Consider the same question but restated in terms of maximum compatibility and binary tree maximum compatibility. Does your answer change?

9. Consider the set of six taxa described by two multi-state characters, $A = (0,0), B = (1,2), C = (0,2), D = (2,1), E = (1,1)$, and $F = (1,0)$, and the tree on the taxa given by: $(((A,B),C),(D,(E,F)))$.
    - Apply the parsimony algorithm to assign states to each node for each of the two characters. What is the parsimony score of this tree?
    - Give two different character state assignments to the nodes to produce the minimum number of changes.

10. Consider the set of sequences (but ignore the tree provided) given for the input given in the previous problem. Find an optimal (unrooted) tree topology T on this set. (Do this without trying to score all possible trees – think about the best achievable score for this specific dataset.) Are either of the characters compatible on T? If not, find an optimal maximum parsimony tree for this input for which at least one character is compatible.

11. Consider the rooted tree given by $((s_1,s_2),(s_3,(s_4,s_5)))$. Suppose that each leaf is labeled by a single nucleotide, with s_1 and s_3 labeled by C, s_2 and s_4 labeled by A, and s_5 labeled by G.
    - Compute the set $A(v)$ for every node v using the Fitch maximum parsimony algorithm on this rooted tree.
    - Find two assignments of nucleotides to the internal nodes of the tree to achieve the best maximum parsimony score, drawing the nucleotide for node v from $A(v)$.
    - Is there a maximum parsimony assignment of nucleotides to the internal nodes of the tree that does *not* always pick the nucleotide for each node v from $A(v)$? (This problem is derived from Felsenstein (2004).)

12. Suppose T and T' are two trees on the same leafset, and T' refines T. Prove that the parsimony score of T' is at most that of T.

13. Suppose the tree is given by $(A,(B,(C,(D,E))))$, and that we have three homoplasy-free characters on these taxa given by:

    - $A = (0,0,1)$
    - $B = (0,1,1)$
    - $C = (0,0,0)$
    - $D = (1,0,0)$
    - $E = (1,0,0)$

    Assume that 0 is the ancestral state and 1 the derived state for each of these characters. Determine the edges in the tree that could contain the root.

14. Suppose **A** is an input matrix for maximum parsimony, so **A** assigns states for each character to all the taxa in a set S. Suppose **A**′ is the result of removing all characters from **A** that are identical on all taxa (i.e., characters c such that $c(s) = c(s')$ for all s, s' in S). Prove or disprove: **A** and **A**′ have the same set of optimal trees under maximum parsimony.

15. Suppose **A** is an input matrix for maximum parsimony and **A**′ is the result of removing all characters from **A** that have different states on every taxon (i.e., characters c such that $c(s) \neq c(s')$ for all $s \neq s'$ in S). Prove or disprove: M and M' have the same set of optimal trees under maximum parsimony.

16. Let **A** be an input matrix to maximum parsimony, and let **A**′ be the result of removing all parsimony uninformative characters from **A**. By the definition of parsimony uninformative, the trees that are returned by an exact maximum parsimony solution for **A**′ will be the same as the maximum parsimony trees returned for **A**. However, suppose you use the characters to define branch lengths in some output tree (as there can be many), as follows. You use maximum parsimony to calculate ancestral sequences, and then you use Hamming distances to define the branch lengths on the tree.

    a. Is it the case that branch lengths you compute on a given tree T must be the same for **A** as for **A**′? (In other words, can branch length estimations change?)
    b. If you use normalized Hamming distances instead of Hamming distances, does your answer to the previous question change?

17. Consider the following input matrix to maximum parsimony:

    - $a = (0,1,0,0,0)$
    - $b = (0,0,1,1,1)$
    - $c = (0,0,2,3,2)$
    - $d = (0,2,0,1,1)$
    - $e = (1,2,0,1,1)$
    - $f = (0,0,3,2,1)$

    Write down all the trees that have the best maximum parsimony scores on this input, and explain how you obtain your answer. Do *not* solve this by looking at all possible trees on $\{a,b,c,d,e,f\}$.

18. Is it the case that maximum compatibility and maximum parsimony always return the same set of optimal trees? If so, prove it; otherwise, find a counterexample. If you find a counterexample, prove it's a counterexample mathematically (i.e., not using software).
19. Find a biological dataset with at least 50 aligned sequences, and try to solve maximum parsimony on the dataset using some good software for maximum parsimony. Since maximum parsimony is NP-hard, the trees you find may be local optima rather than global optima. That said, how many "best" trees does your search return? How different are the trees?

5
Distance-based Tree Estimation Methods

5.1 Introduction

Distance-based tree estimation was briefly introduced in Chapter 1, where we showed that phylogenetic trees could be computed efficiently and accurately using distance-based methods, and that distance-based estimation methods could be proven to be statistically consistent under stochastic models of sequence evolution. In this chapter we go into these issues in greater detail; however, the reader may wish to review Chapter 1 before continuing.

Distance-based tree estimation is really a two-step process, where the first step calculates a distance matrix for the input set, and the second step computes a tree from the distance matrix. Many distance-based tree estimation methods are polynomial time and fast in practice, and so have a computational advantage over most other techniques for computing phylogenetic trees.

In general, the distance matrices that are computed will be symmetric (i.e., $d_{ij} = d_{ji}$ for all i, j) and zero on the diagonal (i.e., $d_{ii} = 0$ for all i). However, these distances generally will not satisfy the triangle inequality, defined as follows:

Definition 5.1 An $n \times n$ matrix **d** is said to satisfy the **triangle inequality** if, for all i, j, k, $d_{ik} \leq d_{ij} + d_{jk}$.

A matrix that is symmetric, zero on the diagonal, and satisfies the triangle inequality is called a "distance matrix." Since phylogenetic distances may not satisfy the triangle inequality, to be completely mathematically rigorous, we should not refer to them as distance matrices, and instead should only refer to them as **dissimilarity** matrices. The literature in phylogeny estimation abuses the term, and refers to dissimilarity matrices as "distance matrices." Therefore, we may sometimes refer to dissimilarity matrices as distance matrices. However, please be aware that the matrices computed in phylogenetic estimation may not satisfy the triangle inequality, and hence any proofs regarding distance-based phylogeny estimation cannot assume this property.

As discussed in Chapter 1, for the approach to have good statistical properties, the estimated distances should be defined so that they converge (as the sequence length increases) to an additive matrix for the true tree. The first step, therefore, depends on the specific assumptions of the evolutionary model.

Distance-based tree estimation can be applied to many types of data, as described in Section 5.12. The main focus of this chapter is the use of distance-based tree estimation where the input is a sequence alignment, and distances between the sequences are computed using an assumed model of sequence evolution. For example, we could compute distances under the Cavender–Farris–Neyman (CFN) model for binary sequences, or under one of the standard DNA sequence evolution models (see Chapter 8).

5.2 UPGMA

One of the original ways of computing trees is UPGMA, which stands for "Unweighted Pair Group Method with Arithmetic Mean" (Sokal and Michener, 1958). UPGMA computes a rooted tree from an input distance matrix in an agglomerative fashion. In the first iteration, it finds a pair of taxa x, y that have the smallest distance, and makes them siblings. These two are then replaced by the cluster $\{x, y\}$, and the distance from $\{x, y\}$ to every other taxon z is defined to be the average of $d(x, y)$ and $d(x, z)$; i.e.,

$$d(\{x,y\}, z) = \frac{d(x,z) + d(y,z)}{2}.$$

In subsequent iterations, the elements are clusters that are either singletons (the original taxa) or sets of two or more taxa. In each iteration, the pair of clusters A and B with the minimum distance is selected to be the next sibling pair and the clusters are merged. The distance matrix is then updated by reducing its dimension by one (removing rows and columns for A and B and replacing it with a row for $A \cup B$) with

$$d(A \cup B, C) = \frac{|A|d(A,C) + |B|d(B,C)}{|A| + |B|}.$$

The process is repeated until all the taxa are merged into a single cluster. Note that this technique produces a rooted tree. Variations of this technique can be considered where other ways of updating the distance matrix are used (e.g., consider the simple variant described in Chapter 1), but the basic algorithmic structure is to find the pair of clusters that are closest and replace them with a new cluster, and then iterate.

We begin with an example of UPGMA applied to a case where the distances obey a strict clock and so produce an ultrametric matrix. Figure 5.1 gives an ultrametric matrix, and Figure 5.2 gives the rooted tree realizing that matrix. If we had applied UPGMA to the matrix in Figure 5.1, we would obtain the tree in Figure 5.2. Hence, UPGMA would have returned the rooted tree realizing the matrix.

Not all distances obey a strict molecular clock, and UPGMA can fail when the input matrix is not sufficiently clocklike. Consider, for example, the tree with lengths on the edges and its matrix of leaf-to-leaf distances (defined by adding up the lengths of the edges on the path between the pair of leaves) given in Figure 5.3. Note that the pair that minimizes

5.2 UPGMA

	B	C	D
A	2	16	16
B		16	16
C			10
D			

Figure 5.1 Ultrametric matrix.

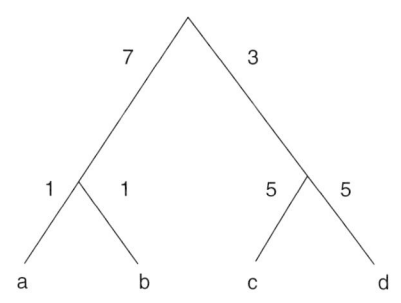

Figure 5.2 Rooted tree realizing the ultrametric matrix from Figure 5.1. Note that the distance from the root to every leaf is the same.

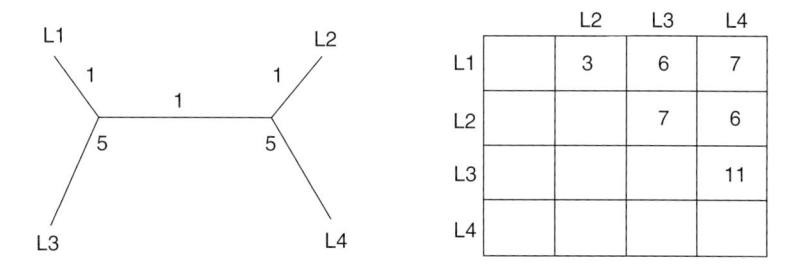

	L2	L3	L4
L1	3	6	7
L2		7	6
L3			11
L4			

Figure 5.3 Additive matrix and its edge-weighted tree.

the distance is L_1, L_2, but that these are not siblings! Thus, when UPGMA is applied to the matrix for this tree, it will produce the *wrong* tree.

UPGMA has a very specific way of updating the distance matrix, and variants of UPGMA can update the matrix using simpler techniques. For example, in Chapter 1, we described a variant where we updated the matrix by just removing the row and column corresponding to one of the two taxa being made into siblings. Clearly the specific technique used to update the distance matrix will impact the final tree. However, *any* agglomerative method that makes the taxon pair with a minimum pairwise distance siblings will fail on an input distance matrix that has its closest pair of taxa not being siblings. Also, for any variant of UPGMA, just knowing that it has that algorithmic design allows you to determine the first sibling pair (up to ties). Therefore, if the dataset has only three leaves, you can determine the rooted tree that will be constructed, and if the dataset has four leaves, then you can determine the *unrooted* tree it will produce. Therefore, for at least some model conditions, proving that UPGMA or some variant of UPGMA produces the wrong tree is generally easy: If the first sibling pair it produces isn't a true sibling pair, the result will be incorrect, no matter what the subsequent steps are.

5.3 Additive Matrices

Given a tree T with non-negative edge weights, and given two leaves x and y in T, we define the distance $D(x,y)$ between x and y to be the sum of the weights of the edges on the path between x and y. It is easy to see that the matrix **D** defined in this manner is a true distance matrix (i.e., it is a dissimilarity matrix that satisfies the triangle inequality). Any distance matrix that can be derived in this manner is said to be **additive**. Furthermore, given any additive matrix, there is a unique tree with *strictly positive* branch lengths that fits the matrix exactly, obtained by collapsing all the edges that have zero length. We will refer to this tree as the *minimally resolved tree* realizing the additive matrix. One of the most well-known theorems about additive matrices was established in Buneman (1974b):

Theorem 5.2 **The Four Point Condition:** *A* n × n *matrix* **D** *is additive if and only if it satisfies the "Four Point Condition" for all four indices* i, j, k, l, *which is that the median and largest of the following three values are the same:*

- $D_{i,j} + D_{k,l}$
- $D_{i,k} + D_{j,l}$
- $D_{i,l} + D_{j,k}$

Only one direction of this theorem is easy, and we will sketch this part of the proof. We will show that if the matrix **D** is additive, then it satisfies the Four Point Condition. The proof of this argument in Chapter 1 was for the special case where the matrix corresponded to a tree with strictly positive edge weights, so here we only have to handle the possibility of zero-weight edges. Since **D** is additive, by definition there is a tree T with non-negative edge weights so that the leaf-to-leaf path distances in T correspond to the entries of **D**. So suppose we have four indices, i, j, k, l. We can assume (without loss of generality) that the tree has one or more edges separating i, j from k, l. If the path separating i, j from k, l has positive weight, then the argument we gave in Chapter 1 applies, and so $D_{ij} + D_{kl}$ is strictly smaller than the other two sums, and the other two sums have the same total score; thus, the Four Point Condition holds. If the path separating i, j from k, l has zero weight, then the three pairwise sums are identical, and the Four Point Condition holds as well.

The other direction requires that we prove that if a matrix satisfies the Four Point Condition for every four indices, then it is additive – and so can be represented by a tree with non-negative branch weights. This argument is non-trivial, and so is not shown here. However, note that it means that any time a matrix satisfies the Four Point Condition, we can use the matrix to construct a tree, and this can be a powerful tool in proving that a method has good statistical properties. Also, proving that a matrix satisfies the Four Point Condition is often easier than proving it corresponds to a tree with non-negative branch weights.

Thus, there are two possible outcomes when we examine an additive matrix and compare the three pairwise sums for four indices, i, j, k, l: All pairwise sums are identical or there are two different values. Can we use this property to compute a tree? Also, since the tree could have zero-length edges, we might want to collapse the zero-length edges, and represent the matrix by a tree with polytomies (i.e., nodes of degree larger than three) in which all edges have strictly positive weight. Such a tree would be the minimally resolved tree realizing the additive matrix. Can we construct such a tree, given an additive matrix?

We answer this question for four leaves, and leave the general problem to the reader. So suppose the matrix $\mathbf{D}$ is only on four indices, i, j, k, l. If the smallest of the three pairwise sums is $D_{ij} + D_{kl}$ and it is strictly smaller than the largest two, then $ij|kl$ is the quartet tree on $\{i, j, k, l\}$. Otherwise, all three pairwise sums are identical, and any tree with strictly positive lengths on the edges that realizes the additive matrix induces a star tree (see Definition 2.25). On the other hand, if zero-length edges are allowed, then the additive matrix can also be realized by a resolved tree on $\{i, j, k, l\}$ in which the internal edge has length 0. Thus, the minimally resolved tree realizing the additive matrix where all three pairwise sums are the same is a star tree.

So, suppose that we have an additive matrix corresponding to a binary edge-weighted tree where all the edge weights are positive. Obviously, the tree can be inferred from its additive matrix, since we can construct the correct quartet tree on every four leaves, and then use the All Quartets Method (see Section 3.4.2) to combine these quartet trees into the tree on the full dataset. However, the All Quartets Method only works with fully resolved quartet trees, so it cannot be applied directly to an additive matrix in which some branches have zero weight. (For that case, a modified version of the All Quartets Method is required so that it can handle unresolved quartet trees.)

Finally, what about estimating trees from dissimilarity matrices that are not additive? Here we show a method that can always estimate a tree on four-leaf datasets, whether or not the dissimilarity matrix is additive.

5.4 Estimating Four-Leaf Trees: The Four Point Method

Given four taxa, i, j, k, l, and given the dissimilarity matrix $\mathbf{d}$ on the four taxa, the Four Point Method (FPM) operates as follows:

- Step 1: Compare the three pairwise sums $d_{i,j} + d_{k,l}, d_{i,k} + d_{j,l}$, and $d_{i,l} + d_{j,k}$, and find the pairwise sums that have the smallest total.

- Step 2: If there are two or more pairwise sums with the same smallest total (i.e., the minimum is not unique), then return a star tree. Else, without loss of generality, assume $d_{i,j} + d_{k,l}$ has the smallest value, and return the quartet tree $ij|kl$.

Theorem 5.3 *Let* **D** *be a* 4×4 *additive matrix corresponding to a binary (i.e., fully resolved) tree* T *on four leaves with strictly positive branch lengths. Then the FPM applied to* **D** *returns tree* T.

Proof Let T have leafset $\{a,b,c,d\}$ with an internal edge separating ab from cd, and let $f > 0$ be the length of the internal edge. The three pairwise sums are all positive, and $D_{ab} + D_{cd} + 2f = D_{ac} + D_{bd} = D_{ad} + D_{bc}$. Hence, the FPM will return $ab|cd$, which is the topology of T. □

We will need the following definitions for the remaining material:

Definition 5.4 Given a binary tree T leaf-labeled by taxon set S, and positive edge weights defined by the function $w : E(T) \to R^+$, we will define E_I to be the set of internal edges in the tree, and $f = min_e\{w(e) : e \in E_I\}$. (Note that $f > 0$ for all trees T with positive branch lengths.)

Definition 5.5 Let **M** and **M**′ denote two $n \times n$ matrices with real values. Then $L_\infty(M,M') = max_{ij} |M_{ij} - M'_{ij}|$.

Lemma 5.6 *Let* T *be an edge-weighted tree with* n ≥ 4 *leaves such that* w(e) $\geq$ f > 0 *for all internal edges* e. *Let* **D** *be the additive matrix for* (T,w), *and let* **d** *be an* n $\times$ n *dissimilarity matrix satisfying* L_∞(d,D) $<$ f$/2$. *Select any four leaves* q $= \{$i,j,k,l$\}$ *and consider the submatrix* **d(q)** *of* **d** *induced by quartet* q. *Then the FPM applied to distance matrix* **d(q)** *will return the tree induced by* T *on leafset* q.

Proof Let **d** be the input dissimilarity matrix, and let $q = \{i,j,k,l\}$ be any four indices. Suppose that T induces quartet tree $ij|kl$, and that **D** is an additive matrix for T under some weighting on the edges. We need to prove that the FPM on q will also return this quartet tree $ij|kl$.

Since T induces $ij|kl$, it follows that the FPM will return $ij|kl$ given matrix **D** as input. Hence, $D_{ij} + D_{kl}$ is the smallest of the three pairwise sums using the matrix **D**. The FPM will return the quartet tree $ij|kl$ for input matrix **d** if and only if $d_{ij} + d_{kl}$ is strictly less than the other two pairwise sums formed using these four indices. Hence, we just need to prove that when $L_\infty(d,D) < f/2$ then $d_{ij} + d_{kl}$ is the smallest of the three pairwise sums.

Note that $D_{ij} + D_{kl} = D_{ik} + D_{jl} - 2F = D_{il} + D_{jk} - 2F$, where F is the length of the path separating the pairs i,j from k,l in the edge-weighted true tree T associated with the additive matrix **D**. Therefore, $F \geq f$, where f is the length of the shortest internal edge in T. Since $L_\infty(d,D) = \delta < f/2$, it follows that the gap between $d_{ij} + d_{kl}$ and the other two pairwise sums can only be reduced by 4δ (adding δ to each of D_{ij} and D_{kl} and subtracting δ from each of the other four entries). Therefore, if $4\delta < 2F$, it follows that the FPM will return $ij|kl$ on input matrix **d**. Since we require that $\delta < f/2$ and $f \leq F$, it follows that $4\delta < 2F$.

Hence, the FPM will return $ij|kl$ for input matrix **d** for every quartet $q = \{i,j,k,l\}$ whenever $L_\infty(d,D) < f/2$. □

5.5 Quartet-based Methods

In the previous section, we presented the FPM and showed that we can use it to construct quartet trees, and in Section 3.4 we presented several methods for combining quartet trees into a tree on the full dataset. The combination of these techniques suggests phylogenetic tree estimation methods based on combining these approaches, the subject of this section.

5.5.1 The Naive Quartet Method

We begin with the simplest quartet-based method for phylogeny estimation. The input is an $n \times n$ dissimilarity matrix **d**, and we assume $n \geq 5$.

Step 1: For every four indices i,j,k,l, use the FPM on matrix **d** restricted to the rows and columns for i,j,k,l. If the output of the FPM for some set of four indices is a star tree (indicating that two or more of the three pairwise sums have the same minimum total), then *return Fail*. Otherwise, the FPM returns a resolved quartet tree for every set of four indices.

Step 2: Apply the All Quartets Method (from Section 3.4.2) to test if the quartet trees are compatible. If they are compatible, then return the tree T that agrees with all the quartet trees; else *return Fail*.

This approach was introduced in Erdös et al. (1999b), where it was called the "Naive Method"; here, we call it the Naive Quartet Method, to emphasize the use of quartet trees.

Theorem 5.7 *(From Erdös et al. (1999b))* Let **d** *be the input* $n \times n$ *dissimilarity matrix and* **D** *be the matrix corresponding to a tree* T *with positive edge lengths. Assume* $L_\infty(d,D) < f/2$, *where* $f > 0$ *is the minimum length of any internal edge of* T. *Then the Naive Quartet Method applied to* **d** *will return* T.

Proof By Lemma 5.6, the FPM will return the correct quartet tree topology on every four leaves in T. Then, since all the quartet trees are correct, the All Quartets Method will return the true tree T from the set of quartet trees computed by the FPM. □

What we have shown is that the Naive Quartet Method will be accurate on a small enough neighborhood of any additive matrix that corresponds to the true tree with positive branch lengths. What does the Naive Quartet Method do when the input matrix **d** is not close enough to an additive matrix corresponding to the true tree? Unfortunately, it is not easy to state the outcome for the Naive Quartet Method given a matrix **d** that is not close to an additive matrix for the true tree. However, since getting even one quartet tree incorrect can produce an incompatible set of quartet trees and the All Quartets Method will return *Fail* whenever the quartet trees are incompatible, the Naive Quartet Method is very vulnerable to errors in the input dissimilarity matrix.

5.5.2 The Buneman Tree and Related Techniques

Any method for combining quartet trees can be used to estimate a tree from the set Q of quartet trees computed using the FPM, or any other technique. Hence, distance-based phylogeny estimation can be addressed through many variations on this theme: Compute quartet trees using the input dissimilarity matrix, and then combine the quartet trees using a quartet amalgamation method.

The Q^* method (Berry and Gascuel, 1997), also known as the Buneman Tree, is the most well known of these methods. The first step of the Q^* method is the calculation of the set Q of estimated quartet trees using the FPM. The second step seeks a maximally resolved tree T, all of whose resolved (i.e., binary) quartet trees are in Q. While it is easy to see that the tree T always exists, since the star tree is a feasible solution, the interesting thing is that there is a unique such tree (Buneman, 1971). The unique maximally resolved tree T satisfying $Q_r(T) \subseteq Q$ is called the Buneman Tree or the Q^* tree. It is easy to see that the Buneman Tree can be calculated in polynomial time using a greedy technique: Arbitrarily order the leaves, and then construct the tree by adding the next leaf into the previous tree, collapsing edges if necessary, to maintain the required property that no edge in the tree be incompatible with any of the quartet trees in Q. In fact, with some care, the Buneman Tree can be computed in $O(n^4)$ time, where n is the number of leaves (Berry and Gascuel, 1997).

The Buneman Tree method is superior to the Naive Quartet Method, in that it is guaranteed to return a tree on every input. However, the Buneman Tree is also generally highly unresolved, since any quartet tree that violates a bipartition will cause that bipartition to not appear. In other words, as the dissimilarity matrix **d** deviates from additivity, the FPM can fail to correctly compute quartet trees, and this will reduce the resolution in the Buneman Tree. Therefore, extensions of the Buneman Tree method (Berry and Bryant, 1999; Bryant and Moulton, 1999; Berry et al., 1999, 2000; Della Vedova et al., 2002; Faghih and Brown, 2010) have been developed that will produce trees with bipartitions that are contradicted by a bounded number of quartet trees in Q.

Yet these quartet-based methods have not generally been as accurate as the better distance-based estimation methods. For example, St. John et al. (2003) evaluated several quartet-based methods, including the Buneman Tree, Quartet Cleaning, and Quartet Puzzling (Strimmer and von Haeseler, 1996), and compared them to neighbor joining (described in the next section), one of the most well-known distance-based methods. On the datasets St. John et al. (2003) examined, they found that neighbor joining had much better accuracy than any of the quartet-based methods. Most importantly, St. John et al. (2003) showed that using neighbor joining to compute a tree on the entire dataset and then using the quartet trees induced by the neighbor joining tree produced more accurate quartet trees than using the FPM to compute quartet trees. This observation suggests that estimating quartet trees independently of the other taxa loses valuable information, and suggests that quartet-based methods for tree estimation may not provide the same level of topological accuracy, compared to methods that analyze the full dataset.

To understand why quartet-based methods have difficulties, consider the CFN model tree with topology *ab|cd* in which the edges to the leaves *a* and *d* are very long (so that the probability of substitution is close to 0.5) and all the other edges are very short (with probabilities of substitution close to 0). Many methods, such as maximum parsimony and UPGMA, will tend to return *bc|ad* (i.e., the wrong tree) given data that evolve down this difficult tree. Other methods, such as neighbor joining and maximum likelihood, will also return the wrong tree unless the sequences are quite long. On the other hand, a much larger tree in which all the edges are relatively short could *include* this difficult quartet tree as a subtree, and could be easier to compute with high accuracy. Hence, quartet-based methods are challenged inherently because the quartet trees they compute can have high error, simply because they are computed in isolation.

5.6 Neighbor Joining

Neighbor joining (Saitou and Nei, 1987) is probably the most well known and the most widely used of all distance-based methods. We will begin with a description of how it operates, using a modification due to Studier and Keppler (1988). As we will see in later sections, neighbor joining is one of the most accurate methods in use today, and there is great interest in understanding why it works well (Steel and Gascuel, 2006).

Neighbor joining is an agglomerative technique, and so operates using iteration, building the tree from the bottom up. The input is an $n \times n$ dissimilarity matrix **d**. In the first iteration, the *n* leaves are all in their own clusters; in subsequent iterations, each cluster is a set of leaves, but the clusters are disjoint. At the beginning of each iteration, the taxa are partitioned into clusters, and for each cluster we have a rooted tree that is leaf-labeled by the elements in the cluster. During the iteration, a pair of clusters is selected to be made siblings; this results in the trees for the two clusters being merged into a larger rooted tree by making their roots siblings. When there are only three subtrees, then the three subtrees are merged into a tree on all the taxa by adding a new node, *r*, and making the roots of the three subtrees adjacent to *r*. Note that this description suggests that neighbor joining produces a rooted tree. However, after the tree is produced, the root is ignored, so that neighbor joining actually returns an *unrooted* tree.

So far, neighbor joining sounds like UPGMA, which we know is statistically inconsistent under some model conditions. Neighbor joining avoids this by deciding which pair of clusters to make into siblings (and how to update the distance matrix) using a more sophisticated strategy, as we now show.

The neighbor joining algorithm. Input: $n \times n$ dissimilarity matrix **d** with $n \geq 4$
Output: Unrooted tree with n leaves labeled $1 \ldots n$

Initialization: Compute the $n \times n$ matrix **Q**, defined by

$$Q_{i,j} = (n-2)d_{i,j} - \sum_{k=1}^{n}(d_{ik}+d_{jk}).$$

While $n > 3$, DO:

Find the pair i,j minimizing $Q_{i,j}$. Without loss of generality, we will call that pair a,b. Make the rooted trees associated with taxa a and b siblings, and call the root of the tree you form u.

Update the distance matrix by deleting the rows and columns for a and b, and including a new row and column for u, and set $d_{u,k} = \frac{d_{ak}+d_{bk}-d_{ab}}{2}$ for all $k \neq u$. Decrement n by 1.

Now $n = 3$; return the star tree with a single internal node v where the roots of the three rooted trees are all adjacent to v.

Theorem 5.8 *Neighbor joining runs in* $O(n^3)$ *time. Furthermore, given an additive matrix* **D** *corresponding to a tree* T *with positive edge weights* w(e), *neighbor joining will return the tree* T. *In addition, if the input is a dissimilarity matrix* **d** *with* $L_\infty(d,D) < f/2$, *where* f *is the smallest weight of an internal edge in* T, *then neighbor joining will return the topology of* T. *Hence, neighbor joining is statistically consistent under the CFN model, when used with CFN distances.*

Proof The running time analysis is easily obtained from Saitou and Nei (1987), but the guarantee that neighbor joining will return the true tree given a dissimilarity matrix that is within $f/2$ (where f is the shortest length of any internal edge in the true tree) of an additive matrix for the true tree comes from Atteson (1999). The proof of statistical consistency then follows since CFN distances converge to the model tree distances as the number of sites per sequence increases. Hence, as the sequence length increases, then $L_\infty(d,D) < f/2$ with probability converging to 1. See also Bryant (2005) and Gascuel (1997) for alternative proofs for the statistical consistency of neighbor joining. □

Note that the proof of statistical consistency for the CFN model can be easily extended to any model in which statistically consistent distance estimation is possible. Hence, neighbor joining is also statistically consistent under all the standard DNA sequence evolution models described in Chapter 8.

5.7 Distance-based Methods as Functions

We have described distance-based tree estimation as operating in two steps, where the first step computes a dissimilarity matrix, and the second step uses that matrix to compute a

5.7 Distance-based Methods as Functions

tree. We now describe the second step as a function that maps dissimilarity matrices to trees with positive branch lengths, and hence to additive matrices:

Definition 5.9 A **distance-based method** Φ is a function that maps dissimilarity matrices to additive matrices, i.e.: the input is an $n \times n$ dissimilarity matrix **d** and the output is an $n \times n$ additive matrix $\mathbf{D} = \Phi(d)$.

In order to prove theorems about statistical consistency for distance-based methods we will generally require the following two properties of the function Φ that maps dissimilarity matrices to additive matrices:

Identity on additive matrices: Φ is the identity on additive matrices (i.e., $\Phi(D) = D$ for all additive matrices **D**).

Continuous on neighborhoods of additive matrices: For all additive matrices **D**, there is some L_∞-neighborhood of **D** on which the distance-based method Φ is *continuous*. That is, for all additive matrices D and any $\delta > 0$, there is an $\varepsilon > 0$ such that for all dissimilarity matrices **d** satisfying $L_\infty(d, D) < \varepsilon$ it follows that $L_\infty(\Phi(d), \Phi(D)) < \delta$.

The next two theorems show that distance-based methods that satisfy these two properties will be statistically consistent under the CFN model, and more generally under any model in which statistically consistent distance-based estimation is possible.

Theorem 5.10 *Consider an* $n \times n$ *additive matrix* **D** *corresponding to a tree* T *with edge-weighting* w. *Let* f > 0 *be the weight of the shortest internal edge in* T, *and let* **D**$'$ *be a different* $n \times n$ *additive matrix such that* $L_\infty(D, D') < f/2$. *Then* **D**$'$ *corresponds to the same tree* T *with some other edge-weighting* w$'$.

Proof Since $L_\infty(D, D') < f/2$, **D** and **D**$'$ induce the same set of quartet trees; hence they define topologically identical trees. □

Theorem 5.11 *Any distance-based method* Φ *that is the identity on additive matrices and continuous on an* L_∞-*neighborhood of every additive matrix is statistically consistent under the CFN model.*

Proof Suppose Φ is continuous on a neighborhood of every additive matrix. Let (T, w) be a CFN model tree (so that w is the branch length function), let **D** be the additive matrix corresponding to the model tree, and let $f > 0$ be the minimum length of any edge in the model tree T. Since Φ is continuous on some neighborhood of every additive matrix, there is some $\varepsilon > 0$ so that whenever matrix **d** satisfies $L_\infty(d, D) < \varepsilon$ then $L_\infty(\Phi(d), \Phi(D)) < f/2$. By Theorem 5.10, $\Phi(D)$ and $\Phi(d)$ are additive matrices for the same tree topology. Since Φ is the identity on additive matrices, $\Phi(D) = D$. Hence, $\Phi(d)$ is an additive matrix that corresponds to an edge weighting of the tree T. Hence, Φ is statistically consistent under the CFN model. □

There is nothing special about the CFN model in this theorem; indeed, the theorem holds for any stochastic model for which the model tree topology can be defined by an additive matrix and all edges in the model tree have positive length (see Chapter 8). Furthermore, most distance-based methods have these two properties – being the identity on additive matrices, and being continuous on neighborhoods of additive matrices. Hence, most distance-based methods can be proven statistically consistent under the CFN model, and other models as well.

5.8 Optimization Problems

What we have described so far are methods that run in polynomial time, and that are not explicitly attempting to solve any optimization problem. However, some other methods have been developed that are explicit attempts to find either optimal solutions or approximate solutions to optimization problems.

A natural optimization problem that has been suggested is to find the additive matrix **D** that is as close as possible to the input dissimilarity matrix **d**, with respect to some way of defining distances between matrices, such as L_1, L_2, and L_∞:

- $L_1(d,D) = \sum_{ij} |d_{ij} - D_{ij}|$
- $L_2(d,D) = \sqrt{\sum_{ij}(d_{ij} - D_{ij})^2}$
- $L_\infty(d,D) = max_{ij} |d_{ij} - D_{ij}|$

Finding an additive matrix that minimizes one of these distances is equivalent to finding an edge-weighted tree whose additive matrix has the minimum distance. Hence, these problems are equivalent to tree estimation problems. The optimization problem based on the L_2 distance is also known as the "ordinary least-squares" (OLS) problem. The "weighted least-squares" (WLS) problem is a variant on this approach, and weights each term in the OLS problem. Hence, the cost function between the input matrix **d** and the additive matrix **D** that is returned is of the form $\sqrt{\sum_{ij} w_{ij}(d_{ij} - D_{ij})^2}$, where the weights w_{ij} depend on additional assumptions (e.g., on how the variance in the distance estimates is modeled). WLS-based approaches were originally suggested in Fitch and Margoliash (1967), who set w_{ij} to be inversely proportional to the estimated distances d_{ij}.

Finding the best tree (i.e., the best additive matrix) with respect to any of these criteria is NP-hard: Day (1987) showed this for L_1 and L_2, and Agarwala et al. (1998) showed this for the L_∞ norm. Agarwala et al. (1998) also showed that finding an arbitrarily close approximation is NP-hard, but provided an $O(n^2)$ time 3-approximation algorithm for the L_∞-nearest tree.

Because the optimization problems are NP-hard, heuristics have been developed that operate by finding good branch lengths on a starting tree with respect to the desired criterion, and then search for trees with better branch weights (i.e., ones that produce better scores). Thus, another natural problem is to find optimal branch weights on a given tree

T with respect to an input dissimilarity matrix **d** and some way of computing distances between dissimilarity matrices and additive matrices:

Input: $n \times n$ dissimilarity matrix **d** and n-leaf tree T.
Output: Weights $w(e)$ on the edges of T so as to produce an additive matrix D minimizing $dist(d, D)$.

The problem depends on how $dist(d, D)$ is defined, and could be (for example) any of the distances we have discussed so far. Fortunately, finding optimal edge weights of a tree with respect to the standard criteria can be performed in polynomial time (Desper and Gascuel, 2005). For example, for the OLS problem, Vach (1989) described an $O(n^3)$ algorithm, and Gascuel (2000) and Bryant and Waddell (1998) gave $O(n^2)$ algorithms. Desper and Gascuel (2005) also provide formulae for optimal branch lengths under the OLS criterion. Bryant and Waddell (1998) also gave $O(n^3)$ and $O(n^4)$ algorithms to find optimal branch lengths under WLS and generalized least squares, respectively.

5.9 Minimum Evolution

The *minimum evolution* (ME) approach to phylogeny, originally proposed in Kidd and Sgaramella-Zonta (1971), encompasses a collection of methods. In its best-known versions, the input is an $n \times n$ dissimilarity matrix **d**, and every tree topology on n leaves is assigned edge weights using a least-squares method. Then, the total length of each tree is computed by adding up the lengths of all the branches in the tree. The tree T with its edge weighting w that minimizes $\sum_{e \in E(T)} w(e)$ is returned. See Desper and Gascuel (2005) for a wonderful discussion about the mathematics involved in these problems, and Felsenstein (2004) and Desper and Gascuel (2005) for the history of these ME approaches.

ME approaches are distinguished generally by how they define the edge weighting of each tree. For example, given a tree T and an input dissimilarity matrix **d**, the optimal weights on the edges could be based on minimizing the L_2 (OLS) or the weighted L_2 (WLS) distances. As discussed above, finding optimal branch lengths of a given tree can be performed exactly in polynomial time for many of these least squares criteria; thus, scoring a given tree topology is fast. However, finding the tree with the minimum total branch length is generally much harder. For example, it is NP-hard under the OLS criterion for integer branch lengths (Bastkowski et al., 2016) and under a variant of the WLS criterion called "balanced minimum evolution" (Pauplin, 2000). See Desper and Gascuel (2002, 2004, 2005); Gascuel and Steel (2006); Fiorini and Joret (2012) for more about the balanced minimum evolution (BME) criterion and its properties.

ME methods, therefore, tend to be heuristics without necessarily provable guarantees about solving their optimization problems. However, we can ask whether an exact solution to an ME method – however it is defined – is statistically consistent. Unsurprisingly, this depends on how the branch lengths are defined; for example, Rzhetsky and Nei (1993) proved that ME used with OLS branch lengths is statistically consistent (if solved exactly);

Gascuel et al. (2001) showed that ME with WLS branch lengths is *not* statistically consistent; and the balanced minimum evolution criterion used in the FastME algorithm *is* statistically consistent (Desper and Gascuel, 2004). The updated FastME software (Lefort et al., 2015) has an improved search strategy over the earlier versions, and runs in $O(n^2 \log n)$ time, where n is the number of leaves in the tree.

5.10 The Safety Radius

A key concept in understanding distance-based methods and being able to prove theoretical properties about the methods (such as being statistically consistent) is the notion of the *safety radius*, which we now define.

Definition 5.12 Let (T, w) be an edge-weighted tree defining additive matrix **A**, and let Φ be a distance-based phylogeny estimation method. Then the **safety radius** of Φ is the largest value x such that whenever the input dissimilarity matrix **d** satisfies $L_\infty(d, A) < xf$, where f is the length of the shortest internal edge in the tree T, then $\Phi(d)$ returns an additive matrix **A'** corresponding to a weighted version of tree topology T.

The importance of the safety radius is that it can be used to prove that a method is statistically consistent, as we now show.

Lemma 5.13 *If Φ is a distance-based method with a positive safety radius under a model* M, *then it is statistically consistent under stochastic model* M *provided that it is possible to estimate distances under* M *in a statistically consistent manner.*

The proof follows from the same arguments as in the proof of statistical consistency for the Naive Quartet Method under the CFN model, and is omitted. However, a few comments are worth noting. First, when a safety radius is established, it typically is independent of the specific model of evolution (e.g., consider the argument for why the Naive Quartet Method is statistically consistent under the CFN model). Second, all the models we have discussed that are identifiable have these statistically consistent distance estimators. Hence, establishing that a method has a positive safety radius will typically mean that you have established it to be statistically consistent under a wide range of models. Finally, the safety radius can be used to compare methods in terms of the conditions that suffice to guarantee accuracy, as we will see.

We begin with a result established in Erdös et al. (1999b) and Atteson (1999), bounding the best possible safety radius for any method.

Lemma 5.14 *(Lemma 2 in Erdös et al. (1999b)) For every additive matrix **A** corresponding to a binary tree in which the shortest internal branch has length* f *and for all* $\varepsilon > 0$ *there is another additive matrix **A'** defining a different tree topology satisfying* $L_\infty(A, A') \leq f/2 + \varepsilon$. *Hence the largest possible safety radius is* $1/2$.

5.10 The Safety Radius

Theorem 5.15 *The safety radii of the Naive Quartet Method, the Buneman Tree method, and the neighbor joining method are all 1/2, so that each is guaranteed to return the correct tree whenever the input distance matrix* d *satisfies* $L_\infty(d,D) < f/2$, *where* f *is the shortest length of any internal edge in the model tree* T *and* D *is the additive matrix for* T. *Since the best possible safety radius is 1/2, these three methods have the optimal safety radius. Hence, all these methods are both statistically consistent under the CFN model of sequence evolution.*

Proof No distance-based method can have safety radius greater than 1/2 (Lemma 5.14). The proof that the Naive Quartet Method has safety radius 1/2 is in Theorem 5.7, and whenever the Naive Quartet Method is accurate then the Buneman Tree method will also be accurate. Hence, the Buneman Tree method has safety radius 1/2. The proof that the neighbor joining method has safety radius 1/2 was provided in Atteson (1999). Hence, all these methods have positive safety radii, and so are statistically consistent under the CFN model of sequence evolution. □

Other methods have also been shown to have an optimal safety radius; for example, Atteson (1999) proved that ADDTREE (Sattath and Tversky, 1977) (a method that is similar to neighbor joining) has the optimal safety radius; Shigezumi (2006) proved that a greedy algorithm for BME has the optimal safety radius; and Pardi et al. (2012) proved that any algorithm finding an exact solution to the BME problem has the optimal safety radius. While these methods have been proven to have an optimal safety radius, the best bounds that have been obtained for some other methods are not as good. For example, Shigezumi (2006) proved that greedy minimum evolution based on OLS has a safety radius at most $\frac{2}{\sqrt{n}}$ where n is the number of leaves in the tree, so that its safety radius approaches 0 as the number of leaves increases.

The safety radius of some well-known methods is still open; for example, the safety radius of the FastME heuristic based on SPR (subtree prune and reconnect) moves is only known to be at least 1/3 (Bordewich et al., 2009). Here we show how to establish some of the easier bounds on the safety radius for some well known methods.

Theorem 5.16 *An exact algorithm for the* L_∞-*nearest tree problem has a safety radius of at least 1/4. Hence, an exact algorithm for the* L_∞-*nearest tree problem is statistically consistent under the CFN model of sequence evolution.*

Proof Suppose Φ is an exact algorithm for the L_∞-nearest tree problem. Suppose that **d** is an $n \times n$ dissimilarity matrix, that **D** is an $n \times n$ additive matrix corresponding to the model tree (T,w), and that $L_\infty(d,D) < f/4$, where $f > 0$ is the minimum length of any internal edge in T. Let $D' = \Phi(d)$. Because Φ is an exact algorithm for the L_∞-nearest tree problem, $L_\infty(D',d) \leq L_\infty(d,D) < f/4$. By the triangle inequality,

$$L_\infty(D',D) \leq L_\infty(D',d) + L_\infty(d,D).$$

Therefore,
$$L_\infty(D,D') < f/4 + f/4 = f/2.$$

By Theorem 5.10, **D** and **D'** are additive matrices for the same tree topology, but with potentially different edge lengths. Hence, Φ has a safety radius of at least $1/4$. □

Theorem 5.17 *Let Φ be an approximation algorithm for the L_∞-nearest tree that has approximation ratio r. Then Φ has a safety radius of at least $\frac{1}{2(r+1)}$, and so is statistically consistent under the CFN model of sequence evolution.*

Proof Let (T,θ) be a model CFN tree, and let **D** denote its additive matrix. Suppose that **d** is a dissimilarity matrix that satisfies $L_\infty(d,D) < \frac{f}{2(r+1)}$ where f is the weight of the shortest internal edge in T. Let $\Phi(d) = \mathbf{D'}$. Since Φ has approximation ratio r, $L_\infty(d,D') \le rL_\infty(d,D) \le \frac{fr}{2(r+1)}$. Then by the triangle inequality,

$$L_\infty(D,D') \le L_\infty(d,D) + L_\infty(D',d)$$
$$< \frac{f}{2(r+1)} + \frac{fr}{2(r+1)} = \frac{f}{2}.$$

Hence, **D** and **D'** are additive matrices for the same tree topology. Therefore, Φ is guaranteed to reconstruct the model tree topology whenever the input matrix **d** satisfies $L_\infty(d,D) < \frac{f}{2(r+1)}$ where **D** is an additive matrix for the model tree, and so has safety radius at least $\frac{1}{2(r+1)}$. □

Note the similarity in the proof to that for Theorem 5.16. As a corollary,

Corollary 5.18 *The 3-approximation algorithm of Agarwala et al. (1998) for the L_∞-nearest tree has a safety radius of at least $1/8$, and is statistically consistent under the CFN model of sequence evolution.*

The reader can verify that the safety radius of the Naive Quartet Method is independent of the specific details of the model of sequence evolution, and hence the Naive Quartet Method is statistically consistent under any model of evolution for which model distances can be estimated in a statistically consistent manner. The same property is true for the other methods discussed above (i.e., neighbor joining, the Buneman Tree, ADDTREE, the Agarwala et al. algorithm, etc.). Since the standard DNA sequence evolution models (described in Chapter 8) have these statistically consistent distance estimators, that means that all these methods are statistically consistent under quite a large range of stochastic sequence evolution models. In particular, the General Markov Model (Steel, 1994b), which contains the Generalized Time Reversible model (Tavaré, 1986), has such a distance estimator. Therefore, by Lemma 5.13,

Theorem 5.19 *The Naive Quartet Method, neighbor joining, the Buneman Tree, an exact or approximation algorithm for the L_∞-nearest tree are all statistically consistent under*

	L2	L3	L4
L1	4	6	8
L2		8	6
L3			12
L4			

M

	L2	L3	L4
L1	4.3	5.2	8.7
L2		7.5	6.4
L3			11.3
L4			

M'

Figure 5.4 On the left is an additive matrix **M**, and on the right is a dissimilarity matrix **M'**. Matrix **M** corresponds to a tree with topology $(L1,(L3,(L2,L4)))$ with internal edge length 2. **M'** is not additive (it does not satisfy the Four Point Condition). Note that $L_\infty(M,M') = 0.8 < 1$. The Four Point Method applied to **M'** returns tree topology $(L1,(L3,(L2,L4)))$, as predicted by the safety radius of the Four Point Method.

any sequence evolution model for which model distances can be estimated in a statistically consistent manner, and in particular under the General Markov Model and all its submodels (e.g., the Generalised Time Reversible Model).

We finish this section with an example.

Example 5.20 Figure 5.4 gives two matrices **M** and **M'**. Matrix **M** is additive and corresponds to quartet tree T with topology $(L1,(L3,(L2,L4)))$ and internal branch length 2. Matrix **M'** is not additive, and satisfies $L_\infty(M,M') = 0.8$. By Theorem 5.15, the Four Point Method has safety radius $1/2$. Since $L_\infty(M,M') < 1/2 \times 2$, the Four Point Method applied to **M'** is guaranteed to return T. An application of the Four Point Method to **M'** shows this is true. Similarly, neighbor joining has safety radius $1/2$ and will return T.

5.11 Comparing Methods

The prior theorems have established that many distance-based methods (the Naive Quartet method, the Buneman Tree, neighbor joining, an exact algorithm for the L_∞-nearest tree problem, and the 3-approximation algorithm by Agarwala et al. (1998)) all have positive safety radii. Can we use this information to infer anything about their performance in practice?

We begin with neighbor joining, the Buneman Tree, and the Naive Quartet Method. By Theorem 5.15, these three methods have the same safety radius; yet these methods perform very differently in practice. For example, the requirement that all quartet trees be correctly computed means that the Naive Quartet Method generally returns *Fail*, except on very small datasets or on datasets with extremely long sequences. Similarly, while the Buneman Tree is basically a better method than the Naive Quartet Method, it tends to produce highly unresolved trees, especially on large datasets (St. John et al., 2003). In contrast, the neighbor joining method generally produces more accurate trees (i.e., ones with lower missing branch rates) than the Buneman Tree, and so is the best of these three methods (Huson et al., 1999a; St. John et al., 2003).

What about the other methods? According to Corollary 5.18, the safety radius of the Agarwala et al. (1998) algorithm might be as small as one-fourth that of the Naive Quartet Method, the Buneman Tree Method, and neighbor joining; if this bound is tight, then the Naive Quartet Method ought to be more accurate than all three of these methods. Yet, Huson et al. (1999a) compared the Agarwala et al. (1998) algorithm to neighbor joining and the Buneman Tree on simulated datasets and found that neighbor joining had the best accuracy, followed by Agarwala et al. (1998), and finally by the Buneman Tree.

In other words, a positive safety radius enables statistical consistency guarantees to be established, but does *not* directly imply how accurate a method will be on data. Similarly, comparing the safety radii of two methods does not directly imply anything about the relative performance on data between the methods. In particular, the safety radius indicates how close the dissimilarity matrix has to be to the model tree additive matrix in order for the method to be guaranteed to produce the true tree without any error, and performance on data inevitably is about partial accuracy rather than exact accuracy.

Studies comparing FastME to neighbor joining have shown that FastME produces topologically more accurate trees under a range of different scenarios (Desper and Gascuel, 2004; Vinh and von Haeseler, 2005; Wang et al., 2006; Vachaspati and Warnow, 2015), suggesting both that the BME criterion is a helpful one and that the search strategy in the FastME software is effective.

It is therefore interesting to consider the inquiry into why neighbor joining works well. While we have described the neighbor joining method as an iterative heuristic that begins with n disconnected leaves and then gradually connects them all, neighbor joining can also be described as a method that begins with a star tree, and then greedily refines around the polytomies, gradually turning a star tree into a binary tree. Steel and Gascuel (2006) used this formulation to show that neighbor joining can be seen as always selecting the next pair of nodes to make siblings on the basis of a BME principle. Other studies also show a connection between neighbor joining and the BME principle (Mihaescu and Pachter, 2008; Haws et al., 2011), which may help explain why neighbor joining does as well as it does in comparison to many other distance-based methods. However, neighbor joining is often not as accurate as FastME. The explanation may be that because neighbor joining uses a greedy heuristic, it cannot undo any steps it makes, while FastME does do a heuristic search to find improved solutions.

5.12 Further Reading

The edge safety radius and other concepts. The original paper introducing the concept of the safety radius is Erdös et al. (1999b), but the concept only became established with Atteson (1999), which established that neighbor joining has a safety radius of $1/2$. However,

Atteson (1999) referred to this as the ℓ_∞ radius, and Gascuel and Steel (2016) referred to this as the "Atteson ℓ_∞ bound." In fact, the term "safety radius" was not used until later.

The safety radius is a condition that guarantees complete accuracy in the reconstructed tree whenever the estimated distance matrix **d** is close enough to the additive matrix **D** defining the true tree. Atteson (1999) also considered a weaker condition, called the "edge safety radius," that addresses the question of which edges will be guaranteed to be recovered when the matrix **d** is not close enough to **D** to guarantee complete accuracy. The edge safety radius is the constant r so that the method is guaranteed to recover any edge whose length is at least w whenever $L_\infty(d,D) < rw$. Clearly $r \leq 1/2$, and the closer r is to $1/2$ the better the edge safety radius. Neighbor joining was established to have edge safety radius of exactly $1/4$ in Mihaescu et al. (2009), and hence does not have an optimal edge safety radius. Two methods based on BME (a greedy heuristic, and also FastME) have edge safety radius of $1/3$, which is strictly better than the edge safety radius for neighbor joining (Bordewich and Mihaescu, 2010).

Another approach to understanding the performance of methods is the "stochastic safety radius" (Gascuel and Steel, 2016). This framework treats distance-based estimation probabilistically, and assumes that the difference between the estimated and true distances are modeled by independent normally distributed random variables with mean 0. In other words, for all pairs x,y of leaves, $d(x,y) = D(x,y) + \varepsilon_{x,y}$, where $d(x,y)$ is the estimated distance between x and y, $D(x,y)$ is the model distance between x and y, and the $\varepsilon_{x,y}$ values are independent normal random variables (Gascuel and Steel, 2016). Under this "random errors model," a method Φ is said to have a b-stochastic safety radius s if Φ will return the true tree with probability at least $1-b$ whenever the estimated distance matrix is close enough to the true distance matrix (where close enough depends on the value of s).

These three ways of modeling the error tolerance of distance-based methods are quite different, and the differences between the stochastic safety radius and the two other models are interesting. First, the stochastic safety radius treatment is probabilistic and so says that when the estimated distance matrix is close enough to the model distance matrix, then with high probability the model tree will be returned; in contrast, the Atteson ℓ_∞ safety radius is deterministic, and says that the true tree *will* be returned whenever the estimated distance matrix is close enough to the true distance matrix. Second, the Atteson ℓ_∞ safety radius does not model the error in estimated distances, but the stochastic safety radius does; hence, the stochastic safety radius approach is best suited to those stochastic models of sequence evolution and techniques for estimating distances for which the error in estimated distances fits the random errors model in the stochastic safety radius approach.

Sequence length requirements for accuracy with high probability. While many distance-based methods are easily shown to be statistically consistent under some models of evolution, it is not as easy to establish the convergence rate, or the sequence length that suffices (or is required) for accuracy with high probability. In particular, given arbitrary upper and lower bounds on the branch lengths, we may wish to know if the method will recover the true tree with high probability from polynomial length sequences, or whether exponential length sequences are needed. Methods that can recover the true tree with high probability from polynomial length sequences are called "absolute fast converging" (*afc*) methods. This definition depends on the specific model of evolution, just as with the definition of statistical consistency; also, if a method is *afc* for a model, then it is necessarily *afc* for all its submodels.

Most of the proofs for methods being *afc* have been for the CFN model; however, some proofs also extend to complex DNA sequence evolution models. Most of the methods that have been established to be *afc* are distance-based (Erdös et al., 1999a,b; Warnow et al., 2001; Nakhleh et al., 2001a; Daskalakis et al., 2006; Roch, 2008, 2010), but maximum likelihood is also *afc* (Roch and Sly, 2016). Interestingly, some methods have also been proven to *not* be *afc*. For example, it is easy to prove that the Naive Quartet Method is not *afc* even for the CFN model, since it requires that every quartet tree be correctly estimated, and these are all estimated independently. Less obviously, neighbor joining, despite its very good performance under some conditions, is also not *afc*, even for simple models (Lacey and Chang, 2006). See Section 8.11 for more on afc methods.

Other types of distances. In this chapter, we have described distance-based tree estimation as having an input set of sequences that evolved down an unknown model tree so that the sequences are all the same length, computing a distance matrix for the sequences, and then computing a tree from the distance matrix. This is the simplest version of distance-based tree estimation. However, distance-based tree estimation occurs in other contexts, and we briefly describe these here.

Distance-based tree estimation is sometimes used for estimating species trees when the input is a set of trees, each estimated for a potentially different subset of the species. Because gene trees can be different from the species tree due to biological processes such as incomplete lineage sorting (ILS), horizontal gene transfer, and gene duplication and loss, methods that take gene tree heterogeneity into account have been developed (see Chapter 10). In particular, when gene tree heterogeneity is due to ILS, then distances between species can be computed using the "average internode distance matrix," which is the average topological distance in the gene trees between every pair of species. NJst (Liu and Yu, 2011) and ASTRID (Vachaspati and Warnow, 2015) are examples of such methods that have been used on biological datasets, and differ only in the methods they use to compute trees after computing the average internode distance matrix.

Distance-based genome-scale trees can also be computed using genomic architectures (e.g., the order and copy number of genes within genomes). For example, statistically consistent methods for computing distances between genomes based on rearrangements (inversions, transpositions, and inverted transpositions) and other types of events (e.g., duplications, fissions, and fusions) have been developed and used to compute genome-scale trees; see Section 10.10 and Wang and Warnow (2006); Wang et al. (2006); Lin et al. (2012b). Similar techniques have also been used to compute phylogenies for manuscripts, such as the Canterbury Tales (Spencer et al., 2003).

Distance-based phylogeny estimation can also be used to compute a tree on whole genomes or long genomic regions, without needing the computation of a multiple sequence alignment. The calculation of distances between genomes can be based on statistical models of genome evolution involving rearrangements, duplications, and other major events (see Section 10.10) or based on distributions of substrings of a fixed length (Vinga and Almeida, 2003; Chan and Ragan, 2013; Chan et al., 2014). Some of the recent approaches, such as Bogusz and Whelan (2017), use sophisticated statistical techniques to compute distances.

Unfortunately, most of these "alignment-free methods" do not have statistical guarantees under sequence evolution models that include indels. However, Daskalakis and Roch (2010) provide an interesting method that they prove is statistically consistent for estimating the tree topology under the model proposed in Thorne et al. (1991), in which sequence evolution includes indels as well as substitutions.

In many of the scenarios we have described, the distance calculation is proven to converge to an additive matrix for the true tree under a statistical model of evolution, so that the overall approach of computing distances and then computing a tree from the distance matrix is statistically consistent under the statistical model of evolution. Hence, once the dissimilarity matrix is calculated, a tree can be computed from the matrix using standard methods (e.g., FastME or neighbor joining). Hence, distance-based tree estimation is a quite general technique that can be applied to many phylogeny estimation problems, and that can have desirable statistical properties.

The impact of missing data. In some cases, the input distance matrix may not have values for all of its entries (i.e., the distance between some pairs of species cannot be computed for some reason). An example of when this happens is in the calculation of distances between species based on gene trees, such as in NJst and ASTRID: if species s_i and s_j are not in any gene tree together, then the internode distance matrix has no value for the distance between s_i and s_j. When the distance matrix has missing entries, then standard distance-based methods, such as neighbor joining and FastME, cannot be applied. Computing trees from matrices with missing entries is very difficult, and only a few methods have been developed for this problem (Criscuolo and Gascuel, 2008). Some supertree methods (see Chapter 7), which construct trees by combining trees on subsets, are also based on distances, and hence also face this kind of challenge. Distance-based tree estimation given missing entries is one of the interesting and basically unsolved problems in phylogenetics.

5.13 Review Questions

1. What is the triangle inequality?
2. What is a dissimilarity matrix?
3. What are p-distances?
4. What is the Cavender–Farris–Neyman distance formula?

5. What is meant by saying that a matrix is additive?
6. What is the Four Point Condition?
7. What is the Four Point Method?
8. What is the Naive Quartet Method?
9. What is the technique for computing the length of the internal branch in the quartet tree on four taxa, given an additive matrix for the four taxa?
10. Are Hamming distances computed on sequences always additive?
11. What is the meaning of safety radius?
12. Suppose **d** and **D** are two $n \times n$ dissimilarity matrices, and that **D** is additive. Give a condition on $L_\infty(d,D)$ under which $NJ(d)$ and $NJ(D)$ will be guaranteed to define the same tree topology.
13. State three distance-based optimization problems and their computational complexity (i.e., are they NP-hard, solvable in polynomial time, or of unknown computational complexity?).
14. Agarwala et al. (1998) presented an approximation algorithm for some optimization problem related to distance-based phylogeny estimation. What problem was that? Is this approximation algorithm statistically consistent under the CFN model?
15. What is meant by alignment-free tree estimation?

5.14 Homework Problems

1. What is the largest CFN distance possible between two binary sequences s and s' of the same length k, under the constraint that they differ in strictly less than $\frac{k}{2}$ positions?
2. Prove or disprove: For all pairs of binary sequences s, s' of the same length k that differ in strictly less than $\frac{k}{2}$ positions, the CFN distance between s and s' is at least the Hamming distance between s and s'.
3. Prove or disprove: For all datasets of binary sequences of the same length k such that all pairs of sequences in the set differ in strictly less than $\frac{k}{2}$ positions, the CFN distance matrix will satisfy the triangle inequality.
4. In the problems preceding this, we have constrained the set of binary sequences of length k to differ pairwise in strictly less than $\frac{k}{2}$ positions. Why did we do this?
5. Consider the matrix in Figure 5.3. Apply UPGMA to the matrix. What is the *unrooted* tree that you obtain? Does it equal the tree given in that figure?
6. Consider the set $\{0, 1, 2, \ldots, 15\}$ and represent them as binary numbers with ten digits (i.e., 8 = 0000001000). Treat these as binary sequences, generated by some CFN model tree. What is the largest CFN distance between any two binary sequences in this set? (Hint: do *not* compute all pairwise distances.)
7. Let T be a CFN model species tree on four leaves A, B, C, and D, with unrooted topology $AB|CD$. Let $\lambda(e)$ denote the CFN branch length of edge e. Let e_I denote the single internal edge in T, and let e_x denote the edge incident with leaf x. Assume that $\lambda(e_I) = 2, \lambda(e_A) = 0.1, \lambda(e_B) = 0.2, \lambda(e_C) = 2.1$ and $\lambda(e_D) = 3.2$.

5.14 Homework Problems

a. Prove or disprove: UPGMA on Hamming distance is statistically consistent for estimating the unrooted tree topology for T.

b. Prove or disprove: UPGMA on CFN distance is statistically consistent for estimating the unrooted tree topology for T.

8. Draw an edge-weighted tree T with at least five leaves and all branches having positive weight. Derive its additive matrix. Check that the Four Point Condition applies for at least two different quartets of leaves.

9. Compute the Hamming distance matrix for the set of four taxa, $\mathscr{L} = \{L_1, L_2, L_3, L_4\}$, given below (each described by four binary characters). Is the distance matrix additive? If you apply the UPGMA method to this distance matrix, what do you get? If you apply the Four Point Method to the matrix, what do you get? What is the solution to maximum parsimony on this input of four taxa? What is the solution to maximum compatibility? Are these characters compatible?

- $L_1 = (0, 1, 0, 1, 0)$
- $L_2 = (0, 0, 0, 0, 0)$
- $L_3 = (1, 0, 0, 0, 0)$
- $L_4 = (1, 0, 1, 0, 1)$

10. Let **D** be an additive matrix corresponding to binary tree T with positive edge weighting w. Let **D'** be the matrix corresponding to the edge-weighted tree obtained by changing the weight of some internal edge to 0. Prove or disprove: **D'** must still satisfy the Four Point Condition.

11. Consider those matrices that correspond to path distances in edge-weighted trees T with positive branch lengths, but where T may not be binary (i.e., the trees can have polytomies). Prove or disprove: Matrices computed this way satisfy the Four Point Condition.

12. Prove or disprove: If C is a set of characters (not necessarily binary) that evolve without any homoplasy on a tree T, then the Hamming distance matrix is additive.

13. Suppose **D** and **D'** are two additive matrices, both corresponding to the same tree topology, but using (perhaps) two different edge weightings. Prove or disprove: For all constants $c > 0$, c**D**+**D'** is also additive.

14. Give an example of two $n \times n$ additive matrices **D** and **D'**, but where **D** corresponds to an edge-weighted binary tree T with positive branch weights, and **D'** is additive and so corresponds to a unique tree T' (which may not be binary) with positive branch lengths where $T \neq T'$, and where $L_\infty(D, D') = f/2$, where f is the minimum length of any internal edge in T.

15. Give an $O(n^2)$ algorithm to compute the $n \times n$ additive distance matrix defined by a tree on n leaves with positive weights on the edges. Hint: use dynamic programming.

16. Suppose $l_1, l_2, \ldots, l_k$ are non-negative integers, and **M** is an $k \times k$ matrix defined by $M_{ij} = 0$ if $i = j$ and otherwise $M_{ij} = l_i + l_j$. Is **M** additive? Prove or disprove.

17. Recall the definition of an additive matrix and the Four Point Method. Is it ever possible for some set of four taxa that the two smallest of the three pairwise sums will be

identical but strictly smaller than the largest pairwise sum, given an additive matrix? If so, give an example, and otherwise prove this cannot happen.
18. Suppose that branch lengths of a tree can be negative. Are there any constraints you can infer about the three pairwise sums in this case?
19. Let T be a binary tree with n leaves and $w: E \to Z^+$ be an edge weighting for T. Give an $O(n^4)$ algorithm to compute the set $Q(T)$ given T and w.
20. Give an $O(n^2 + k)$ algorithm to solve the following problem. The input is an n-leaf tree T with positive edge weights, and a list L of k four-leaf subsets, and the output is the list of quartet tree topologies on every four-leaf subset in L. Hint: use $O(n^2)$ time for the preprocessing to enable each four-leaf subset to be answered in $O(1)$ time.
21. Same as the previous problem, but change the output so that quartet trees have weights on all the edges defined by the additive matrix corresponding to the input edge-weighted tree.
22. Let **D** be an $n \times n$ additive matrix corresponding to binary tree T and positive edge-weighting w with f the minimum length of any internal edge. Let **D'** be an $n \times n$ additive matrix such that $L_\infty(D, D') > f/2$. Can **D** and **D'** define the same tree topology? If so, give an example of such a pair of matrices, or else prove this is impossible.
23. For the additive matrix you produced in the previous problem, compute the tree for every quartet of taxa by applying the Four Point Method. Then apply the Naive Quartet Method to the set of quartets. Verify that you produce the same tree.
24. Take any additive matrix and change one entry. Determine if the new matrix is additive. If not, prove it is not by producing the four leaves for which the Four Point Condition fails. If yes, prove that it is by producing the edge-weighted tree that realizes the new matrix.
25. Prove that every additive matrix uniquely defines a unique tree T (not necessarily binary) with positive branch lengths.
26. Show how to modify the Naive Quartet Method to construct the unique tree corresponding to an edge-weighted (but not necessarily binary) tree with positive branch lengths.
27. Suppose you have an $n \times n$ additive matrix **M** with $n \geq 4$, and you erase the entries corresponding to two symmetric entries M_{ij} and M_{ji} in the matrix. Give an algorithm to infer those two missing values (M_{ij} and M_{ji}) from the remaining data, and prove it correct.
28. Consider the Naive Quartet Method applied to pairwise Hamming distances, and call this the NQM(Hamming) method. Assume the input is a set of binary characters, and characterize those characters that are uninformative for the NQM(Hamming) method.
29. Let $c > 0$ be a constant, and consider the function Φ that maps dissimilarity matrices **d** to c**d** (i.e., $\Phi(d)$ is the matrix **d'** such that $d'_{ij} = cd_{ij}$).
    - Does Φ map additive matrices to additive matrices?
    - Is Φ continuous on some neighborhood of every additive matrix?
    - Is there a value for c such that Φ is statistically consistent for estimating trees under the CFN model? If so, what is the value for c for which this is true? If not, why not?

30. Suppose we know that Φ_1 and Φ_2 are both distance-based methods that satisfy the identity property on additive matrices and are continuous on a neighborhood of every additive matrix. Does the composition of these two methods also satisfy both properties?

31. Suppose Φ maps dissimilarity matrices **d** to **2D**, where **D** is the additive matrix that is closest to **d** under the L_∞ metric. Does Φ satisfy one, both, or neither of the two desired properties (identity on additive matrices, and continuous on a neighborhood of every additive matrix)? If we use CFN distances, is Φ a statistically consistent technique for estimating CFN trees?

32. Prove Theorem 5.17.

33. Suppose Φ is an exact solution to the L_1-nearest tree problem, where the input is a dissimilarity matrix **d** and the output is an additive matrix **D** that minimizes $L_1(d,D) = \sum_{ij} |d_{ij} - D_{ij}|$. Is Φ a statistically consistent method for estimating CFN model trees, if applied to CFN distances? If so, why? If not, why not?

34. Suppose Φ is a $\log n$-approximation algorithm to the L_1-nearest tree problem (see previous problem). Is Φ a statistically consistent method for estimating CFN model trees, if applied to CFN distances? If so, why? If not, why not?

35. Prove the following: If C is a set of binary characters that evolve without homoplasy on a tree T, then the Hamming distance matrix is additive.

36. Consider sequences that evolve down a tree T with n leaves, and let $length(e)$ denote the number of changes that occur on edge e. Consider the $n \times n$ matrix $LENGTH$ defined by this way of defining branch lengths. Prove or disprove: $LENGTH$ is an additive matrix.

37. Consider the case where a set S of binary sequences evolves down a binary tree T without any homoplasy.

    - Prove or disprove: T is an optimal solution to maximum parsimony for this dataset.
    - Prove or disprove: T is an optimal solution to maximum compatibility for this dataset.
    - What is the relationship between T and the strict consensus of the set of all maximum parsimony trees?
    - What is the relationship between T and the strict consensus of the set of all maximum compatibility trees?

38. An edge of a tree on which no changes occur (so that the sequences at the endpoints of the edge are identical) is called a "zero-event" edge. How do zero-event edges impact phylogeny estimation? Can zero-event edges be recovered?

39. Consider the case where a set S of binary sequences evolve down a binary tree T without any homoplasy, and there are no zero-event edges in T with respect to S; hence, every edge has at least one change on it. (See problem 38 for the definition of zero-event edge.)

    - Give an exact characterization of the set of maximum parsimony trees on the set S.
    - Give an exact characterization of the set of maximum compatibility trees on the set S.

- What is the result of applying the Naive Quartet Method to the Hamming distance matrix for S?

40. Consider the case where a set S of binary sequences evolve down a binary tree T without any homoplasy, but the tree T does have zero-event edges with respect to S. Let $E_0 \subseteq E(T)$ denote the zero-event edges in T. Consider T_0, the tree obtained by collapsing all the zero-event edges in T. (See problem 38 for the definition of zero-event edge.)
    - Give an exact characterization to the set of maximum parsimony trees on the set S.
    - Give an exact characterization to the set of maximum compatibility trees on the set S.
    - What is the result of applying the Naive Quartet Method to the Hamming distance matrix for S?

41. Suppose that you have a set S of aligned binary sequences, each of length 100, that you are told have evolved down some unknown CFN model tree. You would like to compute a distance-based tree on S, and so you try to compute the CFN distance matrix. However, when you do this, you find a pair of sequences x, y which differ in 50 or more positions (out of the total 100). What will you do? When will this kind of problem happen? Would you suggest changing the CFN distance calculation to handle such inputs, or would you suggest not using distance-based tree estimation methods? What, in general, would you consider doing to construct a tree under such circumstances? (This is a research question – there are many possible reasonable answers.)

6
Consensus and Agreement Trees

6.1 Introduction

In this chapter we discuss techniques for analyzing sets of trees, which are called **profiles**. Depending on how the set of trees was computed and the objective of the analysis, different techniques are needed.

For example, when a maximum parsimony analysis is performed, many equally good trees may be found, all having the same "best" score (meaning that it is the best score found during the analysis). Similarly, when a Bayesian MCMC (see Section 8.7) analysis is performed, then a random sample of the trees is examined. Sometimes, many different types of methods are used on the same data, and for each analysis a set of trees is obtained. Then, from the full set of trees, each of which has been estimated on the same data, again some kind of point estimate is sought. In each of these cases, a "consensus method" is used to provide a point estimate of the tree from the full set of trees.

An alternative objective might be to find those subsets of the taxon set on which all or most of the trees in the profile agree; this kind of approach does not produce a point estimate of the true tree, but can be used to identify the portion of the history on which all trees agree, and also (potentially) problematic taxa. We refer to this type of approach as an "agreement method."

In this chapter, we discuss consensus and agreement methods, which are methods for analyzing datasets of trees under the assumption that all the trees are estimated species trees with the same leafset S. We discuss supertree methods in Chapter 7, which extend the consensus methods in this chapter to allow for the taxon sets to be different between trees. Finally, methods for analyzing sets of gene trees that can differ from each other due to incomplete lineage sorting, horizontal gene transfer, or gene duplication and loss are discussed in Chapter 10. Many of the methods described in Chapter 10 can also be used as supertree methods.

6.2 Consensus Trees

In general, consensus methods are applied to unrooted trees (and we will define them in that context), but they can be modified so as to be applicable to rooted trees as well.

Here, we will focus on the ones that are the most frequently used in practice: the strict consensus, majority consensus, and greedy consensus. However, we also present three other consensus methods that have interesting properties: the compatibility tree, the Asymmetric Median Tree (Phillips and Warnow, 1996), and the Characteristic Tree (Stockham et al., 2002). We will assume that the input is a set $\mathcal{T} = \{T_1, T_2, \ldots, T_k\}$ of unrooted trees, and that each tree T_i has leafset S with $|S| = n$.

6.2.1 Strict Consensus

The strict consensus tree was defined earlier (see Definition 2.21), and we repeat it here for the sake of completeness. To construct the strict consensus tree, we write down the bipartitions that appear in *every* tree in the input set $\mathcal{T} = \{T_1, T_2, \ldots, T_k\}$. The tree that has exactly that set of bipartitions is the **strict consensus tree**; i.e., the strict consensus of $\mathcal{T}$ is the tree T satisfying $C(T) = \cap_{i=1}^{k} C(T_i)$. Note that the strict consensus tree is unique. Also, the strict consensus is a contraction of every tree in the input; hence, another way of defining the strict consensus tree is to say that it is the maximally resolved tree that is a common contraction of all the trees in $\mathcal{T}$. By construction, the strict consensus tree of $\mathcal{T}$ will not be fully resolved except when all the trees in $\mathcal{T}$ are topologically identical.

6.2.2 Majority Consensus

To construct the majority consensus, we write down the bipartitions that appear in *more than half* the trees in the profile. The tree that has exactly those bipartitions is called the "majority consensus." The majority consensus minimizes the total Robinson–Foulds distance to the input trees, i.e., $\sum_{i=1}^{k} RF(T, T_i)$, and so is also a *median tree* (Barthélemy and McMorris, 1986).

Definition 6.1 Given a set $\{T_2, T_2, \ldots, T_k\}$ of unrooted trees, each on the same leafset, the **majority consensus** tree T is the tree that contains exactly the bipartitions that appear in more than half of the trees.

Lemma 6.2 *For all input sets $\mathcal{T}$ of unrooted trees, there is a unique tree* T *whose bipartition set contains exactly those bipartitions that appear in more than half of the trees in $\mathcal{T}$. Furthermore, the majority tree is either identical to the strict consensus tree or refines it (see Definition 2.15).*

Proof The set of bipartitions that appear in strictly more than half of the trees in $\mathcal{T}$ is pairwise compatible, and so by Theorem 2.20 is compatible as a set. Therefore, by the definition of setwise compatibility, there is a tree T whose bipartition set is exactly that set. Hence, the majority tree always exists and is unique. Since every bipartition in the strict consensus tree appears in the majority consensus tree, if the two trees are not identical then the majority consensus tree is a proper refinement of the strict consensus tree. □

6.2 Consensus Trees 111

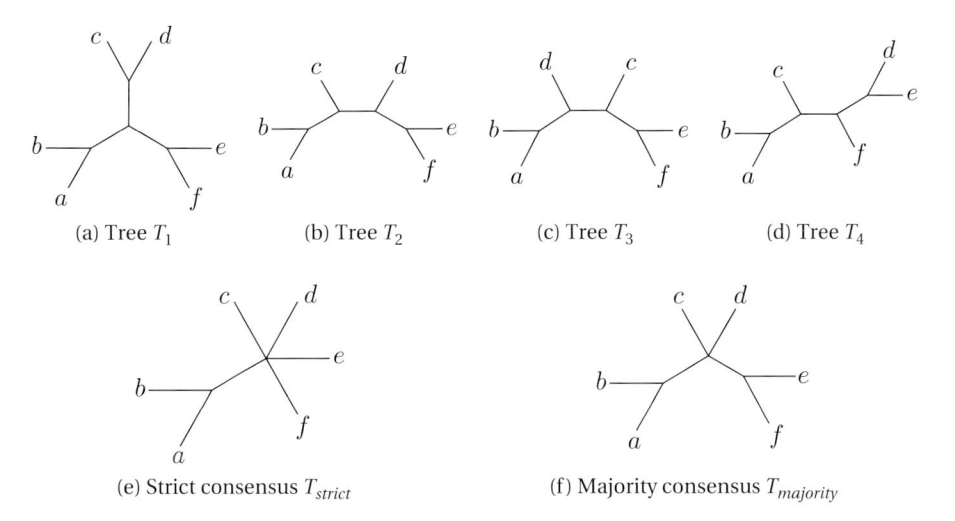

Figure 6.1 (Figure 3.26 in Huson et al. (2010)) We show four unrooted trees (a–d) and their strict consensus (e) and majority consensus (f) trees.

The majority consensus tree may not be fully resolved; however, unlike the strict consensus tree, the majority consensus tree can be fully resolved even when the trees in the profile are different from each other; see Figure 6.1 for one such example.

Majority consensus trees are probably the most commonly computed consensus trees. It is easy to see that the majority consensus tree can be computed in polynomial time, but it can also be computed quickly in practice: A randomized linear time algorithm to compute the majority consensus tree was developed in Amenta et al. (2003).

6.2.3 Greedy Consensus Tree

We now define the greedy consensus tree. To construct the greedy consensus, we order the bipartitions by the frequency with which they appear in the profile. We start with the majority consensus, and then "add" bipartitions (if we can), one by one, to the tree we've computed so far.

When we attempt to add a bipartition $A|B$ to a tree T, we are asking whether we can refine T so that it contains the bipartition $A|B$. If T already contains this bipartition the answer is "Yes," and we do not need to change T at all. If T does not already have the bipartition, we would look for a polytomy in T so that we can refine at this polytomy (by adding just one edge) to create the desired bipartition. As shown in Gusfield (1991), determining if it is possible to refine T to add the bipartition, and if so constructing the unique minimal refinement containing $A|B$, can be performed in $O(n)$ time.

We stop either when we construct a fully resolved tree (because in that case no additional bipartitions can be added), or because we finish examining the entire list. Note that the order in which we list the bipartitions will determine the greedy consensus – so that this particular consensus is *not uniquely defined* for a given profile of trees (we give such an example below). On the other hand, the strict consensus and majority consensus do *not* depend upon the ordering, and are uniquely defined by the profile of trees.

Observation 6.3 The greedy consensus is either equal to the majority consensus or it refines it, since it has every bipartition that appears in the majority consensus. Therefore, the greedy consensus is also called the **extended majority consensus**. While the greedy consensus tree will be at least as resolved as the majority consensus, it may also not be fully resolved.

We can also seek similar consensus trees that are also constructed using a greedy approach as described for the greedy consensus, but which constrain the allowed bipartitions to only those that appear in strictly greater than x percent of the input trees. Thus, the majority consensus is obtained for $x = 50$ and the greedy consensus is obtained for $x = 0$. If we set $x \geq 50$ then a consensus tree always exists that has exactly those bipartitions (all of them, and no others), while for $x < 50$ percent there may not be a consensus tree that has all the bipartitions that appear with that frequency.

6.2.4 Compatibility Trees

Recall that a tree T refines another tree t if t can be obtained from T by collapsing some of the edges in t (Definition 2.15). This definition extends to a set $\mathscr{T}$ of trees, by saying that T is a common refinement of the trees in $\mathscr{T}$ if T refines every tree $t \in \mathscr{T}$.

Definition 6.4 A set $\mathscr{T} = \{T_1, T_2, \ldots, T_k\}$ of trees, all with the same leafset, is *compatible* if there is a tree that refines every tree T_i, $i = 1, 2, \ldots, k$. When $\mathscr{T}$ is compatible, then it has a unique minimal common refinement T, and we refer to T as the **compatibility tree** for $\mathscr{T}$.

We will often be interested in determining whether a set $\mathscr{T}$ of trees is compatible, and in computing a common refinement of the set of trees when they are.

Example 6.5 As an example, consider the trees T_1 and T_2, defined as follows:

- T_1 given by $C(T_1) = \{(abc|defg)\}$, and shown in Figure 6.2(a).
- T_2 given by $C(T_2) = \{(abcd|efg), (abcde|fg)\}$, and shown in Figure 6.2(b).

We can see they are compatible, because T_3, shown in Figure 6.2(c), is a common refinement of each of the trees. Since they are compatible, there should be a minimal common refinement, and we would like to construct it. Note that $C(T_1) \cup C(T_2)$ must be a compatible set of bipartitions, since otherwise there would not be any common refinement. Let T be the tree satisfying $C(T) = C(T_1) \cup C(T_2)$. Note that T is a common refinement of T_1 and T_2, and that every tree that is a common refinement of T_1 and T_2 must refine T. Hence, T is the minimal common refinement of T_1 and T_2 and is the compatibility tree. However, T is different from T_3!

6.2 Consensus Trees 113

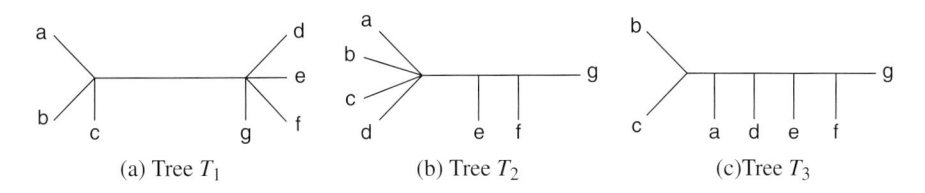

    (a) Tree T_1 (b) Tree T_2 (c) Tree T_3

Figure 6.2 Three trees on the same leafset; tree T_3 is a common refinement of trees T_1 and T_2, and so demonstrates that T_1 and T_2 are compatible. However, T_3 is not a minimal common refinement of T_1 and T_2, and so is not their compatibility tree.

Theorem 6.6 *If a set $\mathscr{T} = \{T_1, T_2, \ldots, T_k\}$ of trees, where each T_i is leaf-labeled by the same set S of taxa, has a compatibility tree T, then $C(T) = \cup_i C(T_i)$. Hence, a set $\mathscr{T} = \{T_1, T_2, \ldots, T_k\}$ of trees on the same leafset is compatible if and only if the set $\cup_i C(T_i)$ is compatible.*

More generally, to see if a set $\mathscr{T}$ of trees is compatible, we write down their bipartition sets, and then we apply the algorithm for constructing trees from bipartitions from Section 2.3.5 to the union of these sets. If $\mathscr{T}$ is compatible, the output will be a tree T that is the compatibility tree; otherwise, the algorithm will fail to return a tree, which will indicate that $\mathscr{T}$ is not compatible.

6.2.5 Asymmetric Median Tree

The asymmetric median tree was proposed in Phillips and Warnow (1996) to address the inability of the median tree approach to construct a compatibility tree when the input set of trees is compatible. For example, consider T_1, T_2, and T_3 on leafset a,b,c,d,e,f, each having only one internal edge. Thus, each tree is defined by its single non-trivial bipartition, as follows: T_1's non-trivial bipartition is $ab|cdef$, T_2's non-trivial bipartition is $abc|def$, and T_3's non-trivial bipartition is $abcd|ef$. The three trees are compatible since $T = (a,(b,(c,(d,(e,f)))))$ is a minimal common refinement of T_1, T_2, and T_3. However, the median tree (i.e., the tree that minimizes the total Robinson–Foulds distance to the three trees) is the star tree (see Definition 2.25), which has total distance of 3. All other trees have a larger total Robinson–Foulds distance. For example, the compatibility tree T has a total distance of 6.

The reason that a compatibility tree is not a median tree is that the Robinson–Foulds distance between the consensus tree and the source trees penalizes edges that are in the consensus tree and that appear in some but not all source trees. Thus, even if a source tree is compatible with all the other source trees, any bipartitions in that source tree that are not in strictly more than half the source trees cannot be in the median tree. That makes the median tree (which is the same as the majority consensus tree) different from the compatibility tree.

The asymmetric median tree addresses this weakness of the median tree. Instead of counting the edges in the consensus tree that don't appear in some source tree, it only counts edges in the source tree that do not appear in the consensus tree. Equivalently, an asymmetric median tree is a tree T on the same leafset as the source trees that minimizes $\sum_{i=1}^{k} |C(T_i) \setminus C(T)|$. Note that the asymmetric median tree is not constrained to be binary. Hence, if T is an incompletely resolved asymmetric median tree for a profile $\mathscr{T}$, then any refinement of T is also an asymmetric median tree for $\mathscr{T}$. Furthermore, if the set $\mathscr{T}$ is compatible, then the compatibility tree is an asymmetric median tree, and every asymmetric median tree refines the compatibility tree.

Finding the asymmetric median tree of k trees is equivalent to solving the maximum independent set on a k-partite graph (Phillips and Warnow, 1996). Hence, finding the asymmetric median tree of two trees can be solved in $O(n^{2.5})$ time, but is NP-hard for even three trees (Phillips and Warnow, 1996). In contrast, the median tree can be computed in polynomial time. Thus, although the asymmetric median tree has some theoretical advantages over the median tree, it is computationally more intensive to compute.

6.2.6 Comparisons Between Consensus Trees

We explore the differences between these consensus trees.

Example 6.7 We give three different trees on the same leafset, defined by the non-trivial bipartitions for each tree:

- T_1 given by $C_I(T_1) = \{(12|3456), (123|456), (1234|56)\}$
- T_2 given by $C_I(T_2) = \{(12|3456), (123|456), (1235|46)\}$
- T_3 given by $C_I(T_3) = \{(12|3456), (126|345), (1236|45)\}$

The bipartitions are:

- $(12|3456)$, which appears three times
- $(123|456)$, which appears twice
- $(1234|56)$, which appears once
- $(1235|46)$, which appears once
- $(1236|45)$, which appears once
- $(126|345)$, which appears once

Using the definition of the strict consensus tree, we see that the strict consensus has only one bipartition, $(12|3456)$, and that the majority consensus has two bipartitions: $(123|456)$ and $(12|3456)$. Note that the greedy consensus tree depends upon the ordering of the remaining four bipartitions (since all appear exactly once, they can be ordered arbitrarily), and so there can be more than one greedy consensus tree. In fact, there are $24 = 4!$ possible orderings of these bipartitions! However, we will only show the results for three of these.

- Ordering 1: $(1234|56), (1235|46), (1236|45), (126|345)$. For this ordering, we see that we can add $(1234|56)$ to the set we have so far, $(12|3456), (123|456)$ to obtain a fully resolved tree. Note that this is equal to T_1.

- Ordering 2: $(126|345), (1236|45), (1234|56), (1235|46)$. For this ordering we see that we *cannot* add the bipartition $(126|345)$ to the set we have so far. However, we can add $(1236|45)$ to obtain a fully resolved tree. This final tree is given by $(1,(2,(3,(6,(4,5)))))$. This is not among the trees in the input.
- Ordering 3: $(126|345), (1235|46), (1234|56), (1236|45)$. For this ordering, we cannot add $(126|345)$, but we can add the next bipartition, $(1235|46)$. When we add this, we obtain a fully resolved tree that is equal to T_2.

Example 6.8 We now consider how the asymmetric median tree and the majority tree can differ. Suppose the input set $\mathscr{T} = \{T_1, T_2, T_3\}$, given by the following bipartitions:

- $C(T_1) = \{(123|456)\}$
- $C(T_2) = \{(12|3456)\}$
- $C(T_3) = \{(1236|45)\}$

The majority consensus of these three trees is the star tree, since there is no bipartition that appears in the strict majority of these trees. However, the three trees are compatible, since their compatibility tree is $(1,(2,(3,(6,(4,5)))))$. Furthermore, the compatibility tree, when it exists, is always the optimal solution to the asymmetric median tree. Hence, the asymmetric median tree on this input is a fully resolved tree, while the majority consensus tree is a completely unresolved tree! Recall that the majority consensus tree is a median tree, using the Robinson–Foulds distance to measure the distance between two trees. Hence, the median tree and the asymmetric median tree can be very different.

6.2.7 The Characteristic Tree

Let T be a consensus tree computed for a set $\mathscr{T}$ of source trees, and let $B(T)$ denote the set of all binary trees that refine T. We can consider $\mathscr{T}$ (and $B(T)$, respectively) as defining a probability distribution on all possible binary trees on S, with every tree in $\mathscr{T}$ (respectively $B(T)$) having the same probability and all other trees having zero probability. In other words, each of these distributions has the uniform distribution on the trees within its specified set, and otherwise has zero probability on the trees outside its set.

Having done this, we can consider how well the distribution defined by $B(T)$ approximates the distribution defined by $\mathscr{T}$. There are many ways of computing distances between distributions, including the L_∞-distance, the L_k-distances ($k = 1, 2$ are popular), and the Kullback–Leibler (KL) divergence (Kullback, 1987). Therefore, we may wish to seek a consensus tree T whose distribution (as defined above) is as close as possible to the distribution defined by $\mathscr{T}$; this is called the **characteristic tree** for $\mathscr{T}$ with respect to the specified distance (Stockham et al., 2002).

As shown in Stockham et al. (2002), the strict consensus tree is the characteristic tree with respect to the KL-distance, and also with respect to the L_1- and L_2-distances. Stockham et al. (2002) also showed that if $|\mathscr{T}|/|B(T)| \leq 0.5$, then the strict consensus tree is the characteristic tree with respect to the L_∞-distance. Conversely, if $|\mathscr{T}|/|B(T)| > 0.5$, then Stockham et al. (2002) showed the characteristic tree can be derived from the strict consensus tree by identifying an edge in the tree which, after being collapsed, produces a tree with the smallest number of binary refinements, and that such an edge can be found in linear time.

6.3 Agreement Subtrees

Agreement subtrees are trees on a subset of the shared taxon set that represent the shared features in the profile. The two main techniques of this type are the *maximum agreement subtree* (MAST) and the *maximum compatibility subtree* (MCST).

6.3.1 Maximum Agreement Subtree

The MAST problem is as follows. Given a set $\mathscr{T}$ of input trees (which may be rooted or unrooted), the objective is to find the largest subset X of the leafset S so that the trees in $\mathscr{T}$ agree on X (i.e., they all induce the same tree on X).

The MAST problem was posed in Finden and Gordon (1995). The first polynomial time algorithm to compute the MAST of two trees (whether rooted or not) was presented in Steel and Warnow (1993), which included an $O(n^2)$ algorithm to compute the MAST of two bounded-degree rooted trees. Subsequently, an $O(n^{1.5} \log n)$ algorithm for computing the MAST of two unbounded-degree trees was found (Farach and Thorup, 1994). Polynomial time algorithms for three or more bounded-degree trees were presented in Amir and Keselman (1994), along with an NP-hardness proof for MAST on three unbounded-degree rooted trees. Thus, the computational complexity of the MAST problem depends on the number of trees and their maximum degree.

6.3.2 Maximum Compatibility Subtree

The objective of the MCST problem (also known as the maximum refinement subtree problem (Hein et al., 1996)) is to find the largest subset X of the leafset S so that the input trees are *compatible* when restricted to X. Thus, the MCST problem is similar to the MAST problem. The motivation for the problem is that many estimated trees have low support branches (see Section 1.7), and when these branches are collapsed the estimated trees have polytomies. These polytomies are considered **soft polytomies**, because what they indicate is uncertainty about the phylogeny as opposed to a statement that the evolution occurred with a multi-way speciation event (which would be a **hard polytomy**). Therefore, if the input source trees are modified by collapsing low support branches, the objective is to find a subset of the taxa on which all the trees have a common refinement.

It is trivial to see that when all the trees in the input are binary, then the MAST of the set is identical to the MCST of the set. However, when the input trees can contain polytomies, then these two trees can be different. Furthermore, the MCST always has at least as many leaves as the MAST. The MCST problem was introduced in Hein et al. (1996), where it was shown to be solvable in polynomial time for two bounded-degree trees, and NP-hard when one of the two trees has unbounded degree. See Ganapathy and Warnow (2001) for polynomial time fixed-parameter algorithms to compute the MCST of three or more trees, and Ganapathy and Warnow (2002) for approximation algorithms for computing the size of the complement of the MCST.

6.4 Clustering Sets of Trees

All the consensus and agreement methods we have described return a single tree from a set of trees. However, when the set of trees potentially indicates multiple different evolutionary hypotheses, then consensus methods that produce a collection of trees, one for each of the possible hypotheses, can provide more insight. One of the interesting approaches is the *reduced consensus* methods, described in Wilkinson (1994, 1995). These methods are designed to output a profile of trees, each of which is a type of agreement tree (i.e., each tree in the profile provides information that is consistent across the set of source trees). Another way to approach this is to divide the set of trees into clusters, and then represent each cluster by its consensus or agreement tree. Approaches such as these can provide insight into alternative phylogenetic hypotheses present in a collection of estimated trees.

Various techniques have been provided to produce these clusters, but the phylogenetic island approach of Maddison (1991) is the most well known. Here, a phylogenetic island is the subset of the input trees that are reachable from each other by TBR (tree bisection and reconnection) moves that only visit the trees within the input set (see Figure 11.3 for a description of the TBR move). Another way of describing a phylogenetic island is that it is a connected component in a graph whose vertices are the input trees, and whose edges connect two trees that are related by a single TBR move. The reason that TBR moves are the basis for the definition of phylogenetic islands is that most heuristic search methods in phylogenetics are based on searching through treespace using TBR moves. Note that the phylogenetic island approach defines disjoint clusters. This definition of phylogenetic island can naturally be extended to allow for different ways of defining neighborhoods of trees, to accommodate different ways of exploring treespace.

An alternative approach based on a statistical optimization is presented in Stockham et al. (2002). Here, the objective is to design the clustering as well as a way of representing each cluster within the collection by its characteristic tree (described earlier) so that the distribution defined by the set of consensus trees is close to the input distribution.

6.5 Further Reading

The consensus tree and agreement subtree methods we have presented are just a small sample of a large number of such methods; Bryant (2003) and Janowitz et al. (2003)

describe a large number of the early consensus methods, but the development of these methods continues.

Some of the recent developments include methods that operate by encoding each source tree as a set of triplet trees (if the source trees are rooted) or quartet trees (if the source trees are unrooted), and then applying methods for combining these subset trees together. An example of this approach is the "local consensus tree" (Kannan et al., 1995), which assumes a rule is provided for combining rooted triplet trees and then constructs a tree, if it exists, consistent with that rule. Quartet-based approaches are generally more popular, since they avoid the problem of rooting the source trees developed; see Chapter 3 for examples of quartet-based tree estimation methods that can be used as consensus methods. Most quartet-based methods are computationally expensive, since they generally require the explicit calculation of all the quartet trees. Some quartet-based methods avoid this computational cost by implicitly, rather than explicitly, computing quartet trees, or else subsample quartet trees, in order to reduce the running time.

6.6 Review Questions

1. Suppose you have ten trees on the same leafset.
   - Define the strict consensus tree, and describe a method for how to calculate the strict consensus tree.
   - Define the majority consensus tree, and describe a method for how to calculate the majority consensus tree.
   - Define the greedy consensus tree, and describe a method for how to calculate the greedy consensus tree.
2. What is the maximum agreement subtree (MAST) problem? What is the computational complexity of MAST on two trees?
3. What is the maximum compatibility subtree (MCST) problem? What is the computational complexity of MCST on two trees?
4. Suppose T and T' are two different trees, each on the same leafset. Can the MCST and MAST of T and T' be identical? Can they be different? Suppose that they have different numbers of leaves; which one must have more leaves? Suppose T and T' are binary; can the MCST and MAST be different?

6.7 Homework Problems

1. Suppose you have a binary tree t on ten leaves and you create two trees T and T' by adding in a leaf x into t in two different places. What is the size of the MAST of T and T'?
2. Suppose you have a collection of binary trees (each of them different), all on the same leafset $\{1, 2, \ldots, n\}$, with $n > 4$. Suppose that T is a compatibility tree for the set. How many trees can be in the collection?

6.7 Homework Problems

3. Consider the following three trees, each on set $S = \{1,2,3,4,5,6\}$.
   - T_1 given by $C(T_1) = \{(12|3456), (123|456), (1234|56)\}$
   - T_2 given by $C(T_2) = \{(12|3456), (123|456), (1235|46)\}$
   - T_3 given by $C(T_3) = \{(12|3456), (126|345), (1236|45)\}$

   Is it possible to order the bipartitions of this set so as to produce T_2 as a greedy consensus? If so, provide one such ordering. If not, explain why not.

4. Suppose you have an arbitrary set $\mathcal{T}$ of trees on the same leafset, and you compute the strict, majority, and greedy consensus trees. For each of the following pairs of trees, suppose they are different; must one of them refine the other? If so, which one, and why?
   - The strict consensus and majority consensus.
   - The greedy consensus and the majority consensus.
   - The strict consensus tree and an arbitrary tree in $\mathcal{T}$.
   - The majority consensus and an arbitrary tree in $\mathcal{T}$.

5. Give two different compatible unrooted trees on the same leafset, and present their minimal common refinement.

6. Give two different trees on the same leafset, neither of which is fully resolved, and which are *not* compatible.

7. Describe a polynomial time algorithm to compute the compatibility tree of two unrooted trees, and implement it. (Remember that the compatibility tree is the minimally resolved tree that is a common refinement of the two input trees.)

8. Let unrooted T_0 given by $(a,(b,(c,((d,e),(f,g)))))$ denote the true tree.

   a. For each unrooted tree below, draw the tree, and write down the bipartitions that are false positives and false negatives with respect to T_0.
      - $T_1 = (f,(g,(a,(b,(c,(d,e))))))$
      - $T_2 = (g,(f,(c,(d,(e,(a,b))))))$
      - $T_3 = (g,(f,(a,(b,(c,(d,e))))))$

   b. Draw the strict, majority, and greedy consensus trees for the three trees $T_1, T_2,$ and T_3. Compute the false negatives and false positives (with respect to T_0) for these consensus trees.

9. Consider an arbitrary unrooted binary true tree and let $\mathcal{T}$ be a set of estimated unrooted trees. Suppose you compute the strict consensus, majority consensus, and greedy consensus of these trees. Now compute the false negative error rates of these three consensus trees, and compare them to each other and also to the false negative error rates of the trees in the set $\mathcal{T}$. What can you deduce? Do the same thing for the false positive error rates.

10. Give two unrooted trees, T_1 and T_2, that are compatible, and their unrooted compatibility tree T_3. Treat T_3 as the true tree, and compute the false negative and false positive rates of T_1 and T_2 with respect to T_3. What do you see?

11. Consider the caterpillar tree T on $s_1, s_2, \ldots, s_6$, and let $\mathscr{T}$ be the set of binary trees produced by adding an additional taxon, x, into T in each of the possible ways (i.e., by subdividing edges and making x a child of the new node).
    - Draw T.
    - Draw $\mathscr{T}$.
    - Draw the maximum agreement subtree for $\mathscr{T}$.
    - Draw the strict consensus tree for $\mathscr{T}$.
    - Draw the majority consensus tree for $\mathscr{T}$.

12. Consider a caterpillar tree T on a set S of n taxa. Consider the set $\mathscr{T}$ that contains all the binary trees formed by adding a new leaf x into T in each of the possible ways (i.e., by subdividing edges and making x a child of the new node).
    1. Express $|\mathscr{T}|$ as a function of n.
    2. Write down the bipartitions of the strict consensus of all the trees in $\mathscr{T}$?
    3. Draw the majority consensus when $n = 4$.
    4. Draw the majority consensus when $n = 5$.
    5. Draw the majority consensus when $n = 6$.
    6. Can you generalize the observations seen here to provide a characterization of the majority consensus trees, as a function of n?

13. Suppose you have a set X of trees, each on the same set of taxa, and you are fortunate enough to be able to label each edge as a true positive or a false positive with respect to the true tree T. Suppose you contract every false positive edge in each tree in X, and consider the resultant set X'. For each of the following pairs of trees T_1 and T_2, indicate if one tree in the pair *must* refine the other tree (and if so, which one). Also indicate if the number of leaves in one tree must be at most the number of leaves in the other one (and if so, which one).
    - Let T_1 be an MCST (maximum compatible subtree) of the trees in X', and T_2 be an MCST of the trees in X.
    - Let T_1 be an MCST of the trees in X, and let T_2 be a MAST of the trees in X.
    - Let T_1 be an MCST of the trees in X', and let T_2 be a MAST of the trees in X'.
    - Let T_1 be a MAST of the trees in X', and let T_2 be the true tree.
    - Let T_1 be an MCST of the trees in X', and let T_2 be the true tree.

14. Suppose you have a set X of trees, each on the same set of taxa, and the MCST and MAST for X have different numbers of taxa. Which one of these has more taxa?

15. Suppose $\mathscr{T}$ is a profile of trees that is compatible. Prove that every asymmetric median tree for $\mathscr{T}$ is a refined compatibility tree.

7
Supertrees

7.1 Introduction

The basic objective of most supertree studies is the assembly of a large species tree from a set of species trees that have been estimated on potentially smaller sets of taxa. Indeed, it is generally believed that construction of the Tree of Life, which will encompass millions of species, will require supertree methods, because no software will be able to deal with the computational challenges involved in such a difficult task. More recently, however, new uses of supertree methods have been discovered, especially in the context of divide-and-conquer strategies. This chapter examines both applications of supertree methods.

Traditionally, supertree methods were used to combine trees computed by different researchers that had already been estimated for different taxon sets. In this case, the person constructing the supertree has no control over the inputs, neither how the different subset trees were constructed nor how the taxon sets of the different subset trees overlap. Furthermore, the person constructing the supertree may not have easy access to the data (e.g., sequence alignments) on which the subset trees were constructed.

A modern and more interesting use of supertree methods is in the context of a divide-and-conquer strategy to construct a very large tree, or to enable a statistically powerful but computationally intensive method to be applied to a larger dataset. In such a strategy, a large set of taxa is divided into overlapping subsets, trees are estimated (using the desired method) on the subsets, and the estimated subset trees are combined into a tree on the full set of taxa using a supertree method.

Divide-and-conquer techniques for constructing large trees have many desirable features: (1) the subsets can be made small enough that expensive methods can be used to construct trees on them; (2) different methods can be used on each subset, thus making it possible to better address heterogeneity within the full dataset; and (3) the subsets can be created so as to have desirable overlap patterns. The first two of these features tend to increase the accuracy of the estimated subset trees, while the third feature can make it easier to construct an accurate supertree from the subset trees (Wilkinson and Cotton, 2006). We will return to the topic of divide-and-conquer strategies and how to use them to construct large trees under a variety of scenarios in Chapter 11. For now, just be aware that supertree methods are more than just ways of assembling large trees from other trees; they are key ingredients in developing methods to enable powerful but expensive methods to run on ultra-large datasets.

Just as in the consensus setting, the input to the supertree method is a set of source trees, referred to as a profile; the distinction between consensus and supertree problems is that in the supertree setting we do not constrain the input trees to have identical leafsets. Hence, any supertree method can be used as a consensus tree method, but the converse is not true. Because of this difference, supertree methods have greater applicability than consensus methods. On the other hand, many optimization problems posed in the supertree setting are NP-hard, whereas the same optimization problems posed in the consensus setting are polynomial time. Seeing supertree method development as an extension to a consensus tree method was first applied by Gordon (1986) in the development of the Gordon's consensus, an extension of the strict consensus to the supertree setting, and then formalized in Cotton and Wilkinson (2007), who developed two supertree methods (both called Majority-Rule Supertrees) that are generalizations of the majority consensus tree.

In the idealized condition, all the source trees are compatible with each other, and the objective will be to construct a compatibility supertree. However, since source trees are estimated trees, they are likely to have some estimation error, and no compatibility supertree will exist. Therefore, rather than trying to find a compatibility supertree (see Section 3.5), the main objective is to find a supertree that is somehow as close as possible, with respect to some criterion, to the input source trees (e.g., minimizing the sum of the distances to the input source trees (Thorley and Wilkinson, 2003)). Examples of optimization problems that have been used to define supertrees include Matrix Representation with Parsimony (MRP) (Baum and Ragan, 2004), Matrix Representation with Likelihood (MRL) (Nguyen et al., 2012), Robinson–Foulds Supertree (RFS) (Wilkinson et al., 2005), Quartet Median Tree, and Maximum Quartet Support Supertree (MQS), all described below. Each of these optimization problems is formulated for inputs containing unrooted source trees, but most have equivalent formulations for inputs of rooted source trees. Supertree software implementing specific approaches (some of which are based on optimality criteria) include MinCut Supertree (Semple and Steel, 2000), Modified MinCut Supertree (Page, 2002), MinFlip Supertree (Chen et al., 2003, 2006), and the MRF Supertree (Burleigh et al., 2004).

Unfortunately, all the optimization problems mentioned above, and in fact nearly all optimization problems that have been posed for supertree construction, are NP-hard. Therefore, heuristics rather than exact algorithms are used to find good solutions to these optimization problems. Thus, supertree methods are, like most problems in phylogenetics, understood by the theoretical properties of the optimization problems on which they are based (such as whether the criteria are biased toward large source trees (Wilkinson et al., 2005)), by the details of their implementations, and by how well they perform on data.

For the rest of this chapter, we will assume that the profile given as input to the supertree problem is $\mathscr{T} = \{t_1, t_2, \ldots, t_k\}$. We will let S_i denote the leafset of tree t_i, and we let $S = \cup S_i$. Most of the supertree problems we will discuss involve profiles of unrooted trees; however, we will also (briefly) describe some methods for computing supertrees when the profile contains rooted trees.

7.2 Compatibility Supertrees

Recall that if T is a compatibility supertree, then T is a minimally resolved tree on leafset $S = \cup S_i$ that is compatible with every tree t_i (Definition 3.1). Equivalently, T is the minimally resolved tree such that one of the following conditions holds: $C(t_i) \subseteq C(T|S_i)$ for every i or $\sum_{i=1}^{k} |C(t_i) \setminus C(T|S_i)| = 0$. Recall also that a refined compatibility supertree is any tree that is compatible with all the trees in the profile, and need not be a minimally resolved such tree.

Now, consider the compatibility supertree problem:

- Input: $\mathcal{T} = \{t_1, t_2, \ldots, t_k\}$, where t_i has leafset S_i and is an unrooted tree.
- Output: A compatibility supertree, if it exists.

Note that even if a profile has a compatibility supertree, it can have more than just one.

Example 7.1 Let $t_1 = ((a,b),(c,d))$ and $t_2 = ((a,b),(c,e))$. Then $(a,(b,(c,(d,e))))$ and $(a,(b,(d,(c,e))))$ are both compatibility supertrees for the profile $\{t_1, t_2\}$.

Determining if a compatibility supertree exists is NP-complete, since if every source tree is a quartet tree, the problem reduces to quartet compatibility, which is NP-complete (Steel, 1992).

7.3 Asymmetric Median Supertrees

In Section 6.2.5, we described a consensus method called the asymmetric median tree that is related to determining if a profile of source trees was compatible. Since consensus methods require that the profile contain trees that have the same set of leaves, we extend that approach to allow for profiles of trees that are on different taxon sets, and we call this the **Asymmetric Median Supertree**:

- Input: Profile $\mathcal{T} = \{t_1, t_2, \ldots, t_k\}$ of unrooted trees, with t_i having leafset S_i.
- Output: Tree T with leafset $S = \cup_{i=1}^{k} S_i$ that minimizes $\sum_{i=1}^{k} |C(t_i) \setminus C(T)|$.

Lemma 7.2 *Let $\mathcal{T}$ be a profile of unrooted source trees, and suppose that $\mathcal{T}$ is compatible so that a compatibility supertree exists. Then* T *is a refined compatibility supertree for $\mathcal{T}$ if and only if* T *is an asymmetric median supertree for $\mathcal{T}$. Furthermore, any refined compatibility supertree* T *satisfies $\sum_{i=1}^{k} |C(t_i) \setminus C(T)| = 0$.*

Computing the asymmetric median supertree is also NP-hard, since a compatibility supertree would be an optimal solution (if it exists), and determining if a compatibility supertree exists is NP-complete. In general, source trees are not required to be binary; however, if we add that constraint, then any compatibility supertree T will satisfy $T|S_i = t_i$ for each $i = 1, 2, \ldots, k$, and so will also satisfy $RF(T|S_i, t_i) = 0$ for each $i = 1, 2, \ldots, k$. This leads to the statement of the Robinson–Foulds Supertree problem, the subject of the next section.

7.4 Robinson–Foulds Supertrees

7.4.1 Problem Formulation

The Robinson–Foulds Supertree (RFS) is a supertree T that minimizes the total Robinson–Foulds (RF) distance between T and the source trees (Cotton and Wilkinson, 2007; Bansal et al., 2010; Chaudhary et al., 2012; Chaudhary, 2015; Kupczok, 2011). Since source trees can be small, we use the extension of the definition of the RF distance from Cotton and Wilkinson (2007) so that it can be used in supertree estimation:

Definition 7.3 Let T have leafset S and let t have leafset $S' \subseteq S$. Then $RF(T,t)$ denotes the RF distance between $T|S'$ and t.

Example 7.4 Let $T = (1,((6,7),(4,(3,(5,2)))))$ and $t = (5,((3,2),(6,7)))$. S', the leafset of t, is $\{2,3,5,6,7\}$. To compute the RF distance between T and t we compute $T|S'$, which is $((6,7),(3,(5,2)))$. The non-trivial bipartitions of $T|S_i$ are $(235|67)$ and $(25|367)$. The non-trivial bipartitions of t are $(235|67)$ and $(23|567)$. Therefore, $RF(T,t) = 2$.

The RF Supertree problem is:

- Input: Profile $\mathcal{T} = \{t_1, t_2, \ldots, t_k\}$ of unrooted source trees.
- Output: Tree T_{RFS} that minimizes the total RF distance to the source trees, i.e.,

$$T_{RFS} = \arg\min_T \sum_{i=1}^{k} RF(T, t_i).$$

The RFS and the asymmetric median supertree are clearly similar, as both are based on bipartitions. However, the asymmetric median supertree does not penalize the supertree for containing bipartitions that do not appear in the source trees, and the RFS does. Thus, an optimal RFS and an optimal asymmetric median supertree can be different. However, the two problems are related to the compatibility supertree problem, as we now show:

Theorem 7.5 Let $\mathcal{T} = \{t_1, t_2, \ldots, t_k\}$ be a set of trees where t_i has leafset S_i, and let $S = \cup_i S_i$. Then T is a compatibility supertree for $\mathcal{T}$ if and only if T has leafset S and $|C(t_i) \setminus C(T|S_i)| = 0$ for all $i = 1, 2, \ldots, k$. Hence, when a compatibility supertree exists, then T is an asymmetric median supertree if and only if T is a refined compatibility supertree. Furthermore, if each t_i is a fully resolved tree, then any refined compatibility supertree T satisfies $RF(T, t_i) = 0$ for all $i = 1, 2, \ldots, k$. Conversely, if T is a tree on leafset $S = \cup_i S_i$ that satisfies $|C(t_i) \setminus C(T|S_i)| = 0$ for all i, then by definition $\mathcal{T}$ is a compatible set of trees, and T is a refined compatibility supertree for $\mathcal{T}$.

We will use this analysis to prove that finding the optimal RFS is NP-hard.

Theorem 7.6 *The Robinson–Foulds Supertree problem is NP-hard.*

Proof Suppose that $\mathscr{A}$ is an algorithm that solves the RFS problem in polynomial time; thus, $\mathscr{A}(\mathscr{T})$ is an RFS for the profile $\mathscr{T}$. We will show that we can use $\mathscr{A}$ to solve the Unrooted Tree Compatibility problem in polynomial time, for the case where the input trees are all binary. Since Unrooted Tree Compatibility is NP-complete even when the input trees are all binary (Theorem 3.11) this will prove that the RFS problem is NP-hard.

Let $\mathscr{T} = \{t_1, t_2, \ldots, t_k\}$ be a profile of unrooted binary trees, with S_i the leafset of t_i for $i = 1, 2, \ldots, k$. We run $\mathscr{A}$ on the profile $\mathscr{T}$ to obtain $T = \mathscr{A}(\mathscr{T})$; thus, T is an optimal solution to the RFS problem. We then compare $T|S_i$ to t_i for each $i = 1, 2, \ldots, k$. If $T|S_i$ refines t_i for every $i = 1, 2, \ldots, k$, then T is a refined compatibility supertree, and otherwise T is not a refined compatibility supertree. This is easily checked in polynomial time. Since we assume that each t_i is binary (i.e., fully resolved), then if $T|S_i$ refines t_i it follows that $T|S_i$ is identical to t_i. If T passes this test for every $i = 1, 2, \ldots, k$, then we will say that $\mathscr{T}$ is compatible, and otherwise we will say that $\mathscr{T}$ is not compatible.

To prove that we have solved the Unrooted Tree Compatibility problem (for the case where every tree in the input profile is fully resolved), we just need to show that the profile $\mathscr{T}$ is compatible if and only if $T = \mathscr{A}(\mathscr{T})$ passes this test for every i. So suppose T passes the test for every $i = 1, 2, \ldots, k$. Then T is a refined compatibility supertree for $\mathscr{T}$, and so the profile $\mathscr{T}$ is compatible. Now suppose that the profile $\mathscr{T}$ is compatible, but that T is not a refined compatibility supertree for $\mathscr{T}$. Since $\mathscr{T}$ is compatible, there is a compatibility supertree, T^*. Note that $T^*|S_i = t_i$ (i.e., $RF(T^*, t_i) = 0$) for $i = 1, 2, \ldots, k$, and so T^* is an optimal solution to the RFS problem. Since $\mathscr{A}$ is an exact algorithm for the RFS problem, T^* and T must both have the same score under the RFS criterion. Hence, $RF(T, t_i) = 0$ for each i as well, and so T is also a refined compatibility supertree for $\mathscr{T}$, contradicting our assumption. Hence, the profile $\mathscr{T}$ is compatible if and only if $T = \mathscr{A}(\mathscr{T})$ is a refined compatibility supertree.

Since $\mathscr{A}$ runs in polynomial time, we have shown that if the RFS problem can be solved in polynomial time then the Unrooted Tree Compatibility problem can also be solved in polynomial time when all the source trees are binary. Since Unrooted Tree Compatibility is NP-complete, even when all the source trees are binary (Steel, 1992), the RFS problem is NP-hard. □

The key part of the proof is that when the input profile is compatible, the compatibility tree would be an optimal solution to the RFS problem. Therefore, the same kind of argument can be used to establish that any supertree problem is NP-hard if a compatibility supertree (if it exists) would be an optimal solution. We summarize this point now, since nearly every supertree problem we discuss henceforth will have this property:

Corollary 7.7 *Let π be a supertree optimization problem where the input is a profile of unrooted source trees. If an exact algorithm for π would return a compatibility supertree* *(when it exists), then π is NP-hard. Hence, in particular, the asymmetric median supertree problem is NP-hard.*

7.4.2 Methods for Robinson–Foulds Supertrees

The RFS problem is a popular approach to supertree estimation, and several methods have been developed to try to find good solutions to the problem. Examples of these methods include PluMiST (Kupczok, 2011), Robinson–Foulds Supertrees (Bansal et al., 2010), MulRF (Chaudhary, 2015), and FastRFS (Vachaspati and Warnow, 2016). Most methods for the RFS problem use heuristic searches to attempt to find good solutions to the optimization problem, but FastRFS (Vachaspati and Warnow, 2016) uses an alternative strategy. FastRFS uses the input source trees $\mathscr{T}$ to compute a set X of allowed bipartitions, and then uses dynamic programming to find an exact solution to the RFS problem within that constrained search space. In other words, the dynamic programming strategy guarantees that the tree that is returned has the best possible score among all trees that draw their bipartitions from X. Furthermore, the running time of FastRFS is $O(nk|X|^2)$, where n is the number of species and k is the number of source trees. Hence, when X is small enough, the algorithm is very fast.

The basic version of FastRFS just uses the bipartitions from the input trees; however, since the input trees may not have all the species, this set needs to be enlarged. The enhanced version of FastRFS uses bipartitions from supertrees computed by other fast supertree methods, and adds them to the bipartition set X. As shown in Vachaspati and Warnow (2016), FastRFS finds better solutions to the RFS than the alternative heuristics, and does so very efficiently.

7.5 Matrix Representation with Parsimony

The Matrix Representation with Parsimony (MRP) (Baum and Ragan, 2004) supertree problem is by far the most well known (and most popular) supertree optimization problem. Although several variants of MRP have been posed that differ depending on whether the input trees are rooted or unrooted, we will discuss the version where the input trees are unrooted.

The input to MRP is a profile $\{t_1, t_2, \ldots, t_k\}$ of unrooted trees where t_i has leafset S_i. From this set of trees, we compute a matrix, called the **MRP matrix**, defined by the concatenation of the matrices obtained for each of the trees in the profile. Thus, to define the MRP matrix for the profile, it suffices to show how we define the MRP matrix for a single tree t_i.

The matrix for source tree t_i on taxon set S_i has a row for every element of $S = \cup_i S_i$ and a column for every internal edge in t_i. To define the column associated to the internal edge e in t_i, we compute the bipartition on the leafset S_i defined by removing e from t_i, and we arbitrarily assign 0 to the leaves on one side of this bipartition and 1 to the other side.

7.5 Matrix Representation with Parsimony

We then assign ? to every $s \in S \setminus S_i$. Thus, for $|S| = n$, each edge in t_i is represented by an n-tuple with entries drawn from $\{0, 1, ?\}$. For each such edge we create a column defined by its n-tuple, and we concatenate all these columns together into a matrix that represents the tree t_i. After computing all the matrices for all the trees, we concatenate all the matrices together into one large matrix, which is called the "MRP matrix." Note that in this matrix, every element $s \in S$ is identified with its row. The number of columns is the sum of the number of internal edges among all the trees. Since each tree can have up to $n - 3$ internal edges, this means that the number of columns is $O(nk)$, where k is the number of source trees.

Under the MRP criterion, we seek the tree that optimizes the maximum parsimony criterion with respect to the input MRP matrix. Since the MRP matrix will in general have ?s, we need to explain how these are handled. Let M be the MRP matrix computed from the profile. Since M may have ?s, we consider the set $\mathcal{M}$ of all matrices that can be formed by replacing the ? entries in M by 0 or 1; hence, if M has p entries that are ?, then $|\mathcal{M}| = 2^p$. In other words, $\mathcal{M}$ is the set of binary matrices that agree with M.

Let $MP(T, M)$ denote the maximum parsimony score of tree T for the matrix M. We denote by $MRP(T, \mathcal{T})$ the MRP score of a tree T with respect to the profile $\mathcal{T}$. Then, $MRP(T, \mathcal{T}) = min\{MP(T, M') : M' \in \mathcal{M}\}$. In other words, we are seeking the *best* way of replacing all the question marks (?s) by zeros and ones so that the result gives us the best possible maximum parsimony score.

Thus, the MRP problem is really the maximum parsimony problem on the MRP matrix, with the understanding of how missing data (as represented by ?s) are handled by maximum parsimony. Because binary character maximum parsimony is NP-hard (Foulds and Graham, 1982), MRP is NP-hard. Here we present a proof of this statement.

Theorem 7.8 *Let $\mathcal{T} = \{t_1, t_2, \ldots, t_k\}$ be an input of unrooted source trees. If $\mathcal{T}$ is compatible, then the MRP matrix defined on this input has a perfect phylogeny, and any optimal solution to MRP will be a refined compatibility supertree. Furthermore, the MRP problem is NP-hard.*

Proof Under the assumption that the source trees are compatible, a compatibility supertree T exists. Then, every column in the MRP matrix (which corresponds to a bipartition in some source tree) will be compatible with T in the sense that the ?s can be replaced by 0s and 1s to correspond to some edge in T. In other words, the columns in the MRP matrix are compatible partial binary characters (where "partial binary characters" are characters with 0s, 1s, and ?s, or equivalently bipartitions on subsets of the taxon set), and T is a perfect phylogeny for the MRP matrix (see Section 4.1). Now suppose T' is an optimal solution to MRP. Then T is also a perfect phylogeny for the input matrix, and so is a refined compatibility supertree. To establish that MRP is NP-hard, we reduce from partial binary character compatibility, which is NP-complete (Bodlaender et al., 1992; Steel, 1992). For each partial binary character (i.e., a bipartition $A|B$ on a subset $S_0 \subseteq S$), we define a tree on leafset S_0 and with one edge defining the same bipartition. The profile of trees defined in this way has a compatibility supertree if and only if the set of partial binary characters is compatible. Hence, MRP is NP-hard. □

7.6 Matrix Representation with Likelihood

Matrix Representation with Likelihood (MRL) was introduced in Nguyen et al. (2012). MRL is nearly identical to the MRP problem, but instead of seeking the maximum parsimony tree for the MRP matrix, the objective is a maximum likelihood tree for the MRP matrix, where maximum likelihood is computed under the CFN model. The CFN model is the symmetric binary model of sequence evolution where the state at the root is equiprobable to be 0 or 1; therefore, the choice of 0 or 1 is randomized to ensure that the state at the root is selected randomly. As with the MRP supertree methods, MRL treats ?s as missing data.

Nguyen et al. (2012) compared leading heuristics for MRP and MRL on simulated datasets, and found that MRL supertrees were generally at least as accurate as MRP supertrees. They also saw that MRL scores were more closely correlated with topological accuracy than MRP scores. Hence, MRL is competitive with MRP, and one of the promising approaches to supertree estimation.

7.7 Quartet-based Supertrees

Quartet amalgamation methods construct trees by combining quartet trees, and are useful in many contexts. Examples of quartet amalgamation methods include the Quartet Puzzling algorithm (Strimmer and von Haeseler, 1996), Weight Optimization (Ranwez and Gascuel, 2001), Quartet Joining (Xin et al., 2007), Short Quartet Puzzling (Snir et al., 2008), Quartets MaxCut (Snir and Rao, 2010), and QFM (Reaz et al., 2014). While some of these methods require a quartet tree on every set of four leaves, others (e.g., Quartets MaxCut) can be used on arbitrary inputs of quartet trees. Therefore, one approach to constructing supertrees is to encode each source tree as a set of quartet trees, and then use one of the appropriate quartet amalgamation methods to assemble a tree from the set of quartet trees.

7.7.1 Maximum Quartet Support Supertrees

A natural optimization problem is to find the supertree that agrees with the largest number of quartet trees defined by the source trees (Wilkinson et al., 2005). We define this problem, as follows. Recall that $Q(t)$ denotes the set of homeomorphic quartet trees induced by a tree t, and that this set can contain star trees (quartet trees that do not have any internal edges) if t is unresolved. Suppose t is a source tree and T is a supertree, and so the leafset of t is a subset of the leafset of T. We define the quartet support of t for T to be $|Q(t) \cap Q(T)|$. Then, the quartet support of a profile $\mathscr{T} = \{t_1, t_2, \ldots, t_k\}$ for the tree T is $\sum_{i=1}^{k} |Q(t_i) \cap Q(T)|$. The Maximum Quartet Support Supertree (MQS) T_{MQS} is the tree

7.7 Quartet-based Supertrees

$$T_{MQS} = \arg\max_{T} \sum_{i=1}^{k} |Q(t_i) \cap Q(T)|.$$

This problem can be rephrased in terms of quartet distance, in the obvious way. Define the quartet distance of t to T to be the number of four-taxon subsets of $\mathscr{L}(t)$ on which T and t induce different quartet trees, and denote this by $d(T,t)$. Given a set $\mathscr{T} = \{t_1, t_2, \ldots, t_k\}$ of gene trees and a candidate species tree T, we define $d(T, \mathscr{T}) = \sum_{i=1}^{k} d(T, t_i)$. Then the tree on leafset $\cup_i \mathscr{L}(t_i)$ that has the smallest total quartet distance to the trees in $\mathscr{T}$ is the Quartet Median Tree, i.e.,

$$T_{median} = \arg\min_{T} \{d(T, \mathscr{T})\}.$$

It is easy to see that any MQS is also a Quartet Median Tree.

Finding the MQS is NP-hard, even if there is a quartet tree on every four leaves (Jiang et al., 2001). However, an easy proof when the set of quartet trees can be incomplete (and so not include a tree on every four leaves) is as follows. A compatibility supertree (if it exists) would be an optimal solution to the MQS problem, and so by Theorem 7.6, the MQS problem is NP-hard.

Since finding a maximum number of the possible quartet trees is NP-hard, approximation algorithms have also been developed; for example, Jiang et al. (2001) developed a polynomial time approximation scheme (PTAS) for the case where the set contains a tree on every quartet.

An interesting variant allows the input quartet trees to be equipped with arbitrary positive weights $w(q)$ and seeks the tree T with the maximum total quartet weight (i.e., maximizing $\sum_{q \in Q(T)} w(q)$). Since the unweighted version is NP-hard, the weighted version problem is also NP-hard. Heuristics for maximum weighted quartet support have been proposed, including Weighted Quartets MaxCut (Avni et al., 2015).

7.7.2 Split-Constrained Quartet Support Supertrees

One way of addressing NP-hard problems is to constrain the search space, and then solve the problem exactly within that constrained space. This kind of approach has been used very effectively with quartet-based optimization problems, and is the subject of this section. Specifically, we define the *Split-Constrained Quartet Support Supertree problem* as follows:

- Input: Function w that assigns non-negative weights to all binary quartet trees on a set S, and set X of bipartitions of S.
- Output: Unrooted binary tree T such that $Q(T)$ has maximum total weight among all unrooted binary trees T' that satisfy $C(T') \subseteq X$. In other words, $T = \arg\max_{T'}$

$\sum_{t \in Q(T')} w(t)$, where the max is taken among all unrooted binary trees T' that draw their bipartitions from X.

Note that when X is the set of all possible bipartitions on S, there is no constraint on the set of trees T that can be considered, and so the Split-Constrained Quartet Support Supertree problem is just the Maximum Quartet Compatibility problem. However, for other settings for X, the constraint on the set of possible trees can be very substantial.

The Split-Constrained Quartet Support problem can be solved in time that is polynomial in the number of species, source trees, and size of $|X|$, using dynamic programming (Bryant and Steel, 2001). To do this, we first define a nearly identical problem (the Clade-Constrained Quartet Support problem) where we seek a rooted tree instead of an unrooted tree, and we constrain the set of clades the rooted tree can have.

Because we are constraining the output tree to be binary, we do not need to consider unresolved (i.e., non-binary) quartet trees. Hence, we will say that quartet tree $ab|cd$ supports the unrooted tree T if $ab|cd \in Q(T)$. Then, since every rooted tree T defines an unrooted tree T_u (see Definition 2.12), we will say that a quartet tree $ab|cd$ supports the rooted tree T if $ab|cd$ supports T_u.

The input to the Clade-Constrained Quartet Support problem is a set C of subsets of S, and a non-negative function w on the set of all possible binary quartet trees of S. The objective is a rooted binary tree T such that $Clades(T) \subseteq C$ and T has maximum quartet support among all rooted binary trees that satisfy this constraint. In other words, letting $R_{C,S}$ denote the set of rooted binary trees on taxon set S that draw their clades from C, then the Clade-Constrained Quartet Support tree is

$$T_{CCQS} = \arg \max_{T \in R_{C,S}} \sum_{t \in Q(T)} w(t).$$

We can construct a rooted binary tree T that has the best Clade-Constrained Quartet Support using dynamic programming. After we find a rooted binary tree with the best Clade-Constrained Quartet Support, we will unroot the rooted tree, thus producing an unrooted binary tree with the best quartet support.

To see how this works, let $\mathcal{T}$ be the set of source trees and X the set of allowed bipartitions given as input to the Split-Constrained Quartet Support Supertree problem. To construct C, we include every half of every bipartition in X, and then we add the singletons (i.e., the elements of S) and the full set S. Note that for every bipartition $A|B$ in X, the set C will contain both A and B.

If $R_{C,S} = \emptyset$, then there are no feasible solutions to the Clade-Constrained Quartet Support problem, and also no feasible solutions to the Split-Constrained Quartet Support problem, and we return *Fail*. Otherwise, suppose $T \in R_{C,S}$ and let $t \in Q(T_u)$. Let v be the (unique) *lowest* node v in T (i.e., the node that is furthest from the root of T) for which at least three of t's leaves are below v. We will say that the *quartet tree* t *is mapped to the node* v *with this property*.

7.7 Quartet-based Supertrees

Let v's children be v_1 and v_2. Since $T \in R_{C,S}$, the sets of leaves below v, v_1, and v_2 are all elements of C. Furthermore, since the set A of leaves below v is an element of C, then if v is not the root of T, then the set $S \setminus A$ is also in C. In other words, the node v defines a *tripartition* of the leafset S into three sets of allowed clades, $(A_1, A_2, S \setminus A)$, where A_i is the set of leaves below v_i for $i = 1, 2$. The root also has an associated tripartition, where the third component is the empty set ($\emptyset$). Just as we talked about mapping quartet trees to nodes, we can say that a quartet tree is mapped to the tripartition associated to the node. We will say that a quartet tree is *induced by* the tripartition it is associated to. Furthermore, the quartet tree $ab|cd$ maps to (U, V, W) if and only if the following properties hold:

1. $ab|cd$ is induced by the tripartition (U, V, W).
2. If the set $\{a, b, c, d\}$ does not split 2–2 among U and V, then two of its leaves are in U and one is in V, or vice versa.

Thus, we can determine if $ab|cd$ maps to a given tripartition just by looking at the tripartition, and we do not need to consider the tree as a whole. Therefore, given any tripartition (U, V, W) of S, we can compute the set of quartet trees that map to the tripartition, and hence the total weight of all quartet trees that map to the tripartition. We will denote this by $QS(U, V, W)$. Since every quartet tree that supports a binary rooted tree T is mapped to exactly one node in T, we can write

$$\text{Support}(T) = \sum_{v \in V(T)} QS(U_v, V_v, W_v),$$

where v defines tripartition (U_v, V_v, W_v). We generalize this by letting $score(T, v)$ be the total quartet support at all the nodes in T at or below v. Then

$$score(T, v) = score(T, v_1) + score(T, v_2) + QS(A_1, A_2, A_3),$$

where v_1 and v_2 are the children of v, A_i is the set of leaves below v_i for $i = 1, 2$, and $A_3 = S \setminus (A_1 \cup A_2)$.

We will use these concepts by computing, for every allowed clade A, the best possible quartet support score achievable on any rooted binary tree T on A, which will be the total support contributed by quartet trees that map to nodes in the tree T. Letting $Qscore(A)$ denote this best possible score, we obtain $QScore(A) = 0$ for any clade A where $|A| \leq 2$. Otherwise, we look over all ways of dividing A into two disjoint sets A_1 and A_2 where each A_i is an allowed clade, and we set

$$QScore(A) = QScore(A_1) + QScore(A_2) + QS(A_1, A_2, A_3),$$

where $A_3 = S \setminus A$. This calculation requires that we compute $QScore(A)$ for each clade A in order of increasing size, and that we precompute the $QS(A_1, A_2, A_3)$ values. In other words, we have formulated a dynamic programming solution to the Clade-Constrained Quartet Support problem, which we now present:

Dynamic Programming Algorithm for the Clade-Constrained Support Problem

- Given set X of allowed bipartitions, compute set C of allowed clades, and include the full set S and all the singleton sets.
- Order the set C by cardinality, from smallest to largest, and process them in this order.
- Compute $QS(U,V,W)$ for all tripartitions (U,V,W) where U, V, and W are each non-empty allowed clades, $U \cup V \cup W = S$, and the three sets are pairwise disjoint.
- For clades $A \in C$ where $|A| \leq 2$, set $QScore(A) = 0$. For all larger $A \in C$, compute $QScore(A)$ in order from smallest to largest, setting

$$QScore(A) = $$
$$max\{QScore(A_1) + QScore(A_2) + QS(A_1, A_2, S \setminus A) : A_i \in C, A = A_1 \cup A_2\}.$$

- Return $QScore(S)$, the maximum quartet support of any $T \in R_{C,S}$. To construct the optimal tree, use backtracking through the dynamic programming matrix.

Lemma 7.9 *For all sets C of allowed clades and all $A \in C$, QScore(A) is the maximum number of quartet trees t that can be satisfied by any rooted tree T on leafset A where at least three of the leaves of t are in A.*

Proof The proof follows by induction, after noting that every quartet tree that is induced in a rooted tree T is mapped to exactly one node in T. □

We summarize this discussion with the following theorem.

Theorem 7.10 *(From Bryant and Steel (2001)) The Clade-Constrained Quartet Optimization problem and the Split-Constrained Quartet Optimization problem can be solved exactly in $O(np + n^4|X| + n^2|X|^2)$ time, where the input has p quartet trees on n taxa and X is the set of constraints (clades or bipartitions).*

A variant of this constrained search problem has an input set $\mathcal{T}$ of k source trees (each leaf-labeled by a subset of S), and sets the weight of a quartet tree $ab|cd$ to be the number of these source trees that induce $ab|cd$. If the algorithm from Bryant and Steel (2001) were applied directly, the running time would be $O(n^5 k + n^4|X| + |X|^2)$. However, an optimal solution to the Split-Constrained Quartet Support Supertree problem for this particular variant can be found in $O(nk|X|^2)$ time, where n is the number of species and k is the number of source trees (Mirarab and Warnow, 2015). This variant of the problem is discussed in greater detail in Chapter 10, due to its relevance to constructing species trees from gene trees in the presence of incomplete lineage sorting.

7.8 The Strict Consensus Merger

7.8.1 Overview of the Strict Consensus Merger

The Strict Consensus Merger (SCM) is a supertree method that is used in the supertree meta-method SuperFine (Swenson et al., 2012b), described in the next section, and

7.8 The Strict Consensus Merger

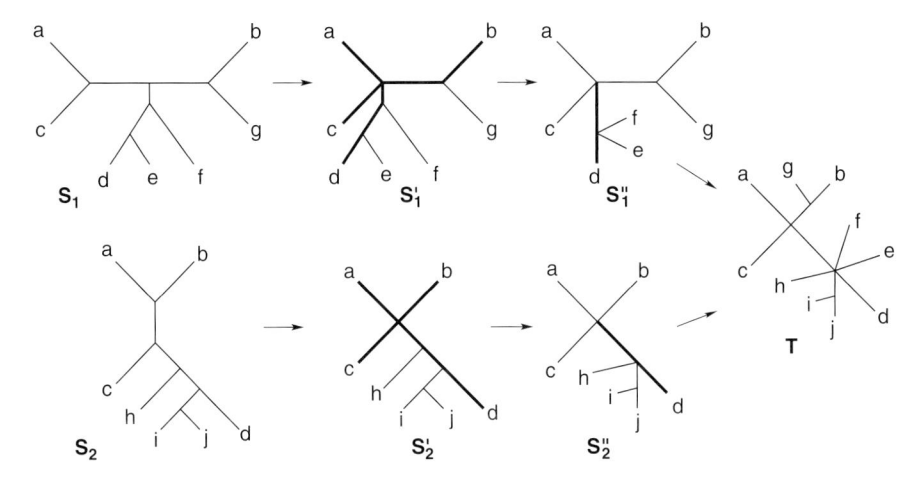

Figure 7.1 (From Swenson et al. (2012b)) Strict consensus Merger (SCM) of two trees S_1 and S_2. Step 1: The two trees both share leafset $B = \{a,b,c,d\}$, but have conflict on this subset. The edges in the two trees that create the conflict are collapsed, producing trees S_1' and S_2'; the backbone (i.e., the strict consensus tree of S_1 and S_2 restricted to B) is shown in bold. Step 2: Both S_1' and S_2' contribute additional taxa to the edge in the backbone leading to the leaf d. The internal edges in the paths corresponding to that edge in S_1' and S_2' are each collapsed, producing trees S_1'' and S_2''; these paths are shown in bold. Finally, the subtrees in S_1'' and S_2'' contributing additional taxa to the edge leading to d are added to the backbone, producing the SCM tree T. Note that the SCM tree is unresolved as the result of both conflict (during Step 1) and a collision (during Step 2).

DACTAL (Nelesen et al., 2012), an iterative divide-and-conquer technique for improving the scalability and accuracy of phylogeny estimation methods, described in Section 11.8. Because of its general utility, we describe the SCM here. For additional details, see Huson et al. (1999a) and Warnow et al. (2001).

The SCM (originally described in Huson et al. (1999a)) is a polynomial time supertree method. The SCM tree is generally highly unresolved, and can in some cases have large polytomies. Hence, unlike most supertree methods, the SCM tree is designed for use as a constraint tree that will need to be subsequently refined into a binary tree.

7.8.2 Computing the Strict Consensus Merger of Two Trees

We begin by describing how the SCM of two trees is performed; see Figure 7.1. The input is two source trees, S_1 and S_2.

- Step 1: We compute B, the common leafset of S_1 and S_2. We then compute the strict consensus tree of $S_1|B$ and $S_2|B$, and refer to this as the "backbone." We modify S_1 and S_2, if necessary, to make them induce the backbone on B; thus, any conflict between $S_1|B$ and $S_2|B$ results in collapsing edges in S_1 and S_2, and introduces at least one polytomy in the SCM tree. After this step completes, we refer to the pair of source trees as S_1' and S_2'.

- Step 2: The taxa that are not in the backbone are now added to it. Each edge in the backbone corresponds to an edge or a path of more than one edge in each of S'_1 and S'_2. We say that S'_i *contributes additional taxa to an edge e in the backbone* if and only if e corresponds to a path of more than one edge in S'_i. When only one of the two trees S'_1 and S'_2 contribute additional taxa to the edge, then those taxa are added directly, maintaining the topological information in the source tree. However, if both trees contribute taxa to the same edge, then this is a "collision," and the additional taxa are added in a more complicated fashion, as we now describe. In the presence of a collision involving edge e, the paths corresponding to edge e are collapsed to a path with two edges in both S'_1 and S'_2, producing trees S''_1 and S''_2. Each S''_i has a single rooted subtree to contribute to the backbone at the edge e. The edge e is subdivided by the addition of a new vertex v_e, and the two rooted subtrees (one from each of S''_1 and S''_2) are attached to vertex v_e, by identifying their roots with v_e. Thus, each collision creates an additional polytomy in the SCM tree.

7.8.3 Computing the Strict Consensus Merger of a Set of Trees

To compute the SCM of a set of trees, an ordering on the set of pairwise mergers must be defined. Different proposals have been made for how to define the ordering, with the objective being that the ordering should maintain as much resolution as possible; thus, collisions should be avoided. One approach that has been used in the SuperFine (Swenson et al., 2012b) supertree method (which uses the SCM in its protocol) is to pick the pair of trees to merge that maximizes the number of shared taxa; however, other approaches can be included.

Given an ordering of pairwise mergers, the SCM of a set of trees proceeds in the obvious way. After each pairwise merger is computed, the set of subset trees is reduced by one, and the process repeats. At the end, a tree is returned that has all the taxa; this is the SCM tree. Hence, there can be more than one SCM tree, depending on the ordering. It is easy to see that the SCM of two trees can be computed in polynomial time, and so the SCM of a set of trees is also polynomial time. We summarize this as follows:

Theorem 7.11 *Let $\mathcal{T}$ be a profile of k trees on set S of n taxa. An SCM of $\mathcal{T}$ can be computed in time that is polynomial in k and n.*

7.8.4 Theoretical Properties of the SCM Tree

If the set of trees is compatible, then there will never be a conflict in any pairwise merger; however, it is still possible for a pair of compatible trees to have a collision, which will result in the SCM tree having at least one polytomy. In other words, the SCM of a set of compatible binary trees may not be fully resolved. In particular, the SCM tree is not guaranteed to solve the Unrooted Tree Compatibility problem. This is not surprising, since the

Unrooted Tree Compatibility problem is NP-hard (Theorem 3.11) and hence a polynomial time algorithm cannot be expected to solve it.

On the other hand, if the source trees are compatible and satisfy some additional constraints, then there is at least one ordering (that can be computed from the data) of the source trees that will produce the unique compatibility supertree (see Theorem 11.8). Thus, the SCM method solves the Unrooted Tree Compatibility problem under some conditions.

By design the SCM is very conservative, and so will collapse edges in the presence of any conflict or collision. As a result, the SCM tends to be highly unresolved. Indeed, the topological error in an SCM tree is high – but the error is in the form of false negatives (i.e., missing branches) instead of false positives.

7.9 SuperFine: A Meta-Method to Improve Supertree Methods

7.9.1 Overview of SuperFine

SuperFine (Swenson et al., 2012b) is a general two-step technique that can be used with any supertree method; see Figure 7.2.

- Step 1: The SCM tree is computed on the profile of source trees.
- Step 2: The SCM tree is resolved into a binary tree using the selected supertree method; see Figure 7.3.

Thus, SuperFine is a meta-method that is designed to work with a selected supertree method, and we refer to SuperFine run with supertree method M by "SuperFine+M." SuperFine has been tested with MRP, MRL, and QMC (Quartets MaxCut) (Swenson et al., 2011, 2012b; Nguyen et al., 2012), and shown to improve the accuracy and scalability of these base supertree methods.

In Section 7.8, we described the SCM and how the SCM tree is computed. We also noted that the SCM tree is typically highly unresolved as a result of collisions (where two source trees can be combined in more than one way) and conflicts (where two source trees are incompatible). Here we show how the SCM tree is refined into a binary tree, one polytomy at a time, using SuperFine.

7.9.2 SuperFine: Refining Each Polytomy in the SCM Tree

The key technique in SuperFine is to recode the topological information in each source tree using a new but smaller taxon set, and then SuperFine runs its selected supertree method on these recoded source trees. The outcome of this process is a supertree on the new taxon set, which is then used to refine the polytomy. This refinement step is applied to each polytomy in turn.

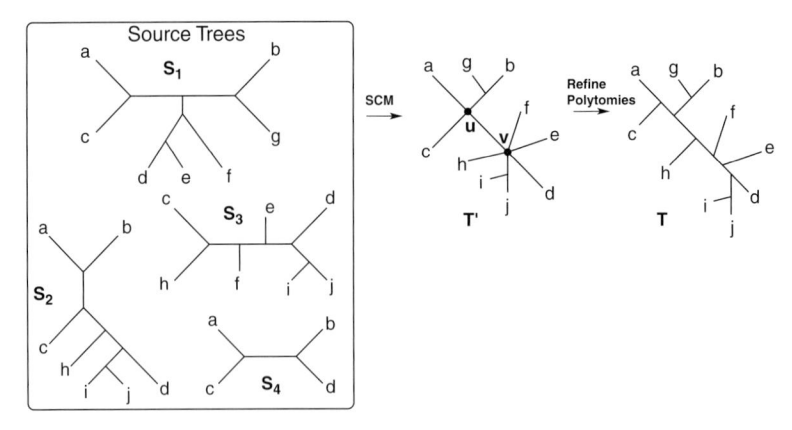

Figure 7.2 (From Swenson et al. (2012b)) Schematic representation of the algorithmic strategy of SuperFine+MRP. Source trees S_1–S_4 are combined pairwise to produce a Strict Consensus Merger (SCM) tree; this process collapses edges, and will only retain internal branches that are compatible with all of the source trees. However, note that it may fail to retain some branches that are compatible with the source trees, as the example shows (for example, *abcefgh|dij* is not in the SCM tree but is compatible with all the source trees). Each polytomy in the SCM tree is then refined by running a preferred supertree method, such as MRP or MRL, on modified source trees.

Here we describe how this recoding is done. Let v be a polytomy in the SCM tree with degree $d > 3$. First, we relabel every leaf in the SCM tree using labels $\{1, 2, \ldots, d\}$, according to which of the d subtrees off the vertex v it belongs to. This produces a relabeling of the leaves of every source tree in $\mathscr{T}$ with $\{1, 2, \ldots, d\}$. Now, if two sibling leaves x, y in a source tree $t \in \mathscr{T}$ have the same label L, we label their common neighbor by L and remove both x and y from t. We repeat this process until no two sibling leaves have the same label. As proven in Swenson et al. (2012b), when this process terminates, the source tree t will have at most one leaf of each label. We refer to this process as "recoding the source trees for the polytomy v," and the modified source trees are referred to as "recoded source trees." Note that each recoded source tree has at most d leaves. Thus, when the polytomy degree is relatively small (compared to the original dataset), this produces recoded source trees that are much smaller than the original source trees.

To refine the SCM tree at the polytomy v, we apply the selected supertree method to the recoded source trees, and obtain a tree $T(v)$ on leafset $\{1, 2, \ldots, d\}$. We then use $T(v)$ to refine the SCM tree at node v. Note that when $d = deg(v)$ is sufficiently small, this step can be quite fast. Furthermore, the refinements around the different polytomies can be performed in parallel (or sequentially, but in any order that is desired), since the different refinements do not impact each other. See Figure 7.3.

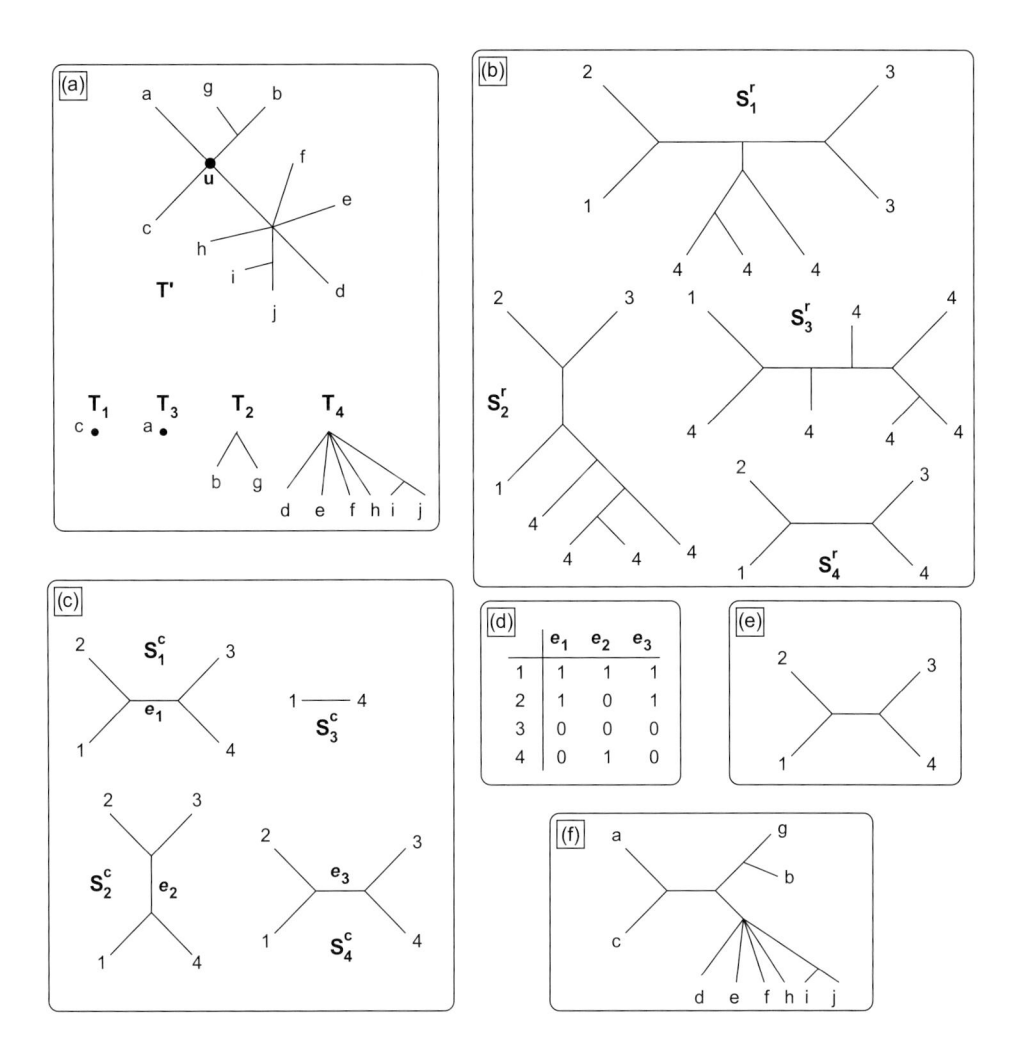

7.9 SuperFine: A Meta-Method to Improve Supertree Methods

Figure 7.3 (From Swenson et al. (2012b)) Schematic representation of the second step of the algorithmic strategy of SuperFine+MRP, in which we refine the SCM tree produced in the first step. The steps here refer to the SCM tree T', polytomy u, and source trees shown in Figure 7.2. (a) The deletion of the polytomy u from the tree T' partitions T' into four rooted trees, T_1, T_2, T_3, and T_4. (b) The leaves in each of the four source trees are relabeled by the index of the tree T_i containing that leaf, producing relabeled source trees S_1^r, S_2^r, S_3^r, and S_4^r. For example, the relabeled version of $S_4 = ac|bd$ is $S_4^r = 12|34$. (c) Each S_i^r is further processed by repeatedly replacing sibling nodes with the same label, until no two siblings have the same label; this results in trees S_1^c, S_2^c, S_3^c, and S_4^c. (d) The MRP matrix is shown for the four source trees, including only the parsimony informative sites; thus, S_3^c does not contribute a parsimony informative site and is excluded. (e) The result of the MRP analysis on the matrix given in (d). (f) The tree resulting from identifying the root of each $T_i, i = 1, 2, 3, 4$, with the node i in the tree from (e).

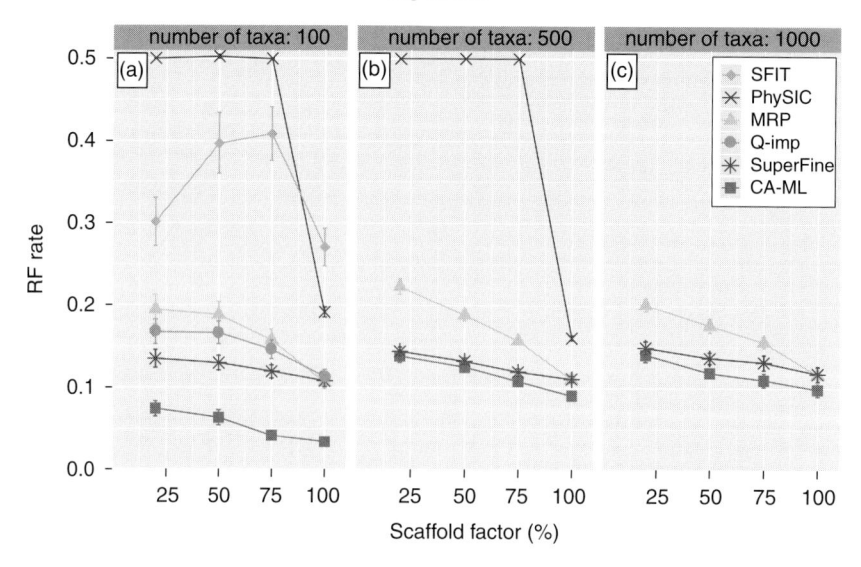

Figure 7.4 (From Swenson et al. (2012b)) Robinson–Foulds (RF) error rates (mean with standard error bars) for supertree methods SFIT, PhySIC, MRP, Q-imputation, SuperFine+MRP, and concatenation using maximum likelihood (CA-ML) for three different taxon sizes, as a function of the scaffold factor. (a) shows results for 100 taxa, (b) shows results for 500 taxa, and (c) shows results for 1000 taxa. Methods not shown in (b) and (c) were unable to complete on these datasets within the allowed timeframe.

7.9.3 Performance in Practice

As shown in Swenson et al. (2011, 2012b) and Nguyen et al. (2012), SuperFine is generally very fast even when run sequentially, and efficient parallel implementations have also been developed (Neves et al., 2012; Neves and Sobral, 2017). Figure 7.4 (from Swenson et al., 2012b) shows a comparison of several supertree methods on estimated source trees, including MRP (the most well-known supertree method), SuperFine using heuristics for MRP as the base supertree method, and concatenation using maximum likelihood (CA-ML). The x-axis shows the "scaffold factor," which is the percentage of the leafset that is in the randomly selected "backbone" source tree. Most methods improve in accuracy as the scaffold factor increases. While CA-ML has the best accuracy (i.e., lowest RF topological error rate), the supertree method with the best accuracy is SuperFine+MRP. Furthermore, although all methods could complete on the 100-taxon datasets, fewer could complete on the 500-taxon datasets, and only CA-ML, SuperFine+MRP, and MRP could complete on the 1000-taxon datasets. A comparison of the running times of these methods is shown in Figure 7.5. Note that the difference in running times is very large on the 500- and 1000-taxon datasets, where CA-ML takes more than 1440 minutes, MRP takes 180–240 minutes, and SuperFine takes 90–180 minutes. Studies evaluating SuperFine with heuristics for MRL (Nguyen et al., 2012) and with QMC (Snir and Rao, 2010) have also shown similar

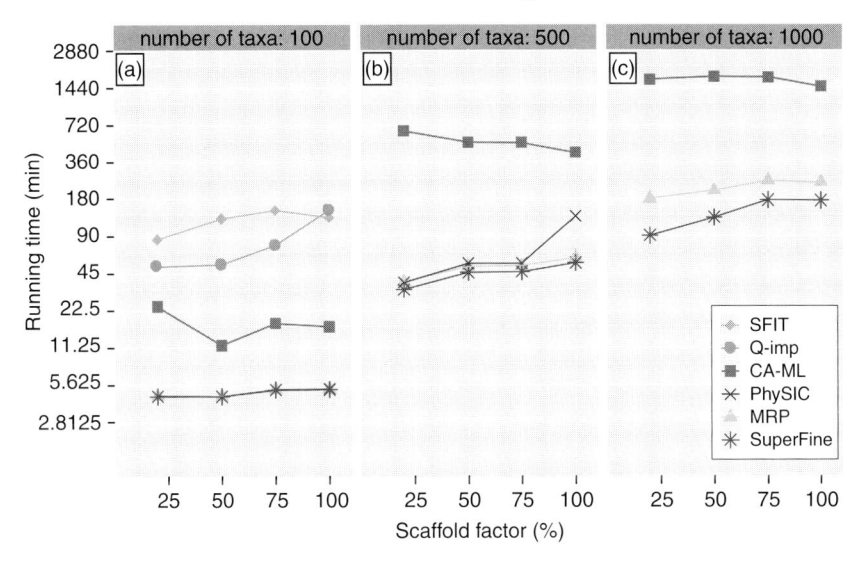

Figure 7.5 (From Swenson et al. (2012b)) Average running times (logarithmic scale), including the time needed to compute source trees, for supertree methods SFIT, Q-imputation, CA-ML, PhySIC, MRP, and SuperFine+MRP for three different taxon sizes, as a function of the scaffold factor. For the supertree methods, running times shown include the time required to generate maximum likelihood source trees using RAxML (Stamatakis, 2006). (a) shows results for 100 taxa, (b) shows results for 500 taxa, and (c) shows results for 1000 taxa. The curves for PhySIC, MRP, and SuperFine+MRP overlap for the 100-taxon datasets. Methods not shown in (b) and (c) were unable to complete on these datasets within the allowed timeframe.

results (Swenson et al., 2011; Nguyen et al., 2012). Thus, SuperFine is a meta-method that uses divide-and-conquer to improve the accuracy and speed of supertree methods. The best supertree analyses come close to the accuracy of a concatenation analysis, and can be much faster.

7.10 Further Reading

Quartet-based supertree methods: Since unrooted trees can be encoded as quartet trees, quartet-based tree construction methods can be used as supertree methods. For example, the QMC method (Snir and Rao, 2010) has been studied as a supertree method in Swenson et al. (2011), and shown to have good accuracy.

Supertrees for rooted source trees: Optimization problems for supertree construction when the input is a set of rooted trees have also been explored. One such approach encodes each rooted source tree as a set of rooted triplet trees (i.e., three-leaf trees), and then seeks a rooted tree that maximizes the number of triplet trees (from the rooted source trees) that agree with it. This is called the **Maximum Triplet Support** problem. Equivalently, this can be formulated as finding a median tree with respect to the triplet distance (Ranwez et al., 2010).

The Maximum Triplet Support problem is NP-hard (Bryant, 1997), and so heuristics have been developed to find good solutions to the problem. The most well known of these heuristics is MinCutSupertree (Semple and Steel, 2000), which the authors describe as "a recursively optimal modification of the algorithm described by Aho et al." In other words, MinCutSupertree is a modification of the ASSU algorithm described in Section 2.9 so that it can be run on incompatible source trees. A modification to the MinCutSupertree method was developed in Snir and Rao (2006), and found to be more accurate than MinCutSupertree. More recent heuristics for this problem include the SuperTriplets method (Ranwez et al., 2010) and Triplets MaxCut (Sevillya et al., 2016).

Distance-based supertree estimation: Another class of supertree methods takes advantage of the fact that the source trees for the supertree problem typically define distance matrices of some type, so that supertree estimation methods can use the distance matrices to infer the final supertree. For example, estimated phylogenetic trees are often based on maximum likelihood, and so come with branch lengths that reflect the expected number of changes on the branch; thus, a maximum likelihood tree defines an additive matrix for its taxon set. The same is true for trees computed using distance-based methods such as neighbor joining or FastME, when the distances are based on statistical sequence evolution models. Even trees computed using maximum parsimony come with branch lengths, although these cannot be interpreted in quite the same way. Thus, all these methods can be used to produce additive matrices, and in some cases these matrices are even ultrametric.

As Willson (2004) observed, the use of these "distance matrices" is a valuable source of information in supertree construction, since they naturally enable the estimation of branch lengths in the supertree, something that is not possible when just using the source trees alone.

Here we describe a classical distance-based supertree problem, **Matrix Representation with Distances** (MRD) from Lapointe and Cucumel (1997). The input is a set of k dissimilarity matrices, each on a subset of the taxon set S. We seek an additive matrix on the full species set that *minimizes the total distance to the input matrices*; in other words, we seek a median tree with respect to some way of measuring distances between matrices. Common ways of measuring distances use the L_∞, L_1, and L_2 norms, but other norms can also be used. Note also that we require that the output be an additive matrix, so that we can use it to define the supertree topology and branch lengths. If we wish to compute a rooted supertree with branch lengths reflecting elapsed time, then we will require that the output be an ultrametric matrix.

Distance-based supertree estimation is not new (e.g., Lapointe and Cucumel, 1997; Lapointe et al., 2003; Willson, 2004; Criscuolo et al., 2006), but none of the current methods has become widely used. However, Build-with-Distances (Willson, 2004) came close to MRP and in some cases was more accurate (Brinkmeyer et al., 2011), leading the authors to conclude that distance-based supertree estimation might have the potential to replace MRP.

As discussed in Chapter 5, there is a substantial literature on related problems, where the input is a single dissimilarity matrix M and the objective is an additive matrix or an ultrametric matrix that is optimally close to *M*, under some metric between distance matrices. Some of these problems are solvable in polynomial time, others are NP-hard but can be approximated, and some are hard to approximate; see Ailon and Charikar (2005); Fakcharoenphol et al. (2003); Agarwala et al. (1998) for an entry to this literature.

Statistical properties of supertree methods: The supertree methods presented so far are heuristics for NP-hard optimization problems. As such, understanding their statistical properties is quite challenging, since characterizing the conditions under which the heuristics are guaranteed to find globally optimal solutions to their criteria is difficult. However, suppose that each of the optimization problems could be solved exactly – i.e., suppose that globally optimal solutions could be found. Could we say anything about the probability of recovering the true supertree? To answer this question, we pose this as a statistical estimation problem in which the (unknown) true supertree is used to generate a sequence of source trees under some random process.

Suppose that the source trees are on subsets of the full taxon set, and are generated by a random process defined by a model species tree on the full set of taxa. For example, the model could assume that a source tree is the tree induced by the model species tree on a random subset of the taxon set. Under this model of source tree generation, all the source trees are compatible, and the true species tree is a compatibility supertree. Since the random process generates all subtrees with non-zero probability, the model species tree is identifiable (i.e., the model species tree has non-zero probability, and no other tree on the full taxon set has non-zero probability). Furthermore, any method that is guaranteed to return a compatibility supertree for the input set is statistically consistent under this model. Thus, exact solutions for many supertree optimization problems (e.g., MRP, RFS, and MQS) will be statistically consistent methods for species tree estimation under this model.

When the source trees are estimated species trees, then species tree estimation error is part of the generative model. Similarly, when the source trees are gene trees, then gene tree heterogeneity due to biological factors such as incomplete lineage sorting and gene duplication and loss is also part of the generative model. Finding a supertree that maximizes the probability of generating the observed source trees is the **Maximum Likelihood Supertree problem**, proposed in Steel and Rodrigo (2008). Several variants of the generative model have been proposed, and maximum likelihood and Bayesian methods have been developed under these models (Ronquist et al., 2004; Steel and Rodrigo, 2008; Bryant and Steel, 2009; Cotton and Wilkinson, 2009; De Oliveira Martins et al., 2016).

Under some conditions, the RFS problem will provide a good solution to the maximum likelihood supertree under one of the exponential models described in Steel and Rodrigo (2008) (see discussion in Bryant and Steel (2009)). Therefore, heuristics for the RFS discussed in Section 7.4.2 (e.g., PluMiST (Kupczok, 2011), RFS (Bansal et al., 2010), MulRF (Chaudhary, 2015), and FastRFS (Vachaspati and Warnow, 2016)) can be used as heuristics for the Maximum Likelihood Supertree problem.

7.11 Review Questions

1. Define the MRP problem (what is the input and what is the output?).
2. Explain how to write down the MRP matrix.
3. What is the difference between the MRP and MRL optimization problems?
4. Suppose you have a set of 100 binary trees, each tree has ten species, and the total number of species is 500. How many rows and columns are in the MRP matrix?
5. Define the Compatibility Supertree problem (what is the input and what is the output?).
6. Suppose that $\mathscr{T}$ is a set of source trees and it has a compatibility supertree T. Now suppose that T' is the solution to the Robinson–Foulds Supertree problem given input $\mathscr{T}$. Must T' also be a compatibility supertree for $\mathscr{T}$?
7. Define the Split-Constrained Quartet Support Supertree problem (what is the input and what is the output?).
8. Let $T = ((a,f),(b,(c,(d,e))))$ and $T' = (g,(a,(c,(b,(h,d)))))$. Let T^* be the SCM of T and T', and let T^{**} be the homeomorphic subtree of T^* restricted to the leafset of T. Report the RF distance between T^{**} and T.

7.12 Homework Problems

1. Suppose X is a set of compatible unrooted trees on different sets of leaves. What can you say about the solution space to MRP on input X?
2. Suppose you have 1000 trees, each with 100 leaves, and 5000 taxa overall. How big is the MRP matrix?
3. Let T_1 and T_2 be binary trees with leafset S, and let $\mathscr{T}$ be a set of source trees with leaves drawn from S, with $S = \cup_{t \in \mathscr{T}} C(t)$. Let $X = C(T_1) \cup C(T_2)$. Let T be an optimal solution to the Split-Constrained Quartet Support Supertree problem given the set $\mathscr{T}$ of source trees and constraint set X. Must T be one of T_1 or T_2? If so prove it, and otherwise give a counterexample.
4. Prove Lemma 7.2.
5. Under the assumption that the input trees are fully resolved (i.e., binary), is it always the case that every optimal solution to the Maximum Quartet Support Supertree problem is fully resolved?
6. Prove that the Quartet Median Tree and the Maximum Quartet Support Supertree are the same.
7. Recall that the Split-Constrained Quartet Support Supertree problem requires that the output tree be binary. Consider the relaxed version of this problem where the supertree is not required to be binary, and in fact can have unbounded degree. Would the problem still be solvable in polynomial time?

Part II

Molecular Phylogenetics

8
Statistical Gene Tree Estimation Methods

8.1 Introduction to Statistical Estimation in Phylogenetics

Phylogeny estimation from biomolecular sequences is often posed as a statistical inference problem where the sequences evolve down a tree via a stochastic process. Statistical estimation methods take advantage of what is known (or hypothesized) about that stochastic process in order to produce an estimate of the evolutionary history. When we consider phylogeny reconstruction methods as statistical estimation methods, many statistical performance issues arise. For example: Is the method guaranteed to construct the true tree (with high probability) if there are enough data (i.e., is the method statistically consistent under the model)? How much data does the method need to obtain the true tree with high probability (i.e., what is the sample complexity of the method)? Is the method still relatively accurate if the assumptions of the model do not apply to the data that are used to estimate the tree (i.e., is the method robust to model misspecification)?

Markov models of evolution form the basis of most computational methods of analysis used in phylogenetics. The simplest of these models are for characters with two states, reflecting the presence or absence of a trait. However, the most common models are for nucleotide (four-state characters) or amino acid (20-state characters) data. They can also be used (although less commonly) for codon data, in which case they have 64 states (Goldman and Yang, 1994; Yang et al., 2000; Kosiol et al., 2007; Anisimova and Kosiol, 2009; De Maio et al., 2013).

In Chapter 1, we described the Cavender–Farris–Neyman (CFN) model of binary sequence evolution, and a simple method to estimate the CFN tree from binary sequences. In this chapter we focus on sequence evolution models that are applicable to molecular sequence evolution. As we will see, the mathematical theorems and algorithmic approaches are very similar to those developed for phylogeny estimation under the CFN model.

Statistical identifiability is an important concept related to Markov models. We say that a parameter (such as the tree topology) of the Markov model is *identifiable* if the probability distribution of the patterns of states at the leaves of the tree are always different for two Markov models that are different for that parameter. Thus, some parameters of a model may be identifiable while others may not be. For example, the unrooted tree topology is identifiable under the CFN model, but the location of the root is not.

Statistical consistency is another important concept, but is a property of a method rather than of the model. That is, a method is statistically consistent under a model if the method converges to the correct answer as the amount of data increases. Thus, if a method is statistically consistent under a model, then the model is identifiable. Conversely, if the model is not identifiable, then no method can be statistically consistent under the model. However, if a (supposedly) good method is proven to be inconsistent under a model, it does not follow that the model is not identifiable. Furthermore, it is possible for a model to be identifiable, but for no statistically consistent method to exist (Devroye and Győrfi, 1990)!

Note that statistical consistency is a question about the probability that the method will return the true tree as the amount of data increases. Hence, a method can be statistically consistent and still do poorly in simulations – because the simulation may not consider performance with large enough amounts of data.

Statistical consistency or inconsistency of a method, and similarly identifiability or non-identifiability, cannot be established using a simulation; instead a formal proof is required. As we describe these models, we will address whether the model parameters (tree topology and possibly numeric parameters) are statistically identifiable, and for those parameters that are identifiable, we will ask which methods are statistically consistent techniques for estimating these parameters.

8.2 Models of Site Evolution

8.2.1 Nucleotide Site Evolution Models

We will describe site evolution in general terms, allowing for any number of character states: two for presence/absence characters, four for nucleotides, and 20 for amino acids. Also we refer to "sites" instead of "characters," to emphasize that these are generally for positions within a multiple sequence alignment.

Markov models of site evolution used in phylogenetics are stochastic evolutionary processes operating on a rooted binary tree T. The state at the root is drawn from a distribution given by π (so that π_x is the probability of state x, and is assumed to be strictly positive for all states), and then the site evolves down each edge incident with the root (i.e., (r,a) and (r,b), where a and b are the two children of r). This process produces states at a and b, and the process continues on the edges below a and b. Eventually, this process produces states at every node in the tree. (Note the description was in general terms, but for nucleotide evolution the states would be one of the four possible nucleotides.)

To complete this description, we need to show what happens on each edge. The process on each edge is described by a continuous-time Markov chain that is defined by a transition matrix **P**. Furthermore, **P** is parameterized by branch length t, so that $P(t)_{xy}$ is the probability of moving from character state x to character state y given branch length t, and $P(t)_{xx}$ is the probability of staying in state x for branch length t. (We use branch length instead of time, since evolutionary change is not always proportional to time.) Equivalently, the random process operating on edge *e* can be described by a substitution probability matrix $M(e)$, which is the transition matrix for edge *e* evaluated at branch length *length(e)*.

This process repeats for each site, and each site evolves independently of the other sites, and under exactly the same process. This is expressed by saying the sites evolve under independent and identically distributed (*i.i.d.*) processes. When site evolution is independent, then different sites can have different states at the nodes of the tree.

This algorithmic description of the evolutionary process implies the **Markov property** of site evolution, because the probability of seeing a particular state at the head of an edge *e* is determined by the state at the tail of the edge and the substitution probability matrix $M(e)$ (equivalently, by the state at the tail of the edge, the branch length, and the transition matrix), and not by anything else.

The importance of parameterizing the evolutionary process on each edge by branch length is that it enables us to rewrite the transition matrix **P** in terms of a rate matrix **Q**, so that **P** is related to **Q** by matrix exponentiation:

$$P(t) = e^{Qt} = \sum_{n=0}^{\infty} Q^n \frac{t^n}{n!}.$$

Nearly all Markov models of evolution in use in phylogenetics assume that there is a *single rate matrix that governs all the edges in the tree*. As a result, these Markov models are **stationary**, which means that the transition rate matrices do not depend on time. Standard Markov models are also designed to be **time-reversible**, so that the probability distribution of site patterns (where a **site pattern** is one assignment of states that appear at the leaves of the tree) is unchanged by changing the location of the root.

As we will see, the unrooted tree topology for standard Markov models is identifiable, and maximum likelihood estimation is a statistically consistent estimation of the unrooted tree topology. The assumption of a single rate matrix has the benefit of reducing the number of parameters that have to be estimated in a maximum likelihood analysis, since instead of estimating the branch lengths and a 4×4 rate matrix for every edge in the tree, we only need to estimate one 4×4 rate matrix and the branch lengths in the tree. In other words, we reduce the total number of parameters we need to estimate quite dramatically.

The different models are thus characterized by how they constrain the branch lengths, the rate matrices, and the probability distribution π of states at the root. Many (but not all) models allow the distribution π of states at the root to be fairly arbitrary, and only require that $\pi_x > 0$ for all states x (i.e., they require that all character states have positive probability at the root). Similarly, branch lengths are normally constrained to be strictly positive, but sometimes zero-length branches are allowed. The main distinction between the models, therefore, is how they constrain the rate matrices.

The most constrained of the time-reversible stationary four-state models is the **Jukes–Cantor model**, presented in Jukes and Cantor (1997). Under this model (referred to as the **JC69** model), all nucleotides are equiprobable at the root, and if there is a nucleotide substitution on an edge, then all changes are equiprobable. In other words, the JC69 model is the DNA version of the CFN model. We could therefore represent a JC69 model tree either with a substitution probability $p(e)$ for edge e or with a branch length $\lambda(e)$ on edge e indicating the expected number of changes on edge e, just as we did for the CFN model. In that formulation, $\lambda(e) = -\frac{3}{4} \ln\left(1 - \frac{4}{3}p(e)\right)$. However, we can also write this in terms of a rate matrix and branch lengths (using the same lengths as given above), where $Q_{xy} = Q_{uv}$ for $x \neq y$ and $u \neq v$. We also need to specify the distribution of states at the root, which is given by $\pi_x = 1/4$ for all nucleotides x. Thus, the JC69 model is an example of a model that can be expressed in terms of a common rate matrix across the tree.

The **Generalised Time Reversible** (i.e., **GTR**) model (Tavaré, 1986) makes the fewest constraints on the rate matrix of all the time-reversible stationary models, and is the most commonly used model for phylogenetic inference on DNA sequences. Intermediate models, some of which are shown in Figure 8.1, can be obtained by relaxing the constraint given in the JC69 model in various ways. The models in the figure are all identifiable, and estimation under these models is generally computationally feasible. These are all examples of standard DNA site evolution models, with GTR the most complex of the standard models; see Hillis et al. (1996), Li (1997), and Yang (2014) for more information.

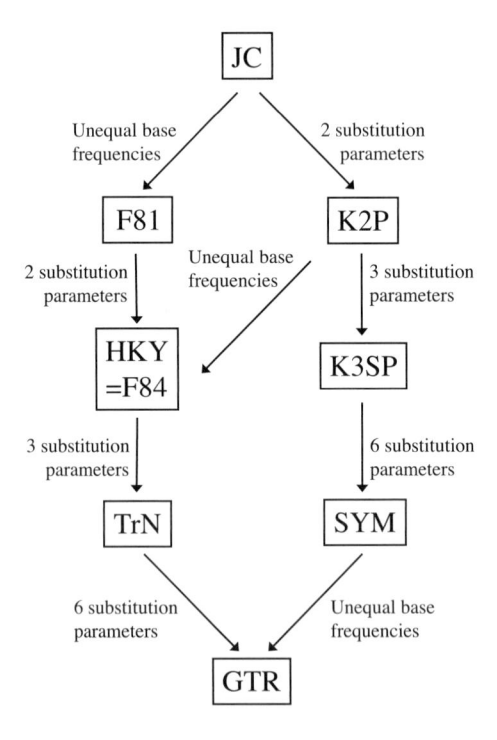

Figure 8.1 (Figure 3.9 in Huson et al. (2010)) The hierarchical relationships between standard DNA site evolution models. A directed edge from one model to another indicates that the first model is a submodel of the second. Thus, the Jukes–Cantor model at the top is the most simplified, and the Generalised Time Reversible (GTR) model at the bottom is the most complex.

In all of these models, phylogeny estimation operates as follows. Given the observed sequences, the distribution π of states at the root is estimated by the frequency of each state in the observed sequences. Then, during the phylogenetic analysis, the branch lengths (and sometimes also the rate matrices) are estimated using various techniques. The result is an unrooted tree T, typically with branch lengths and a 4×4 rate matrix. The rate matrices vary in terms of the number of free parameters, so that once the branch lengths are fixed the estimation of the rate matrix can range from being easy (i.e., for the JC69 model) to challenging (e.g., for the GTR model).

Yet, the assumption that there is one common rate matrix that governs all the sites and the entire tree is not always realistic, because the relative probability of the substitutions can depend on evolutionary pressures that can change over the evolutionary tree, as well as along the sequence. More general models have been proposed in order to relax the assumption of a common rate matrix operating across the tree. Some of these generalizations have resulted in non-identifiable models, so that estimation under these more general models can be unappealing. The **No Common Mechanism** model (Tuffley and Steel, 1997), described in Section 8.13, is an example of such a model.

However, some more general sequence evolution models *are* identifiable, and the **General Markov model** (Steel, 1994b) is a primary example of such a model. The General Markov (GM) model makes very few constraints on the evolutionary process. The only constraint on the distribution of states at the root is that $\pi_x > 0$ for all states. For every edge e there is a 4×4 substitution probability matrix $M(e)$ where $M(e)_{xy}$ is the probability of changing from character state x to state y on edge e, and the only constraint on these matrices is that $det(M(e)) \neq 1, -1, 0$ for all edges e. Therefore, if we rewrite the matrices in terms of branch lengths and rate matrices, the rate matrices on the different edges may be arbitrarily different from each other. One of the consequences of this is that the GM model is not necessarily stationary.

It is easy to see that every GTR model tree is a GM model tree, so that the GTR model is a special case of the GM model. Since the GTR model is the most general of the time reversible stationary models, this implies that the GM model contains all these (standard) site evolution models as special cases. Also, as we will see, the unrooted GM model tree topology is identifiable, and the same is true for all its submodels.

8.2.2 DNA Sequence Evolution Models

The stochastic models we have seen so far have addressed the evolution of a single site; to enable the analysis of a collection of sites, we assumed that all the sites evolve independently down the same tree under the same process. Yet, it is clear that some sites evolve quickly and others evolve slowly, and so there is rate variation across sites.

For those models that are based on rate matrices (i.e., the GTR model and its submodels), the assumption that there is one common rate matrix that governs the tree makes it possible

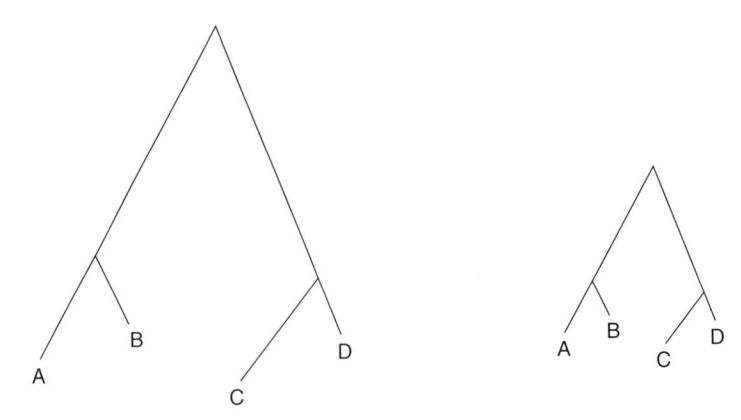

Figure 8.2 Two model trees that are scaled versions of each other. A model in which each site picks one of these trees with some probability is a rates-across-sites model of site variation.

to include rate variation across sites (i.e., where some sites evolve faster or slower than others) by allowing each site to pick a rate from a distribution. For example, consider a JC69 model tree, and let $\lambda(e)$ denote the expected number of changes of a random site on edge e. Under the assumption that the distribution of rates has mean 1, then a random site will still have $\lambda(e)$ expected number of changes on edge e. Now suppose the rate drawn by site i is 2; then the expected number of changes of site i on edge e will be $2\lambda(e)$. In other words, site i will evolve twice as quickly as a random site. Therefore, under a **rates-across-sites** model, if site i is twice as fast as site j on edge e, then site i is twice as fast as site j on every edge! Figure 8.2 shows an example of rates-across-sites by presenting the model trees for two sites; note that the branch lengths remain proportionally the same, and that the tree on the left is a scaled-up version of the tree on the right. Note that under the rates-across-sites assumption, the sites still evolve under an *i.i.d.* model. Finally, the most commonly used rates-across-sites model is the gamma distribution, which means that each site draws its rate from a common gamma distribution.

8.2.3 Amino Acid Site Evolution Models

Amino acid sequence evolution is also modeled similarly, but with 20×20 rate matrices instead of 4×4 matrices. Unlike the nucleotide case, however, the 20×20 rate matrices are typically not estimated from the data, and instead are pre-computed based on biological datasets; in other words, the usual amino acid rate matrices have no free parameters. The distribution π of states at the root is sometimes fixed but can be estimated from data.

The earliest amino acid models were developed in Dayhoff et al. (1978), and were based on empirical frequencies observed in a relatively small database of aligned amino acid sequences. More recent amino acid models, such as the JTT (Jones et al., 1992) and WAG (Whelan and Goldman, 2001) models, have been developed using much larger databases of aligned amino acid sequences.

8.3 Model Selection

In order to construct a phylogeny on a given set of aligned sequences using a model-based approach, the statistical model underlying the dataset must be selected. For example, the tree that is estimated using GTR+Gamma+I (whereby some proportion of sites are invariable, and the others draw their rates from a gamma distribution) could be different from the tree that is estimated under the JC69 model without any site variation. While GTR+Gamma+I is typically the most general model that is considered for phylogeny estimation from gene trees, it is not always the case that accuracy is maximized by analyzing datasets under this model; sometimes a more restricted model might lead to a more accurate estimate of the gene tree topology (Hoff et al., 2016). Hence, selecting the best sequence evolution model is one of the important steps in a phylogenetic analysis.

The usual approach is to take the estimated alignment and compute a tree on the alignment; then, for this tree and alignment, a pair of statistical models (e.g., K2P vs. HKY, or GTR vs. HKY+Gamma) is compared for their fit to the data. Some of these comparisons will involve nested models, but others may not. In particular, standard nucleotide site evolution models (as shown in Figure 8.1) are nested, but the protein site evolution models are not.

The likelihood ratio test (LRT) can be used to compare models that are nested, and this was the initial approach used for model selection in phylogenetics (Frati et al., 1997; Huelsenbeck and Crandall, 1997; Sullivan and Swofford, 1997; Posada and Crandall, 1998). However, when models are not nested then other methods such as the Akaike information criterion (AIC) (Akaike, 1974) and Bayesian information criterion (BIC) (Schwarz, 1978) can be used.

The BIC was developed in order to reduce the tendency, present in both the AIC and the LRT, to prefer increases in model complexity (Schwarz, 1978). A comparison between the AIC and BIC scores will demonstrate this difference, and will be helpful for understanding model selection. The AIC score is given by

$$AIC = -2\ell + 2p,$$

and the BIC score is given by

$$BIC = -2\ell + p \log n,$$

where ℓ is the maximum log likelihood score achieved for the dataset on the given tree, p is the number of free parameters in the model, and n is the length of the multiple sequence alignment (more generally, n is the sample size). Increases in the maximum likelihood score will decrease the AIC and BIC scores, and increases in the number of parameters in the model will increase the AIC and BIC scores. Finally, the sequence length only impacts the BIC score, and increases in the sequence length increase the BIC score. AIC and BIC scores are positive (log likelihood scores are negative). When using either the AIC or the BIC to compare two sequence evolution models, the model with the smallest score is preferred.

A comparison between these formulae shows that a small increase in p will be tolerated more by the AIC than by the BIC, so that the AIC will tend to pick more complex models than the BIC. The LRT also tends to favor more complex models than the BIC (Schwarz, 1978). There is also a modification to the AIC criterion called the "corrected AIC" (Sugiura, 1978), and written AIC_c, which is designed to reduce the tendency of the AIC to favor complex models. All these techniques (LRT, AIC, BIC, and AIC_c) respond differently to increases in model complexity, and may pick different models as a result for a given dataset.

There is continued research into which approach provides the best results for phylogeny estimation (Posada and Crandall, 2001; Posada and Buckley, 2004; Sullivan and Joyce, 2005; Kosiol et al., 2006), and in practice multiple tests are often used. Several software packages are available to choose between different sequence evolution models using these techniques. For example, ProtTest (Darriba et al., 2011) is frequently used for amino acid models, and ModelTest is frequently used for nucleotide models (Posada and Crandall, 1998; Darriba et al., 2012). A new method for efficiently selecting between a large set of nucleotide models is PLTB (Hoff et al., 2016).

Different parts of a single gene may evolve under different site evolution models, and so partitioning of the dataset is often needed; PartitionFinder (Lanfear et al., 2012) is software that combines techniques to partition a nucleotide dataset with model selection. In addition, different parts of the tree may require different sequence evolution models, even for a single gene (Kosiol et al., 2006).

8.4 Distance-based Estimation

Distance-based tree estimation has two steps: First a matrix of pairwise distances is computed, and then a method is used to compute a tree for the distance matrix. In Chapter 5, we discussed methods for the second step, and we note that some of these methods (such as neighbor joining) have the nice property that they are guaranteed to return a tree T whenever the distance matrix **d** is close enough to an additive matrix **D** defining the tree T.

In this chapter, we show how to compute a matrix of pairwise distances given a set of aligned sequences. These computations are done using formulae that are specifically designed for the models of sequence evolution we have discussed above, and have the guarantee that as the sequence length increases, the matrix of estimated pairwise distances will converge to an additive matrix defining the model tree. Therefore, under the assumption that the sequences are generated under the same model of evolution, the two-step process is statistically consistent. Furthermore, as we will see, the calculation of these distances is polynomial time for standard sequence evolution models.

8.4.1 Statistical Methods for Computing Distances

In Chapter 1, we showed how to compute distances under the CFN model so that these distances would converge to an additive matrix for the model tree. Here we address the same issue, but for the DNA sequence evolution models we have discussed.

Computing distances for the Jukes–Cantor model: Let $S = \{s_1, s_2, \ldots, s_n\}$ be a set of sequences, each of length k, that have evolved down a JC69 model tree T. We let $\lambda_{i,j}$ denote the expected number of changes of a random site on the path in T between s_i and s_j. Thus, letting λ_e denote the expected number of changes on edge e, it is easy to see that λ is an additive matrix for T. We estimate the model distance between s_i and s_j using

$$\hat{\lambda}_{i,j} = -\frac{3}{4} ln\left(1 - \frac{4}{3}\frac{H(i,j)}{k}\right),$$

where $H(i,j)$ denotes the Hamming distance between s_i and s_j. Furthermore, $\hat{\lambda}_{ij}$ converges to λ_{ij}, as $k \to \infty$. This is called the *Jukes–Cantor distance correction*.

Computing distances for the General Markov model: The calculation of pairwise distances for the GM model is more complicated than for the JC69 or CFN models. For a given pair of sequences s_i and s_j that have evolved down a GM model tree T, let $\hat{f}_{ij}(\alpha, \beta)$ denote the relative frequency of s_i having state α and s_j having state β. Then let $\mathbf{F}_{ij}$ denote the 4×4 matrix (with rows and columns indexed by the four possible character states) using the $\hat{f}_{ij}(\alpha, \beta)$ values for its entries. Let $d_{ij} = -log|det\mathbf{F}_{ij}|$. Then the matrix d converges as the sequence length increases to an additive matrix for T. This is called the **logdet** distance calculation (Steel, 1994b). One of the important things to realize is that the matrix of pairwise distances that this logdet distance calculation converges to is not necessarily the matrix of expected numbers of changes for a random site between each pair of sequences (see discussion in Steel (1994b)). Thus, although it will converge to an additive distance matrix for the model tree, it does not enable the estimation of a branch length that can be interpreted in terms of the expected number of changes on the edge. This makes logdet distances different from other distances we've seen, such as Jukes–Cantor distances.

The logdet distance correction can be used to compute distances for any of the submodels of the GM Model. Hence, it can be used with sequences that evolve down GTR model trees, JC69 model trees, etc. However, model-specific distance corrections are also available.

8.4.2 Statistical Properties of Distance-based Methods

In Chapter 5, we presented a collection of methods that can construct trees when given distance matrices as input. For example, we showed that the Naive Quartet Method, neighbor joining, the Agarwala et al. method for 3-approximating the L_∞-nearest tree, and FastME are robust – to varying extents – to error in the estimated distance matrix in the following sense. For each of these methods, there is a positive safety radius δ that depends on the model tree, so that when the input dissimilarity matrix d satisfies $L_\infty(d, \mathbf{D}) < \delta f$ where $\mathbf{D}$ is an additive matrix defined by an edge-weighting of the model tree T with minimum internal branch length f, then the method is guaranteed to return the tree T (see Section 5.10).

As discussed in Section 5.10, when a distance-based method has a positive safety radius (i.e., when $\delta > 0$), then the method is statistically consistent under any sequence evolution model for which estimated pairwise distances converge to the model pairwise distances as the sequence length increases. Thus, since the estimated distances computed using the CFN distance calculation converge to the model CFN distances, each of these methods is statistically consistent under the CFN model. The same statement is true for the GM model, since logdet distances converge to additive matrices for the GM model tree. And, since the GM model contains all the standard DNA sequence evolution models (including the JC69, Kimura 2-parameter, and other models), this means that the distance-based methods are statistically consistent under the standard DNA sequence evolution models. We summarize this as follows:

Observation 8.1 To prove that a distance-based method is statistically consistent at estimating the unrooted tree topology under a model of sequence evolution, it suffices to establish

- for every model tree, the topology is defined by an additive matrix A,
- the additive matrix A can be estimated in a statistically consistent manner, and
- the distance-based method has a positive safety radius, and so is tolerant of some error in the estimated distances.

The first two conditions (existence of an additive matrix for the model tree, and a statistically consistent technique for estimating the additive matrix) depend on the model of sequence evolution, but hold for all the standard sequence evolution models. The third condition depends on the distance method.

8.5 Calculating the Probability of a Set of Sequences on a Model Tree

The models of evolution that we have been working with all assume that the sites evolve independently and identically down some model tree. Hence, the probability of a set of sequences in a gap-free multiple sequence alignment is just the product over all the sites of the probability of the pattern for that site. Furthermore, nearly all models used in phylogenetics are time-reversible, so that the location of the root has no impact on the result of the calculation. Hence, to calculate the probability that a given model tree (T, θ) would generate a sequence dataset S, we can root the tree T anywhere we like (including at a leaf), compute the probability of the site pattern at the leaves for each site, and then multiply all these probabilities together.

8.5.1 The Brute Force Approach

We now show how to calculate the probability of a set S of aligned sequences given a JC69 model tree (T, θ), where T is a rooted binary tree and θ provides the probability of change on each edge. More complex models require additional parameters, such as a 4×4 DNA

8.5 Calculating the Probability of a Set of Sequences on a Model Tree

rate matrix governing the tree (for the GTR model) or 4×4 substitution matrices for every edge in the tree (for the GM model). Since these are single-site models, we will calculate the probability of the sites one at a time, and then multiply the probabilities of the sites together.

When we do the calculation of a site pattern at the leaves, we have to consider all the possible assignments of states to the internal nodes of the tree. We can do this using a brute force approach, which enumerates all the possible assignments of states to the internal nodes, and then calculates each of these scenarios independently, as follows.

For a binary model tree with n leaves, there are $n-1$ internal nodes in the tree (since the model tree is rooted and binary). Assuming the site has r states (e.g., $r = 2$ for the CFN model and $r = 4$ for the JC69 and GTR models), there are r^{n-1} ways of assigning states to each of the internal nodes for a single site. For each of these r^{n-1} possible assignments, the probabilities of all the state changes on the edges must be calculated, and then they are multiplied together; each of these individual calculations produces the probability of observing exactly those state changes. The probability of the root state must also be included in the calculation, and is based on the parameter π rather than the transition matrices and branch lengths. Finally, each of these r^{n-1} values must be added together. Overall, this approach, although it works, is computationally expensive, even for the simplest case where $r = 2$, once n is not small. We demonstrate this with brute force technique on a small (four-leaf) CFN tree.

Example 8.2 Let (T, θ) be the CFN model tree given by $(r, (a, (b, c)))$, and rooted at leaf r. Assume the substitution probabilities on the external edges are all 0.1, and the substitution probability on the internal edge is 0.4. We are given four aligned sequences:

- r = 100
- a = 000
- b = 101
- c = 111

To compute the probability of this set of four sequences, we will compute the probability of each site pattern, and then multiply them together.

We begin with the first site pattern, $r = 1, a = 0, b = 1, c = 1$. The probability that $r = 1$ is 0.5, since r is the root and the two states are equiprobable under the CFN model. We label the internal nodes of T by u and v, with v the parent of b and c, and u the parent of a and v. There are four possible assignments of states to u and v, and for each one we have to compute the probability of the state changes on the tree.

- $u = 0, v = 0$. In this case, there are changes on edges $(r,u), (v,b)$, and (v,c), and no changes on the other edges. We examine each edge in turn and note the probability of each of the implied events (change on the edge or no change on the edge). For example, there is a change on edge (r,u) since $r = 1$ and $u = 0$. The substitution probability on that edge is 0.1, so that event has probability 0.1. Similarly, there is no change on edge (u,a), and that event has probability 0.9. We multiply these probabilities together and

get $0.1^3 \times 0.9 \times 0.6$. Since the root state has probability 0.5, the total probability of this assignment of states to the nodes of the tree is $0.5 \times 0.1^3 \times 0.9 \times 0.6 = 0.00027$.
- $u = 0, v = 1$. The same type of calculation gives $0.5 \times 0.9^3 \times 0.4 \times 0.1 = 0.01458$.
- $u = 1, v = 0$. The same type of calculation gives $0.5 \times 0.1^3 \times 0.9 \times 0.4 = 0.00018$.
- $u = 1, v = 1$. The same type of calculation gives $0.5 \times 0.1 \times 0.9^3 \times 0.6 = 0.02187$.

After we compute these probabilities we add them up, and obtain

$$\Pr(r = 1, a = 0, b = 1, c = 1|(T, \theta)) = 0.0369.$$

Note the care that needs to be taken in computing the probability of a single site.

We do this for each site. Each calculation involves computing the probability of the four possible assignment of states to all the nodes in the tree, and then adding up the four values. The same type of calculation for the second site ($r = a = b = 0, c = 1$) yields 0.0369 (note it has the same probability of the first site, which makes sense because of the symmetry in this model tree). The third site ($r = a = 0, b = c = 1$) has probability 0.1361. Hence, the joint probability of the three sites is $0.0369^2 \times 0.1361 = 0.00018517896$.

Note that this calculation not only tells you the probability of the model tree generating the four sequences, it lets you know *which site* of the three has the highest probability, and for that matter what states are most likely at each internal node for each site pattern. Given two different model trees for the same sequence dataset, you would be able to compare the probabilities of generating the sequence dataset, and pick the model tree that has the higher probability of generating the data. In other words, the **likelihood** of the model tree is the probability of the data given the tree, and the problem of seeking the model tree that has the highest probability of generating the observed data is the **maximum likelihood** problem.

8.5.2 Felsenstein's Pruning Algorithm

The problem with the brute force approach from the previous section is how onerous it is to do, since the calculation is done explicitly for all the ancestral states, and there are r^{n-1} assignments of states to the internal nodes. Hence, this is an *exponential time algorithm*. The same problem can be solved using a much more efficient approach, as we now show.

Felsenstein (1981) introduced a dynamic programming algorithm to compute the probability of a single site on a model tree, much along the lines of the dynamic programming algorithm used to compute the maximum parsimony score of a fixed tree. This algorithm, referred to as "Felsenstein's Pruning Algorithm" (and sometimes as "Felsenstein's Peeling Algorithm") runs in $O(r^2 n)$ time to compute the probability of a single site for a given model tree with n leaves and where the sites have r states, under the assumption that the calculation of the probability of any state transition can be performed in $O(1)$ time. Hence, for k sites, the running time is $O(knr^2)$. This means calculating the probability of binary sequences on a given CFN model tree requires $O(kn)$ time, and so does calculating the probability of DNA sequences on a given GTR model tree, since for each of these cases the number r of states is treated as a constant.

The input is a set S of gap-free aligned sequences, along with the rooted model tree (T, θ). While θ depends on the model, for each of the standard models (e.g., GTR), for any edge $e = (a, b)$ in the tree (with a the tail and b the head), we can efficiently compute the probability of any state–state transition on the edge.

Since the sites evolve identically and independently, we will show how to compute the probability of the site pattern for a single site. Then, by multiplying the probabilities across the sites we can obtain the probability of the set of sequences we observe at the leaves of the rooted tree T. Since the usual stochastic models of sequence evolution are time-reversible, we can calculate the probabilities by rooting the tree at any node, including at a leaf.

We will use the basic algorithmic approach in the Sankoff dynamic programming algorithm (see Section 4.3.2) for weighted maximum parsimony in order to calculate the probability of the sequence data given the model tree. In the parsimony dynamic programming algorithm we had subproblems $Cost(v, x)$ for each node v in the rooted tree and character state x, where $Cost(v, x)$ denotes the minimum parsimony cost of the tree T_v under the assumption that we label v by state x.

To compute the probability of the site pattern at the leaves, we will let $FPA(v, x)$ (FPA stands for "Felsenstein's Pruning Algorithm") denote the probability of the site pattern at the leaves in T_v given the model tree (T, θ) (which includes all the substitution probabilities on the edges of T_v), assuming that the state at v is x. We calculate each $FPA(v, w)$ from the bottom up, as we now show. Using this formulation, the boundary cases (where v is a leaf) have $FPA(v, x) = 1$ if the state at the leaf v is x, and otherwise $FPA(v, x) = 0$. We let Σ denote the state space for the site patterns (e.g., $\Sigma = \{A, C, T, G\}$ for DNA sequences). For internal vertices v with children w_1 and w_2,

$$FPA(v,x) = \sum_{a \in \Sigma} [\Pr(v=x|w_1=a) \times FPA(w_1,a)] \times \sum_{a \in \Sigma} [\Pr(v=x|w_2=a) \times FPA(w_2,a)].$$

Finally, when v is the root of the tree, the probability of the observed data at the leaves of the tree is given by $\sum_{a \in \Sigma}(\pi_a \times FPA(v,a))$. Hence, if the root of the tree is one of the leaves (and so the state is given), then this last expression simplifies to $\pi_b \times FPA(v,b)$, where b is the state at the leaf acting as the root.

8.6 Maximum Likelihood

Maximum likelihood (ML) phylogeny estimation is a very popular technique used for phylogeny estimation in systematics. We present this approach in the context of estimating trees under the JC69 model.

Recall that a JC69 model tree consists of a rooted binary tree T and the numerical parameters for the model, which are just the probabilities of change on the edges of the tree. Thus, a JC69 model tree is described as a pair (T, θ), where θ is the set of substitution probabilities on the edges of the tree. Under the JC69 model, ML estimation takes as input a set S of sequences, each of the same length, and seeks the JC69 model tree (T^*, θ^*) that maximizes $\Pr(S|T^*, \theta^*)$, the probability of generating the observed sequence data S. Because the JC69 model is time-reversible, moving the root from one edge to another edge within a tree T does not change the probability of generating the observed data; therefore, the result of an ML analysis is an *unrooted* model tree. (This is also the reason that other phylogeny estimation methods, except for a few distance-based methods that assume a molecular clock, return unrooted trees.)

Note that we have described ML under the JC69 model, but that the description applies therefore to any of the statistical models of evolution we have discussed so far. Thus, ML under the GTR model would seek the GTR model tree that maximized the probability of generating the observed sequence data. The only thing that changes as we substitute one model for another in these different formulations is what we mean by θ. Even more complex models (that go beyond the GM model) can be considered, and ML can be extended to address estimation under these models.

Since ML requires the estimation of all the model parameters, as the model becomes more parameter-rich, ML estimation becomes more computationally intensive. Indeed, there is also the possibility of *over-fitting* if there are too many model parameters. Furthermore, while the ML scores may be distinct for different trees, the differences in scores between different trees are sometimes small enough to not be considered significant; for this reason, the problem of finding the optimal as well as the near-optimal trees is sometimes the objective. In addition, although ML is a natural optimization problem, it does not follow that ML estimation will be, by necessity, statistically consistent. For example, as we will see in Section 8.13, ML is not statistically consistent under the No Common Mechanism model.

Computational complexity of maximum likelihood. ML, if run exactly (so that optimal solutions are found), is statistically consistent under the JC69, GTR, and GM models (Chang, 1996). However, finding an optimal ML tree is an NP-hard problem (Roch, 2006), and so heuristics are used instead of exact solutions. As with any NP-hard phylogeny estimation problem, because the number of trees on n leaves is exponential in n, examining all tree topologies is infeasible, and heuristic searches are used to explore treespace. However, heuristic searches for ML are more complicated than heuristic searches for maximum parsimony, because even scoring a tree is computationally intensive! That is, while Felsenstein's Pruning Algorithm provides a way to compute the probability of a set of sequences given a model tree, this assumes that we know the set θ of numeric parameters for the model tree. As it turns out, the estimation of numeric parameters on a tree is also addressed heuristically, and may not find global optima (Steel, 1994a). Hence, ML phylogeny estimation is complicated and computationally intensive, even for the simplest sequence evolution models.

8.7 Bayesian Phylogenetics

Bayesian methods are similar to ML methods in that they also calculate likelihoods of trees based upon explicit parametric mathematical models of evolution. However, while ML methods attempt to find the model tree that has the largest probability of generating the observed data, Bayesian methods try to sample from the set of model trees with frequency proportional to their likelihoods, given the observed data. The result is that Bayesian methods produce a set of trees rather than a single tree, and they use this set to evaluate the statistical support for different evolutionary hypotheses.

Bayesian methods are usually implemented using Markov Chain Monte Carlo (MCMC) techniques. In this type of approach, a Bayesian method performs a random walk through the space of model trees. Every time it visits a model tree, it computes the probability of the observed data, using Felsenstein's Pruning Algorithm. The probability of accepting the new model tree depends on whether the probability has increased or decreased; for example, in the Metropolis–Hastings (Hastings, 1970) version of MCMC, the new tree may be accepted even if the probability has decreased. If the new tree is accepted, the MCMC chain continues from the new model tree.

The Bayesian MCMC walk is guaranteed to converge to the stationary distribution on model trees, as long as the MCMC walk satisfies the **detailed balance property**, which essentially says that $\pi(i)p_{ij} = \pi(j)p_{ji}$, where $\pi(x)$ is the equilibrium probability of state x and p_{xy} is the probability of moving to state y from state x (here, x and y (and so also i and j) represent model trees). Thus, the design of Bayesian methods requires some care, both in terms of how model trees are visited and the probabilities of accepting a new model tree, in order for the detailed balance property to hold. However, if the detailed balance property holds and if the MCMC walk lasts long enough, then it will start sampling model tree topologies proportionally to their likelihoods, as defined by the sequence evolution model and observed data.

After the MCMC chain has run long enough to reach stationarity, a sample of the model trees it visits is then taken from the trees visited after some "burn-in" period (i.e., after the chain has run long enough to reach the stationary distribution). If a point estimate of the tree topology is desired, then a summary of the distribution may be computed using several different techniques. For example, a majority consensus tree of the sampled trees can be computed. In a Bayesian analysis, the frequency with which a tree topology appears in the sample is its posterior probability; hence, another common summary statistic is the **maximum *a posteriori* (MAP)** tree, which is the tree topology that appears the most frequently in the sample (Rannala and Yang, 1996).

A third approach assembles a set of tree topologies, starting with the most frequently observed trees and then decreasing, until a desired threshold is achieved (e.g., where 95 percent of the probability distribution is in the set), and then computes a point estimate based on this subset of trees. Each point estimate can then be used as an estimated

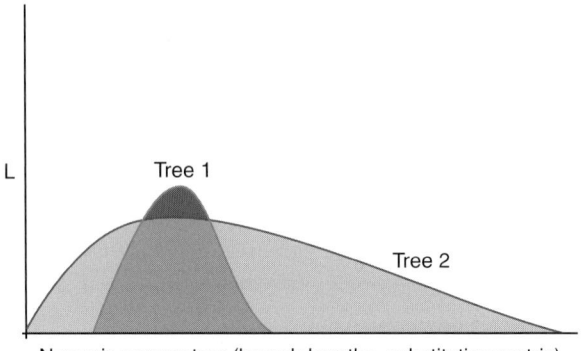

Figure 8.3 The *x*-axis indicates the numeric parameters for the GTR model trees that can equip the different tree topologies, Tree 1 and Tree 2; the *y*-axis indicates the likelihood score. Note that the likelihood curve for Tree 1 has a larger peak than that for Tree 2, but Tree 2 has a larger area under its curve than Tree 1. Therefore, GTR ML would pick Tree 1, but marginalizing over the GTR numeric parameters (as in a Bayesian analysis) would favor Tree 2.

phylogeny, and the support for the branches of the phylogeny can similarly be computed based on the observed distribution. Finally, if the Bayesian method is run properly (i.e., so that it reaches the stationary distribution), then its point estimates will be statistically consistent under the JC69 model as well as under more general models, such as the GTR model (Steel, 2013).

As a result, there is an important distinction between ML and Bayesian estimation. As seen in Figure 8.3, Tree 1 has a higher ML score than Tree 2, because for at least one set of numeric parameters Tree 1 has higher probability of generating the input sequence alignment than Tree 2. However, the total probability of producing the input sequence alignment is higher for Tree 2 than Tree 1, when all the possible numeric parameters are taken into account. Hence, for this input, Tree 2 would be preferred over Tree 1 in a Bayesian framework.

There are several challenges in using Bayesian methods. For example, Bayesian methods require priors, and the choice of priors can affect the accuracy of the resultant trees; hence, the selection of priors requires great care (Alfaro and Holder, 2006; Yang, 2009). The second is that Bayesian MCMC methods need to be run until they have reached the stationary distribution, and it is not at all straightforward to assess whether stationarity has been reached. Furthermore, on large datasets, it can require very long running times to reach stationarity. As a result, Bayesian MCMC analyses are typically not used for very large sequence datasets (e.g., those containing 1000 or more sequences), although some ML methods such as FastTree-2 (Price et al., 2010) can complete analyses on 1000 sequences in under an hour (Liu et al., 2012a), and on much larger datasets (even 1,000,000 sequences) in days (Mirarab et al., 2015a; Nguyen et al., 2015b).

8.8 Statistical Properties of Maximum Parsimony and Maximum Compatibility

We have already shown that distance-based methods can be statistically consistent under standard stochastic models of evolution, including the GM model, provided that the distances between sequences are computed using appropriate statistical methods. We noted (although we did not prove) that ML and Bayesian methods are also statistically consistent under these standard models. We now turn to the maximum parsimony and maximum compatibility problems, and their statistical properties under stochastic models of sequence evolution. Are these statistically consistent under any of these models of evolution?

Here the story is not so positive. Felsenstein (1978) gave several examples of very simple four-leaf model trees on which maximum parsimony is inconsistent. Worse, on these model trees, parsimony converges to the wrong tree as the number of sites increases. This is expressed by saying that maximum parsimony is **positively misleading**, an even stronger statement than just being inconsistent. Since maximum compatibility and maximum parsimony choose the same tree for four-leaf datasets, this result also applied to maximum compatibility, and hence establishes that on these model trees maximum compatibility is also positively misleading. These negative examples provide a cautionary note about maximum parsimony and maximum compatibility, and have resulted in many evolutionary biologists choosing to use other methods.

In Section 4.7, we discussed parsimony informative characters, and showed that the parsimony informative characters for any four-sequence dataset are those that split 2:2. We also showed that the tree returned by maximum parsimony would be the tree with the bipartition associated with the 2:2 split that appears the most frequently in the input. Hence, to analyze maximum parsimony on a dataset with four sequences, we only need to note the parsimony informative sites and see which one appears the most frequently; the tree yielding that bipartition is the tree that parsimony will return.

We will use this type of analysis to prove that maximum parsimony (and hence maximum compatibility) is positively misleading on some CFN model trees.

Example 8.3 We present a model CFN tree in which parsimony (and hence compatibility) will be positively misleading. This is an example of a **Felsenstein Zone Tree**, in recognition of Felsenstein's discovery. Let T be a rooted CFN model tree with topology $((a,b),(c,d))$ and assume that the two edges incident with a and d have substitution probabilities equal to $1/2 - \varepsilon$ and that the other edges have substitution probabilities ε. We also assume that $0 < \varepsilon < 0.25$, so that $\varepsilon < 1/2 - \varepsilon$; thus, all the internal edges are short (substitution probabilities of ε) and only the external edges incident with a and d are long (substitution probabilities of $1/2 - \varepsilon$). In fact, in our analysis, we will assume that ε is very close to 0, and so the short edges are *much* shorter than the long edges.

There are three parsimony informative site patterns, and so we only need to determine the site pattern that has the highest probability of the three. These site patterns are:

- P1: $a = b \neq c = d$, corresponding to unrooted tree with bipartition $ab|cd$.
- P2: $a = c \neq b = d$, corresponding to unrooted tree with bipartition $ac|bd$.
- P3: $a = d \neq b = c$, corresponding to unrooted tree with bipartition $ad|bc$.

Site pattern P1 matches the model tree, while site patterns P2 and P3 differ from the model tree. So for parsimony to be statistically consistent, the probability of P1 must be strictly larger than the other two site patterns. We will show that for small enough ε, not only is this not the case, but the probability of P3 is strictly greater than the probability of the other two site patterns. In other words, for small enough ε, maximum parsimony will return the wrong tree with probability converging to 1 as the sequence length increases.

We will prove that maximum parsimony will converge to $((b,c),(a,d))$, which in essence is putting b and c as siblings, because b and c are connected by a path of only short branches together and so are nearly always the same state. However, it is also the case that maximum parsimony puts the two leaves that are on the long branches together as siblings. As a result, this phenomenon is referred to as **long branch attraction**, or **LBA**.

In the analyses below, we let r be the root of the model tree and let u and v be the children of r. The children of u are a and b, and the children of v are c and d.

Calculating the probability of P1, $a = b \neq c = d$. This case can be divided into two subcases, $a = b = 0, c = d = 1$ and $a = b = 1, c = d = 0$; by the symmetry of the CFN model, these are equiprobable. We analyze $a = b = 0, c = d = 1$; the probability of site pattern P1 will be twice that of this subcase. Because $b \neq c$, there must be at least one change of state on the path from b to c; since each edge on that path has substitution probability ε, the probability of each of these character state changes is less than ε. There are eight possible assignments of character states to r, u, and v, and each root state has probability $1/2$. Hence, no matter what the state assignment is it cannot have probability more than $\varepsilon/2$; therefore, the total probability of $a = b = 0, c = d = 1$ is strictly less than 4ε, and the probability of $a = b \neq c = d$ is strictly less than 8ε.

Calculating the probability of P2, $a = c \neq b = d$. Because $b \neq c$ here as well, the same analysis shows that site pattern P2 also has probability strictly less than 8ε.

Calculating the probability of P3, $a = d \neq b = c$. By symmetry of the CFN model, the probability of site pattern $a = d \neq b = c$ is twice that of the site pattern $b = c = 0, a = d = 1$. The probability of site pattern $b = c = 0, a = d = 1$ is the sum of the probabilities of each of the eight possible states that u, v, and r can take. Hence, it is strictly greater than the probability of any one of these eight different ways of assigning states to u, v, and r. We will obtain a lower bound on the probability that $b = c = 0, a = d = 1$ by setting the state assignments for u, v, and r all to 0. Thus,

$$\Pr(b = c = 0, a = d = 1) >$$
$$\Pr(b = c = r = u = v = 0, a = d = 1) =$$
$$\Pr(b = c = u = v = 0, a = d = 1 | r = 0) \times \Pr(r = 0) =$$
$$\frac{1}{2}\Pr(b = c = u = v = 0, a = d = 1 | r = 0).$$

8.8 Statistical Properties of Maximum Parsimony and Maximum Compatibility

To calculate $\Pr(b = c = u = v = 0, a = d = 1 | r = 0)$, we note that this assignment of states to nodes of the tree has changes only on the two long edges; hence

$$\Pr(b = c = u = v = 0, a = d = 1 | r = 0) = (1-\varepsilon)^4 (\frac{1}{2} - \varepsilon)^2.$$

Therefore,

$$\Pr(b = c \neq a = d) > \frac{1}{2}(1-\varepsilon)^4 (\frac{1}{2} - \varepsilon)^2.$$

Comparing site pattern probabilities for small ε. We have established that

$$(1) \Pr(a = b \neq c = d) < 8\varepsilon,$$

$$(2) \Pr(a = c \neq b = d) < 8\varepsilon,$$

$$(3) \Pr(a = d \neq b = c) > \frac{1}{2}(1-\varepsilon)^4 (\frac{1}{2} - \varepsilon)^2.$$

As $\varepsilon \to 0$, $\frac{1}{2}(1-\varepsilon)^4 (\frac{1}{2} - \varepsilon)^2$ converges to 0.125 from below. Hence, for small enough ε, we can make the probability of site pattern P3 as close to 0.125 as we wish. Since the probabilities of P1 and P2 are bounded from above by 8ε, this means we can also make P1 and P2 have probabilities that are as close to 0 as we want. Hence, for small enough ε the probability of P3 is strictly greater than the probability of P1 and P2, achieving the result we claimed. For example, when $\varepsilon = 0.01$, the probability of site pattern P3 is greater than $0.5 \times 0.99^4 \times 0.49^2 > 0.115$, and the probabilities of P1 and P2 are both less than 0.08. Hence for this setting of ε, P3 has strictly greater probability than either P1 or P2. Obviously, any smaller value for ε will produce the same inequality, and some bigger values for ε will also work.

Finally, since maximum compatibility and maximum parsimony are identical on four-leaf trees, if maximum parsimony is positively misleading then so is maximum compatibility. Hence, we have proven the following:

Theorem 8.4 *There are CFN model trees on which maximum parsimony and maximum compatibility are positively misleading and so will return the wrong tree topology as the number of sites increases, with probability converging to* 1.

The theorem is not restricted to just CFN model trees and not just for four-taxon trees. Maximum parsimony can be inconsistent even under the strict molecular clock (Zharkikh and Li, 1993) and on large trees (Kim, 1996). More generally, the conditions in which maximum parsimony is inconsistent are not limited to just a few rare cases (Kim, 1996). However, that doesn't mean that maximum parsimony is statistically inconsistent on *all* CFN model trees. Indeed, it is not hard to prove that maximum parsimony is statistically consistent for four-leaf CFN trees with sufficiently long internal edges and sufficiently short external edges. Therefore, statistical inconsistency is only a statement that there are *some* model conditions in which the method is not statistically consistent.

The model tree used in the proof of Theorem 8.4 is an example of LBA, where two long branches placed near very short branches may "attract" each other. The impact of LBA on statistically consistent methods is different, since they will recover the true tree with probability converging to 1 as the sequence length grows. Even so, maximum likelihood and other methods will also be impacted by LBA, since they will tend to return the wrong tree until the sequences are long enough (imagine, for example, what ML or neighbor joining would do with a single site evolving down the model tree in the proof of Theorem 8.4). This observation suggests that adding additional loci can reduce error in phylogenetic analyses (Lecointre et al., 1994; Cummings et al., 1995), provided that the loci evolve down the same model tree (see Chapter 10 for conditions under which loci evolve down different trees). However, while adding loci that evolve down the same model tree will improve accuracy for statistically consistent methods, it does not address the issue of statistical inconsistency, which is an asymptotic statement.

8.9 The Impact of Taxon Sampling on Phylogenetic Estimation

One of the questions that has been raised is whether the Felsenstein Zone can be avoided, perhaps by adding taxa to break up long branches (Hendy and Penny, 1989). Thus, increasing the density of the *taxon sampling* might lead to model trees for which maximum parsimony is consistent rather than inconsistent. Or, even if increasing the taxon sampling doesn't change the issue of statistical consistency, it might lead to reduced topological error rates.

The idea of increasing taxonomic sampling to improve accuracy, originally suggested in Hendy and Penny (1989), led to a provocative simulation study (Hillis, 1996), which showed that many methods, including maximum parsimony, might be able to produce highly accurate trees on a very large dataset, and that larger trees might be easier than small trees.

Hillis' paper led to a vigorous exchange in the journal *Trends in Ecology and Evolution* about the nature of statistical consistency, accuracy, how to evaluate methods, and the value of model-based modes of phylogenetic analysis (Kim, 1996; Rannala and Yang, 1996; Hillis, 1997; Purvis and Quicke, 1997a,b; Yang and Goldman, 1997). Nevertheless, over the years the observation that increasing taxonomic sampling often leads to improved accuracy has been confirmed. There is a large literature on this subject; see Graybeal (1998), Pollock et al. (2002), Zwickl and Hillis (2002), Hillis et al. (2003), Poe (2003) for some of the earliest studies; Heath et al. (2008) for an interesting survey on the early literature; and Leebens-Mack et al. (2005), Brandstetter et al. (2016), and Wang et al. (2016) for some analyses of biological datasets exploring the same basic questions.

The basic insight into the (generally) beneficial impact of increased taxon sampling is that sequences that have evolved down a common tree are related, and so analyses of all the data together assist in the inference of their common history. As a simple example, consider a Jukes–Cantor caterpillar tree where every edge is short (i.e., the substitution probability is very low) and that has 100 leaves, $s_1, s_2, \ldots, s_{100}$, ordered as you go from left to right in the tree. For low enough substitution probabilities, the quartet tree on $s_1, s_{50}, s_{51}, s_{100}$ is in the Felsenstein Zone, and so maximum parsimony will be inconsistent, and even statistically consistent methods like ML, neighbor joining, and the Four Point Method will make mistakes unless the sequences are very long. Now, consider what happens when we try to construct a tree on the entire set of 100 sequences. The Naive Quartet Method will probably fail, since if it fails on even one quartet it will fail to return a tree (and we have already noted that the quartet tree $s_1 s_{50} | s_{51} s_{100}$ is difficult to construct correctly). However, the entire tree may be much easier to compute using ML!

Studies evaluating quartet-based methods have also tended to suggest a similar beneficial impact of taxon sampling, but somewhat indirectly. For example, the simulation study reported in St. John et al. (2003) showed that quartet trees estimated independently by neighbor joining were *less* accurate than quartet trees computed in a two-step process: First compute a neighbor joining tree on the entire dataset, then look at the induced quartet trees it defines. As a result, quartet-based methods for constructing trees are often less accurate than methods that analyze the entire dataset together (St. John et al., 2003).

Whether it is more beneficial to add taxa or loci is another question that studies have addressed using both simulated and real datasets, and evidence suggests that generally both are beneficial. Yet, neither adding taxa nor adding loci is necessarily always benign. For example, if you add too many taxa, then short branches can become too short, which will increase error (since potentially there will not be enough sites to even detect the shortest branches). Another problem with adding too many taxa and/or loci is that evolution can become more heterogeneous (a subject we return to in Chapter 10) as you enlarge the dataset in either direction, leading to the need to consider more parameter-rich models, and hence more computationally intensive methods. Furthermore, as the dataset becomes larger, computational issues become increasingly important.

Despite the challenges involved in analyzing large datasets, the move to large datasets is clearly seen as potentially highly beneficial in terms of accuracy, provided that the model complexity and computational burden can be addressed. See Chapter 11 for algorithmic approaches to improve accuracy and scalability to large trees.

8.10 Estimating Branch Support

A very common concern is figuring out how reliable each branch is within the tree T computed on a set of aligned sequences. Depending on the method for performing the phylogenetic analysis, different approaches can be used to assess branch support. Perhaps the most common approach that can be used with any phylogenetic estimation method is *non-parametric bootstrapping*. This technique was introduced in Felsenstein (1985), and subsequently elaborated on in Efron et al. (1996) and Holmes (2003).

In its simplest form, the input sequence alignment is used to generate a large number (e.g., 100 or more) of **bootstrap replicate** datasets. A bootstrap replicate is a data matrix with the same dimensions as the original matrix, but where the columns of the bootstrap replicate are obtained by sampling with replacement from the original data matrix. As a result, some columns from the original matrix will appear not at all, some will appear exactly once, and others will appear more than once. After the bootstrap replicate datasets are obtained, a phylogeny is estimated on each bootstrap replicate dataset using the same method as was used to estimate a tree on the original dataset. This produces a set of bootstrap trees (one per replicate dataset).

The most common use of bootstrapping is to characterize the support for each edge in the tree T. Specifically, if an edge e in T defines a bipartition $A|A'$, then we look at the bootstrap trees and determine the fraction of those trees that also have edges defining the same bipartition; that fraction is the bootstrap support for the edge e.

The interpretation of bootstrap support values is complicated, since high bootstrap support may not indicate high probability of accuracy. For example, if bootstrapping is performed for a maximum parsimony analysis and the model condition is one where maximum parsimony is positively misleading, then there can be very high support (even 100 percent support) on branches that are not present in the model tree. Hence, high bootstrap support can be misleading when the estimation method is not statistically consistent. However, even with statistically consistent methods the interpretation of bootstrap support is not easy. That said, edges with low support values (below 50 percent) are generally considered unreliable, and edges with support values above 95 percent are considered reliable. In the middle region, where support values are between 50 percent and 95 percent, opinions differ as to the reliability of edges (Swofford et al., 1996).

When the phylogenetic analysis is a Bayesian MCMC method, then another technique is typically used. Recall that the Bayesian MCMC methods operate by performing a random walk through treespace, and a random sample of the model trees that are visited (after "burn-in") is saved. That set of trees is then used to produce a distribution on tree topologies (i.e., what fraction of the model trees in the set have a particular tree topology), as well as a distribution on bipartitions (i.e., what fraction of the model trees in this set have a particular bipartition). If a single-point estimate tree is desired, then typically the MAP tree is returned; however, a consensus tree (e.g., a majority consensus or a greedy consensus) of the sampled trees is also sometimes returned. The branch supports on the tree are obtained by using the percentage of the trees in the set that induce the same bipartition, and are called the *posterior probabilities* for each edge.

Note the similarity between how branch support is computed for both techniques – bootstrapping and Bayesian MCMC – the only difference is how the set of trees is computed. However, this difference is actually quite important, and the interpretation of branch support derived from the two techniques is correspondingly substantially different. In general,

Bayesian support values (posterior probabilities) tend to be higher than bootstrap support values, so interpreting branch support needs to take this into consideration (Rannala and Yang, 1996; Douady et al., 2003; Holmes, 2003, 2005; Alfaro and Holder, 2006).

8.11 Beyond Statistical Consistency: Sample Complexity

Saying that a method is statistically consistent says nothing in essence about how well it will perform on data, since you don't know how much data it needs to return the true tree with high probability. Thus, another issue of practical importance is the amount of data that a method needs to reconstruct the true tree with high probability. In the statistical literature, this is referred to as the "sample complexity" of a statistical estimation method.

An interesting example of a simulation study examining this issue is the "Hobgoblin of Phylogenetics" paper (Hillis et al., 1994), which presented model conditions and sequence lengths under which some statistically inconsistent methods were more accurate than some statistically consistent methods. In addition, some studies have provided mathematical analyses of the amount of data that are needed to obtain good accuracy with high probability, and expressed these in terms of the model tree parameters, the subject of the next section.

8.12 Absolute Fast Converging Methods

The model tree used to establish that maximum parsimony is statistically inconsistent provides strong evidence that the accuracy of a phylogenetic estimation method depends on the model tree branch lengths. Furthermore, we have already seen that several distance-based methods are guaranteed to be correct when the estimated distance matrix is within $f/2$ of the model distance matrix, where f is the length of the shortest internal edge in the model tree; hence, branch lengths also impact the theoretical guarantees of statistically consistent methods. The number of sequences (i.e., leaves in the model tree) obviously impacts running time; perhaps not so surprisingly, it can also affect accuracy. The impact of these different model parameters (number of sequences and branch lengths) on method accuracy, and on the sequence lengths that suffice for accuracy with high probability, is the subject of this section.

Consider a Jukes–Cantor tree (T, θ) with n leaves, where for each edge e the length of e (i.e., the expected number of changes of a random site on e) is denoted by $\lambda(e)$. Suppose that f is the length of the shortest edge and g is the length of the longest edge in T. Both f and g impact the sequence length requirements of a phylogeny reconstruction method, as we will see. Short edges are less likely to have changes, so to recover them we will need longer sequences. Hence, as $f \to 0$, the sequence lengths that suffice for exact reconstruction with high accuracy will increase. Long edges also increase the sequence length requirements, but for a different reason. Consider the impact of having a long edge e in the tree defining bipartition $A|B$. As the length of e increases, it becomes easier and easier to detect the bipartition $A|B$; for example, the Naive Quartet Method would have no trouble in correctly reconstructing the quartet trees of the form that split 2:2 around e. On the other hand, when e is very long, every two sequences $a \in A$ and $b \in B$ will be nearly as different as two randomly selected sequences. What this means is that even if the subtrees on A and on B could be reconstructed exactly, it could be very difficult to find the correct way to attach the two trees to each other. Thus, both f and g impact the sequence length requirements for phylogeny estimation methods.

Starting with Erdös et al. (1999a), a substantial mathematical framework has been developed to understand the sequence lengths that suffice for accuracy with high probability for different phylogeny estimation methods. We introduce a somewhat simple approach to this question, as presented in Warnow et al. (2001).

For a given stochastic model of evolution M (e.g., M could be the JC69 model, the GTR model, the GM model, etc.), we will define $M_{f,g}$ to be the set of model trees (with any number of leaves) where $f \leq \lambda(e) \leq g$, and where $f > 0$ and g are fixed but arbitrary. We will assume that Φ is a statistically consistent method, and we wish to bound the sequence lengths that suffice for accuracy with high probability given any model tree in $M_{f,g}$. Since Φ is statistically consistent, for every $\varepsilon > 0$ and for every $(T, \theta) \in M_{f,g}$, there is some sequence length K such that $Pr(\Phi(S) = T) > 1 - \varepsilon$ if S is a set of sequences of length K generated on (T, θ). We express the sequence lengths that suffice for accuracy with high probability as a function that depends on n (the number of leaves), ε, f, g, and Φ. Treating $f, g,$ and ε as fixed (but arbitrary), this means we can express the sequence length requirement for Φ as a function only of n, the number of leaves in the model tree.

We continue the discussion under the JC69 model of DNA sequence evolution, since it's the easiest to understand. Hence, we will talk about the $JC_{f,g}$ family of JC69 model trees.

Definition 8.5 The phylogeny estimation method Φ is **absolute fast converging** (*afc*) for the JC69 model if for all f, g, ε and all model trees $(T, \theta) \in JC_{f,g}$, there exists a polynomial $p(n)$ such that $Pr(\Phi(S) = T) > 1 - \varepsilon$ when given sequences of length at least $p(n)$ that have evolved down (T, θ).

Erdös et al. (1999a) presented the Dyadic Closure Method, which is the first method that was proven to be afc under a stochastic model of sequence evolution. In Section 3.4.4, we presented a method to compute the dyadic closure $cl(Q)$ of a set Q of binary quartet trees, and we showed that if T and T' two binary trees with the same leafset as Q, then if $cl_2(Q) = Q(T)$ and $Q \subseteq Q(T')$, then $T = T'$. Based on this observation, the Dyadic Closure Method is as follows:

The Dyadic Closure Method: Given a set S of n sequences,

- Compute the $n \times n$ Jukes–Cantor distance matrix **d** for S.
- For each $q \in d_{ij}$,

- Use the Four Point Method to compute a tree on every four sequences whose pairwise distances are at most q, and denote this set of quartet trees by $\mathscr{T}_q$.
- Compute the dyadic closure of $\mathscr{T}_q$ (i.e., $cl_2(\mathscr{T}_q)$). If $cl_2(\mathscr{T}_q) = Q(T)$ for some tree T, then set $T_q = T$, and otherwise set T_q to be the null tree.

- If for any q, T_q is not the null tree, return it; otherwise return Fail.

Theorem 8.6 *The Dyadic Closure Method is statistically consistent under the JC69 model.*

Proof We will show that for any $\varepsilon > 0$ then for long enough sequences, there will be a value q such that with high probability T_q is the model tree, and that if $T_{q'}$ is not the null tree then it is identical to the model tree as well. Let (T, θ) be a JC69 model tree, and let $qmax = max_{ij}\{d_{ij}\}$. Since the Four Point Method is statistically consistent, for any $\varepsilon > 0$, there is a $K > 0$ such that $\mathscr{T}_{qmax} = Q(T)$ with probability at least $1 - \varepsilon$ given sequences of length at least K that evolve down (T, θ). By Corollary 3.9, when $\mathscr{T}_{qmax} = Q(T)$, then for any $q \leq qmax$ either $\mathscr{T}_q$ is the null tree or $\mathscr{T}_q = Q(T)$. Hence, for any $\varepsilon > 0$, there is a sequence length K such that the Dyadic Closure Method returns the model tree T with probability at least $1 - \varepsilon$ given sequences of length at least K. □

This theorem only establishes that the Dyadic Closure Method is statistically consistent, but does not bound its sequence length requirements. However, as shown in Erdös et al. (1999a), the Dyadic Closure Method will return the true tree with high probability from polynomial length sequences!

Theorem 8.7 *For any JC69 model tree (T, θ) for which $f \leq \lambda(e) \leq g$ for all edges e, the Dyadic Closure Method will reconstruct the model tree topology T with probability at least $1 - \varepsilon$ given sequences of length $C_1 n^{C_2 g}$, where C_1 is a constant that depends on f and ε, C_2 is a constant that does not depend on f, g, or ε, and n is the number of leaves in the tree. Hence, the Dyadic Closure Method is absolute fast converging under the JC69 model.*

Proof We sketch the argument given in Erdös et al. (1999a), and express it in terms of the JC69 model. First, suppose that (1) q is large enough that $\mathscr{T}_q$ contains trees on all the short quartets of T, and (2) all trees in $\mathscr{T}_q$ are correctly computed using the Four Point Method. By Theorem 3.8, $cl_2(\mathscr{T}_q) = Q(T)$, where T is the model tree topology, and so $T_q = T$. Now let $q' \neq q$. If $\mathscr{T}_{q'}$ contains any incorrect quartet tree, then $T_{q'}$ will be the null tree; note also that if this happens then $q' > q$. If $q' < q$ then $\mathscr{T}_{q'}$ cannot contain incorrect quartet trees; in that case, if $cl_2(\mathscr{T}_{q'})$ contains a tree on every quartet of leaves then $T_{q'} = T_q = T$. Otherwise, $cl_2(\mathscr{T}_{q'})$ does not have a tree on every quartet of leaves and so $T_{q'}$ is the null tree. Hence, when these two conditions hold then for all q', either $T_{q'}$ will be the null tree or identical to T_q. Finally, the sequence length that suffices for (1) and (2) to be true with probability at least $1 - \varepsilon$ is at most $C_1 n^{C_2 g}$, where C_1 is a constant that depends on ε and f, and C_2 is a constant that does not depend on f, g, or ε. Hence, the Dyadic Closure Method is afc under the JC69 model. □

Much theoretical work has been done since Erdös et al. (1999a) about the sequence length requirements of different methods. For example, we now know that the neighbor joining method (and its variants) are *not* afc (Lacey and Chang, 2006), and that ML *is* afc (Roch and Sly, 2016).

The literature on the topic of sequence length requirements is quite rich, and is surveyed in Roch and Sly (2016). The theory about sequence length requirements has also led to innovations in algorithm design for phylogeny estimation (e.g., Huson et al., 1998, 1999a; Nakhleh et al., 2001a; Warnow et al., 2001; Brown and Truszkowski, 2012), some of which are described in Chapter 11.

8.13 Heterotachy and the No Common Mechanism Model

In most of the models of DNA sequence evolution we have discussed so far, there is a single rate matrix **Q** that governs the entire tree. To model site variability, sites can pick rates of evolution from a distribution, so that some sites will evolve twice as quickly as others, some will evolve one-third as quickly, etc. However, even rates-across-sites models do not really model site heterogeneity, since they only permit sites to be scaled versions of each other.

Real variation between sites is known to be more heterogeneous than this, and is referred to as **heterotachy** (Lopez et al., 2002). Figure 8.4 presents two model trees with the same underlying tree topology but with different branch lengths, so that neither tree can be obtained from the other by multiplying all the edges by a constant. Hence if site *i* evolves down one tree and site *j* evolves down the other, this would not fit a rates-across-sites assumption; thus, this is an example of heterotachy. See Lopez et al. (2002), Kolaczkowski and Thornton (2004), Lockhart et al. (2006), Taylor et al. (2006), and Zhou et al. (2007) for studies about heterotachy and its impact on tree estimation.

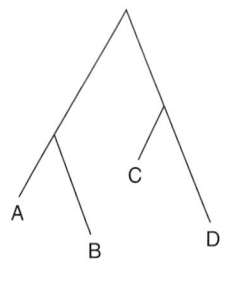

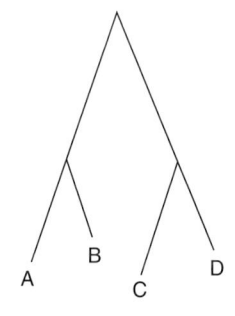

Figure 8.4 Two model trees with the same underlying tree topology, but with different branch lengths on the edges. The two trees are not scaled versions of each other; hence, this pattern of site variation is not consistent with a rates-across-sites model.

8.13 Heterotachy and the No Common Mechanism Model

Tuffley and Steel (1997) provide a model of sequence evolution that addresses heterotachy, by assigning an independent substitution matrix to each combination of edge and site. Under this model, the evolutionary process on every edge and site is independent of what happens on any other edge and site. Under the assumption that all character state changes are equiprobable (as in the CFN and JC69 models) and that the state at the root is selected at random, the evolutionary process is fully specified by parameters $p_{e,i}$ that specify the probability of a change on edge e for site i, with $0 \leq p_{e,i} \leq \frac{r-1}{r}$. Tuffley and Steel called this the **No Common Mechanism model** (NCM).

Theorem 8.8 *(From Tuffley and Steel (1997)) Let S be a set of r-state sequences. Then tree* T *is an optimal maximum likelihood tree under the r-state No Common Mechanism model if and only if* T *is an optimal solution to maximum parsimony.*

Proof We sketch the proof; see Tuffley and Steel (1997) for additional details. Let T be an arbitrary binary tree topology with leaves labeled by S. The maximum probability of generating S on T is obtained by assigning sequences at each of the internal nodes in order to optimize maximum parsimony, and then setting $p_{e,i} = 0$ if site i does not change on e and otherwise setting $p_{e,i} = \frac{r-1}{r}$. Under this setting for θ, $Pr(S|(T,\theta)) = (\frac{1}{r})^{MP(T,S)+1}$. (This calculation uses the fact that if a site changes on an edge, it changes to each of the other $r-1$ states with equal probability.) Hence, T is a maximum parsimony tree for S if and only if T is a maximum likelihood tree for S under the r-state NCM model. □

The consequence of this theorem is substantial! Let (T,θ) be a model CFN tree on which maximum parsimony is positively misleading (e.g., the Felsenstein Zone Tree from Example 8.3). Since the NCM model for binary sequences contains the CFN model as a special case, (T,θ) is an NCM model tree. If we compute ML trees on data generated by (T,θ) under the NCM model, then on every dataset we examine we will return the maximum parsimony tree. Since maximum parsimony will converge to a tree other than the model tree as the sequence length increases, ML under the NCM will converge to the same wrong tree. In other words, ML will be positively misleading on (T,θ) as well. We summarize this observation in the following theorem:

Theorem 8.9 *(From Tuffley and Steel (1997)) For every* r $\geq$ 2, *ML under the* r*-state NCM model is not statistically consistent and can be positively misleading.*

The proof we provided was for $r = 2$, but can be easily extended to larger values for r, and is given as a homework problem. Thus, statistical consistency depends on the model of evolution – so that with sufficiently complex models, even good techniques (such as ML) may not have the desired property of being statistically consistent. This is an important thing to realize, since it is easy to assume that the only time ML won't be consistent is if the assumed model is not the model that generated the data. This theorem clearly shows that this is not the case!

8.14 Further Reading

This chapter has only touched on statistical models of evolution, a rich and deep subject in its own right, and many textbooks have been written that provide excellent coverage of this area (Page and Holmes, 1998; Grauer and Li, 2000; Nei et al., 2003; Yang, 2009; Grauer, 2016). In addition, we have focused on methods, such as maximum likelihood, that provide point estimates of the tree topology, rather than on Bayesian approaches to phylogenetics, which output a distribution of the tree topology. For a deeper look at Bayesian phylogenetics, see Holder and Lewis (2003), Alfaro and Holder (2006), and Chen et al. (2014).

One of the interesting subjects we didn't cover in substantial detail is phylogenetic estimation when there is variation between sites, and even establishing statistical identifiability under these models is non-trivial. When all the sites evolve down a GTR model tree under the same rate, or with rates drawn from a gamma distribution, the tree is identifiable (Allman et al., 2008). However, the tree may not be identifiable under more general distributions of rates across sites. Mossel and Roch (2013) provide a good survey of the literature in the area, and also show that phylogenetic trees are generally identifiable for many rates-across-sites models, provided the trees are large enough. Mossel and Roch (2013) also provide efficient statistically consistent algorithms to compute the trees under these models. The inference of branch lengths under sufficiently general rates-across-sites models is also compromised, and not always identifiable (Evans and Warnow, 2005). The problem seems to be that the gamma distribution has some nice properties that allow many parameters to be estimated, but more general models may not. The question is therefore why the gamma model is used, rather than some other model? The answer may be the ease with which calculations can be performed (Felsenstein, 2004):

There is nothing about the gamma distribution that makes it more biologically realistic than any other distribution, such as the lognormal. It is used because of its mathematical tractability.

This observation leads to basic concerns about the reliability of dates at ancestral nodes, since these depend very much on branch length estimations and hence on the details of the model for rate variation. Thus, one direction for future research is to evaluate the impact in practice on estimates of dates resulting from violations of the assumed model of site variation.

Another aspect of molecular evolution that we have not discussed is *cladogenesis*, i.e., "lineage branching," and so refers to models for tree shape. For example, to model the speciation process, cladogenesis would address the rate at which species divide into subspecies or create new species, and perhaps die. Similarly, cladogenesis could model gene duplication and loss processes (see Chapter 10). The Yule model (Yule, 1924) is the earliest cladogenesis model, and models lineage branching as a pure birth process with a constant rate. The Yule model was subsequently extended to the birth–death model (Kendall, 1948), which allows for species to go extinct. The birth–death model is more general than the Yule model, but it makes the simplifying assumption that speciation and extinction occur at constant rates, which is not entirely realistic (Mooers and Heard, 1997).

However, the shape of the tree and the branch lengths are impacted also by taxon sampling strategies, and cladogenesis and taxon sampling both impact phylogenetic inference and downstream analyses (Mooers, 2004; Heath et al., 2008). Thus, better models of tree shape and branch lengths would be helpful in designing simulation studies to evaluate the performance of phylogeny estimation methods.

Better cladogenesis models are important for other reasons as well. For example, they can provide insights into the factors that impact speciation rates, and how speciations adapt to different environments. Bayesian methods require a prior on model trees, and so more realistic models of tree shape and branch length could be informative for these priors. Finally, distinguishing between trees that are close in terms of likelihood scores could be assisted by better cladogenesis models. Felsenstein (2004), Blum and Francois (2006), Heath et al. (2008), Phillimore and Price (2008), and Morlon et al. (2010) provide a good entry into the literature on cladogenesis models; see also Raup et al. (1973), Raup (1985), Guyer and Slowinski (1991), Gould et al. (1997), Aldous (1991, 2001), McKenzie and Steel (2001), and Heard and Mooers (2002) for a sample of the early work in this area.

8.15 Review Questions

1. Suppose we have a CFN model tree with branch substitution probabilities given by $p(e)$, as e ranges over the branches of the tree. Show how to define $\lambda(e)$.
2. What is a JC69 model tree? What is the JC69 distance correction?
3. What is a Generalised Time Reversible (GTR) model tree?
4. What is a General Markov (GM) model tree?
5. What is the relationship between the GM, GTR, and JC69 models?
6. What is the safety radius of neighbor joining?
7. What is the logdet distance correction? What models is it applicable to?
8. What is meant when we say that a method M is statistically consistent for estimating a GTR model tree?
9. Can statistical consistency be proven using a simulation study?
10. What is meant when we say that a method M is statistically inconsistent for estimating a GTR model tree?
11. Can statistical inconsistency be proven using a simulation study?
12. What is meant by saying that a method M is positively misleading for estimating a GTR model tree?
13. Suppose you are given a JC69 model tree and a set of sequences at the leaves. What is the computational complexity of computing the probability of the sequences at the leaves?
14. What is the JC69 Maximum Likelihood problem? What is its computational complexity?
15. Comment on the use of ML under the GTR model to analyze data that have evolved under the JC69 model. Is this statistically consistent?

16. Is maximum parsimony statistically consistent under the JC69 model? If not, what does it mean to say this?
17. What is the No Common Mechanism (NCM) model for binary sequences? Is maximum parsimony statistically consistent under the binary sequence NCM model? Is ML statistically consistent under the binary sequence NCM model?
18. What is the relationship between the NCM and CFN models?
19. What techniques are used to compute branch support on estimated trees?

8.16 Homework Problems

1. Prove or disprove: Every GTR model tree is a JC69 model tree.
2. Prove or disprove: Every JC69 model tree is a GTR model tree.
3. Prove or disprove: Every CFN model tree is a JC69 model tree.
4. Prove or disprove: Every GM model tree is a GTR model tree.
5. Prove or disprove: Every GTR model tree is a GM model tree.
6. True or false? If a method M is statistically consistent under the JC69 model, then it is also statistically consistent under the GTR model.
7. Consider the following algorithm for estimating JC69 model trees from sequence data. Given a set of sequences, we compute JC69 distances for the sequences. We then check to see if the distance matrix is additive; if it is, we return the tree T corresponding to the additive distance matrix, and otherwise we return a random tree. Prove or disprove: This is a statistically consistent method under the JC69 model.
8. Suppose we are given sequence dataset S generated by an unknown JC69 model tree. We compute logdet distances, and then run neighbor joining on the distance matrix we obtain. Is this a statistically consistent method?
9. Suppose we are given sequence dataset S generated by an unknown GTR model tree. We compute Jukes–Cantor distances, and then run neighbor joining on the distance matrix we obtain. Is this a statistically consistent method?
10. Suppose you have the CFN tree T with topology $((A,B),(C,D))$ with every edge having $p(e) = 0.1$, and rooted at A (note that in this tree, the root has only one child).
    - Compute the probability that $B = C$.
    - Compute the probability that $A = C$.
11. For the same CFN tree as in the previous problem, compute the probability that $A = B = C = D = 0$.
12. Consider a CFN model tree T with topology $((A,B),(C,D))$. Treat this as a rooted tree, with A being the root, and thus having five edges. Suppose the internal edge is labeled e_I, and we set $p(e_I) = 0.4$, and $p(e) = 0.001$ for all the other edges. Compute the probability of the following events:
    - $A = B = 0$ and $C = D = 1$
    - $A = C = 0$ and $B = D = 1$
    - $A = D = 0$ and $B = C = 1$

13. In this problem we will define a set of different CFN model trees on the same tree topology, $((A,B),(C,D))$ but with different edge parameters. We let e_I be the internal edge separating A,B from C,D, and let e_x be the edge incident with leaf x (for $x = A,B,C,D$). The trees are then defined by the edge parameters $p(e)$ for each of these edges, with these $p(e)$ given as follows:

- For T_1, we have $p(e_A) = p(e_C) = 0.499$, and $p(e) = 0.0001$ for the other edges e.
- For T_2, we have $p(e_I) = 0.499$ and $p(e) = 0.01$ for the other edges e.

Think about what kinds of character patterns you would see at the leaves of the trees, and answer the following questions;

a. Of the three parsimony informative character patterns, identify which one(s) would appear most frequently for tree T_1.
b. Of the three parsimony informative character patterns, identify which one(s) would appear most frequently for tree T_2.
c. For each of these model trees, do you think maximum parsimony would be statistically consistent? Why?
d. For each of these model trees, do you think UPGMA on CFN distances would be statistically consistent? Why?
e. For each of these model trees, do you think neighbor joining on CFN distances would be statistically consistent? Why?

14. Consider CFN model trees, all with the same tree topology, $((A,B),(C,D))$, but with different edge parameters. We let e_I be the internal edge separating A,B from C,D, and let e_x be the edge incident with leaf x (for $x = A,B,C,D$). The trees are then defined by the edge parameters $p(e)$ for each of these edges, with these $p(e)$ given as follows:

- For T_1, we have $p(e_A) = p(e_C) = 0.499$, and $p(e) = 0.0001$ for the other edges e.
- For T_2, we have $p(e_I) = 0.499$ and $p(e) = 0.01$ for the other edges e.
- For T_3, we have $p(e) = 0.499$ for all edges e.
- For T_4, we have $p(e) = 0.0001$ for all edges e.

a. Suppose one of these CFN trees generated a dataset of four sequences, and you had to guess which one generated the data. Suppose the dataset consisted of four sequences A,B,C,D of length 100 that were all identical; which would you choose?
b. Same question as above, but suppose the dataset consisted of four sequences A,B,C,D of length 10, where
    - $A = 0100100111$
    - $B = 0000000000$
    - $C = 0010101001$
    - $D = 0000000000$

15. Suppose we are given sequence dataset S generated by an unknown JC69 model tree, and we analyze the sequences using GTR maximum likelihood (solving the problem exactly). Will this be a statistically consistent method? (More to the point, if we estimate the tree under GTR using a statistically consistent method for GTR, such as

ML, but the data are generated by a JC69 model tree, is this a statistically consistent method?)

16. Recall the Cavender–Farris–Neyman (CFN) model, and consider three methods: maximum likelihood under CFN, maximum parsimony, and UPGMA on CFN distances.

   a. Consider invariant characters (i.e., characters that assign the same state to all the taxa). For each of the methods given above, say whether the invariant characters are informative, and explain your reasoning.

   b. Consider characters that are different on every taxon. For each of the methods above, say whether these characters are informative, and explain your reasoning.

17. Consider a CFN model tree T given by $((A,B),(C,D))$. Treat this as a rooted tree, with A being the root, and thus having five edges. Suppose the internal edge is labeled e_I, and we set $p(e_I) = 0.4$, and $p(e) = 0.001$ for all the other edges. Would maximum parsimony be statistically consistent on this model tree? Why?

18. Consider the following type of character evolution down a rooted binary tree T, in which every node is labeled by a unique integer (which may be positive, negative, or zero); note this means that in a tree with n leaves, there are $2n-2$ distinct labels. We do not assume that the label of a node is larger or smaller than its parent node, but we do assume that the label at the root is 0. The state of the character at the root is always 0. Every edge e in the tree T has a substitution probability $p(e)$ with $0 < p(e) < 1$. On an edge $e = (x,y)$, with x the parent of y, the character changes its state with probability $p(e)$; if it changes state, then the new state is y. As with other models we've studied, if there are multiple sites that evolve down the same tree, we assume that the substitution probabilities $p(e)$ govern all the sites, but can differ between edges. We also assume that the labels at the nodes are part of the model tree, and so are the same for all characters that evolve down the tree.

   a. Suppose the rooted model tree T has topology $(a,(b,c))$. Let the parent of b and c be labeled by 3, and let a be labeled by 5, b be labeled by 2 and c be labeled by 4. Recall that the root is always labeled by 0.

      - Suppose that a character evolves down this model tree but *never changes its state*. What are the character states at the leaves (a,b,c) for this character?
      - Suppose that the character evolves down this model tree and changes exactly once – on the edge from the root to a; what are the character states at the leaves for this character?
      - Suppose the character evolves down this model tree and changes exactly once – on the edge from the root to the parent of b and c. What are the character states at the leaves for this character?
      - Suppose the character evolves down this model tree and changes state on every edge of the tree. What are the character states at the leaves of the character?

   b. Suppose the following four sequences evolve down some unknown model tree of this type:

- $u = (3,0,1)$
- $v = (3,0,5)$
- $w = (0,8,2)$
- $x = (0,8,4)$

What is the tree topology, and what are the labels at the nodes of the tree? (Recall we already know that the root label is 0.)

c. Suppose the following six sequences evolve down some unknown model tree of this type:

- $A = (4,2,0,3,1)$
- $B = (4,2,0,3,6)$
- $C = (0,2,0,3,7)$
- $D = (0,0,0,3,8)$
- $E = (0,0,5,5,9)$
- $F = (0,0,5,5,10)$

What is the tree topology, and what are the labels at the nodes of the tree?

d. Suppose the following three sequences are given to you. Is it possible that they evolve down some unknown model tree of this type?

- $A = (4,0)$
- $B = (4,2)$
- $C = (0,2)$

If so, present the tree; otherwise prove this cannot be the case.

e. Describe a polynomial time statistically consistent method to infer the model tree topology from the site patterns. What is the running time of your algorithm? (Don't just say "polynomial.") What is your justification for saying it is statistically consistent under this model?

19. Prove that for all $r > 2$, there are model trees under the r-state No Common Mechanism model for which maximum likelihood is positively misleading.

20. Find a biological dataset with at least 50 aligned sequences, and try to solve maximum parsimony on the dataset using good software for maximum parsimony. How many "best" trees do you obtain? Now do the same for maximum likelihood. What do you find? If there are more "best" trees under one criterion than the other, what is your explanation for this? You might also want to repeat this experiment on a simulated dataset.

9
Multiple Sequence Alignment

9.1 Introduction

Phylogeny estimation generally begins by estimating a multiple sequence alignment on the set of sequences. Once the multiple sequence alignment is computed, a tree can then be computed on the alignment (Figure 9.1). Not surprisingly, errors in multiple sequence alignment estimation tend to produce errors in estimated trees (Ogden and Rosenberg, 2006; Nelesen et al., 2008; Liu et al., 2009a; Wang et al., 2012) and other downstream analyses. Hence, multiple sequence alignment is an important part of phylogeny estimation.

As we have seen, there are many methods for estimating trees from gap-free data. However, because multiple sequence alignments almost always contain gaps, represented as dashes, phylogeny estimation methods must be modified to be able to analyze alignments with dashes. Typically this is performed by treating the dashes as missing data (i.e., missing data means there is an actual nucleotide or amino acid, but it is not known). Alternatively, the dashes are sometimes treated as an additional state in the sequence evolution model, thus producing five states for nucleotide alignments or 21 states for amino acid alignments. Finally, sometimes sites (i.e., columns in the multiple sequence alignment) containing dashes are eliminated from the alignment before a tree is computed. The different treatments of sequence alignments can result in quite different theoretical and empirical performance.

Multiple sequence alignments are computed for different purposes, including phylogeny estimation and protein structure prediction, and the definition of what constitutes a correct alignment depends, at least in part, on the purpose for the alignment. For some biological datasets, curated alignments, typically based on experimentally confirmed structural features of the molecules (e.g., secondary structures or tertiary structures of RNAs and proteins), are used as benchmarks for evaluating alignment methods. Examples of such benchmarks for evaluating large amino acid alignments include HomFam (Sievers et al., 2011), BAliBASE (Thompson et al., 1999), and the 10AA collection (Nguyen et al., 2015b), while the Comparative Ribosomal Website (CRW) provides benchmarks for RNA alignment (Cannone et al., 2002). Evolutionary alignments, on the other hand, are defined by the evolutionary history relating the sequences.

First align, then construct the tree

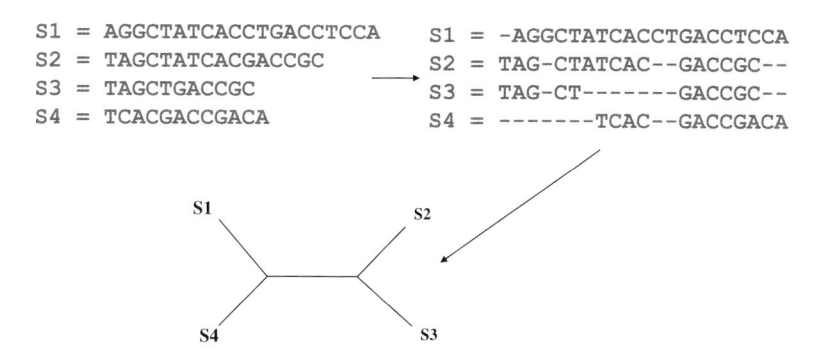

```
S1 = AGGCTATCACCTGACCTCCA        S1 = -AGGCTATCACCTGACCTCCA
S2 = TAGCTATCACGACCGC            S2 = TAG-CTATCAC--GACCGC--
S3 = TAGCTGACCGC          →      S3 = TAG-CT-------GACCGC--
S4 = TCACGACCGACA                S4 = -------TCAC--GACCGACA
```

Figure 9.1 Two-phase phylogeny estimation. In the standard two-phase approach, a multiple sequence alignment is first computed, and then a tree is computed on the alignment.

In either case, whether the multiple sequence alignment is based on a known evolutionary history (as in a simulation) or an established 3D structure for a protein family, the alignment serves as a reference, and the columns (i.e., sites) within the alignment define the "homologies." Therefore, homology can be based on structural features or evolutionary histories, leading to the opposing concepts of "structural homology" (also called "analogy") and "evolutionary homology" (also called **positional homology**). Thus, evolutionary homology for two letters or two sequences indicates descent from a common ancestor, so that saying that two sequences are *positionally homologous* is a strong statement. In contrast, while structural alignments are expected to be close to the true (evolutionary) alignment, convergent evolution may create conditions where the best structural alignment puts nucleotides or amino acids in the same site (thus implying homologies), even though these specific homologies are not present in the true evolutionary alignment (Iantomo et al., 2013). In other words, structural homology may not be identical to evolutionary homology (Reeck et al., 1987).

The main focus of this book is on phylogeny estimation, and so we will address multiple sequence alignment from that perspective. Hence, when we say that two letters or sequences are homologous, we will be referring specifically to positional homology. However, alignment methods can be evaluated using either type of benchmark – phylogenetic sequence alignment benchmarks with known evolutionary alignments (generally as a result of simulations) and structural sequence alignment benchmarks with known or estimated structures. In the rest of this chapter, we describe the various techniques that are used to produce multiple sequence alignments.

9.2 Evolutionary History and Sequence Alignment

Sequences evolve under processes that include events such as insertions and deletions (jointly called **indels**) that change the length of the sequences, and must be accounted for in a phylogenetic analysis. To explain how this is accounted for, we begin by showing how a pair of sequences that are related by evolution can be aligned. Figure 9.2 shows how one sequence evolves into another sequence through a combination of insertions, deletions, and substitutions, and the pairwise alignment that reflects the evolutionary history. There can be more than one true pairwise alignment relating two sequences, however, as the following example shows.

Example 9.1 Suppose sequence ACAT evolves into AGAT by deleting the C (thus creating AAT) and then inserting the G between the two As. The evolutionary history defines which pairs of nucleotides (one from each sequence) are homologous, and so any pairwise alignment that produces the precise set of pairwise homologies matching the evolutionary history is valid. However, for this pair of sequences with this evolutionary history, there is more than one representation as a pairwise alignment, each of which accurately reflects the true homologies. The first pairwise alignment is:

```
A C - A T
A - G A T
```

and the second pairwise alignment is

```
A - C A T
A G - A T
```

The only difference between these two alignments is the order in which the second and third sites appear, and so they define the same set of homologies. For this reason, they are equivalent alignments.

The true multiple sequence alignment of a set of DNA sequences (up to the ordering of sites) is defined by the evolutionary history relating the sequences. That is, every column in the true alignment should contain nucleotides that all have a common ancestor from which they have evolved via substitutions. To construct such an alignment, therefore, we would need to know the true history relating the sequences – including the sequences at each node of the phylogeny relating the sequences, and how each sequence evolved from its parent sequence. If we knew this history, then we would be able to define the true pairwise alignment on each edge of the tree, and then combine all these pairwise alignments into a multiple sequence alignment via transitivity.

9.3 Computing Differences Between Two Multiple Sequence Alignments

In this section we will describe ways of calculating differences between two multiple sequence alignments of the same underlying sequence dataset, and we will show how these measures can be used to quantify the error in a multiple sequence alignment estimation.

9.3 Computing Differences Between Two Multiple Sequence Alignments 181

```
        Deletion   Substitution
    ...ACGGTGCAGTTACCA...
                    Insertion          ...ACGGTGCAGTTACC-A...
    ...ACCAGTCACCTA...                 ...AC----CAGTCACCTA...
```

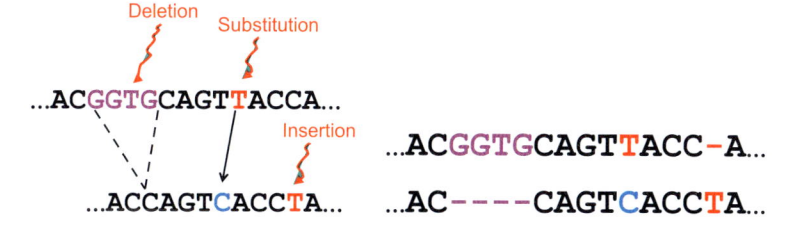

The true multiple alignment
- **Reflects historical substitution, insertion, and deletion events**
- **Defined using transitive closure of pairwise alignments computed on edges of the true tree**

Figure 9.2 Evolution and the true multiple sequence alignment. The top sequence evolves into the bottom sequence via the deletion of the substring GGTG, the substitution of a T for a C, and the insertion of a T. This corresponds to the pairwise alignment on the right. Note that two letters are placed in the same column only when they have a common history. Thus, the substring GGTG in the top string is above dashes in the bottom string, and indicates that deletion event. Similarly, the red T is above the blue C, to indicate that they have a common history. (The use of color here is only to help illustrate the points; nucleotides don't otherwise have colors!)

As with phylogeny estimation, the evaluation of multiple sequence alignments on biological data is challenging due to the inherent difficulty in knowing the true evolutionary history that relates a set of sequences. Simulation studies provide an alternative way to evaluate alignments, since then we can know the true alignment. The challenge in using simulations is that the models used to generate the data need to be sufficiently realistic that performance on the simulated data should be similar to performance on real data. This is a high standard, and may not be possible for standard simulations, which assume that the sites evolve under *i.i.d.* models (Morrison et al., 2015). Whether the evaluation is based on simulation or biological benchmarks, the reference alignment (i.e., true alignment in the case of a simulation, or structural alignment in the case of a biological benchmark) enables the error in the estimation to be quantified. Alternatively, the estimated alignment can be evaluated in terms of its impact on downstream analyses. For example, if an alignment is used to predict a protein structure, then how well the protein structure is predicted can be used as the metric to evaluate the alignment. Similarly, if the alignment is used to estimate a phylogenetic tree, then the error in the estimated tree computed using that alignment can be the basis for the evaluation. For these and other reasons, alignment evaluation is a complicated issue (Iantomo et al., 2013; Morrison et al., 2015). In what follows, we will focus on standard alignment estimation criteria, each of which is based on shared **homology pairs**, a concept we now make precise in the context of a nucleotide alignment.

S=	-	A	C	A	T	T	A
S'=	T	-	-	A	C	-	A

Table 9.1 S = ACATTA *evolves into* S' = TACA *by (1) deleting the first two letters (A and C) and the second T (thus producing ATA), (2) changing the remaining T into a C (thus producing ACA), and (3) putting a T at the front (thus producing TACA). The alignment shown is the true alignment of S and S' reflecting this evolutionary history.*

S=	-	A	C	A	T	T	A
S'=	T	A	C	A	-	-	-

Table 9.2 *An estimated alignment between S and S'. If we compare the alignment in this table to the pairwise alignment given in Table 9.1, we will note that both have two columns occupied by two As. Yet the two columns in the true alignment with two As (i.e., columns four and seven) correspond to different pairs of these nucleotides than the two columns in the estimated alignment with two As (i.e., columns two and four). For example, column four in the true alignment aligns the second A from sequence S and the first A from sequence S', and neither of the columns in the estimated alignment with two As use this specific pair of nucleotides. The same is true of column seven in the true alignment. In fact, the alignment given in this table has no homology pairs in common with the true alignment given in Table 9.1*

Each site (i.e., column) in a multiple sequence alignment puts the nucleotides in the column in relationship to each other, and is an assertion of homology. Thus, if there are five nucleotides in a given site, then there are ten (five choose two) pairs of nucleotides in the column, and all of these pairs are considered to be homology pairs. Hence, the set of homology pairs for a given nucleotide multiple sequence alignment contains all pairs of nucleotides that appear together in the same column within the alignment. It is important to note that we distinguish between different occurrences of the same nucleotide, based on where it appears in the sequence. This distinction will become clear when we evaluate the estimated alignment in Table 9.2 with respect to the true alignment given in Table 9.1.

Just as with techniques to quantify the error in tree estimation, we consider two types of error – false negatives and false positives. We show by example how we calculate false negatives and false positives for two alignments of the same pair of sequences $S = ACATTA$ and $S' = TACA$. Suppose that the true alignment between S and S' is given in Table 9.1, and we have an estimated alignment shown in Table 9.2. Both alignments have a column with two As, but they aren't using the same copies of this nucleotide. Hence, the two homology pairs (one from each alignment) are not identical.

The number of true positive homology pairs is the total number of homology pairs that both alignments share. The false positive homology pairs in an estimated alignment are the ones it produces that aren't in the true alignment, and the false negative homology pairs are the ones in the true alignment that are missing from the estimated alignment. The number of these true positives, false positives, and false negatives can be turned into *rates* by dividing by the appropriate number of homology pairs. Although this is described in terms of two sequences within a pairwise alignment, the definition extends to multiple sequence alignments, but still depends on homology pairs. These false negative and false positive rates are then values between 0 and 1, and are measures of *error*. Since they depend on the sum-of-pairs scores, they are referred to as the SPFN and SPFP rates, respectively (Mirarab and Warnow, 2011).

Measures of alignment accuracy (as opposed to error) are also commonly used, with the total column (TC) score counting the number of columns in the true and estimated alignment that are exactly identical (some authors have defined TC scores differently, for example by removing sites that have any gapped entries). This is a popular technique for evaluating alignment accuracy, but since it depends on matching everything in a column the TC score can be very low – especially for large datasets, or datasets that are highly heterogeneous. Two other terms are also used to discuss alignment accuracy: the modeler score and the SP-score. The modeler score is the same as precision, and is given by $1 - SPFP$, while the SP-score is the same as recall, and is given by $1 - SPFN$; see Appendix B for these statistical terms. We summarize this as follows:

Definition 9.2 Let $\mathscr{A}$ be the reference multiple sequence alignment, and $\mathscr{A}'$ an estimated multiple sequence alignment, both of the same set S of unaligned sequences. Each alignment can be represented by its set of pairs of homologous letters, and these sets can be compared.

- The **SPFN rate** is the fraction of the truly homologous pairs (in $\mathscr{A}$) that are not present in $\mathscr{A}'$; this is also called the sum-of-pairs false negative rate. The **SP-score** is equivalent to $1 - \text{SPFN}$, and is a measure of recall.
- The **SPFP rate** is the fraction of the homologous pairs in $\mathscr{A}'$ that are not present in $\mathscr{A}$; this is also called the sum-of-pairs false positive rate. The **modeler score** is $1 - \text{SPFP}$, and is a measure of precision.
- The **TC** score is the **number** of columns that are identical (including gaps) in the two alignments; this is also called the total column score.

Thus, the SPFN and SPFP rates represent error rates, and the TC score, SP-score, and modeler score represent accuracy measures. All these measures (SPFN, SPFP, SP-score, modeler score, and TC score) can be calculated in linear time using FastSP (Mirarab and Warnow, 2011).

Computing "distances" or "similarities" between two alignments. These measures, SPFN, SPFP, etc., are defined by comparing an estimated alignment to a true or reference alignment on the same dataset. However, these can also be used to compute "distances"

9.4 Edit Distances and How to Compute Them

One way to compute a pairwise alignment of two sequences is to define a cost for a substitution and a cost for an insertion or deletion (which typically depends on the length of the insertion or deletion), and then use these costs to define the minimum cost of an edit transformation. This minimum cost is called the "edit distance," and any pairwise alignment achieving that minimum cost is called an "optimal alignment." As we will see, there are cases where there is more than one optimal alignment, and so the optimal alignment may not be unique.

The actual edit distance depends on the specific costs for each event. For example, in Table 9.3, we show a pairwise alignment of two sequences $S = AACT$ and $S' = CTGG$ that is optimal when each indel has cost 1 and each substitution has cost 3. On the other hand, the alignment shown is far from optimal when indels have cost 4 and substitutions have cost 1!

Finding an optimal alignment between two sequences is sometimes straightforward. Under the Levenshtein distance, substitutions and insertions and deletions of single nucleotides all have unit cost; hence, if the input pair of sequences is $S = AAT$ and $S' = CAAGG$, there are several transformations of S into S' of minimum cost, and it is not hard to find one of the minimum cost transformations. For example, we could insert C at the beginning of $S = AAT$, obtaining $CAAT$. We could then change the T into a G, obtaining $CAAG$. Finally, we could add one more G, obtaining $CAAGG$, which is S'. Since each step had unit cost, the cost of this transformation of S into S' is 3. With a little effort, you will be able to see that there is no transformation of S into S' that uses fewer than three steps, and so this is a minimum cost transformation of S into S'. Since this transformation also defines an evolutionary process relating S and S', it has a pairwise alignment associated to it, shown in Table 9.4.

Finding the minimum cost transformation of S into S' was easy for this case, but what if the sequences are much longer? Say, for example, $S = AATTAGATCGAATTAG$ and

S=	A	A	C	T	-	-
S'=	-	-	C	T	G	G

Table 9.3 *Alignment of two sequences with four indels and two matches (no substitutions). This is an optimal pairwise alignment of the two sequences for some settings for the relative cost of indels and mismatches. For example, if each indel cost 1 and each substitution costs 3, then this alignment has total cost 4, and no other alignment can have lower cost. However, if each indel costs 4 and each substitution costs 1, then this alignment would have total cost 16, and an indel-free alignment would have total cost 4.*

S=	-	A	A	T	-
S'=	C	A	A	G	G

Table 9.4 *An optimal pairwise alignment of two sequences when all substitutions and single-letter indels have unit cost.*

S' = *CATTAGGATTGAACATTAGTACA*? You can quickly convince yourself that finding a minimum cost transformation (where each insertion, deletion, and substitution has unit cost) by trial and error is painful, and that even if you could find one for this particular pair, you wouldn't want to do this if the two sequences had hundreds of nucleotides in them. Fortunately, there are fast methods to find minimum cost transformations between two sequences.

Needleman–Wunsch. The dynamic programming algorithm in Needleman and Wunsch (1970) presents the alignment problem in terms of maximizing similarity. However, algorithms for maximizing similarity can be turned into algorithms for minimizing distances, and vice versa. Hence, we will describe the Needleman–Wunsch algorithm in terms of minimizing edit distances, and for only its simplest version.

Gap costs. There are multiple ways of defining the cost of a gap of length L, which can be used within the Needleman–Wunsch dynamic programming algorithm, including:

- Simple gap penalties: The cost of a gap of length L is cL for some constant $c > 0$.
- Affine gap penalties: The cost of a gap of length L is $c_0 + c_1 L$, where $c_0 > 0$ and $c_1 > 0$.

The simple gap penalties are sometimes referred to as linear gap penalties, but the term "linear" actually encompasses affine gap penalties as well. Other gap penalties (e.g., Miller and Myers, 1988; Gu and Li, 1995; Altschul, 1998; Qian and Goldstein, 2001; Chang and Benner, 2004) have also been proposed that may provide better fits to biological data (though see Cartwright, 2006), but affine gap penalties are currently the most commonly used.

Edit distances under a simple gap model. We begin with the simplest case of computing the edit distance between two DNA sequences (i.e., strings over the alphabet $\Sigma = \{A, C, T, G\}$) under a simple gap penalty model. The simple gap penalty is not particularly realistic, but it's a good starting point for understanding how algorithms work to compute these optimal pairwise alignments.

We will assume that the cost of an indel of a single letter is 1 and each substitution has constant cost C that may be different from 1. Under the assumptions of a simple gap model, the minimum cost edit transformation can be calculated efficiently using dynamic programming. But before we introduce the dynamic programming solution to this problem, we will define a *canonical edit transformation* defined by a pairwise alignment.

Canonical edit transformation, given a pairwise alignment. Suppose you are given two sequences S and S' in a pairwise alignment. There are many edit transformations that are consistent with the given pairwise alignment, of course, because the *order* of the events changes the transformation. However, if all indels are always of length 1, then modulo the order of events that occur, there is a canonical edit transformation that is implied by the pairwise alignment – start at the left end of the alignment and perform the events implied by the sites, from left to right, until you reach the right end of the pairwise alignment. We continue with some notation that will make this exposition easier to follow.

Definition 9.3 Let $S = s_1 s_2 s_3 \ldots s_n$, so that s_i is the ith letter in S. The ith prefix of S is the string $S_i = s_1 s_2 \ldots s_i$, and we let $S_0 = \lambda$ denote the empty string, which is (vacuously) a prefix of S.

Hence, the first j positions in a pairwise alignment $\mathcal{A}$ between S and S' define an alignment between the corresponding prefixes of S and S'. Furthermore, if $\mathcal{A}$ is an optimal pairwise alignment of S and S', then the first j positions of $\mathcal{A}$ define an optimal alignment of the corresponding prefixes of S and S' specified by the first j columns of $\mathcal{A}$.

Now suppose we are working with a simple gap penalty, so that each indel of length L costs L times an indel of a single letter. Once we have specified the cost for each substitution and for each single indel, then given any transformation of S into S' and the associated pairwise alignment, the cost of the transformation is the sum of the costs of the sites in the alignment, as we will now show. For example, suppose that each substitution and each indel has unit cost. Now, given the pairwise alignment

```
- A A T -
C A A G G
```

the sites indicate the following events that transform sequence *AAT* into *CAAGG*:

- the first site indicates an insertion of a *C* before the string, and has cost 1;
- the second and third sites indicate no changes (no indels and no substitutions), and so have cost 0;
- the fourth site indicates a substitution of *T* by *G*, and has cost 1;
- the fifth site indicates the insertion of a *G* at the end of the string, and has cost 1.

Thus, the cost of this transformation and its associated pairwise alignment is 3. It is easy to see that this is an optimal pairwise alignment of the two strings. Under the assumption of a simple gap model, each optimal pairwise alignment of two strings is obtained by extending an optimal alignment of prefixes of the two strings. On the other hand, the following is also a pairwise alignment of AAT and CAAGG achieving the same score:

```
- A A - T
C A A G G
```

Thus, the optimal pairwise alignment may not be unique.

We will use these ideas and terminology to define a dynamic programming algorithm to compute the edit cost between any two strings A and B, and then use those calculations to find a minimum cost transformation of A into B and its corresponding pairwise alignment. We begin with the simple gap cost function, but make it even simpler by constraining each single substitution and single letter indel to have unit cost. Later, we will show how to modify the dynamic programming to allow for more general cost functions.

Definition of the subproblems. The input is two strings $A = a_1 a_2 \ldots a_m$ and $B = b_1 b_2 \ldots b_n$. We let $F(i,j)$ denote the edit distance between A_i and B_j, $0 \leq i \leq m$ and $0 \leq j \leq n$, where A_i refers to the ith prefix of A and B_j refers to the jth prefix of B. We let $H(a_i, b_j)$ denote the Hamming distance between a_i and b_j, so that $H(a_i, b_j) = 0$ if $a_i = b_j$ and $H(a_i, b_j) = 1$ otherwise.

The base case. The base case is $i = j = 0$, which denotes the cost of transforming an empty string into an empty string; it is easy to see that $F(0,0) = 0$.

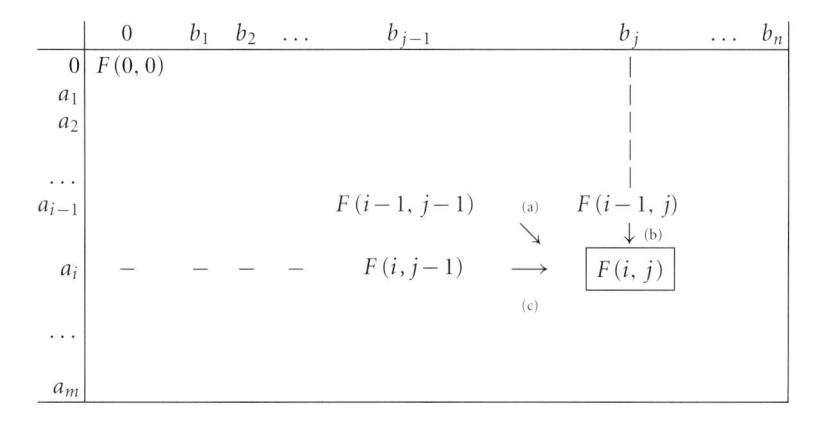

	0	b_1	b_2	...	b_{j-1}		b_j	...	b_n
0	$F(0,0)$								
a_1									
a_2									
...									
a_{i-1}					$F(i-1, j-1)$	(a)	$F(i-1, j)$		
					↘		↓ (b)		
a_i	—	—	—	—	$F(i, j-1)$	→	$F(i, j)$		
						(c)			
...									
a_m									

Figure 9.3 (Figure 2.4 in Huson et al. (2010)) The Needleman–Wunch dynamic programming approach to computing the minimum cost of any pairwise alignment (i.e., edit distance) between two sequences, $A = a_1 a_2 \ldots a_m$ and $B = b_1 b_2 \ldots b_n$, under a simple gap cost model. The $F(i,j)$ entry in the $m \times n$ matrix $\mathbf{F}$ indicates the cost of an optimal pairwise alignment between A_i and B_j, where A_i is the ith prefix of string A and B_j is the jth prefix of string B. The order in which the matrix is filled in requires that $F(i-1, j-1), F(i, j-1)$ and $F(i-1, j)$ all be computed before $F(i, j)$ is computed (e.g., it can be filled in from top to bottom, or from left to right). Case (a) corresponds to a pairwise alignment in which the last site matches a_i with b_j; case (b) corresponds to a pairwise alignment in which the last site has only a_i, and case (c) corresponds to a pairwise alignment where the last site has only b_j. The three cases each have a corresponding total cost (defined by the cost of aligning the two prefixes and then the additional cost of a single indel for cases (b) and (c), or the cost of "substituting" a_i with b_j (i.e., no cost if $a_i = b_j$), and the minimum of these three costs is then placed in $F(i,j)$; see text for details.

The recursive definition. Suppose we want to compute $F(i,j)$ and we have computed all "smaller" subproblems. Hence, in particular, we have computed $F(i,j-1)$, $F(i-1,j)$, and $F(i-1,j-1)$. Thus, we know the edit distance between A_i and B_{j-1}, between A_{i-1} and B_j, and between A_{i-1} and B_{j-1}. Now, imagine you have an optimal edit transformation of A_i into B_j and the corresponding pairwise alignment. The final site of the pairwise alignment must take one of the following forms:

- Case 1: a_i and b_j are aligned together in the final site. In this case, the other sites (before this last site) define a pairwise alignment of A_{i-1} and B_{j-1}.
- Case 2: a_i is aligned to a dash in the final site. In this case, the other sites (before this last site) define a pairwise alignment of A_{i-1} and B_j.
- Case 3: b_j is aligned to a dash in the final site. In this case, the other sites (before this last site) define a pairwise alignment of A_i and B_{j-1}.

For Case 1, the pairwise alignment of A_i and B_j either involves a match (when $a_i = b_j$) or a mismatch (when $a_i \neq b_j$). Cases 2 and 3 each involve an indel (so the second case involves a deletion of a_i and the third case involves an insertion of b_j). The costs of these events are as follows:

- Case 1: The cost implied by the last site is 0 if $a_i = b_j$ and otherwise the cost is 1. Hence, the total cost is $F(i-1,j-1) + H(a_i,b_j)$.
- Case 2: The cost of the last site is 1. Hence, $F(i,j) = F(i-1,j) + 1$.
- Case 3: The cost of the last site is 1. Hence, $F(i,j) = F(i,j-1) + 1$.

Although we don't yet know the optimal pairwise alignment, we do know that it takes one of these forms. Hence, if we have already computed $F(i,j-1), F(i-1,j)$, and $F(i-1,j-1)$, we can set $F(i,j)$ to be the *minimum* of the three possible costs (using the above analysis). In other words, we set

$$F(i,j) = min\{F(i-1,j-1) + H(a_i,b_j), F(i-1,j) + 1, F(i,j-1) + 1\}.$$

Filling in the DP matrix. We need to compute $F(i,j)$ for all $0 \leq i \leq m$ and $0 \leq j \leq n$. We can compute these entries in any order we like, as long as we don't try to compute $F(i,j)$ before we compute the values on which it depends. Hence, we can fill in the matrix row-by-row, column-by-column, or even in a diagonal way. We'll do this (for simplicity's sake) row by row. Thus:

For all $0 \leq i \leq m$ and $0 \leq j \leq n$, $F(0,j) = j$ and $F(i,0) = i$
For $i = 1$ to m DO

    For $j = 1$ to n DO

        $F(i,j) = min\{F(i-1,j-1) + H(a_i,b_j), F(i-1,j) + 1, F(i,j-1) + 1\}$

Return $F(m,n)$

Note that how $F(i,j)$ is defined depends on whether i or j is 0, since these require special treatment, and that the final answer is located in $F(m,n)$.

Figure 9.3 presents a visualization of how $F(i,j)$ is computed during the dynamic programming algorithm; the arrows going into $F(i,j)$ indicate the values that need to be examined during the calculation of the $F(i,j)$ entry. For this simple cost function where all substitutions and single indels have unit cost, the calculation is quite easy. We examine a_i and b_j: If they are the same, then the optimal cost will be equal to $F(i-1,j-1)$. Otherwise, $F(i,j)$ will be exactly $1 + min\{F(i,j-1), F(i,j-1), F(i-1,j-1)\}$, because the last site in the pairwise alignment of A_i and B_j will have to incur unit cost.

Finding the optimal alignment from the DP matrix. This algorithm computes the edit distance between two strings under the assumption that all events (indels and substitutions) have unit cost, but does not compute any edit transformation (or pairwise alignment) achieving that cost. We will show how we can use backtracking to output at least one such minimum cost transformation. (In fact, the backtracking can provide an implicit representation of *all* the minimum cost transformations!)

When you set $F(i,j)$, you are finding the minimum of three values. Whichever entry (or entries, if there are more than one) gives you the smallest value, put an arrow from the box for $F(i,j)$ to the box that gave you the smallest value. For example, if $F(i,j)$ was set to be $F(i-1,j-1)+1$, then put an arrow from $F(i,j)$ to $F(i-1,j-1)$. Then, at the end of the computation, these arrows will define at least one path from $F(m,n)$ all the way back to $F(0,0)$, and each such path will define a minimum cost edit transformation and its pairwise alignment.

Running time. It is easy to see that the algorithm takes $O(1)$ time to compute $F(i,j)$ for each i,j, given that you compute the values in an appropriate order. Thus, the running time of the Needleman–Wunsch algorithm is $O(nm)$, where the input is a pair of strings, one that has length m and the other that has length n.

Modifying the DP algorithm to handle arbitrary substitution cost matrices. Suppose that the cost of substituting one nucleotide by another is symmetric, but depends on the pair of nucleotides. In other words, we are given a 4×4 symmetric substitution cost matrix M where $M[x,y]$ is the cost of substituting nucleotide x by nucleotide y.

It is straightforward to modify the dynamic programming algorithm to address this case; you just need to modify how you define $F(i,j)$ to account for different costs for mismatches in the final site. In other words, you would replace $H(a_i,b_j)$ with $M(a_i,b_j)$.

Modifying the algorithm to account for gap costs where a single indel costs C is also easy: just change $+1$ for the increase in the cost incurred by an indel in the last site to $+C$. More extensive modifications are needed to modify the algorithm to account for more general indel cost functions. While affine gap penalties can be handled without an increase in computational complexity, more general gap penalty models may require the examination of more entries in the DP algorithm in order to compute each entry. Even so, arbitrary gap cost functions can be accommodated within polynomial time (i.e., cubic instead of quadratic time).

Maximizing similarity instead of minimizing distance. A variant of this problem is obtained by defining the similarity between two strings, and then seeks the pairwise alignment yielding the maximum pairwise similarity score. It is not hard to modify the Needleman–Wunsch algorithm appropriately so that it is described in those terms; however, note that instead of penalizing for mismatches, the algorithm must explicitly favor matches.

9.5 Optimization Problems for Multiple Sequence Alignment

Now that you have seen how to compute an optimal pairwise alignment using edit distances, we can consider multiple sequence alignment. The first question is how to define the cost of a multiple sequence alignment. There are two very natural ways to define the cost of a multiple sequence alignment that can be considered, both of which are extensions of the cost of a pairwise alignment.

9.5.1 Sum-of-Pairs Alignment

The first optimization problem we present is called "Sum-of-Pairs" (SOP), and is defined as follows. We assume we can compute the cost of any pairwise alignment (and we allow quite general formulations for this cost). The cost of a given multiple sequence alignment of three or more sequences is defined by summing the costs of its set of induced pairwise alignments. The SOP optimization problem is then as follows: Given input set S of sequences and the function for computing the cost of any pairwise alignment, find an alignment $\mathscr{A}$ on S such that the sum of the induced pairwise alignments is minimized.

Some examples should make this clear. As an example, consider three sequences s_1, s_2, s_3 in a multiple sequence alignment given in Table 9.5. Suppose that every insertion, deletion, and substitution has unit cost. Then, examining the multiple sequence alignment in Table 9.5, we compute the total SOP cost as follows. The induced pairwise alignment between s_1 and s_2 has cost 2, the induced pairwise alignment between s_1 and s_3 has cost 3, and the induced pairwise alignment between s_2 and s_3 has cost 3. Hence, the multiple sequence alignment has SOP cost 8. However, a multiple sequence alignment of these three sequences is achievable that has SOP score 7; find it!

s_1	A	-	-	C
s_2	A	T	A	C
s_3	C	-	A	G

Table 9.5 *A multiple sequence alignment on three sequences with SOP score 8, where each indel or substitution has unit cost.*

9.5 Optimization Problems for Multiple Sequence Alignment

The SOP optimization problem asks us to find a multiple sequence alignment with minimum SOP cost. Finding the best multiple sequence alignment for a set of n sequences, based on a given cost function c, is NP-hard (Wang and Jiang, 1994). An extension of the dynamic programming algorithm described above can be used to find an optimal multiple sequence alignment (under the SOP criterion) in time that is exponential in the number n of sequences (i.e., $O(k^n)$ time, where each sequence has at most length k); this makes the calculation of optimal alignments feasible when n is very small. However, in general, optimal multiple sequence alignments under the SOP criterion are not generally attempted.

9.5.2 Tree Alignment

In the Tree Alignment problem (Sankoff, 1975), the input is an unrooted tree T with leaves labeled by the set S of n sequences, $S = \{s_1, s_2, \ldots, s_n\}$, and the function c that determines the edit distance between two sequences. Suppose we have also assigned sequences to every node in the tree; the total cost of the tree will be the sum over all the edges of the edit distances between the sequences labeling the endpoints of the edge. The Tree Alignment problem then seeks a way of assigning sequences to the internal nodes of T so that the total cost of the tree is minimized. As we have already seen, given the sequences at the nodes of the tree, we can compute the minimum transformation on every edge, and hence also a pairwise alignment for every edge; therefore, given the sequences at the nodes of the tree, we can also define the multiple sequence alignment associated with the minimum transformations on the edges. Thus, the assignment of sequences to the internal nodes that achieves the minimum total cost for the tree also defines a multiple sequence alignment on the sequences at the leaves, which we call an **optimal tree alignment**.

Example 9.4 Suppose the set of sequences is $S = \{s_1, s_2, s_3\}$, where $s_1 = ATA$, $s_2 = AAT$, and $s_3 = CAA$, and suppose indels have a very large cost (say 100) while substitutions have unit cost. Now let T be the unrooted tree with the sequences in S at the leaves. What is the best label of the internal node you can find, and what is the cost for this tree? You should be able to find that the sequence AAA gives a cost of only 3, and is the best possible sequence that could be obtained for this tree. However, if we had been constrained to pick the label only from the sequences at the leaves, then the cost would have been larger. Thus, better scores may be achievable when the internal nodes can be labeled differently from the sequences at the leaves.

Example 9.5 Suppose the input is $s_1 = AC$, $s_2 = ATAC$, and $s_3 = CAG$, and suppose we are given an unrooted tree T with an internal node X, and s_1, s_2, and s_3 at the leaves. Suppose insertions, deletions, and substitutions each have unit cost. If we set $X = s_1$, the cost of the tree would be the sum of the costs of the optimal pairwise alignments between s_1 and the other two sequences. The cost of the optimal alignment between s_1 and s_2 is 2, and the cost of the optimal alignment between s_1 and s_3 is 2; hence, the total cost would be $2 + 2 = 4$. What would the cost have been if we used $X = s_2$? It would have been the cost of the pairwise alignment between s_1 and s_2, plus the cost of the pairwise alignment between s_2 and s_3, and so $2+3=5$. Finally, if we used $X=s_3$, then the cost would have been the cost of the pairwise alignment between s_1 and s_3, plus the cost of the pairwise alignment between s_2 and s_3, or $2+3=5$. Hence, if we restricted X to be one of the input sequences, then the best result would have cost 4, and would be obtained by setting $X=s_1$. On the other hand, if we are not constrained to selecting from among the input sequences, we could let X be some other sequence. Can you find a better solution for this problem than $X=s_1$?

The Tree Alignment problem (Sankoff, 1975; Sankoff and Cedergren, 1983) allows the internal nodes of the input tree to be labeled by arbitrary strings, and is NP-hard even for simple gap penalty functions (Wang and Jiang, 1994; Wareham, 1995). In contrast, the maximum parsimony problem (which assumes a fixed alignment is given) on a fixed tree can be solved exactly in polynomial time using dynamic programming (see Section 4.3). Thus the Tree Alignment problem is *harder* than the fixed tree maximum parsimony problem.

Some special cases of Tree Alignment can be solved in polynomial time; for example, when the number of sequences is three, then the Tree Alignment problem amounts to finding the median of the three sequences, which can be computed using dynamic programming (Sankoff and Cedergren, 1983). Tree alignment can also be solved exactly on larger trees, but the running time grows exponentially with the number of leaves (Sankoff and Cedergren, 1983). While exact solutions on large datasets may be intractable, approximation algorithms that have bounded error have been developed (Wang et al., 1996, 2000; Wang and Gusfield, 1997). For example, Wang and Gusfield (1997) showed that an optimal "lifted-alignment" approach (which roots the tree, and then assigns sequences to the internal nodes by always picking a sequence from one of its two children) would be a two-approximation to the optimal alignment. In other words, the best lifted alignment would have a total treelength that was no more than twice that of the optimal alignment.

9.5.3 Generalized Tree Alignment

The Tree Alignment problem assumes that the tree is given as input, and the objective is the best labels to the internal node to minimize the total cost of all the induced pairwise alignments on the edges of the tree. However, if the tree is not given in the input, then the input would be a set S of unaligned sequences and the output would be a tree T with internal nodes labeled by sequences over the same alphabet of minimum total cost. This is called the *Generalized Tree Alignment* problem.

The Generalized Tree Alignment is a special case of the Steiner Tree problem, one of the most well-studied computational problems. The input to the Steiner Tree problem is a set S of points, drawn from a metric space, and the objective is to find a set X of additional points (called the "Steiner points") in the metric space and a spanning tree on $S \cup X$ that has minimum cost, where the cost is the sum of the distances on the edges of the tree.

9.5 Optimization Problems for Multiple Sequence Alignment

It is not surprising that the Generalized Tree Alignment problem is also NP-hard, but the Generalized Tree Alignment problem is even MAX SNP-hard (Wang and Jiang, 1994), so that arbitrarily good approximation algorithms are unlikely to be computationally efficient.

The 2-approximation to Generalized Tree Alignment. Although the Generalized Tree Alignment problem is MAX SNP-hard, it does have an efficient 2-approximation algorithm, which we now describe. The input is a set S of sequences and a function for computing the edit distance between any two sequences. Hence, given any graph G with sequences labeling every node, we define the weight of an edge (s, s') to be the edit distance between s and s', and $w(G)$ to be the total of the weights of the edges in G.

- Step 1: Compute the edit distance between every pair of sequences in S,
- Step 2: Construct the weighted complete graph G with one vertex for each sequence in S, and where the weight of the edge (s, s') is the edit distance between s and s'.
- Step 3: Find a minimum spanning tree T for G (that is, a connected acyclic subgraph T of G that includes every node in G and that has minimum total weight) using, for example, Kruskal's algorithm (Kruskal, 1956).
- Step 4: We now make a new tree T', as follows. We begin by copying T. Then, for every vertex v in T that is not a leaf, we add a vertex v' that is adjacent to v and give it the same sequence label. Thus if T has n leaves and n' internal nodes, then T' has $n + 2n'$ leaves and n' internal nodes. Return T'.

It is easy to see that $w(T) = w(T')$, and that every sequence in S appears as a leaf in T'; hence, T' is a phylogeny for S with the same weight as T. Since we have computed the optimal pairwise alignment for every two sequences that are connected by an edge in T', we can also compute a multiple sequence alignment that agrees with these pairwise alignments by transitivity.

The running time to compute T' depends on how long it takes to compute the edit distance between two sequences (which is polynomial under the models we have discussed) and the time to compute the minimum spanning tree (which is also polynomial). Hence, this is a polynomial time algorithm. We now prove that T' is a 2-approximation to the Generalized Tree Alignment problem.

Theorem 9.6 *(From Takahashi and Matsuyama (1980); Gusfield (1993)) Let S be any non-empty set of sequences, and let* T' *be a tree constructed by the algorithm described above. Then* w(T') < 2w(T^{opt}), *where* T^{opt} *is an optimal solution to the Generalized Tree Alignment problem on input S.*

Proof Since $w(T) = w(T')$, it suffices to show that $w(T) < 2w(T^{opt})$. Consider the graph G^* obtained by doubling every edge in T^{opt}; since every node G^* has even degree, G^* has a circuit C that covers every edge exactly once (i.e., G^* is Eulerian). Hence, $w(C) = w(G^*) = 2w(T^{opt})$. Let P be a path obtained by deleting one edge from C, and note that $w(P) < w(C) = 2w(T^{opt})$. Since P will generally have multiple appearances of a given vertex, we construct a path P' from P that contains exactly one copy of each vertex by only including the initial occurrence of each vertex, and skipping over the remaining occurrences. For example, if $P = v_1, v_2, v_3, v_1, v_4, v_3, v_2, v_4$, then P' would be v_1, v_2, v_3, v_4. Since edit distances satisfy the triangle inequality, it follows that $w(P') \leq w(P)$. Hence, $w(P') \leq w(P) < w(C) = 2w(T^{opt})$. Because C was an Eulerian circuit, P' is a spanning tree for G^*, and so $w(T) \leq w(P')$ (since T is a minimum spanning tree for G^*). Hence, $w(T') \leq w(P') < 2w(T^{opt})$, and the theorem is proved. □

There have been many follow-ups to this theory, including algorithms with improved performance guarantees or improved performance in practice; see Du et al. (1991), Gusfield (1993), Karpinski and Zelikovsky (1997), and Schwikowski and Vingron (1997, 2003) for an entry into this literature.

9.6 Sequence Profiles

One of the major developments in molecular sequence analysis was the development of profile hidden Markov models in the early 1990s (Brown et al., 1993; Haussler et al., 1993; Krogh et al., 1994). These profile HMMs are statistical models that can be constructed from multiple sequence alignments, and then used to construct alignments for additional sequences. However, profile HMMs are also used in other bioinformatics analyses, including the detection of remote homology, classification of proteins into protein families, inference of structures of proteins and RNAs, and identification of genes during genome annotation. Because profile HMMs are somewhat complicated, we postpone the discussion of profile HMMs until Section 9.7, and begin here with a simpler (and yet also important) concept: *sequence profiles* (Gribskov et al., 1987).

Given a multiple sequence alignment, we can compute the frequency of each letter (nucleotide or amino acid) in a given position within the multiple sequence alignment. For example, given a DNA sequence alignment, for each position in the alignment we can associate a four-tuple giving the proportion of the sequences in which each of the four nucleotides appears. No site in a multiple sequence alignment is entirely gapped (or, rather, if such a site is produced for some reason, it is then deleted as it is meaningless). Hence, a nucleotide alignment with k sites defines a profile of length k, with the ith position of the profile occupied by the four-tuple for the distribution of nucleotides in the ith site. Similarly, an amino acid alignment with k sites defines a profile of length k where the ith position is occupied by a 20-tuple. As an example, see Table 9.6, which presents a single multiple sequence alignment with five sequences and its associated profile.

The values in the profile can be used in several ways. For example, they can be used to define a similarity score between a sequence and the profile, in which case the profile can be used to search a database for sequences that have high similarity to the sequences used to construct the profile (see Section 9.20). Another use of a profile is as a generative model, as we now show.

9.6 Sequence Profiles

Alignment				
$s_1 =$	A	-	-	C
$s_2 =$	A	T	A	C
$s_3 =$	C	C	A	G
$s_4 =$	A	T	G	C
$s_5 =$	G	A	G	G

Profile				
A	0.6	0.25	0.5	0.0
C	0.2	0.25	0.0	0.6
T	0.0	0.50	0.0	0.0
G	0.2	0.0	0.5	0.4

Table 9.6 *A multiple sequence alignment on five sequences, and its associated (unadjusted) sequence profile.*

Using sequence profiles as generative models Using the profile given in Table 9.6, we create the graphical model with a begin state, an end state, and four match states M_1, M_2, M_3, and M_4, one for each position. The match states emit nucleotides with the probabilities given by their frequencies for each position. This creates a generative model with topology shown in Figure 9.4.

Note that we can generate DNA strings of length four by picking a nucleotide for the first position with probabilities equal to the observed relative frequencies, then picking a nucleotide for the second position, etc. As a result, the two most probable sequences for that profile would be *ATAC* and *ATGC*, each of which has probability $0.6 \times 0.5 \times 0.5 \times 0.6 = 0.09$. It should also be obvious that any sequence with a *T* in the first position or a *G* in the second position cannot be generated by this profile, and has probability 0.

In this case, allowing entries in the profile to be zero (as in the first position of the profile shown here) results in the profile being unable to generate some sequences that have the same length as the profile. For example, when we try to use this profile to generate additional sequences, we will not be able to generate any sequences that begin with

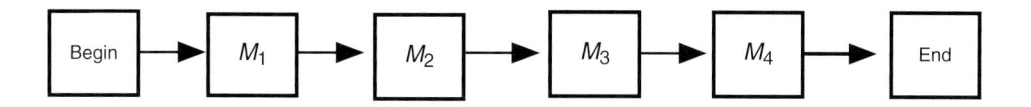

Begin → M_1 → M_2 → M_3 → M_4 → End

Figure 9.4 The graphical model for a sequence profile representing an alignment of length four, adapted from a figure in Durbin et al. (1998). The graphical structure of a sequence profile, with a begin state (Begin) and end state (End), and four match states, M_1, M_2, M_3, and M_4. There is a transition edge from Begin to M_1, from M_i to M_{i+1} for $i = 1, 2, 3$, and from M_4 to End; each of these transition edges has probability 1. Each match state emits a single symbol from its alphabet from a probability distribution.

T. Therefore, it is common to modify the profile so that all entries are strictly positive, typically accomplished using mixtures of observed frequencies and *pseudo-counts*, which are very small numbers often based on Dirichlet mixture priors (Sjölander et al., 1996). However, so that we can keep the exposition simple, we will work with profiles based purely on the observed frequencies (i.e., no adjustment will be made to ensure that all the entries are positive), and we'll call these **unadjusted sequence profiles**.

Given a profile P associated with an alignment, we can construct a probabilistic graphical model to generate sequences with the same distribution as P. Figure 9.4 shows the graphical model when the profile has length 4, but the process can be generalized to any length profile in the obvious way. We include a *start state* (also called a "begin" state) and an *end state*. In between the vertices for these two states, we include one vertex for each position in the alignment; these vertices are called *match states*. There is a transition edge from the start state to the match state for position 1, then a transition edge from the match state for position 1 to the match state for position 2, etc. Finally, there is a transition edge from the match state for position 4 to the end state. The graphical model produced in this fashion is a directed graph in which all edges move from left to right; thus, this is a directed acyclic graph (DAG).

Because of the special structure of this graph, every path through this model starts at the begin state, visits every match state (in turn), according to the sites in the alignment, and then ends at the end state. The order in which the states are visited is determined by the transition edges, which are always followed (i.e., the transition probability on every edge is 1).

All states other than the start and end states are associated with positions in the alignment, and so are associated with a distribution on the nucleotides at that position. Thus, each match state has a vector of emission probabilities for the different nucleotides, using the probability distributions we obtained from the profile. When we visit one of these states, we select a nucleotide using the distribution for the state, and write it down. In this way, following a path through the graphical model produces a DNA sequence of length four.

This probabilistic graphical model is Markovian, because the emission probabilities and transition probabilities for each state depend only on the state, and not on the path taken to reach the state. This is a *gap-free* model, so that if you know the topology of the model (i.e., the graphical structure), you will be able to know the path you took through the model for any given sequence that can be emitted by the model.

Although a profile can be seen as a generative model, it's limited to generating sequences of the same length as the profile. If we include gaps in a profile, so that the probability vector associated with the ith site has an extra entry for a gap (i.e., a vector of length five for nucleotide sequences, or a vector of length 21 for amino acids), then we can generate sequences that are shorter than the profile length, but we still cannot generate sequences that are longer. Thus, a more general model than profiles is needed to enable sequence length heterogeneity. Profile hidden Markov models (profile HMMs) are graphical models that were developed to address this issue.

9.6 Sequence Profiles

$$S_1 = A\ T\ G$$
$$S_2 = A\ T\ C$$
$$S_3 = A\ A\ G$$
$$S_4 = \text{-}\ \text{-}\ G$$

Figure 9.5 Four sequences in an alignment with a gap in sequence S_4.

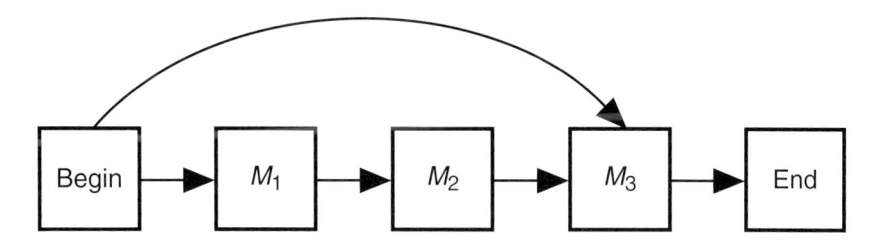

Figure 9.6 The graphical structure of a generative model with three match states for the alignment given in Figure 9.5. Note the deletion edge from the begin state to the third match state to represent the gap in sequence S_4. Figure adapted from Durbin et al. (1998: 105).

A slightly more complex model. To motivate the design of a profile HMM, we will begin with the challenge of modeling an alignment that has gaps, each representing either an insertion or a deletion event. Figure 9.5 shows an alignment of four sequences in which one sequence has a gap, and we interpret the gap as a deletion event. How can we model this alignment?

Figure 9.6 shows how we modify the sequence profile to allow us to generate sequence S_4; since S_4 is missing nucleotides in the first two positions, we put an edge from the begin state directly to the third match state. The resultant figure has only one extra edge, which we refer to as a deletion edge. We would need to set the transition probabilities on the edges leaving the begin state so that they summed to 1; since there are four sequences in the alignment, we would set the probability of following the deletion edge to 0.25, and the probability of following the edge *Begin* $\to M_1$ to 0.75. This is a very simple case of a sequence alignment with a single gap (here, of length two), and a very simple model that can be used to represent the alignment.

This approach can be generalized by including a deletion edge for every gap we see in the given multiple sequence alignment, but if the alignment has many gaps, then the resultant model can have many deletion edges, as shown in Figure 9.7.

There are several limitations to using deletion edges to model all gaps. One limitation, which we've already noted, is that when the alignment is large and has many gaps, then

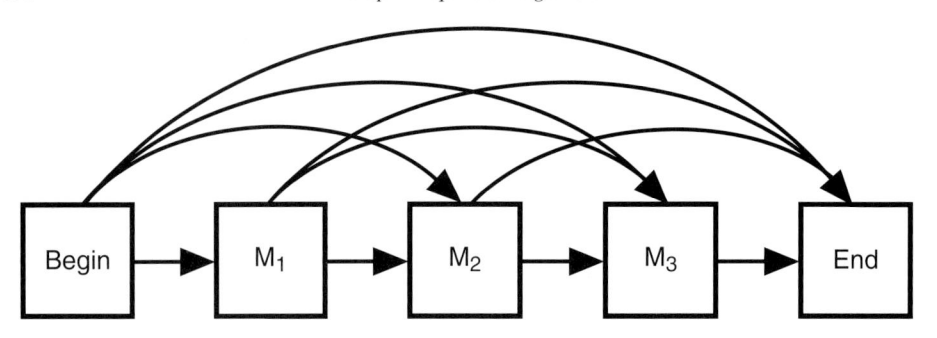

Figure 9.7 The graphical structure of a generative model with three match states, in which there is a deletion edge from every state to all states to its right other than the immediate successor. Figure adapted from Durbin et al. (1998: 105).

the model can become quite large, with potentially $O(k^2)$ deletion edges for an alignment of length k. However, another limitation is that the generative model created in this way *cannot* generate sequences that are longer than the multiple alignment. In the next section, we will show another approach to modeling sequence alignments that addresses both of these limitations – profile hidden Markov models.

9.7 Profile Hidden Markov Models

9.7.1 The Graphical Model

Profile hidden Markov models are statistical models that are quite similar to these profiles, but that have greater flexibility in modeling sequence variation, and that can be used as generative models as well. However, in many cases, given a sequence generated by a profile HMM, you may not be able to tell what path was used to produce the sequence. That is why they are referred to as "hidden" Markov models.

Suppose we are given an alignment that has many different gaps of different lengths and in different locations. One option is to have many deletion edges and insertion nodes to model the variation in the sequences. However, this could result in a large number of transition edges, with the consequence that inferring the model could become difficult. Instead, to keep the graphical model small (so that it has a linear number of transition edges and states), we will use nodes rather than edges to represent deletion events, and we will connect the deletion states together to model long deletions. We will also allow self-loops on the insertion states, so that arbitrarily long insertions can be modeled.

Figure 9.8 shows the generic graphical structure for a profile HMM, in which states M_j, I_j, and D_j indicate the jth match, insertion, and deletion states, respectively. The model has an insertion state I_0 between the begin state and the first match state, and all insertion states have self-loops. As a result, insertions of any length can be produced, and the model can generate sequences of any length. This figure shows the graphical structure but does

9.7 Profile Hidden Markov Models

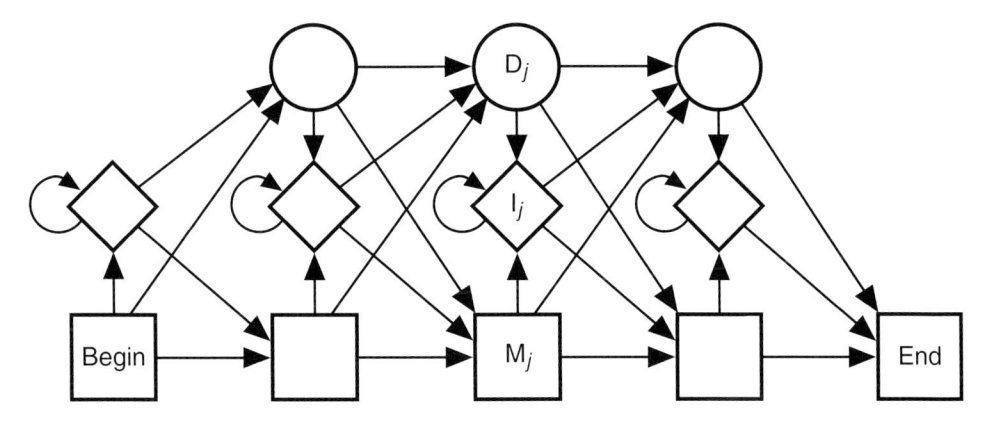

Figure 9.8 (Figure 5.2 in Durbin et al. (1998)) The graphical structure of a profile hidden Markov model (HMM), where the nodes of the profile HMM correspond to states of the model and directed edges represent the transitions between states. Each profile HMM has a begin state and an end state, and three other kinds of nodes: match states (indicated by squares), insertion states (indicated by diamonds), or deletion states (indicated by circles). Nodes representing insertion states can have self-loops, but no other nodes can; other than these self-loops, the underlying directed graph is acyclic. Not shown in this figure are the numeric parameters: the directed edges in the profile HMM are associated with transition probabilities, with the sum of the probabilities on the outgoing edges of any node being 1. Every match state and every insertion state emits a single symbol from its alphabet, and has an associated emission probability; the deletion states are "silent" and do not emit anything.

not show the numeric parameters of the model, which are the transition probabilities on the edges and the emission probabilities for the match and insertion states. Recall that the transition probability on an edge is the probability that the edge will be taken, and so for every node v, the transition probabilities on the outgoing edges from v must sum to 1. Similarly, the emission probabilities of the different symbols in the underlying alphabet must sum to 1 for any match or insertion state. Note that the alphabet depends on the type of data; thus, the alphabet would be $\{A,C,T,G\}$ for DNA sequence, or $\{A,C,U,G\}$ for RNA sequences, or the alphabet of 20 amino acids for amino acid sequences.

9.7.2 Strings Generated by a Profile HMM

For a given profile HMM, every path from the start state to the end state defines a set of strings over the alphabet that can be generated by the path. The empty string λ (i.e., the string of length 0) can only be generated by the path that avoids all the match and insertion states and instead uses the deletion states.

Consider a profile HMM M with three match states, using the graphical model shown in Figure 9.8. Since deletion states do not emit any symbols, the path $Begin \to D_1 \to D_2 \to D_3 \to End$ will produce the empty string, but all other paths will produce non-empty strings. The path $Begin \to M_1 \to M_2 \to M_3 \to End$ generates strings of length three, but cannot generate strings that are shorter or longer than three. Furthermore, depending on the emission probabilities, some strings may not be able to be generated by any path through a given model. For example, if the match and insertion states have zero probability of emitting the nucleotide A, then no DNA string with any As can be generated by the model.

Some strings can be generated by more than one path through one of these models. For example, for the model M described above, suppose that the probability of emitting each nucleotide is non-zero for every insertion or match state, and the transition probabilities from the match state M_i to I_i is strictly between 0 and 1. In this case, every non-empty sequence can be generated by multiple paths through the model. For example, the string AAA can be generated by the path $Begin \to M_1 \to M_2 \to M_3 \to End$ and also by $Begin \to M_1 \to I_1 \to D_2 \to M3 \to End$. Thus, one of the interesting consequences of using a profile HMM as a generative model is that even if we know the entire probabilistic graphical model (i.e., the underlying directed graph, the emission probabilities for the match and insertion states, and the transition probabilities on all the directed edges), we will typically not be able to infer the path through the model that generated a given sequence. Although we've already noted some cases (e.g., the empty string) where it is possible to know the path that was taken through the model to generate the string, the general case is that many strings can be generated by two or more paths through one of these models. Whenever a Markov model has the property that some strings can be generated by two or more paths through the model, we will say that the model has "hidden states," and we will refer to these models as *hidden* Markov models. Furthermore, because these models are based on sequence profiles (with match states corresponding to sites in a multiple sequence alignment), these are called "profile hidden Markov models," or more simply profile HMMs.

9.7.3 Probability Calculations on Profile HMMs

Although it is not possible to know the true path in a profile HMM that generated a given string s, the probabilistic framework allows us to answer several questions that will be useful to us, such as:

- Question 1: Given a profile HMM M and a string s that can be generated by M, what path through M is most likely to have produced s?
- Question 2: Given a profile HMM M and a string s that can be generated by M, what is the probability that M generated s?
- Question 3: Given two profile HMMs M and M' and a string s, which profile HMM is more likely to have generated s?

As we will show, each of these questions can be addressed using dynamic programming algorithms.

Question 1: Computing the most probable path through a profile HMM. Question 1 (finding the path in the model that is most likely to have generated a given string) can be answered in polynomial time, using a dynamic programming algorithm called the *Viterbi*

Algorithm, after Andrew Viterbi who developed it in the context of coding theory. Because insertion states have self-loops, each insertion state can be used several times. Therefore, the path is actually a *walk* (since the term "walk" allows for vertices to repeat, and the term "path" doesn't generally allow repeated appearance of a vertex). However, in the HMM literature, this distinction between paths and walks is not followed, and the objective of the Viterbi algorithm is described in terms of finding the best path. For that reason, we will continue to abuse the term in this chapter, and refer to the search for an optimal path.

We assume we are given a string x and a profile HMM H, and we wish to find a path π^* through H that *could* have generated the string x, and among all such paths the one that maximizes the probability of generating x. We will use $P(x,\pi)$ to denote the probability that the string x is generated by a path π that begins at the start state and ends at the end state of H. Every insertion or match state that π uses will generate a symbol from Σ; hence, the number of insertion or match states used by the path must be the length of the string x, or else $P(x,\pi) = 0$.

We will use the following notation:

- Σ is the alphabet (e.g., all the symbols that can be generated by any state in H).
- a_{vw} is the transition probability for the directed edge vw (i.e., the probability of moving directly to w from v, given that you start at v).
- $e_v(b)$ is the probability of emitting symbol $b \in \Sigma$ when in state $v \in V$, under the assumption that v is not a deletion state.
- The string $x = x_1 x_2 \ldots x_n \in \Sigma^n$, and its ith prefix $x_{1\ldots i}$ is $x_1 x_2 \ldots x_i$; by convention, the 0th prefix is the empty string.
- V will denote the set of states of H, and will include a start state v_0 and an end state v_L.
- Every path π we will examine will begin at the start state v_0 but may terminate at some state other than v_L, the end state for H.

The subproblems we will compute will have the form $MPP(v,i)$, and will denote the *maximum probability of any path* through the profile HMM H that begins at v_0 and ends at v and that generates the ith prefix of x. If we can compute all these subproblems correctly, then $MPP(v_L, n)$ will give us the value that we are seeking – the maximum probability of any path through H that begins at the start state and ends at the end state, and that generates x.

The boundary condition is where $v = v_0$, and for these we have $MPP(v_0, 0) = 1$ and $MPP(v_0, i) = 0$ for $i \geq 1$. The recursion for $MPP(v,i)$ depends on whether v is a state that can emit symbols or not.

If v is a deletion state or the end state, then it cannot emit any symbols. Any path π beginning at v_0 and ending at v and that generates a string $x_{1\ldots i}$ can be written as a path $\pi' v$, where π' is the subpath of π beginning at v_0 and ending at a node w immediately preceding v in π. Hence wv will be a directed edge in H. Therefore, when v is a deletion state or the end state, $MPP(v,i) = max_w \{MPP(w,i) a_{wv}\}$.

If v is a match state or an insertion state, then it must emit the symbol x_i. Furthermore, π', the subpath of π as described above, must generate the string $x_{1\ldots i-1}$. Therefore, when v

emits symbols, then $MPP(v,i) = max_w\{MPP(w,i-1)a_{wv}e_v(x_i)\}$. Putting this together, we have the following recursion:

$$MPP(v,i) = \begin{cases} 1 & v = v_0 \text{ and } i = 0 \\ 0 & v = v_0 \text{ and } i > 0 \\ max_w\{MPP(w,i-1)a_{wv}e_v(x_i)\} & v \text{ is a match state or an insertion state} \\ max_w\{MPP(w,i)a_{wv}\} & \text{otherwise} \end{cases}$$

The order in which we compute the subproblems must ensure that we have computed all $MPP(v,i)$ before we compute any $MPP(w,i)$ where wv is a directed edge in H. Hence we only need to specify an ordering of the states in the HMM. Examining the layout of the generic profile HMM given in Figure 9.8, we note that no directed edges go from right to left; hence, we will start at the left and move to the right in listing the states. We compute the subproblems associated with the start state first (these are the boundary conditions). Then, we compute subproblems associated with the insertion state above the start state. Now we can go to the first column of three states immediately to the right of the start state (i.e., the states $M_1, I_1,$ and D_1). The match state and the deletion state both have directed edges to the insertion state and so must precede the insertion state but can appear in either order. Hence, we can order these three states as $M_1, D_1,$ and I_1. We then move to the next column and put the three states in the same relative ordering, M_2, D_2, I_2. We repeat this until all the match, deletion, and insertion states are included, and then finish with the end state.

Because we can order the subproblems so that each subproblem depends only on subproblems that appear before it in the ordering, a dynamic programming approach can correctly compute all the subproblems. The Viterbi algorithm returns $MPP(v_L, n)$, which is the maximum probability of generating string x of length n among all paths in H that begin at the start state and end at the end state. Backtracking can then be used to return a maximum probability path through the profile HMM (and hence to align the sequence to the profile).

The running time analysis is straightforward. There are $O(kn)$ subproblems, where k is the number of states in H and n is the length of x. Computing an ordering on the subproblems takes $O(k)$ time. There are at most three other states w that can immediately precede any vertex v in the H. To compute $MPP(v,i)$ for any vertex v requires that we examine at most three other states w, look up previously stored values and multiply them, and take the maximum of the resultant three products. Hence, to compute $MPP(v,i)$ takes $O(1)$ time, and so the running time for the Viterbi algorithm is $O(kn)$. Enabling backtracking does not change the computational complexity, and so finding a maximum probability path can also be performed in $O(kn)$ time.

The algorithm we have described computes the maximum probability of any path through the model. However, in many implementations, the objective is not the maximum probability of any path but rather the maximum logarithm of the probability of any path. Switching to using logarithms of probabilities rather than probabilities does not change the set of paths that are optimal, but can improve numerical aspects of the implementation.

Note also that when logarithms are used, the values are summed rather than multiplied (i.e., $log(ab) = log(a) + log(b)$).

Question 2: Computing the probability that a profile HMM generates a string x. The Viterbi algorithm is a dynamic programming algorithm to find the maximum probability of any path through a given profile HMM to generate a given string. However, as discussed above, sometimes we want to know the total probability, over all possible paths through the given model, of generating a given string x. A very similar dynamic programming algorithm can be used to compute that total probability. The key idea is to define subproblems $TP(v,i)$ to denote the total probability of all paths through the model that begin at the start state and end at state v and that generate the ith prefix $x_{1...i}$. Setting up the boundary conditions and the recursion is then straightforward. The algorithm for this problem is called the *Forward Algorithm*, and is described in detail in Durbin et al. (1998) and Ewens and Grant (2001).

Question 3: Comparing two models. The third question we asked was how to determine which of two models is more likely to have generated a given string. Now that Question 2 can be answered, it is easy to do: Given string x and two profile HMMs H_1 and H_2, we just compute $P(x|H_1)$ and $P(x|H_2)$, and return the model that had the larger probability.

9.7.4 Building Profile HMMs

Building a profile HMM from an input multiple sequence alignment has two basic steps: First the graphical model is computed (which essentially amounts to determining which columns in the alignment correspond to match states and which correspond to insertion states), and then the numeric parameters on the graphical model are computed. The determination of which columns correspond to match states and which do not is typically based on the percentage of ungapped sites, with some methods automatically making all columns that are at least 50 percent gapped into insertion states.

Once the graphical model is constructed, the numeric parameters can be set to maximize the joint probability of the sequences in the multiple sequence alignment; this calculation is enabled by noting that for each such sequence, the path through the model is given by the multiple sequence alignment. The only wrinkle here is ensuring that no sequence has zero probability, which can happen if the set of sequences never use some state; in this case, the estimations of the parameters are adjusted using appropriate pseudocounts (for example using Dirichlet mixture priors (Sjölander et al., 1996). When the input sequences are unaligned, computing a profile HMM turns out to be much more challenging, even if the graphical model is given; one of the popular approaches is the Baum–Welch method (Baum, 1972).

9.7.5 Using Profile HMMs to Align Sequences

Profile HMMs are used in multiple ways, including searching databases for homologous sequences, annotating sequences, etc. However, profile HMMs can also be used to align sets of sequences. We describe this in the context of amino acid sequence alignment for established protein families, under the assumption that each protein family has full-length proteins. However, some databases, such as Pfam (Bateman et al., 2002), only model single domains; this complicates this process because the result is not an alignment of the entire protein.

We assume we are given a database of established protein families and their curated alignments. Then, we build a profile HMM for each of the curated multiple sequence alignments. Then, given a collection S of amino acid sequences that are all part of a particular protein family F, we can produce a multiple sequence alignment for S as follows:

- Note the reference alignment A and the profile HMM M for the family F.
- For each sequence $s \in S$, find a maximum probability path through M. Once the path is computed, it associates each amino acid in s to either a match state or an insertion state. The amino acids that are associated with match states are added to the site for the match states, indicating that homology has been inferred by the alignment. The treatment of amino acids that are associated with insertion states is more complicated; because they have not been inferred to be homologous to anything in the reference alignment, each such amino acid *should* be added to a new site where every other entry is gapped. In practice, some alignment methods will place all such amino acids into the same single insertion column, on top of each other; such an approach has the advantage of saving space, but because it implies homology, it is also potentially misleading. (Note, the amino acids might actually *be* homologous, but the inference technique does not establish this!)
- The multiple sequence alignment for the set S is the transitive closure of the alignments that have been computed for each sequence in S. Equivalently, the sites in the new alignment contain all those amino acids from the sequences in S that align to a given match state; all other amino acids are insertions, and are not considered homologous to any other amino acid.

The accuracy of the output multiple sequence alignment depends on several factors, including the accuracy of the reference alignment A and the accuracy of the alignments of the sequences in S to A produced by using a maximum likelihood path through the HMM M. Note that this type of approach cannot be used to align datasets for which there is no reference alignment; in other words, this is not a *de novo* alignment strategy.

9.8 Reference-based Alignments

In the approach we have just described, the multiple sequence alignment is aided by the use of a previously computed *reference alignment* (for example, the Pfam alignments for each protein family).

PAGAN (Löytynoja et al., 2012) is another approach that uses a reference alignment, but it uses it in a very different manner. In addition to the reference alignment, a tree on the alignment is given as part of the input. PAGAN then computes optimal sequences for the internal nodes of the tree, and then adds the input sequences to the reference alignment by finding the most similar sequence in the tree for each input sequence, and aligning the input sequence to that tree sequence. Löytynoja et al. (2012) showed that PAGAN compared favorably to an HMM-based method using the popular HMMER (Eddy, 2009; Finn et al., 2011) suite of tools, especially under model conditions with high rates of evolution. However, PAGAN sometimes failed to align sequences, while HMMER aligned all sequences.

9.9 Template-based Methods

Another type of reference alignment method uses external alignments that come with either structural or biochemical annotations, and uses that information in guiding the multiple sequence alignment. For example, there are some homologous protein sequences whose sequences have solved three-dimensional structures, and statistical models that provide structural information about the underlying sequence alignment have been developed and used to good advantage in protein alignment (Kemena and Notredame, 2009). This type of approach also benefits from the fact that sequences with known structures can be more accurately aligned. Statistical models that incorporate additional features, especially ones related to structure or biochemical properties, are called "templates," and many multiple sequence alignment methods have been developed that take advantage of templates. Morrison (2006) provides an interesting discussion of early template-based methods (up to 2006), but see also Preusse et al. (2007), DeSantis et al. (2006), Neuwald (2009), Gardner et al. (2012), Nawrocki (2009), Nawrocki et al. (2009), and Löytynoja et al. (2012).

Some of these template-based methods use curated alignments based on structure and function of well-characterized proteins or rRNAs; for example, the protein alignment method MAPGAPS in Neuwald (2009) and the rRNA sequence alignment method in Gardner et al. (2012) use curated alignments. COBALT (Papadopoulos and Agarwala, 2007), 3DCoffee (O'Sullivan et al., 2004), PROMALS (Pei and Grishin, 2007), and the method in Kececioglu et al. (2010) similarly use external information like structure and function, but then use progressive alignment techniques (see Section 9.12) or other techniques to produce the final alignment. Clustal-Omega (Sievers et al., 2011) also has a version, called "External Profile Alignment," that uses external information (in the form of alignments) to improve the alignment step. Several studies (Zhou and Zhou, 2005; Pei and Grishin, 2007; Neuwald, 2009; Deng and Cheng, 2011; Gardner et al., 2012; Ortuno et al., 2013; Sievers et al., 2013) have shown that alignment methods that use high-quality external knowledge can surpass the accuracy of some of the best purely sequence-based alignment methods.

9.10 Seed Alignment Methods

In the previous three sections we discussed methods that use external reference alignments in order to align a collection of sequences. These methods use various techniques to add sequences into the reference alignment (e.g., profile HMMs or inferring ancestral sequences) and can provide good accuracy. Furthermore, because the input sequences can be aligned independently, they also provide excellent scalability. However, because these methods depend on external reference alignments, they are not *de novo* alignment methods. In this section, we describe how the basic strategies of these reference alignment methods can be extended to enable *de novo* alignment. We refer to these methods as "seed alignment methods" because they operate by *computing* a seed alignment, rather than using a precomputed seed alignment.

The input to a seed alignment method is a set S of unaligned sequences. A seed alignment method has two basic steps:

- A subset S_0 of the input set S is selected, and an alignment $\mathscr{A}$ is computed for S_0; this alignment is then referred to as the "seed alignment."

- The remaining sequences are added to $\mathscr{A}$.

For example, profile HMMs can be used in a seed alignment method. After the alignment $\mathscr{A}$ is computed for the subset S_0, a profile HMM is built for $\mathscr{A}$, and used to add all the sequences in $S \setminus S_0$ to $\mathscr{A}$ using the Viterbi algorithm. Note that how the set S_0 is selected and aligned will have an impact on the downstream alignment.

One of the main advantages of this kind of *de novo* multiple sequence alignment method is its tremendous scalability to large datasets. If the number of sequences selected for the seed alignment is small, then even computationally intensive multiple sequence alignment methods can be used to produce the seed alignment. In addition, the addition of the non-seed sequences to the seed alignment is fast (polynomial time using the Viterbi algorithm) and each sequence can be aligned independently.

One of the applications where this approach has been useful is phylogenetic placement, in which the objective is to add sequences into a tree constructed for a reference alignment. EPA (Berger et al., 2011) and pplacer (Matsen et al., 2010) are two phylogenetic placement methods that operate by representing the given sequence alignment by a profile HMM, add the sequences into the alignment using the Viterbi algorithm, and then place the aligned sequence into the tree optimally with respect to maximum likelihood. As shown in Berger et al. (2011) and Matsen et al. (2010), such approaches can be highly accurate and very fast. In Section 9.17, we will show an elaboration on the seed alignment approach in which the seed alignment is represented by a collection of profile HMMs instead of a single HMM.

9.11 Aligning Alignments

There are many techniques that have been developed to merge pairs of alignments. One approach seeks a merged alignment that minimizes the SOP score; this is NP-hard (Ma et al., 2003), but can be solved exactly for moderate-sized datasets (Kececioglu and Starrett, 2004).

More recent techniques for merging pairs of alignments have generally used two types of techniques: represent each alignment by a profile and then align the profiles (i.e., profile–profile alignment) (Gotoh, 1994; Kececioglu and Zhang, 1998; Yona and Levitt, 2002; Edgar and Sjölander, 2003; Sadreyev and Grishin, 2003; Edgar and Sjölander, 2004; Wheeler and Kececioglu, 2007) or represent each alignment by a profile HMM and then align the two profile HMMs (i.e., HMM–HMM alignment) (Söding, 2005). Profile–profile alignment and HMM–HMM alignment are each accomplished using dynamic programming techniques similar to those used to align two sequences (e.g., the Needleman–Wunsch algorithm) or to add a sequence into an alignment by finding an optimal path through a profile HMM (e.g., the Viterbi algorithm). Given an alignment of the two profiles (or of the two profile HMMs), this defines an association between pairs of match states in each alignment, and hence a way of merging the two alignments together. Two alignments can also be merged together using a two-level HMM, where one level models sequence evolution and the other level models structural evolution (Löytynoja and Goldman, 2008a). Note that these techniques for merging two alignments do not change the alignments they operate on.

Example 9.7 We show how a profile–profile alignment technique can be used to merge two multiple sequence alignments on disjoint sequences into an alignment on the full dataset. Suppose alignments $\mathscr{A}_1$ and $\mathscr{A}_2$ are given as input; see Table 9.7.

$\mathscr{A}_1$:
$s_1 =$	-	-	T	A	C
$s_2 =$	-	A	T	A	C
$s_3 =$	C	A	-	-	G
$s_4 =$	C	A	A	T	G
$s_5 =$	C	-	T	-	G

$\mathscr{A}_2$:
$s_6 =$	C	T	-	-	A	C
$s_7 =$	C	-	A	T	A	C
$s_8 =$	G	-	A	-	A	T

Table 9.7 *Two multiple sequence alignments on disjoint sets of sequences. Tables 9.8 through 9.10 show the steps performed to merge these two alignments together using profile–profile alignment techniques.*

		a_1	a_2	a_3	a_4	a_5
	A	0	1	$\frac{1}{4}$	$\frac{2}{3}$	0
P_1:	C	1	0	0	0	$\frac{2}{5}$
	G	0	0	0	0	$\frac{3}{5}$
	T	0	0	$\frac{3}{4}$	$\frac{1}{3}$	0

		b_1	b_2	b_3	b_4	b_5	b_6
	A	0	0	1	0	1	0
P_2:	C	$\frac{2}{3}$	0	0	0	0	$\frac{2}{3}$
	G	$\frac{1}{3}$	0	0	0	0	0
	T	0	1	0	1	0	$\frac{1}{3}$

Table 9.8 *Unadjusted profiles* P_1 *and* P_2 *for the alignments* $\mathscr{A}_1$ *and* $\mathscr{A}_2$, *respectively, from Table 9.7.*

We compute profiles P_1 and P_2 for $\mathscr{A}_1$ and $\mathscr{A}_2$, respectively (see Table 9.8). We will use very simple profiles, where the ith position is just the frequency vector of the four nucleotides A, C, G, T, for the ith site in the alignment. For example, for $\mathscr{A}_1$, the first position is 100 percent Cs, so its frequency vector will be $[0, 1, 0, 0]^T$ (i.e., these are unadjusted profiles). We will denote the ith vector in P_1 by a_i, and the ith vector in P_2 by b_i. Note that each profile has the same length as its associated alignment; thus, P_1 has the length five and P_2 has length six.

While there are several ways to align two profiles, we'll use one that is motivated by the idea of minimizing the expected edit distance, under the assumption that each indel and each substitution has unit cost. Consider a_i in P_1 and b_j in P_2, and treat them as generative models. Hence, $P(x|a_i)$ is the probability of observing x in position i in P_1, and similarly $P(y|b_j)$ is the probability that y appears in position j for profile P_2.

We define the cost of putting a_i and b_j in the same column to be the expected cost of aligning a randomly generated nucleotide in the ith position of P_1 and the jth position of P_2. Hence,

$$\text{cost}(a_i, b_j) = \sum_{x \neq y} P(x|a_i) P(y|b_j).$$

For example, $\text{cost}(a_1, b_1) = 1/3$, $\text{cost}(a_2, b_1) = 1$, $\text{cost}(a_2, b_3) = 0$, and $\text{cost}(a_5, b_6) = 11/15$. Note also that the cost of aligning a non-empty column against an entirely gapped column is 1.

After we compute $\text{cost}(a_i, b_j)$ for all i, j, we use those as the cost of each "substitution" of a_i by b_j, and we run the Needleman–Wunsch algorithm to find a minimum cost alignment of the two profiles. If we apply this technique to the two profiles in Table 9.8, we obtain the pairwise alignment given in Table 9.9, which has total cost $1/3 + 1 + 0 + 1/4 + 1/3 + (2/15 + 3/5) = 2.65$.

Note that the pairwise alignment of P_1 and P_2 defines a merger of alignments $\mathscr{A}_1$ and $\mathscr{A}_2$, and hence a multiple sequence alignment of $\{s_1, s_2, \ldots, s_8\}$ (shown in Table 9.10) that

| P_1 | a_1 | - | a_2 | a_3 | a_4 | a_5 |
| P_2 | b_1 | b_2 | b_3 | b_4 | b_5 | b_6 |

Table 9.9 *An optimal alignment of profiles P_1 and P_2 from Table 9.8, where a_i and b_i represent the unadjusted frequency vectors for the ith positions in P_1 and P_2, respectively.*

s_1	-	-	-	T	A	C
s_2	-	-	A	T	A	C
s_3	C	-	A	-	-	G
s_4	C	-	A	A	T	G
s_5	C	-	-	T	-	G
s_6	C	T	-	-	A	C
s_7	C	-	A	T	A	C
s_8	G	-	A	-	A	T

Table 9.10 *The final multiple sequence alignment $\mathscr{A}$ of $\{s_1, s_2, \ldots, s_8\}$ obtained by aligning $\mathscr{A}_1$ and $\mathscr{A}_2$ from Table 9.7, computing their profiles P_1 and P_2 (see Table 9.8), and then aligning the profiles (see Table 9.9). Recall that $\mathscr{A}_1$ is an alignment of $\{s_1, s_2, s_3, s_4, s_5\}$, and that $\mathscr{A}_2$ is an alignment of $\{s_6, s_7, s_8\}$. Note that $\mathscr{A}$ agrees with $\mathscr{A}_1$, but includes an all-gap column. Similarly, $\mathscr{A}$ agrees with $\mathscr{A}_2$.*

is consistent with $\mathscr{A}_1$ and $\mathscr{A}_2$. Hence, during this profile–profile alignment procedure, we never changed the alignments $\mathscr{A}_1$ and $\mathscr{A}_2$. This is a property of profile–profile alignment and also HMM–HMM alignment strategies: the underlying alignments that are represented by profiles or profile HMMs are not modified during the process.

9.12 Progressive Alignment

9.12.1 Using a Guide Tree to Inform the Alignment

Nearly all the popular multiple sequence alignment methods rely on progressive alignment, to lesser or greater extents. Progressive alignment techniques, introduced in Hogeweg and Hesper (1984) and Feng and Doolittle (1987), begin by computing a rooted binary tree (called a merge tree or a **guide tree**) from the input set of sequences; then, the multiple sequence alignment is built from the bottom up, starting at the internal nodes that have only leaves as children and moving toward the root. In each step, the alignments below a node are merged into a new alignment on a larger set of sequences. For example, assume that v_1 and v_2 are two vertices in the guide tree with a common parent, v, and that you have already computed an alignment A_1 of the sequences at the leaves below v_1 and A_2 of the sequences below v_2. To produce an alignment of the leaves below v, then, we only need to merge A_1 and A_2 together. These pairwise mergers are performed in a variety of

(a) Input

a:	CAGGATTAG
b:	CAGGTTTAG
c:	CATTTTAG
d:	ACGTTAA
e:	ATGTTAA

(b) Pairwise distances

	a	b	c	d	e
a	0	1	3	4	4
b	1	0	2	4	4
c	3	2	0	5	5
d	4	4	5	0	1
e	4	4	5	1	0

(c) Guide tree

a: CAGGATTAG
b: CAGGTTTAG
c: CATTTTAG
d: ACGTTAA
e: ATGTTAA

(d) Progressive alignment

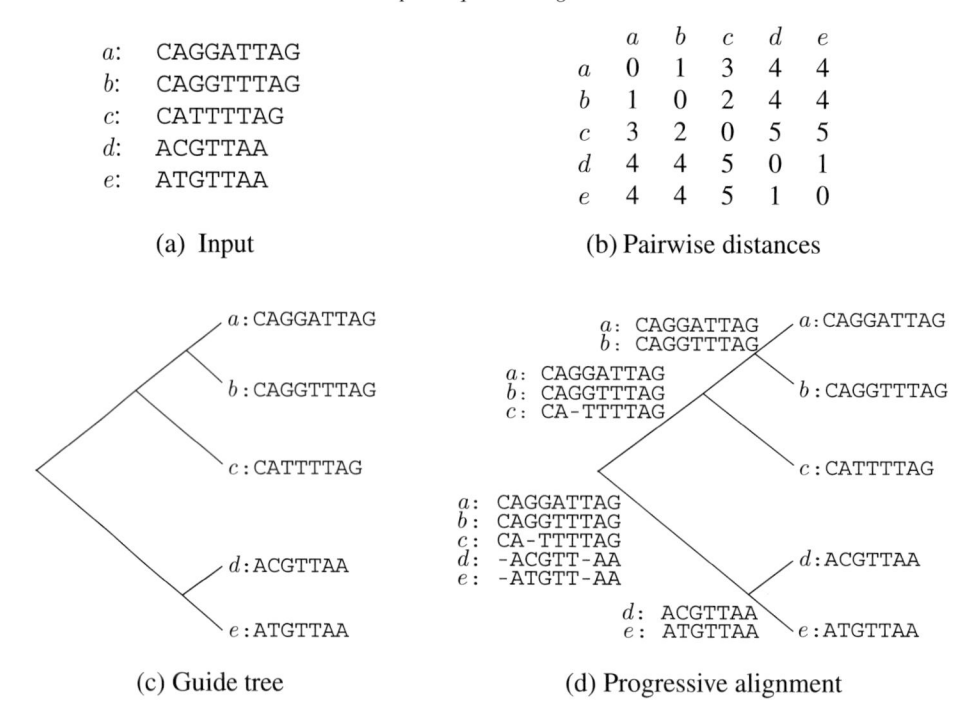

a: CAGGATTAG
b: CAGGTTTAG

a: CAGGATTAG
b: CAGGTTTAG
c: CA-TTTTAG

a: CAGGATTAG
b: CAGGTTTAG
c: CA-TTTTAG
d: -ACGTT-AA
e: -ATGTT-AA

d: ACGTTAA
e: ATGTTAA

a: CAGGATTAG
b: CAGGTTTAG
c: CATTTTAG
d: ACGTTAA
e: ATGTTAA

Figure 9.9 (Figure 2.7 in Huson et al. (2010)) A progressive alignment of a set of five sequences, a,b,c,d,e. The input of unaligned sequences is shown in (a). In (b) we show a calculation of the pairwise edit distances between the input sequences, where all substitutions and indels have unit cost. Based on this matrix of pairwise distances, we use UPGMA to compute a rooted guide tree, which is shown in (c). Then, the sequences are aligned using the guide tree, from the bottom up, using Needleman–Wunch. First, a and b are aligned, and their pairwise alignment A_{ab} is placed at the parent node of a and b. Then, d and e are aligned, and their pairwise alignment is placed at the parent node of d and e. Then the pairwise alignment A_{ab} between a and b is aligned to c; this is performed by representing A_{ab} as a profile, and aligning c to the profile using a modification of Needleman–Wunch. The three-way alignment A_{abc} of a,b,c is placed at the MRCA (most recent common ancestor) of a,b,c. Finally, the two alignments A_{abc} and A_{de} are aligned using a profile–profile alignment technique, and the five-way alignment of all sequences is placed at the root. Note that during a progressive alignment, the alignments on the subsets are never modified, except through the introduction of gapped sites.

ways (discussed in the previous section) that do not change A_1 or A_2. When the root is reached, the resultant alignment contains all the sequences at the leaves of the tree, and so is a multiple sequence alignment of the entire dataset.

The first progressive alignment method to gain prominence was Clustal (Higgins and Sharp, 1988), which gave birth to a large family of related multiple sequence alignment methods including Clustal-W (Thompson et al., 1994) and Clustal-Omega (Sievers et al., 2011), all building on previous techniques and improving in accuracy and scalability. Now, most popular multiple sequence alignment methods (e.g., Muscle (Edgar, 2004a,b); Prank (Löytynoja and Goldman, 2005); and MAFFT (Katoh et al., 2005)) use some form of progressive alignment as part of their alignment strategy.

Progressive alignment methods differ in many ways, including how the guide tree is computed and how the alignment is built using the guide tree; a very simple one is described in Figure 9.9. Some of the most interesting developments in progressive alignment strategies treat the guide tree as a surrogate for the true phylogeny, and then use that assumption to explicitly inform the alignment and to ensure that the alignment that is produced is feasible from an evolutionary perspective. This type of method is referred to as "phylogeny-aware." Prank was the first of these phylogeny-aware methods, and PAGAN (its successor) has a variant that builds on Prank. As shown in Löytynoja and Goldman (2008b), when the true tree is given as input, Prank produces more accurate patterns of insertions and deletions than leading alternative alignment methods (even when also given the true tree). However, both Prank and PAGAN are impacted by the choice of guide tree, so that errors in the input guide tree can reduce accuracy in the final alignment.

A recent improvement on both Prank and PAGAN is Canopy (Li et al., 2016), another phylogeny-aware technique that co-estimates alignments and trees in an iterative fashion; the use of iteration enables Canopy to be much more robust to errors in the input guide tree, so that Canopy improves on the accuracy of both Prank and PAGAN.

9.12.2 Techniques to Improve Progressive Alignments

Progressive alignment methods have several vulnerabilities. For example, many studies have shown that the guide tree can have an impact on the resultant alignment and also on trees estimated on the resultant alignment (Nelesen et al., 2008; Liu et al., 2009a; Penn et al., 2010; Wang et al., 2012; Capella-Gutiérrez and Galbadón, 2013; Smith et al., 2013; Toth et al., 2013).

While many alignment methods use quick-and-dirty techniques to compute guide trees, the increasing evidence that guide trees can have a high impact on the final alignment on some datasets has led many alignment methods to be redesigned to infer their guide trees more carefully. As an example, Yamada et al. (2016) presents a new variant of MAFFT that carefully computes a guide tree using a computationally intensive technique, and obtains improved accuracy as a result. Similarly, Nelesen et al. (2008) showed that using a maximum likelihood tree on a MAFFT alignment as the guide tree improved ProbCons (Do et al., 2005) and several other multiple sequence alignment methods. Prank (Löytynoja and Goldman, 2005) is another alignment method that is highly sensitive to the guide tree it uses (Capella-Gutiérrez and Galbadón, 2013), and Liu et al. (2009a) and Toth et al. (2013) have shown that improved accuracy can be obtained by the use of carefully computed guide trees (maximum likelihood on good alignments) instead of Prank's default guide tree.

For those alignment methods that benefit from using guide trees that are close to the true tree in terms of their topologies, iteration between alignment estimation and tree estimation can result in improved accuracy (Capella-Gutiérrez and Galbadón, 2013). The first suggestion of using iteration for co-estimation of alignments and trees may be attributable to Hogeweg and Hesper (1984), but iteration between alignment estimation and tree estimation is increasingly used. For example, the use of iteration between alignment and tree estimation is why Canopy improves on its predecessor methods, PAGAN and Prank. Iteration between alignment and tree estimation is also used in the SATé family of methods (Liu et al., 2009a, 2012b; Mirarab et al., 2015a), described in Section 9.16.3. However, not all alignment methods improve in accuracy when using guide trees that are close to the true tree, and so the choice of guide tree clearly depends on the method, and perhaps also on properties of the dataset.

Another issue that impacts progressive alignment methods is that errors made early in the progressive alignment persist, since subsequent mergers never change the alignments they merge together. One strategy to deal with this is to attempt to correct errors in the resultant alignment; examples of such techniques include *polishing* (Berger and Munson, 1991; Hirosawa et al., 1995), which breaks the set of sequences into subsets and then re-aligns the induced sub-alignments. Another approach seeks to reduce the number of errors made early in the progressive alignment. One of the most powerful techniques to reduce errors in early steps of a progressive alignment is called "consistency."

9.13 Consistency

Consistency was initially proposed in Gotoh (1990), and the approach was elaborated on in a sequence of early papers (Vingron and Argos, 1991; Kececioglu, 1993; Morgenstern et al., 1996, 1998; Notredame et al., 1998). The basic idea in consistency is to use an input library of alignments (typically just pairwise alignments, but multiple sequence alignments can also be used) on the input sequences to produce a support value for each of the possible homology statements, where a **homology statement** is an assertion that a particular pair of nucleotides (or amino acids), one from each of two different sequences, are homologous. Once these support values are computed, they can be used within progressive alignment methods to determine how to merge two alignments together.

Given the library of pairwise alignments, there are many ways to compute these support values. For example, the support values *could* just be the frequency with which the given pair of nucleotides (or amino acids) are in the same column in one of the input alignments. This is a promising approach, but it is not what is meant by the term **consistency**, which refers to the use of intermediate sequences in the calculation of support values, as we now explain.

Suppose x and y are two nucleotides in sequences A and B. If there is a third sequence C with a nucleotide z such that the pairwise alignment between A and C aligns x and z and the pairwise alignment between B and C aligns y and z, then by transitivity of homology, it follows that x and y should be homologous. Therefore, our confidence in the homology of x and y should increase. Furthermore, the larger the number of intermediate sequences that provide this evidence of x and y being homologous, the larger our confidence should be.

This simple observation is the main common idea in consistency-based methods. However, algorithmically there are key issues that affect how well the consistency-based approach works: (1) how the input library is computed; (2) how the support values are computed for each possible homology statement using the input library; and (3) how the support values are used to inform the construction of the multiple sequence alignment method.

Many techniques have been proposed for the computation of the input library. For example, while a single alignment might be computed for a given pair of sequences, better results might be obtained by varying the technique to calculate the pairwise alignment. In addition, when Bayesian methods are used to compute pairwise alignments, then a sample from the posterior distribution on pairwise alignments can be used.

The support values for each homology statement can be computed using various techniques, ranging from very simple ones (for example, just counting the number of intermediate sequences that provide the extra support for the homology statement) to more sophisticated ones based on probabilistic models.

Once the support values are computed, the problem of finding the best multiple sequence alignment optimizing the total support can be considered. One such optimization problem is the NP-hard Maximum Weight Trace alignment problem (Kececioglu, 1993), which constrains the multiple sequence alignment to take its homology statements from a specified set of allowed pairs, and then seeks the multiple sequence alignment having the maximum total support. Because the Maximum Weight Trace problem is NP-hard, greedy approaches have been employed; for example, T-Coffee (Notredame et al., 2000; Taly et al., 2011) uses the support values within a progressive alignment strategy.

Consistency-based alignment methods often provide substantial improvements in accuracy over standard alignment methods, especially when the support values are based on probabilistic models, so that the support value is explicitly an estimate of the probability that the homology statement is true (Kemena and Notredame, 2009; Pais et al., 2014).

The sequence alignment method most closely associated with consistency is probably T-Coffee (Notredame et al., 2000; Taly et al., 2011). However, many other methods have used consistency to great advantage, including Di-Align (Morgenstern et al., 1996, 1998; Subramanian et al., 2008), FSA (Bradley et al., 2009), MAFFT (Katoh et al., 2005; Yamada et al., 2016), MUMMALS (Pei and Grishin, 2006), PCMA (Pei et al., 2003), Pecan (Paten et al., 2009), Probalign (Roshan and Livesay, 2006), ProbCons (Do et al., 2005), and PROMALS (Pei and Grishin, 2007).

9.14 Weighted Homology Pair Methods

Some alignment methods use a technique that is very similar to consistency in that they compute support values for each homology pair, and then compute the output multiple sequence alignment using those support values. Some of these techniques use sophisticated statistical methods to sample either pairwise or multiple sequence alignments in order to derive their support values. In many ways these techniques are similar in flavor and objective to the consistency methods, but because these techniques may not use the

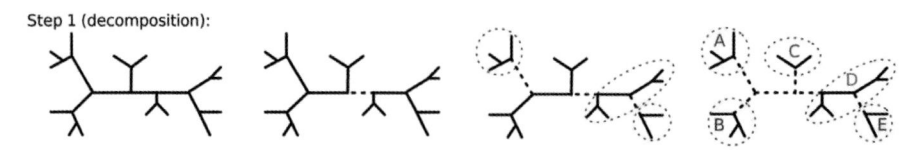

Figure 9.10 (From Mirarab et al. (2015a)) A centroid-edge decomposition strategy used to divide a sequence dataset into disjoint subsets based on a given tree. In each iteration, a **centroid edge** is deleted, where a centroid edge is one whose removal splits the leafset in half (or as close to half as possible). Once all the subtrees are small enough (below the user-specified threshold), the decomposition terminates. Note that subset D is not a clade in the input tree. This is the decomposition strategy used in the multiple sequence alignment methods SATé-2 (Liu et al., 2012b), PASTA (Mirarab et al., 2015a), and UPP (Nguyen et al., 2015b). It is also used in the phylogenetic placement method SEPP (Mirarab et al., 2012), in the metagenomic taxon identification method TIPP (Nguyen et al., 2014), and in the protein family classification method HIPPI (Nguyen et al., 2016).

intermediate sequence approach that is normally associated with the consistency approach, we will instead refer to them as *weighted homology pair* methods.

A new approach that uses a weighted homology pair approach is PSAR-Align (Kim and Ma, 2014), which aims to improve an input multiple sequence alignment. The basic approach is to use an input multiple sequence alignment to generate a set of new suboptimal alignments by sampling from the set of possible alignments under a statistical model. Once the set of suboptimal alignments is generated, it is used to define probabilities for each homology statement. Once the support values are computed, PSAR-Align uses a sequence annealing algorithm in FSA (Bradley et al., 2009) to compute a revised multiple sequence alignment for the input sequences. PSAR-Align is able to improve multiple sequence alignments for gene sequences (Kim and Ma, 2014) as well as for whole genomes (Earl et al., 2014).

9.15 Divide-and-Conquer Methods

Some multiple sequence alignment methods use a divide-and-conquer strategy in which the sequence set is divided into subsets, the sequences in each subset are aligned, and then the subset alignments are merged into a final alignment. Methods that use this strategy include the mega-phylogeny method developed in Smith et al. (2009), SATé (Liu et al., 2009a), SATé-2 (Liu et al., 2012b), PASTA (Mirarab et al., 2015a), SATCHMO-JS (Hagopian et al., 2010), PROMALS (Pei and Grishin, 2007), and MAPGAPS (Neuwald, 2009). Of these methods, SATCHMO-JS, PROMALS, and MAPGAPS can only be used on proteins, but mega-phylogeny, SATé, SATé-2, and PASTA can be used on both nucleotides and protein sequences. The MAPGAPS method is a bit of an outlier in this set because the user provides the dataset decomposition, but we include it here for comparative purposes.

While the methods differ in some details, they use similar strategies to estimate alignments. Most estimate an initial tree, and then use the tree to divide the dataset into subsets.

The method to compute the initial trees differs, with SATCHMO-JS, SATé, SATé-2, and PASTA using methods that first compute a multiple sequence alignment and then a tree, PROMALS using a UPGMA tree on k-mer distances, and mega-phylogeny using a reference tree and estimated alignment.

The subsequent division into subsets is performed in two ways. In the case of mega-phylogeny, SATé, SATCHMO-JS, and PROMALS, the division into subsets is performed by breaking the starting tree into clades or complements of clades so as to limit the maximum dissimilarity between pairs of sequences in each set. In contrast, SATé-2 and PASTA remove centroid edges from the unrooted tree, recursively, until each subset is small enough (where the maximum size is a parameter that can be set by the user); see Figure 9.10 for a graphical example of this decomposition. Note that SATé-2 and PASTA do not attempt to limit the maximum dissimilarity between sequences in each set, and that the sets produced by the SATé-2 decomposition may not form clades or complements of clades in the tree.

Alignments are then produced on each subset, with PROMALS, SATé, SATé-2, PASTA, and mega-phylogeny re-estimating alignments on each subset, and SATCHMO-JS using the alignment induced on the subset by the initial MAFFT alignment. These alignments are then merged together into an alignment on the full set, with each method using somewhat different techniques. PASTA uses a combination of profile–profile alignment methods and transitive closure, SATCHMO-JS uses HMM–HMM alignment techniques, and PROMALS uses external knowledge about protein structure to help determine how to merge the subset alignments together.

MAPGAPS shares many features with these four methods, but has some unique features that are worth pointing out. MAPGAPS requires the user to provide a dataset decomposition and also a manually curated seed alignment and associated template reflecting structural and functional features of the protein family. The algorithm operates by estimating alignments on the subsets using standard alignment methods (either Muscle or PSI-BLAST (Altschul et al., 1997)), and then uses the template to merge the subset-alignments together. Of the methods described here, PASTA and MAPGAPS have the best scalability, and both can analyze ultra-large datasets, even those with one million sequences.

9.16 Co-estimation of Alignments and Trees

Multiple sequence alignment can be seen as the inference of positional homology between letters in the input sequences, and so is a question about the evolutionary history relating the sequences. On the other hand, phylogeny estimation tends to be phrased as the inference of the branching structure (i.e., the tree or network) relating the sequences. Thus, multiple sequence alignment and phylogeny estimation address complementary aspects of the inference of evolutionary history, and it would make sense to try to co-estimate them rather than do one first and then the other.

One common co-estimation strategy is to extend maximum parsimony to allow for indels, and seek the tree and alignment that gives the minimum total cost (i.e., co-estimation based on the Generalized Tree Alignment problem (Section 9.5.3). Another approach is to co-estimate the alignment and tree under a model of sequence evolution that includes indels; this type of method is called "statistical co-estimation." Finally, some methods have been developed that co-estimate the alignment and tree but are not based on any optimization criterion; we refer to these as "heuristic co-estimation," to emphasize that these methods are not based on any optimality criterion.

9.16.1 Parsimony-style Co-estimation

Recall that the input to the Generalized Tree Alignment problem is a set S of sequences, and the output is a tree T with internal nodes labeled by sequences so as to minimize the total treelength across the tree, where the length of an edge is the edit distance between the sequences labeling the endpoints of the edge. Any such node-labeled tree, combined with the implied pairwise alignment on every edge of the tree, defines a multiple sequence alignment for the input set S. Hence, any algorithm for the Generalized Tree Alignment problem can be seen as a method to co-estimate the alignment and the tree. This approach is sometimes referred to as "direct optimization," to emphasize that it estimates the tree directly from unaligned sequences. However, when the alignment is considered part of the output, it is also a parsimony-style co-estimation technique.

Several methods for the Generalized Tree Alignment problem (Vingron and von Haeseler, 1997; Wheeler et al., 2006; Varón et al., 2007; Liu and Warnow, 2012) have been developed (see also Vingron and von Haeseler (1997) for a related problem), but the software POY (Wheeler et al., 2006; Varón et al., 2007) is the most well known. The accuracy of the alignments and trees produced by POY, or by any heuristic for Generalized Tree Alignment, depends on the specific heuristics used to find local optima (Ogden and Rosenberg, 2007), and also on the choice of cost function (Liu and Warnow, 2012). In particular, how the cost of a gap of length L is defined (e.g., simple gap penalty, affine gap penalty, or something else) impacts the accuracy of the final alignment and tree produced by the heuristic search, with affine gap penalties tending to produce more accurate trees than simple gap penalties (Liu and Warnow, 2012). However, even with affine gap penalties, direct optimization methods have not been shown to be competitive with the leading two-phase methods (e.g., maximum likelihood on MAFFT alignments) in terms of either tree or alignment accuracy (Liu and Warnow, 2012). On the other hand, to date the set of gap penalties that have been explored in this context is quite limited, and it is possible that more complex ways of computing editing distances will enable direct optimization methods to provide improved accuracy compared to two-phase methods.

9.16.2 Statistical Co-estimation Methods

Several statistical models of sequence evolution have been proposed that show sequences evolving down a tree with insertions, deletions, and substitutions. The earliest such model was from Bishop and Thompson (1986), which led to several other models, of which TKF91 (Thorne et al., 1991) and TKF92 (Thorne et al., 1992) are the most well known.

Advances have been made to improve the biological realism of these models, to reduce the computational burden of inferring under these models, or to enable estimation of alignments and/or trees (Allison et al., 1992a,b; Hein et al., 2000; Holmes and Bruno, 2001; Lunter et al., 2003a,b; Metzler, 2003; Miklós, 2003; Miklós et al., 2004; Lunter et al., 2005a,b; Redelings and Suchard, 2005; Suchard and Redelings, 2006; Novák et al., 2008; Rivas and Eddy, 2008; Bouchard-Côté and Jordan, 2013; Rivas and Eddy, 2015; Ezawa, 2016).

One of the exciting advances in this area is the methods that co-estimate alignments and trees under statistical models, such as StatAlign (Novák et al., 2008; Herman et al., 2014) and BAli-Phy (Redelings and Suchard, 2005; Suchard and Redelings, 2006). However, the computationally most efficient of the co-estimation methods, BAli-Phy, is limited to at best moderate-sized datasets, and even these can be very computationally intensive (i.e., weeks of analysis to analyze 100 or 200 sequences). The running time issues are often caused by reliance on Bayesian MCMC methods to sample from the joint distribution of alignments and trees, since convergence to the stationary distribution can require a very long time.

If the true tree is known, or a good estimate of the tree is available, then phylogeny-aware methods like Prank, PAGAN, and Canopy that are also based on an explicit statistical model of sequence evolution can be used. In fact, Blackburne and Whelan (2013) refer to Prank as a "heuristic to full statistical alignment." Furthermore, Blackburne and Whelan (2013) showed that Prank alignments cluster with point estimates of alignments produced by BAli-Phy (using either the maximum *a posteriori* or posterior decoding alignment), while alignments produced by the usual alignment methods (ClustalW, Dialign, MAFFT, MUMMALS, Muscle, ProbAlign, ProbCons, and T-Coffee) cluster together. Based on this pattern, Blackburne and Whelan (2013) partition alignment methods into two sets – those that are "evolution-based" and those that are "similarity-based" – and they group Prank in the evolution-based set. It would seem likely that the other phylogeny-aware methods would group with Prank, and hence with the evolution-based set.

The statistical phylogenetics community regards the development of statistical models of sequence evolution and the inference of alignments (and co-estimation of alignments and trees) under these models to be tremendously important, especially if they are based on sound biological models (Ezawa, 2016). With the increased interest in large datasets (with several hundred or even thousands of sequences), the development of statistical co-estimation methods that can analyze large datasets in reasonable time frames is possibly one of the most important outstanding estimation problems in phylogenetics.

9.16.3 Heuristic Co-estimation Methods

Co-estimation of alignments and trees can also be performed without a specific optimization problem (as in direct optimization and in statistical estimation), and we refer to this class of methods as "heuristic co-estimation" methods, to emphasize the lack of optimality criterion. The first of these methods may have been the iterative technique that alternated between alignment estimation and tree estimation provided in Hogeweg and

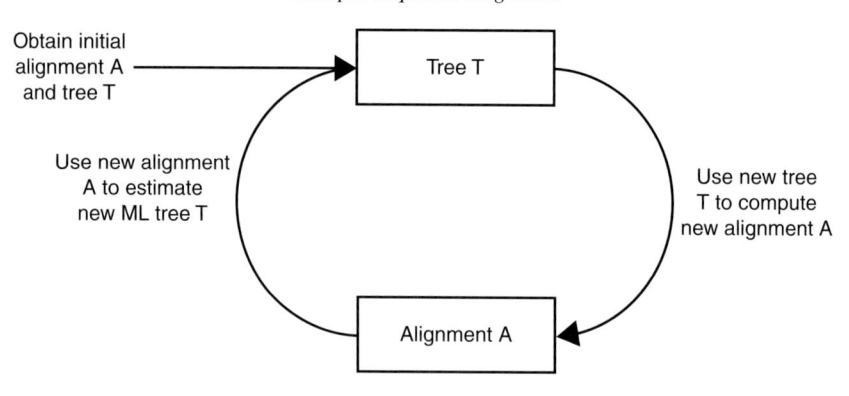

Figure 9.11 (From Liu et al. (2009a)) SATé, SATé-2, and PASTA all use the same basic iterative strategy to compute an alignment and a tree from unaligned sequences. In the first iteration, an initial alignment and tree are computed, typically based on some fast techniques. In each iteration, a new alignment is computed using a divide-and-conquer strategy on the tree from the previous iteration, and a maximum likelihood tree is computed on the new alignment. The process repeats until a stopping criterion is triggered, which may be a fixed number of iterations or a maximum amount of time. The output is either the last alignment/tree pair, or the pair optimizing some criterion (e.g., the maximum likelihood score).

Hesper (1984), but several other methods (e.g., SATé, SATé-2, PASTA, and Canopy, each discussed earlier) have been developed that also use iteration between alignment and tree estimation. Each of these, therefore, can be seen as a heuristic co-estimation method. Similarly, although SATCHMO-JS does not iterate between tree estimation and alignment estimation, it is also a heuristic co-estimation method. PASTA has the best scalability of these three methods (e.g., it can analyze datasets with 1,000,000 sequences), and its algorithmic design provides some insight into how to design methods for large-scale alignment estimation.

PASTA is the most recent member in the SATé family of methods that use iteration plus divide-and-conquer to co-estimate alignments and trees. Figure 9.11 shows the iterative approach, and Figure 9.12 shows how PASTA uses divide-and-conquer to re-align sequences during each iteration. In the first iteration, an alignment and tree are computed on the unaligned sequences using any preferred fast method, and the tree is decomposed by repeatedly deleting a centroid edge until all the subsets are small enough. The default maximum subset size in PASTA is 200, but the user can set this parameter value to suit their dataset. In Step 2, a spanning tree is computed on the subtrees using the tree, so that two subtrees that are near in the tree are made adjacent in the spanning tree. In Step 3, the leafsets of each subtree are then aligned using a preferred external multiple sequence alignment method; the default is MAFFT, but other methods can be used as well (see Nute and Warnow (2016) for the use of BAli-Phy within PASTA). This produces a set of alignments on subsets that have no shared sequences. In Step 4, adjacent alignments are merged together using OPAL (Wheeler and Kececioglu, 2007) or Muscle; this produces

9.16 Co-estimation of Alignments and Trees

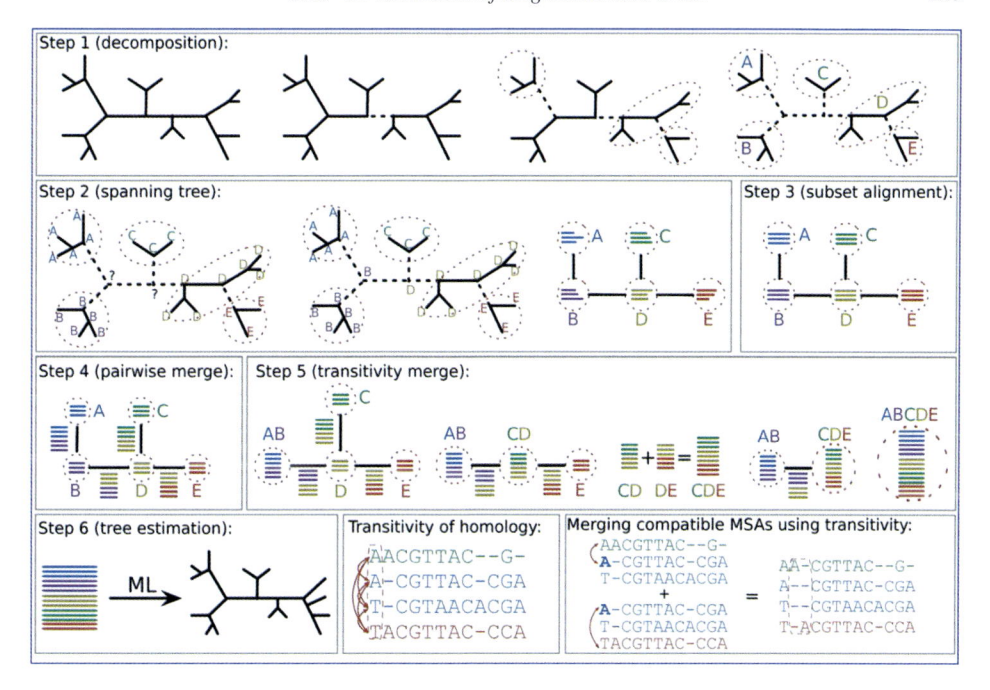

Figure 9.12 (From Mirarab et al. (2015a)) In each iteration, PASTA uses the tree from the previous iteration to divide the sequence set into disjoint subsets, aligns each subset using a preferred external method, such as MAFFT, merges the alignments together into an alignment on the full dataset, and then computes a new tree from the alignment. In general, PASTA iterates several times, with the largest improvement obtained in the first iteration, and most of the advantages obtained by the third iteration. See text for details.

a set of alignments that may overlap, and agree wherever they overlap. In Step 5, these subset alignments are then merged together using transitivity, to define a multiple sequence alignment on the full set of sequences. Step 6 then computes a maximum likelihood tree for this multiple sequence alignment. If desired, this process can repeat, starting from this newly estimated tree.

PASTA is designed to work with a preferred multiple sequence alignment method (the "base method"), but only applies the base method to small subsets. PASTA also depends on external codes to align two alignments and to compute maximum likelihood trees. Thus, PASTA (like SATé and SATé-2) is a framework rather than a single method. As shown in Mirarab et al. (2015a), when the base method is MAFFT (run using the -l-ins-i variant), then PASTA produces alignments on large datasets that are generally at least as accurate as those produced by MAFFT, and maximum likelihood trees on PASTA alignments are generally more accurate than maximum likelihood trees on MAFFT alignments. The use of BAli-Phy within PASTA (Nute and Warnow, 2016) provided improved scalability, enabling BAli-Phy to be used in computing alignments and trees on very large datasets (even some containing 10,000 sequences); furthermore, the use of BAli-Phy as the base method produced improved alignment and tree accuracy compared to default UPP, which uses MAFFT -l-ins-i.

PASTA shares the same basic iteration plus divide-and-conquer strategy as SATé and SATé-2. SATé-2 and PASTA have the same decomposition strategy, which differs from that of SATé and enables SATé-2 and PASTA to analyze larger datasets and also have better accuracy. However, PASTA uses a different technique to merge alignments together, and this enables it to be even more accurate and analyze much larger datasets. SATé-2 uses OPAL or Muscle to hierarchically merge all the alignments into a single alignment; PASTA restricts the use of OPAL or Muscle to only aligning pairs of the initial subset alignments, and then uses transitivity to complete the construction of the full multiple sequence alignment. PASTA can align 1,000,000 sequences with high accuracy, but SATé-2 can only analyze about 50,000 sequences, and SATé is limited to about 30,000 sequences.

Figure 9.13 shows a comparison of the running times for PASTA and SATé-2 on simulated datasets ranging from 10,000 to 200,000 sequences; there is a dramatic reduction in running time largely as a result of greatly reducing the running time needed to merge all the subset alignments together.

Table 9.11 (from Nguyen et al. (2015b)) compares UPP (Nguyen et al., 2015b), PASTA, MAFFT, Muscle, and Clustal-Omega (each run in default mode) on a collection of large biological and simulated datasets. The ROSE (Stoye et al., 1998), RNASim (Guo et al., 2009), and Indelible (Fletcher and Yang, 2009) datasets are simulated, the CRW (Comparative Ribosomal Website) datasets are RNA datasets with structural alignments (Cannone et al., 2002), and the rest are amino acid datasets with structural alignments. Generally MAFFT, PASTA, and UPP have the best accuracy on these datasets, with each sometimes the best performing for a given criterion and dataset.

Note that the three criteria (alignment SP-error, alignment TC score, and delta-FN tree error) each rank methods differently; for example, on the Indelible 10K dataset, Muscle is substantially worse than MAFFT with respect to delta-FN (tree error) score (32.5 percent vs. 24.8 percent), and alignment SP-error (62.4 percent vs. 41.4 percent) but substantially better with respect to TC score (18.3 percent vs. 7.8 percent). This is not an uncommon observation; Liu et al. (2012b) also showed similar inversions. Why these inversions occur is not known, but one obvious conclusion is that the standard alignment criteria are not very well correlated with phylogenetic accuracy. The development of alignment criteria that are predictive of phylogenetic accuracy is a topic for future research.

9.17 Ensembles of HMMs

The use of single HMMs to model a sequence family and hence enable multiple sequence alignment has been well documented. In this section, we describe some elaborations on this approach that use collections of HMMs instead of a single HMM to model a multiple sequence alignment.

9.17 Ensembles of HMMs

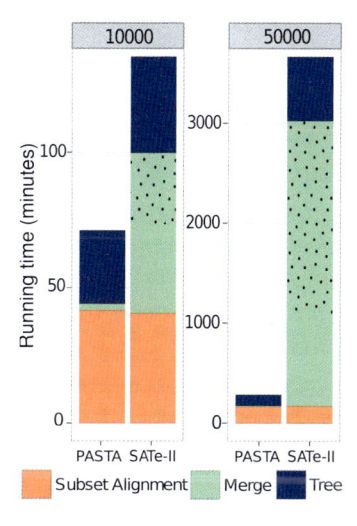

Figure 9.13 (From Mirarab et al. (2015a)) Running time profiling of PASTA and SATé-II on one iteration on RNASim datasets with 10,000 and 50,000 sequences; the dotted region indicates the last pairwise merger.

UPP is a recent multiple sequence alignment method that was developed to provide high accuracy and scalability to large datasets (Nguyen et al., 2015b). UPP, which stands for "Ultra-large alignments using Phylogeny-aware Profiles," is very similar to the seed-alignment strategy described above, except that instead of using a single profile HMM to represent the seed alignment, it uses an "Ensemble of HMMs." As shown in Nguyen et al. (2015b), using the ensemble of HMMs provides improved accuracy, especially under conditions where the input dataset has high heterogeneity or a large number of sequences.

Figure 9.14 shows the algorithmic steps in UPP, starting from the input set S of unaligned sequences. A random subset S_0 of the input S (which we call the backbone) is selected, and an alignment A and phylogenetic tree T are computed on the subset using PASTA. Because PASTA (like most methods) is affected by sequence length heterogeneity, the backbone is constrained to contain only those sequences that are considered full-length. Then, the tree T is used to compute a hierarchy of subsets of S_0, as follows. The first subset is the full set S_0. Then, a centroid edge is found and removed from the tree; the two subsets that are formed are then added to the hierarchy. The decomposition repeats recursively on each subtree until the subset sizes are below some specified threshold provided by the user; the default in UPP is 10 percent of the full set S_0. This creates a collection of subsets, with some subsets being quite small, and each subset has an alignment, as defined by the alignment A restricted to the set S_0. For each of these subset alignments, we build a profile HMM, thus producing a collection of profile HMMs that we refer to as an "Ensemble of HMMs." Now for each $s \in S \setminus S_0$, we add s to the backbone alignment A by finding the best-fitting profile HMM in the ensemble of HMMs for s, and then use the alignment of s to A defined by the maximum probability path for s through that best profile HMM.

Table 9.11 *(From Nguyen et al. (2015b))* We report the average alignment SP-error (the average of SPFN and SPFP error) (top), average ΔFN error (middle), and average TC score (bottom), on the collection of biological and simulated datasets ranging from 1000 sequences (ROSE NT) to 93,681 sequences (HomFam). All scores represent percentages, and so are out of 100. Results marked with an "X" indicate that the method failed to terminate within the time limit (24 hours on a 12-core machine). Muscle failed to align two of the HomFam datasets; we report separate average results on the 17 HomFam datasets for all methods and the two HomFam datasets for all but Muscle. We did not test tree error on the HomFam datasets (therefore, the ΔFN error is indicated by "NA"). The tier ranking for each method is shown parenthetically. MAFFT was run in three ways: MAFFT-L-INS-i was used within PASTA and UPP on subsets, and on all datasets with at most 1000 sequences; for most of the larger datasets MAFFT was run in default mode (–auto); the exceptions were the RNASim 100K dataset, three replicates from the Indelible 10K M3 dataset, and the CRW 16S.B.ALL dataset, where we used MAFFT-PartTree (because the default setting failed to run). All MAFFT variants included the –ep 0.123 parameter.

Method	ROSE NT	RNASim 10K	Indelible 10K	ROSE AA	CRW	10 AA	HomFam (17)	HomFam (2)
Average alignment SP-error								
UPP	7.8 (1)	9.5 (1)	1.7 (2)	2.9 (1)	12.5 (1)	24.2 (1)	23.3 (1)	20.8 (2)
PASTA	7.8 (1)	15.0 (2)	0.4 (1)	3.1 (1)	12.8 (1)	24.0 (1)	22.5 (1)	17.3 (1)
MAFFT	20.6 (2)	25.5 (3)	41.4 (3)	4.9 (2)	28.3 (2)	23.5 (1)	25.3 (2)	20.7 (2)
Muscle	20.6 (2)	64.7 (5)	62.4 (4)	5.5 (3)	30.7 (3)	30.2 (2)	48.1 (4)	X
Clustal	49.2 (3)	35.3 (4)	X	6.5 (4)	43.3 (4)	24.3 (1)	27.7 (3)	29.4 (3)
Average ΔFN error								
UPP	1.3 (1)	0.8 (1)	0.3 (1)	1.8 (1)	7.8 (2)	3.4 (2)	NA	NA
PASTA	1.3 (1)	0.4 (1)	<0.1 (1)	1.3 (1)	5.1 (1)	3.3 (1)	NA	NA
MAFFT	5.8 (2)	3.5 (2)	24.8 (3)	4.5 (3)	10.1 (3)	2.3 (1)	NA	NA
Muscle	8.4 (3)	7.3 (3)	32.5 (4)	3.1 (2)	5.5 (1)	12.6 (3)	NA	NA
Clustal	24.3 (4)	10.4 (4)	X	4.2 (3)	34.1 (4)	3.5 (2)	NA	NA
Average TC score								
UPP	37.8 (1)	0.5 (2)	11.0 (3)	2.6 (2)	1.4 (1)	11.4 (1)	47.3 (1)	40.3 (3)
PASTA	37.8 (1)	2.3 (1)	48.0 (1)	5.4 (1)	2.3 (1)	12.1 (1)	46.1 (2)	50.0 (1)
MAFFT	31.4 (2)	0.4 (2)	7.8 (4)	0.6 (3)	0.7 (2)	12.1 (1)	45.5 (2)	46.9 (2)
Muscle	9.8 (3)	<0.1 (2)	18.3 (2)	2.7 (2)	0.7 (2)	10.5 (2)	27.7 (4)	X
Clustal	5.7 (4)	0.2 (2)	X	3.1 (2)	0.1 (2)	11.8 (1)	38.6 (3)	31.0 (4)

As shown in Nguyen et al. (2015b), using an ensemble of HMMs to represent a backbone alignment compared to a single HMM substantially improves the accuracy of the alignment and subsequent phylogenetic trees, as computed using maximum likelihood heuristics. Furthermore, while the accuracy of UPP increases with the number of sequences in the subset S_0, improved accuracy compared to most other multiple sequence alignment methods was

9.17 Ensembles of HMMs   223

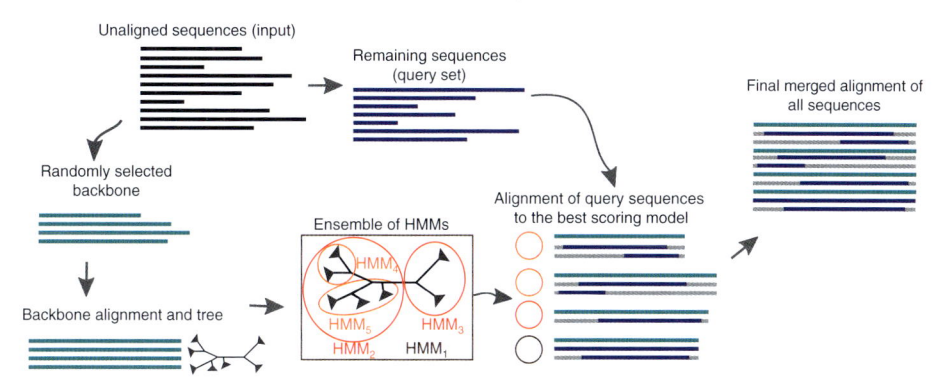

Figure 9.14 (Figure from Nguyen et al. (2015b)) UPP's algorithmic design. The input is a set S of unaligned sequences. In the first step, a random subset S_0 of the sequences is selected from S, where S_0 is supposed to be full-length or close to full-length; all other sequences are called "query sequences." An alignment A and tree T is computed on S_0 (default is PASTA), and referred to as the "backbone" alignment and tree henceforth. The backbone tree is used to compute a hierarchical collection of subsets of S_0 ranging from all of S_0 down to subsets of size at most B (where B is a parameter determined by the user, but the default is $|S_0|/10$). Each subset X in this collection has an alignment $A|X$ obtained by restricting the backbone alignment A to the sequences in X. A profile HMM is built on each alignment $A|X$; this is the ensemble of HMMs for the backbone alignment A. Every query sequence s is then compared to every profile HMM in the ensemble of HMMs, and the profile HMM with the best bit score for s is used to add s into the backbone alignment A.

found even for small subsets; for example, a 1,000,000-sequence synthetic dataset was aligned with very high accuracy using a backbone of only 100 sequences. UPP also scales linearly in terms of running time and is trivially parallelizable.

The key observation in UPP is that a collection of profile HMMs computed using this technique is a better model for a multiple sequence alignment of a large, heterogeneous sequence dataset than a single profile HMM. UPP is just one of a set of methods, each based on similar phylogenetically defined ensembles of profile HMMs, and aimed at different bioinformatics problems (Mirarab et al., 2012; Nguyen et al., 2014, 2016). In addition, other ways of representing multiple sequence alignments by collections of profile HMMs have been developed, using somewhat different techiques than those used in UPP.

The decomposition strategy used in UPP is very similar to the one used in PASTA, with the distinction that UPP's decomposition includes all the subsets created during the decomposition, whereas PASTA's decomposition only used the smallest subsets. Both decompositions can produce subsets that are not clades or complements of clades in the tree; however, requiring the decomposition to produce clades resulted in a small increase in the error rate and a substantial increase in running time.

Further research into how to build these ensembles is needed to provide benefits beyond multiple sequence alignment; other applications where these ensembles might provide advantages over existing approaches that rely on single HMMs includes orthology detection, remote homology detection, protein family and superfamily classification, and protein structure prediction.

9.18 Consensus Alignments

Since there are many different ways of computing multiple sequence alignments, one of the challenges is determining which alignment method to use. Or, if time permits, many alignments can be computed, and then one of them can be selected based on some criterion that is believed to be correlated with accuracy (DeBlasio and Kececioglu, 2015).

Another approach, and the subject of this section, is to use some technique to generate a set of possible multiple sequence alignments, and then use that set to compute a consensus alignment (also called a summary alignment), in the hope that the consensus alignment would be more accurate than a randomly selected alignment from the set.

The set of sequence alignments can be generated in many different ways, for example by including several different alignment methods, using only one method but varying the parameter settings (e.g., varying the gap penalty function or changing the substitution matrix), using different guide trees, sampling from the posterior distribution when using Bayesian methods such as BAli-Phy or StatAlign for estimating alignments, sampling from alignments produced by iterative methods such as PASTA, etc.

Once the set of multiple sequence alignments is computed, a consensus alignment can be computed. While consistency-based methods such as M-Coffee (Wallace et al., 2006) can obviously be used to construct consensus alignments (if necessary, first turning all the multiple sequence alignments into collections of pairwise alignments), the consensus alignment can be constructed using techniques that are not consistency-based (i.e., that do not explicitly use intermediate sequences in order to compute the support value for each homology statement).

MergeAlign (Collingridge and Kelly, 2012) and WeaveAlign (referred to as MinRisk in Herman et al. (2015)) are two such consensus alignment methods. Both of these methods compute a DAG (directed acyclic graph) so that paths through the DAG represent valid multiple sequence alignments. Although the DAGs used in MergeAlign and WeaveAlign are similar, there are some differences. In both methods, the nodes of the DAG represent the columns in at least one of the input alignments, but the directed edges in the DAG for MergeAlign represent only those pairs of columns that appear adjacent in at least one of the input alignments, while the directed edges in the DAG for WeaveAlign correspond to any pair of columns that could appear next to each other in a valid multiple alignment. MergeAlign then computes a consensus alignment that maximizes the average weight of its columns, where the weight of a column is the number of the input alignments that have that column, while WeaveAlign optimizes statistical criteria based on decision theory. Thus, the two methods differ in that they use somewhat different DAGs and they optimize different criteria.

WeaveAlign's approach assumes that the input set of alignments is sampled proportionally to their probabilities, but can be used with any input set of alignments. Because the effective sample size defined by the WeaveAlign DAG is typically larger than the original set of sampled alignments, the DAG representation typically provides a better estimation of the posterior probability distribution of alignments for the input set of sequences than the original sample. The usual dynamic programming techniques can then be applied to the DAG to construct several summary alignments in polynomial time (in fact, in linear time), including the MAP (maximum *a posteriori*) alignment.

Herman et al. (2015) studied WeaveAlign as a method to compute the maximum expected accuracy alignment (Holmes and Durbin, 1988; Do et al., 2005; Roshan and Livesay, 2006; Bradley et al., 2009; Paten et al., 2009; Hamada and Asai, 2012) with respect to several alignment accuracy measures, and evaluated these alignments on collections of sampled multiple sequence alignments for different simulated and biological datasets analyzed using StatAlign. The WeaveAlign method takes a parameter g that "penalizes longer alignments by a factor proportional to the penalty on false positives" (Herman et al., 2015). For appropriate choices of g and the optimality criterion that depended on the dataset, the WeaveAlign alignments were typically *more accurate* than the average alignment in the StatAlign sample, thus achieving the basic goal of getting a better result from the summary alignment than a randomly picked alignment in the sample.

Herman et al. (2015) also compared WeaveAlign to T-Coffee and MergeAlign on StatAlign alignments computed on a collection of simulated and biological datasets, and observed that for many choices of the alignment accuracy measure and parameter g, WeaveAlign produced summary alignments that were typically more accurate than those obtained using T-Coffee or MergeAlign. Furthermore, on the datasets they explored, WeaveAlign was much faster than T-Coffee and MergeAlign.

The excellent performance of WeaveAlign in comparison to T-Coffee and MergeAlign is very encouraging, and suggests several directions for further research. For example, which optimization criterion should be used, and how should the g parameter be set? Results shown in Herman et al. (2015) suggest that the settings that are best probably depend on dataset properties, but this needs to be explored. Another question is whether the improvement in accuracy obtained using WeaveAlign compared to MergeAlign is due to the change in optimality criterion or that the DAG used by WeaveAlign represents a larger set of multiple sequence alignments.

The scalability of this approach to larger datasets has not yet been sufficiently evaluated. Since proper sampling of alignments using StatAlign is limited to about 30 sequences (Herman et al., 2015), "approximate" techniques for generating the set of sampled alignments must be used whenever the number of sequences exceeds this bound. One such approximate sampling technique explored in Herman et al. (2015) performed well on the largest dataset they explored, which had 122 sequences, but additional exploration will be needed to develop good approximate techniques for generating the sampled alignments for much larger datasets.

Overall, the results in Herman et al. (2015) suggest that computing consensus alignments optimizing statistical criteria based on decision theory for a carefully selected set of sampled alignments can provide high accuracy. Hence, the development of such statistical consensus alignment methods, and good approximation techniques for sampling alignments on large sequence datasets, are two highly promising directions for further research.

9.19 Discussion

Several different techniques for estimating alignments have been described in this chapter:

- Using statistical models, such as sequence profiles, profile HMMs, and ensembles of HMMs, to represent multiple sequence alignments. These models enable the development of *reference-based* alignment methods and *seed alignment* methods that either use reference alignments or else build multiple sequence alignments on a small subset of the input sequences, and then use those seed alignments to align the remaining sequences. Both reference-based and seed alignment methods are highly scalable, and some can analyze ultra-large datasets.
- Methods for aligning alignments, typically based on sequence profiles or profile HMMs.
- Progressive alignment techniques that compute an alignment from the bottom up on the basis of a guide tree, using methods for aligning alignments.
- Divide-and-conquer, where the sequence dataset is divided into subsets, often on the basis of an estimated phylogeny, the subsets are aligned using a selected multiple sequence alignment method, and then merged together using methods for aligning alignments. Like seed alignment methods, divide-and-conquer methods are also highly scalable.
- Consistency, which uses a set of alignments (perhaps given as input, or perhaps computed) in order to inform the alignment.
- The estimation of the alignment (or co-estimation of the alignment and a tree) under a phylogenetic statistical model of sequence evolution. Bayesian methods for statistical co-estimation of alignments and trees are particularly popular, despite how computationally intensive they are.
- Consensus alignment methods that can combine information in a set of different multiple sequence alignments on the same set, and produce more accurate alignments as a result.

Popular alignment methods combine several of these techniques; for example, using progressive alignment as well as consistency. Some of these techniques, such as seed-alignment methods and divide-and-conquer methods, are specifically geared toward large datasets, and have enabled alignments of datasets with as many as 1,000,000 sequences (Neuwald, 2009; Mirarab et al., 2015a; Nguyen et al., 2015b). On the other hand, the highest accuracy may well come from methods that are based on phylogenetic models of sequence evolution, such as BAli-Phy (Redelings and Suchard, 2005; Suchard and Redelings, 2006). The integration of statistical methods into divide-and-conquer methods or seed alignment methods may enable the analysis of very large sequence datasets. An example of one such approach is the integration of BAli-Phy into PASTA and then into UPP, explored in Nute and Warnow (2016).

9.20 Further Reading

More general statistical models. Much of this chapter has examined the use of profile hidden Markov models (HMMs) for representing multiple sequence alignments, and then techniques for using those profile HMMs to infer alignments. However, other statistical models have also been proposed. For example, Markov Random Fields (Ma et al., 2014) could be used instead of HMMs, and would have the advantage of being able to represent correlations between sites within an alignment.

Local alignments. The sequence alignment methods we described produced alignments in which every sequence appeared in its entirety. This is called a *global alignment*. However, in some conditions (though not usually for phylogenetic purposes), a different kind of alignment is desired. For example, we may want to find a close match of a short string S to a longer string S'. The application for this is where we are searching a database $\mathscr{D}$ of "full-length" strings for a match to a short string S; hence, we only want to focus on the substrings within long strings that give good matches to S. Thus, given a full-length string S' in $\mathscr{D}$, we would search for a pair of indices i,j so that the substring of S' between indices i and j gave a very good match to S. More generally, we may have two full-length strings, and may wish to find substrings of each that are highly similar. This is the *local pairwise alignment* problem. The Needleman–Wunsch algorithm doesn't solve this problem, since it finds a global alignment rather than a local alignment. However, Smith and Waterman (1981) showed how to modify the dynamic programming approach so that the optimal local alignment could be found.

Phylogeny+HMM models. Profile HMMs and phylogenetic models of sequence evolution, such as the one used in BAli-Phy, can both be seen as models of sequence evolution, but they are very different and have different strengths and weaknesses. Profile HMMs have an advantage over phylogenetic models in that they allow different sites within an alignment to evolve very differently from other sites (e.g., with different substitution matrices); furthermore, the distribution of gaps within the alignment can be modeled with greater care. However, profile HMMs have no underlying phylogenetic tree, although they can be seen as models of "star-like" evolution; hence they have a disadvantage over phylogenetic models that better model diversity due to evolutionary distances.

On the other hand, phylogenetic models, such as the one in BAli-Phy, inherit the basic limitations of the sequence evolution models that they are based on; most of these assume a single substitution rate matrix governing all the sites (i.e., even if there is rate variation, the substitution rate matrix does not change), and many treat all the sites as *i.i.d.* down a single tree. In contrast, profile HMMs can model site variation in a more general and probably more realistic way.

Profile HMMs and the standard phylogenetic sequence evolution models are orthogonal: profile HMMs model sequence variation spatially (moving from left-to-right), and phylogenetic sequence evolution models address the dimension of time. Thus, each type of model provides important insights into sequence variation, which the other lacks, suggesting the possibility of novel models that are combinations of both types of models.

The first attempts to merge these two approaches to modeling sequence variation were by Yang (1995); Felsenstein and Churchill (1996); Mitchison and Durbin (1995), with elaborations in the papers on tree-HMMs (Mitchison, 1999), evolutionary HMMs (Holmes and Bruno, 2001), and phylo-HMMs (Siepel and Haussler, 2004).

These "Phylogeny+HMM" models, which are typically constructed after a multiple sequence alignment and tree are computed for a set of sequences, have been able to provide better fits to nucleotide sequence data. In addition, they have been used to predict secondary structures, detect recombination events, and predict membership in gene families (Qian and Goldstein, 2003; Siepel and Haussler, 2004; Domelevo-Entfellner and Gascuel, 2008). Phylogeny+HMM models could also be used to compute multiple sequence alignments and/or phylogenies for inputs of unaligned sequences. The challenge here is not mathematical so much as computational, and no current approach of this type has yet been shown to be feasible in practice for large datasets (e.g., hundreds of sequences).

In a sense the technique used in the UPP alignment method, described in Section 9.17, is an approximation to a Phylogeny+HMM model. UPP models a multiple sequence alignment using a collection of profile HMMs, where each profile HMM is based on a subset of the sequences, and the subsets of the sequences are computed using an estimated phylogeny for the multiple sequence alignment. However, other methods have also used techniques to construct a phylogenetically defined collection of profile HMMs to represent a multiple sequence alignment, and then used that more elaborate model to improve phylogenetic placement (Mirarab et al., 2012), identification of metagenomic reads (Nguyen et al., 2014), protein family classification (Brown et al., 2007; Nguyen et al., 2016), protein function prediction (Krishnamurthy et al., 2007), orthology detection (Afrasiabi et al., 2013), and protein domain identification (Bernardes et al., 2016).

"Phylogeny-aware" methods. The term "phylogeny-aware" appears in several methods. For example, Prank, PAGAN, and Canopy are methods that are referred to as phylogeny-aware, and UPP is another alignment method that is also referred to as phylogeny-aware. Yet, the meaning of the term "phylogeny-aware" differs between the two methods.

As discussed in Löytynoja and Goldman (2005) and Löytynoja et al. (2012), Prank and PAGAN are designed for use with externally computed guide trees, which are presumed to be the true tree. Hence, the term "phylogeny-aware" indicates the explicit interpretation of the guide tree as a true phylogeny so as to ensure that the alignments that are constructed are phylogenetically consistent with the guide tree. Prank and PAGAN have very high accuracy when the input is the true tree; hence, in a sense they can be seen as "post-tree" analysis methods. This distinguishes them from methods where the objective in estimating the alignment is to enable a highly accurate estimate of the true tree. However, Canopy – the newest of the phylogeny-aware methods developed by this group – uses iteration, and hence is robust to its guide tree, since it functions only as a starting tree.

In contrast, UPP ("Ultra-large alignment using Phylogeny-aware Profiles") is not phylogeny-aware in the same sense; it uses the tree to perform the division of the sequence set into subsets, and there is no need for the tree to be extremely close to the true tree for this decomposition strategy to be helpful. Instead, the term "phylogeny-aware" in the name for UPP refers to its use of the current tree to inform the decomposition of the dataset into subsets, so as to design the ensemble of hidden Markov models.

Similarly, all progressive alignment methods use guide trees, and SATé and PASTA use estimated trees to perform the divide-and-conquer step. Although these methods are based on trees, they are not considered phylogeny-aware in the sense that Prank, PAGAN, and Canopy are.

Masking alignments. One of the common post-alignment steps is to *mask* those sites in the alignment that seem to either be poorly aligned or that have too many gaps, in the hope that removing the "noise" from the alignment will improve the downstream analyses, such as tree estimation or the detection of sites undergoing positive selection. Two well-known methods for identifying noisy and/or unreliable sites are probably Guidance (Penn et al., 2010) and GBlocks (Talavera and Castresana, 2007), but various methods have been developed to identify these poorly aligned sites (Tan et al., 2015).

One of the main uses of masking is to improve the estimation of the phylogeny. However, while many of the early studies showed that tree estimation could be improved as a result of masking noisy sites, some recent studies have suggested otherwise. More specifically, masking alignments for the estimation of gene trees, as opposed to concatenation-based species tree estimation, seems to often make gene trees less accurate (Tan et al., 2015).

Genome alignment. Because whole genomes are increasingly being used for phylogenetic analysis (e.g., Jarvis et al., 2014), accurate whole genome alignment (WGA) has become of increasing importance (Earl et al., 2014). Genomes evolve with events such as inversions, duplications, and rearrangements, and hence go beyond the events modeled by multiple sequence alignment methods. As a result, WGA requires a different type of method (see Section 10.10). Duplication events can produce multiple genomic regions within a single genome that are homologous to each other, and hence potentially also homologous to multiple regions within other genomes; these present particular challenges to WGA methods. Rearrangement events (i.e., transpositions, inversions, etc.) are also challenging, because the detection of homology requires the ability to detect the locations within the genome where these events occurred; this is difficult, even if duplications are not present. Therefore, alignment methods that by design produce collinear alignments are unsuitable for WGA, whenever there are duplications and/or rearrangements.

Whole genome alignment also presents computational challenges due to the lengths of the genomes being aligned. One approach to dealing with the computational issues is to pick one of the genomes to serve as a reference, and then align all the remaining genomes to this reference genome using pairwise WGA methods. Reference-based multiple WGA has the advantage of being much faster than methods that compare all pairs of genomes, but can have reduced accuracy. For example, if some genes are missing in the reference genome but present in other genomes, it will not be possible to infer homologies for this gene between the non-reference genomes. More generally, based on evidence from multiple sequence alignment methods and their performance compared to pairwise alignment methods, reference-based multiple WGA is not expected to be as accurate as multiple WGA that can infer homologies through multi-way comparisons. However, reference-based WGAs can be highly accurate for closely related genomes, and provide improved speed.

Some of the WGA methods aim for speed, and so explicitly avoid dealing with rearrangements or duplications; as a result, they output WGAs that may only cover a subset of the sites in each of the genomes. However, even WGA methods that can handle rearrangements may not handle duplications, and so will also generally output WGAs that only cover some of the sites within each of the genomes. Finally, in order to avoid false positive homologies, many WGA methods may simply not include various genomic regions within their output alignments. These and other factors result in WGAs often being incomplete, in the sense that they miss some genomic regions, and so do not produce global multiple alignments.

The development of methods for pairwise and multiple WGA is an area of very active research, involving a combination of novel data structures, algorithmic approaches, and algorithmic engineering (including HPC implementations); for an entry into this literature, see Hohl et al. (2002), Brudno et al. (2003a), Kent et al. (2003), Brudno et al. (2003b), Blanchette et al. (2004), Raphael et al. (2004), Schwartz et al. (2003), Darling et al. (2004), Ma et al. (2006), Phuong et al. (2006), Bray and Pachter (2004), Dewey (2007), Paten et al. (2008), Dubchak et al. (2009), Paten et al. (2009), Darling et al. (2010), Angiuoli and Salzberg (2011), Paten et al. (2011), Kim and Ma (2014), and Frith and Kawaguchi (2015).

A competition between some of the recent WGA methods on simulated whole genome datasets is reported in Earl et al. (2014). This competition included WGA methods that are used in major genome browsers and stand-alone methods. While many methods did well in the primate simulation (i.e., evaluating accuracy for aligning closely related genomes), there were substantial differences between methods on the mammalian simulation (i.e., evaluating accuracy for aligning more distantly related genomes). The best-performing method in this competition was Cactus (Paten et al., 2011), but TBA (Blanchette et al., 2004), MULTIZ (Blanchette et al., 2004), PSAR-ALIGN (Kim and Ma, 2014) used to modify the MULTIZ alignment, and VISTA-LAGAN (Dubchak et al., 2009) also had good accuracy. Some of the observations were expected; for example, reference-based WGA methods decreased in accuracy with the evolutionary distances to the reference genomes, and accuracy decreased as duplications and rearrangements increased. However, the study also showed that some algorithmic design choices impacted accuracy more than others. In particular, several methods operated by first determining synteny blocks, and then aligned the sequences within the synteny blocks, and the algorithm used to define the synteny blocks had a greater impact on the final performance than the algorithm used to align the sequences within the blocks. Finally, the study showed that duplications were not handled well by most WGA methods.

9.21 Review Questions

1. List three criteria you can use to evaluate an estimated multiple sequence alignment for accuracy with respect to a reference multiple sequence alignment.
2. Express the modeler score and SP-score in terms of SPFN and SPFP.
3. What does the Needleman–Wunsch algorithm solve?
4. What is the running time for the Needleman–Wunsch algorithm, when the input is a pair of sequences, one of length L and the other of length M?
5. What is the difference between local and global pairwise alignment?
6. Describe two optimization problems for multiple sequence alignment (i.e., specify the input and the output).
7. What is a minimum spanning tree?
8. What is an Eulerian graph?
9. What does it mean to say that the states in a profile HMM are hidden?
10. What are the numeric parameters in a profile HMM?
11. How many free parameters are there in a profile HMM having the network topology shown in Figure 9.8? Express this number as a function of the number k of match states.
12. Briefly describe what a progressive alignment does, and one of the issues involved in using a progressive alignment strategy.
13. Briefly describe what "consistency" is, and give an example of how it might be used in an alignment method.
14. What is POY, and what optimization problem does it try to solve?

9.22 Homework Problems

1. Let s and s' be two sequences, where $s = GGATT$ and s' evolves from s by substituting the A in s by a C, and then appending AT to the end of s.
   - What is s'?
   - What is the true pairwise alignment of s and s'?
2. Let $s = AACT$ and $s' = CTGG$ be two sequences, and consider the pairwise alignment of s and s' given in Table 9.3. Describe an evolutionary history relating s and s' for which the given alignment would be the true pairwise alignment.
3. Let an indel and a substitution each have unit cost, and let $s = AATTAAG$ and $s' = TTAAGC$. Use the dynamic programming algorithm to compute the minimum edit distance between these two sequences (i.e., the entries of the matrix should always be non-negative, and should represent the least cost of any transformation of the associated prefixes). Show all entries in the matrix you compute using this dynamic programming algorithm, and the optimal pairwise alignment that you obtain.
4. Let S and S' be two DNA sequences, with S of length L and S' of length L'. Give a polynomial time dynamic programming algorithm to determine the length of the longest common subsequence of S and S'. (Note that a common subsequence is not

the same thing as a common substring; for example, *AAA* is a common subsequence of $S = ATTGATA$ and $S' = TAGGATCA$, but *AAA* is not a substring of either S or S'.)

5. Let $s_1 = AC, s_2 = ATAC$, and $s_3 = CAG$. Suppose that insertions, deletions, and substitutions each have unit cost, and let T be a tree with s_1, s_2, and s_3 at the leaves, and an internal node X.

   - Draw the tree alignment produced by setting $X = s_1$. What is the SOP cost of this alignment? What is its tree alignment cost?
   - Pick another sequence for the internal node, and compute the tree alignment you obtain for that other sequence. What is the tree alignment cost? What is the SOP cost?
   - Prove or disprove: there is only one optimal solution to tree alignment for this input, and it has $X = s_1$.

6. Let $s_1 = ATA$, $s_2 = AAT$, and $s_3 = CAA$. Suppose all insertions, deletions, and substitutions have unit cost. Find an optimal solution to the Tree Alignment problem on the tree T with one internal node and three leaves (i.e., find the best sequence to label the internal node). Is your solution unique? If so prove it, or else show another sequence with as good a score.

7. Let S be an arbitrary set of sequences and assume that insertions, deletions, and substitutions have unit cost. Let T be a tree with one internal node and all the sequences in S at the leaves. Let M be the optimal tree alignment on S obtained by assigning the best possible sequence to the internal node of T. Prove that for all sets S, the SOP cost of M is at least the tree alignment cost of M.

8. Let S be an arbitrary set of sequences and assume that insertions, deletions, and substitutions have unit cost. Let T be a tree with one internal node and all the sequences in S at the leaves. Let M be the optimal tree alignment on S obtained by assigning the best possible sequence to the internal node of T. For what sets S is the SOP cost of M guaranteed to be exactly the same as the tree alignment cost of M?

9. Let S be an arbitrary set of sequences and assume that insertions, deletions, and substitutions have unit cost. Let T be a tree with one internal node and all the sequences in S at the leaves. Let M be the optimal tree alignment on S obtained by assigning the best possible sequence to the internal node of T. For what sets S is the SOP cost of M guaranteed to be exactly twice that of the tree alignment cost of M?

10. Suppose you have two sequences x and y of the same length, and the gap-free alignment of x and y. Let $P_{x,y}$ denote the associated unadjusted profile for the gap-free alignment of x and y, and let $S_{x,y}$ denote the set of all sequences that could be generated by $P_{x,y}$ (i.e., the set of all sequences that have non-zero probability of being generated by $P_{x,y}$).

    - Prove that $\{x, y\} \subseteq S_{x,y}$ (i.e., that x and y always have non-zero probability of being generated by $P_{x,y}$.
    - Prove or disprove: For all x, y, $S_{x,y} = \{x, y\}$ (i.e., every other sequence has zero probability of being generated by $P_{x,y}$).

9.22 Homework Problems

- Prove or disprove: $\exists x, y$ such that $|S_{x,y}| > 2$.
- Prove or disprove: $\exists x, y$ such that $|S_{x,y}| = 2$.

11. Consider the gap-free alignment given for sequence dataset $s_1 = AACTAAG$, $s_2 = AATATAG$, $s_3 = ATAAAAG$, $s_4 = TTATTAG$, and $s_5 = TATATAG$.

    - Write down the unadjusted profile hidden Markov model that represents this multiple sequence alignment, and that doesn't include any insertion or deletion states.
    - What are the most likely sequences to be generated by this model? (If there is only one, say so – and otherwise give them all.)
    - What is the probability of generating sequence *AACTAAG*?
    - What is the probability of generating sequence *CTAAAAG*?

12. Suppose you have a profile HMM M where the match states only emit letter A and the insertion states only emit letter T.

    - Suppose M has no deletion states, and the insertion states have no self-loops. Are any of the states of M actually hidden? In other words, if you know M (the graph and its associated numerical parameters) and you are given a sequence s that could be generated by M, can you determine the path taken through M to generate s? Does the answer to your question depend on s or the details about M?
    - Same question as above, but now suppose the insertion states have self-loops.
    - Same question as above, but now suppose that the insertion states have self-loops and that the model has deletion states.

13. Recall that PASTA uses a recursive centroid edge decomposition to decompose its dataset into disjoint subsets, where a centroid edge is an edge whose deletion creates two subsets that are as close as possible in size to each other. Design a polynomial time algorithm to find a centroid edge and its bipartition in an input binary tree, and analyze the running time as a function of the number of leaves in the input tree.

14. An alternative decomposition strategy that has been used in PASTA is a recursive decomposition on the longest branch in the tree, until each subset is small enough. Design a polynomial time algorithm to find the longest edge and its bipartition in the input tree, and analyze the running time as a function of the number of leaves in the input tree.

15. Another decomposition strategy that could be used in PASTA is to decompose on the midpoint of the longest leaf-to-leaf path in the input tree, under the assumption that the tree has positive branch lengths. Design a polynomial time algorithm to find this "midpoint" edge and its bipartition in the input tree, and analyze the running time as a function of the number of leaves in the input tree.

10
Phylogenomics: Constructing Species Phylogenies from Multi-Locus Data

10.1 Introduction

One of the fascinating challenges in estimating the evolutionary history of a set of species is that different regions within the genomes can evolve differently due to various biological phenomena (Maddison, 1997; Mallo and Posada, 2016; Posada, 2016). One of the most obvious causes for this difference is horizontal gene transfer (Syvanen, 1985), whereby DNA is transferred from the genome of one species into that of another. Horizontal gene transfer (HGT) is especially frequent among prokaryotes (Gogarten et al., 2002), but occurs in other organisms as well. While it can be argued that a species tree is useful as a model (Mindell, 2013) and can make sense in the presence of HGT (even for prokaryotes) if it is based on genes that are resistant to HGT, a full depiction of evolutionary history when HGT has occurred requires a more general graphical model called a "phylogenetic network" (Morrison, 2014a).

Hybrid speciation, where two different species have viable offspring, is another biological process that requires a phylogenetic network. Well-known examples of hybridization include mules and hinnies (which are hybrids of horses and donkeys) and ligers and tigons (which are the hybrid offspring of lions and tigers). Hybrid speciation has long been known to be common in plants (Rieseberg, 1997), and its extent in other organisms is increasingly apparent (Pennisi, 2016). Like HGT, a proper representation of the evolutionary history of a dataset in which hybridization appears requires a phylogenetic network, rather than a tree.

There are also biological processes that cause different parts of the genomes to evolve differently, but where the species history is still correctly modeled as a tree. For example, under a gene duplication and loss model (Ohno, 1970), a gene evolves within the branches of a species tree with duplication events (that increase the number of copies of the gene within a lineage) and loss events (that reduce the number of copies of a gene), so that a given species can have multiple copies of the same gene within its genome.

Another process that can create discordance with the species tree is the **multi-species coalescent** (MSC) model (Maddison, 1997; Pamilo and Nei, 1998) (i.e., the multi-species version of the coalescent process), which models how alleles segregate into populations, and so reflects population-level processes. Under the MSC, the genes evolve within the species tree, and can be different from the species tree in the presence of "incomplete lineage sorting" (Maddison, 1997).

The relative contribution of each of these biological causes for gene tree heterogeneity is debated, and in any event will depend on the particular dataset. Yet in general, heterogeneity in the evolutionary history across the genome is increasingly appreciated as an interesting source of insight into how species evolve, and one that requires new methods for phylogenetic analysis. Indeed, while the default model is still a species tree, some have argued that events such as HGT and hybridization are natural and sufficiently frequent that the default model for evolutionary histories ought to be a phylogenetic network rather than a tree (Morrison, 2014a).

This chapter covers the estimation of species phylogenies (i.e., trees and networks, both), addressing multiple causes for gene tree heterogeneity. While phylogenetic network estimation is clearly important, it is a subject in its own right, with several textbooks devoted to it (Huson et al., 2010; Morrison, 2011; Gusfield, 2014); therefore, we mainly focus on species tree estimation. A large part of this chapter is focused on methods for species tree estimation when gene trees can be incongruent due to incomplete lineage sorting (ILS). We also include some (brief) material about species tree estimation in the presence of gene duplication and loss, as well as in the estimation of the underlying species tree in the presence of random but bounded HGT. Finally, we also include a discussion of the two main types of phylogenetic networks that have been developed – evolutionary networks and data-display networks.

10.2 The Multi-Species Coalescent Model (MSC)

The presentation in this chapter aims for an intuitive understanding of the models that we analyze, rather than a full understanding with all the parameters and model assumptions. The reader who is interested in obtaining a deeper understanding of these models should see the survey of the coalescent in Hudson (1991), or one of the excellent textbooks on population genetics (Ewens, 2000; Durrett, 2008; Wakeley, 2009).

The MSC model treats each species as a population of individuals, with each individual having a set of alleles for each gene (one allele for haploid organisms, two for diploid, etc.). This basic perspective – of treating a species at a given point in time as a population of individuals – is an important and powerful perspective that leads to substantial insights into the process of evolution, and that also introduces new computational problems.

Over time, different alleles assort into different populations, so that speciation events can lead to different species having different sets of alleles among their individuals. When this happens, gene trees defined using a single allele from each selected individual can be different from the species tree and from each other. The forward process we have described is called "lineage sorting." If this process results in gene tree discordance with the true species tree, then it is called **incomplete lineage sorting** or "deep coalescence." Lineage sorting is mathematically equivalent (in terms of the probability distribution it produces on gene trees) to a backward process called the multi-species coalescent.

In the MSC, the lineages at the leaves trace their history backward in time. Under this backward "coalescent" process (Kingman, 1982), any two lineages are equally likely to be the first to "coalesce" on an edge. This is the only property we will use to establish the theoretical guarantees of the methods we develop.

Coalescence is easy to understand using the Wright–Fisher model (Fisher, 1922; Wright, 1931). Each leaf in the tree represents one allele of a gene in a particular individual. Each lineage picks its parent at random from the prior generation, and coalescence occurs when two or more lineages pick a common parent. Over time, all the lineages will eventually coalesce into a single lineage, thus creating a gene tree that fits inside the species tree.

For any two lineages, the first opportunity they have to coalesce is on the edge above their most recent common ancestor (MRCA); deep coalescence is the event that occurs when two or more lineages fail to coalesce on that first possible edge. When deep coalescence occurs, three or more lineages enter the next edge (i.e., the one further back in time, and closer to the root), and any two of the lineages have equal probability of coalescing first. Thus, deep coalescence creates the potential for the gene tree to be different from the species tree, but the gene tree can still match the species tree even if there is deep coalescence. The terms "incomplete lineage sorting" and "deep coalescence" have the same meaning, but one expresses it in terms of the forward process and the other expresses the meaning in terms of the backward process.

Figure 10.1 shows four gene trees that evolved within the same species tree under the MSC. Note that only one of the gene trees differs from the species tree in terms of topology; however, the branch lengths of the gene trees differ from the species tree, even when the tree topologies match. Note also that deep coalescence can occur without a change in topology.

Under the MSC, the probability of two lineages coalescing on a branch, given that they both enter the branch, depends on the effective population size (each selects their parent at random from the individuals in the previous generation, so increases in the effective population size decrease the probability of coalescence) and the number of generations represented by the branch (more generations increase the probability of coalescing). Thus, each branch e of the species tree is associated with two parameters: the number $t(e)$ of generations over the time period represented by the branch and the effective population size $pop(e)$ on the branch (or twice that if the organism is diploid). Under the MSC, the probability of coalescence depends only on the ratio between these two parameters, which is referred to as the length $l(e)$ of the edge e in coalescent units:

$$l(e) = \frac{t(e)}{pop(e)}.$$

Under the coalescent model, the waiting time to coalescence for two lineages is exponentially distributed; hence, the probability that a particular pair of lineages coalesce on edge e is $1 - e^{-l(e)}$ (Hudson, 1983). Hence, as $l(e) \to \infty$, the probability of two lineages coalescing

10.2 The Multi-Species Coalescent Model (MSC)

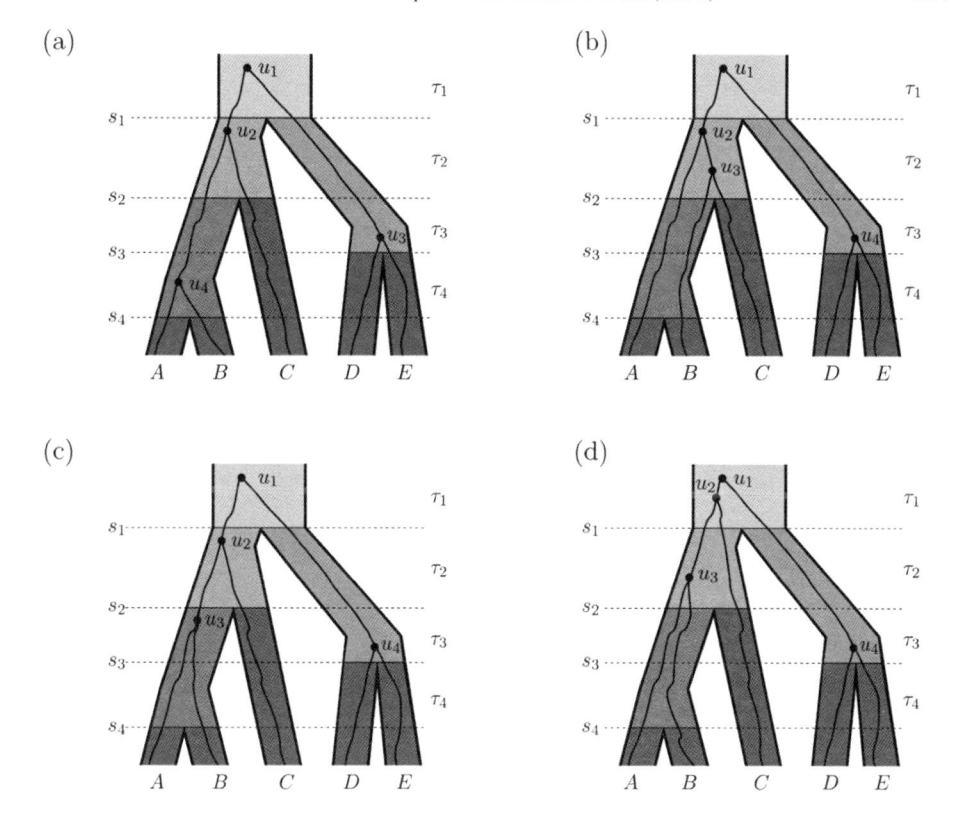

Figure 10.1 (Figure 1 in Stadler and Degnan (2012)) Gene evolution within a species tree under the multi-species coalescent model. The figure shows four of the different possible scenarios (there are others) for a gene tree within the species tree $(((A,B),C),(D,E))$. Of the four different gene trees (a–d), only gene tree (b) has a different tree topology from the species tree, resulting from incomplete lineage sorting (also known as deep coalescence). Interestingly, gene tree (d) also displays a deep coalescence, as the lineages from A and B do not coalesce on the edge above the MRCA of A and B, and yet the gene tree matches the species tree. Reproduced from *Algorithms for Molecular Biology* 7:7 (2012) under the Creative Commons Attribution License.

on edge e approaches 1, while as $l(e) \to 0$ the probability of two lineages coalescing on e approaches 0.

The following lemma from Hudson (1983) (see also Nei (1986)) about the gene tree distribution defined by three-leaf rooted species trees under the MSC will turn out to be key to proving identifiability of rooted species trees under the MSC, and for developing statistically consistent methods for estimating rooted species trees under the MSC.

Lemma 10.1 *(From Hudson (1983); Nei (1986)) Let* $T = (c, (a, b))$ *be a rooted species tree and let* L *be the length in coalescent units of the internal branch in* T *(i.e., the branch above the MRCA of* a *and* b*). Then the probability that a rooted gene tree matches the* *rooted species tree is* $1 - \frac{2}{3}e^{-L} > 1/3$, *and the probability of the other two gene trees are both* $\frac{1}{3}e^{-L} < 1/3$.

In other words, under the MSC, for any model species tree with three leaves, the most probable rooted gene tree will match the species tree, and the other two rooted gene trees will have strictly smaller and equal probabilities. We delay a proof of this lemma until later.

A model MSC species tree with more than three leaves is also described by the pair (T, θ), where T is a rooted binary tree with leaves labeled by a set of species, and θ is the set of coalescent unit branch lengths of the tree. Species trees with very short branches will produce gene trees that conflict with the species tree with greater probability than species trees with very long branches. In fact, a model species tree defines a probability distribution on the gene tree topologies, and – as we will show – is identifiable from that distribution. Furthermore, while calculating the probability of a large gene tree is complicated, it turns out to be fairly straightforward to calculate the probability of small gene trees, and to reason about the species tree from the observed distribution of small gene trees.

For example, under the MSC, every gene tree has some positive probability of being generated. That is, no matter what the species tree topology and branch lengths (in coalescent units) are, there is strictly positive probability that no two lineages coalesce on any edge of the tree, so that all lineages enter the branch above the root. At that point, every tree is possible, since every order of coalescent events has some non-zero probability. In other words, every gene tree topology will appear with probability converging to 1 as the number of sampled genes increases. Therefore, we should not be surprised to see gene tree heterogeneity, and instead should expect it.

10.3 Using Standard Phylogeny Estimation Methods in the Presence of ILS

How accurate are the species tree topologies estimated using standard phylogeny estimation methods, in the presence of gene tree heterogeneity resulting from ILS? The evidence from simulation studies is mixed, with some studies showing that some standard methods can be (fairly) accurate, even in the presence of moderate levels of ILS (Bayzid and Warnow, 2013), and others showing that standard methods can have high error and produce highly supported incorrect trees (Kubatko and Degnan, 2007). What about the theoretical properties? Are any of the standard methods statistically consistent?

We answer this question for concatenation using maximum likelihood, which is possibly the most frequently used method for estimating trees from multi-locus data. The input is a set of multiple sequence alignments, with one alignment for every locus. A concatenation analysis begins by concatenating all the alignments into one large "super-alignment" (also called a supermatrix), and then a tree is estimated on the super-alignment using maximum likelihood.

The simplest version of concatenation analysis using maximum likelihood (CA-ML) seeks the maximum likelihood model tree (T, θ), where T is a binary tree and θ is the set of numeric model parameters. For example, under the Jukes–Cantor model, θ contains the branch lengths for every edge in T. This is called an *unpartitioned maximum likelihood analysis*. Note that under this analysis, all the sites of all the loci are assumed to evolve down the same Jukes–Cantor model tree.

Under a more general approach, the different loci are allowed to evolve down different model trees, but the assumption is that all the model trees share the same topology, and hence only differ in their numeric parameters. Concatenation analyses under this assumption require the estimation of numeric parameters for each locus, and are therefore more computationally intensive. These analyses are called *fully partitioned maximum likelihood analyses*.

Roch and Steel (2015) proved that unpartitioned maximum likelihood is statistically inconsistent (and even positively misleading) under the MSC model, thus establishing that this common way of estimating phylogenies is not a statistically rigorous way of estimating species trees from multi-locus data. However, they left open the question of whether fully partitioned maximum likelihood (in which all numeric parameters are independently estimated for each of the loci) is statistically inconsistent or consistent under the MSC; see discussion in Warnow (2015) and Mirarab et al. (2015b).

To date, no standard phylogeny estimation method has been shown to be statistically consistent under the MSC. For example, standard consensus methods applied to gene trees are not statistically consistent (Degnan et al., 2009). Yet, some new methods have been designed that *are* statistically consistent, as we will see.

10.4 Probabilities of Gene Trees under the MSC

Recall that a species tree, with its branch lengths in coalescent unit, defines a probability distribution on rooted gene trees. Can we expect that the rooted gene tree matching the species tree should appear more often than any other topology? The answer is *No*: There are conditions in which the most probable gene tree may be topologically different from the species tree! When this happens, the most probable rooted gene tree is called an **anomalous gene tree** (AGT), and the model species tree is said to be in the **anomaly zone** (Rosenberg, 2002, 2013; Degnan and Rosenberg, 2006; Degnan, 2013). However, by Lemma 10.1, for any rooted three-leaf species tree, the most probable rooted gene tree is topologically identical to the species tree; in other words, there are no AGTs on three species. Here we provide a proof of this lemma.

Proof of Lemma 10.1. Let T be an arbitrary rooted model species tree with topology $(c,(a,b))$. Let e be the edge from the root of T to the parent of a and b, and let L be the length of the edge e in coalescent units. Let p_0 denote the probability that the lineages from a and b coalesce on edge e; then under coalescent theory, $p_0 = 1 - e^{-L}$ (see Hudson (1983) for the derivation of this formula). Note that $1 > p_0 > 0$, since $L > 0$. Consider how a gene tree is formed under the MSC as the lineages coalesce from the leaves of the species tree toward the root. If the lineages coming from a and b coalesce on edge e, then the gene tree t is topologically identical to the species tree T. To obtain a gene tree with a different topology from the species tree, therefore, there must be no coalescent event on the edge e. Hence, the three lineages (one from a, one from b, and one from c) will all "enter" the edge above the root, at which point any two of them will have equal probability of coalescing first. If the first pair to coalesce comes from a and b, then we obtain a gene tree with topology equal to that of T; otherwise, we will obtain a different tree topology. Putting this together, letting t denote the gene tree,

$$\Pr(t = ((a,b),c)) = p_0 + \frac{1-p_0}{3} = 1 - \frac{2}{3}e^{-L}.$$

Furthermore, since any two lineages have equal probability of coalescing, the probabilities of the other two gene tree topologies are equal, and so

$$\Pr(t = ((a,(b,c))) = \Pr(t = (b,(a,c))) = \frac{1-p_0}{3} = \frac{e^{-L}}{3}.$$

Finally, since $L > 0$, it follows that

$$\Pr(t = ((a,b),c)) > \frac{1}{3},$$

and so

$$\Pr(t = (a,(b,c))) = \Pr(t = (b,(a,c))) < \frac{1}{3}.$$

In other words, the probability of generating a rooted gene tree with exactly the same tree topology as the species tree is *strictly greater* than the probability of producing either of the other two rooted gene trees, and the other two rooted gene trees have equal probability of being generated. As a consequence, under the MSC, for any model species tree with three leaves, the *most probable* rooted gene tree topology is the rooted species tree topology.

What about unrooted gene trees? It is not too hard to show that the most probable *unrooted* gene tree is topologically identical to the unrooted species tree, no matter what the four-leaf species tree and branch coalescence probabilities are (see Allman et al. (2011) and Degnan (2013) for proofs). Thus, there are no anomalous unrooted four-leaf gene trees (AUGTs).

However, for four or more species there are anomalous rooted gene trees, and for five or more species there are anomalous unrooted gene trees (Degnan, 2013; Rosenberg, 2013). Hence, these positive results are restricted to small datasets. Fortunately, these positive results are sufficient to enable algorithm development. We summarize this point as a theorem, which we will use when we analyze methods for estimating species trees.

Theorem 10.2 *For all rooted three-leaf species trees with branch lengths in coalescent units, the most probable rooted gene tree is topologically identical to the rooted species tree. For all rooted four-leaf species trees with branch lengths in coalescent units, the most probable unrooted gene tree is topologically identical to the unrooted four-leaf species tree.*

We will show how to use these results to design species tree estimation methods that can handle gene tree incongruence due to incomplete lineage sorting, and so are statistically consistent under the MSC.

10.5 Coalescent-based Methods for Species Tree Estimation

Several methods have been developed to estimate species trees in the presence of ILS, many of which are statistically consistent under the MSC. A major assumption is that there is no recombination within loci and free recombination between loci; this is referred to by saying that the loci are coalescent genes, or just **c-genes**. These methods also assume that discord between true gene trees is solely due to ILS, and so do not address other biological sources of gene tree discord (i.e., hybridization, horizontal gene transfer, gene duplication and loss, gene flow following speciation, etc.). As we will show, these assumptions make it possible to use the input data to estimate the distribution of gene trees defined by the unknown true species tree, and hence estimate the species tree.

We will assume that the species tree estimation method either directly outputs an estimate of the species tree, or that its output can be used to produce a point estimate of the species tree; the latter is necessary when using Bayesian methods that produce a distribution of the species tree topology rather than a single-point estimate (see Section 8.7 for how to obtain point estimates of the species tree from the estimate of the posterior distribution obtained by a Bayesian MCMC analysis).

- Summary methods. The input is a set of gene trees, from which a species tree is estimated. For some methods, the gene trees must be rooted, but some methods are designed for unrooted gene trees. Some summary methods require numeric model parameters (e.g., branch lengths), but most operate just using the tree topologies. Summary methods are the most popular because they tend to be reasonably fast.

- Co-estimation methods. The input is a set of sequence alignments (one for each locus), and the gene trees and species tree are co-estimated from the alignments. The advantage of co-estimation methods is potential improvement in accuracy, but at the expense of running time.

- Site-based methods. The input for these methods – as for co-estimation methods – is a set of sequence alignments. Therefore, co-estimation methods are also site-based methods since they take the same input and both estimate species trees, but some site-based methods do not attempt to reconstruct gene trees. The advantage of such site-based methods is that they do not need to estimate gene trees, and so the species tree reconstruction can be done even when accurate gene tree estimation is very difficult.

10.5.1 Summary Methods

Recall that summary methods estimate the species tree by combining gene trees. There are many different types of summary methods, but at a top level they can be distinguished by whether they do or do not require rooted gene trees.

The careful reader will note that the proofs of statistical consistency for summary methods assume that the input is a set of true gene trees (whether rooted or not). Most summary methods only require that the topology is given, and do not assume that the numeric parameters (e.g., branch lengths) are available. These summary methods can be applied to estimated gene trees, which will in general have some error, and so the proofs of statistical consistency we provide will not be relevant to the case where gene trees have estimation error. The impact of gene tree estimation error on summary methods can be substantial, and is discussed in Section 10.5.2.

Summary methods that require rooted gene trees. We begin with the case of estimating the rooted species tree from a set of rooted gene trees. If there are only three species, we can estimate the true species tree by counting the number of times each of the three possible rooted gene trees occurs, and return whichever one appears the most frequently. By Theorem 10.2, as the number of genes increases, with probability converging to 1, the rooted true gene tree that appears the most frequently will be the true rooted species tree. Thus, we have a statistically consistent method for estimating any rooted species tree on three leaves from rooted true gene trees.

Now, suppose we are given a set $\mathcal{T} = \{t_1, t_2, \ldots, t_k\}$ of rooted gene trees, and each t_i has the same leafset S with $|S| \geq 4$. How can we estimate the rooted species tree T given $\mathcal{T}$? We will use the fact that there are no anomalous rooted three-leaf trees to estimate all the rooted three-leaf species trees, and then combine them.

To estimate the rooted species tree on a given set A of three leaves, we will just examine each rooted gene tree to see what homeomorphic subtree it induces on A. After we complete this, we can determine which of the three rooted triplet trees on A appears the most frequently among the gene trees. By Theorem 10.2, the most probable rooted gene tree on A will be topologically identical to the rooted true species tree on A. Hence, as the number of true rooted gene trees goes to infinity, for every set A of three species, the probability that the most frequent rooted gene tree on A is the rooted true species tree on A will converge to 1.

We do this for every set A of three species, and thus assemble a set of rooted three-leaf trees (one tree for every three species). If all these rooted three-leaf trees are correct (i.e., equal to the true rooted species tree on the three species), then we can construct the true species tree T using the ASSU algorithm described in Section 3.3. If the ASSU algorithm returns a tree, this is our estimate of the species tree; else, the ASSU algorithm rejects the dataset, saying the three-leaf trees are not compatible, and we return "Fail." While failure is always possible, as the number of genes increases the probability of failure goes to 0, and the probability of returning the true species tree goes to 1.

In other words, we have described a very simple algorithm for inferring the rooted species tree from rooted gene trees. We call this algorithm **SRSTE**, for a "simple rooted species tree estimation."

SRSTE: a simple algorithm to construct rooted species trees under the MSC model. The input is a set of rooted gene trees, each on the same set S of $n > 3$ species, and the output is either an estimated tree T or "Fail."

- Step 1: For all three leaves a, b, c, determine the most frequently induced gene tree on a, b, c, and save it in a set $\mathcal{T}$. (If there are ties, pick any most frequent tree topology at random.)
- Step 2: Apply the ASSU algorithm to the set $\mathcal{T}$ of rooted three-leaf trees, and return its output.

Note that if $\mathcal{T}$ is compatible, then the ASSU algorithm outputs a tree T that agrees with all the rooted triplets. If $\mathcal{T}$ is not compatible, then the ASSU algorithm returns *Fail*, and there is no tree that agrees with the trees in $\mathcal{T}$.

Theorem 10.3 *SRSTE is a statistically consistent method polynomial time algorithm for estimating the rooted species tree from true rooted gene trees under the MSC model.*

The proof of statistical consistency follows from the derivation of the algorithm, and the running time is easily seen to be polynomial. Thus, the SRSTE method is a very simple statistically consistent method for estimating rooted species trees from rooted gene trees under the MSC model, and it runs in polynomial time. Unfortunately, because SRSTE uses the ASSU algorithm, it requires all triplet trees that it computes to be compatible; hence, if even a single rooted triplet tree is incorrectly computed, SRSTE will fail to return the true species tree. In fact, the most likely outcome of using SRSTE is that it will not return any tree at all. Hence, SRSTE is a theoretical construct rather than a useful tool.

Many coalescent-based summary methods have been developed for constructing species trees from rooted gene trees that have better empirical performance than SRSTE, and that are also statistically consistent under the MSC. For example, instead of using the ASSU algorithm to combine the rooted three-leaf trees, we could use the R^* consensus method to combine the rooted three-leaf trees, as described in Degnan et al. (2009). In fact, many methods estimate species trees from rooted gene trees and have been proven to be statistically consistent under the MSC, including STELLS (Wu, 2012), MP-EST (Liu et al., 2010), STEM (Kubatko et al., 2009), STAR (Liu et al., 2009b), GLASS (Mossel and Roch, 2011), and iGLASS (Jewett and Rosenberg, 2012).

Methods that are designed to construct supertrees can also be statistically consistent under the MSC. For example, the SuperTriplets method (Ranwez et al., 2010) computes a tree from a set of rooted trees, under a criterion that depends on the frequency of the induced triplet trees, and is statistically consistent under the MSC.

Summary methods that use unrooted gene trees. Although gene trees are often presented as rooted trees, the correct location of the root is often very difficult. For example, a standard technique is to root the gene tree at an outgroup species; however, when gene trees can differ from the species tree due to ILS, the correct rooted version of the gene tree may not be at the branch leading to the outgroup taxon or taxa. Rooting is relatively easy when the strict molecular clock holds, but otherwise can be challenging. Hence, the locations of roots in estimated rooted gene trees are often wrong. Therefore, methods that seek to estimate the species tree by combining information from rooted gene trees must address the impact of error in these estimated rooted trees on the resultant estimated species tree.

Alternatively, the estimated gene trees can be treated as unrooted trees, and the unrooted species tree can be estimated from unrooted gene trees. As we will see, summary methods based on unrooted gene trees can also be designed that have excellent performance.

We begin with a very simple method to estimate species trees by combining unrooted gene trees. This method, which we call **SUSTE** (for simple unrooted species tree estimation), is the unrooted equivalent of SRSTE.

SUSTE: a simple quartet-based method to estimate species trees under the MSC model.
Recall Theorem 10.2, which said that there are no anomalous four-leaf unrooted gene trees, so that for all unrooted four-leaf model species trees (tree topologies and coalescent unit branch lengths), the most probable unrooted four-species gene tree is identical to the unrooted species tree. Hence, a simple method that is very similar to SRSTE suffices to estimate the unrooted species tree.

The input to SUSTE is a set of unrooted gene trees, each on the same set S of $n \geq 4$ species, and the output is either an estimated tree T or "Fail." For a given set A of four species, there are three possible unrooted gene tree topologies, and in a given set $\mathcal{T}$ of gene trees, there will be a most frequently observed unrooted tree topology. We will refer to an unrooted tree topology on A that appears the most frequently as a *dominant quartet tree* for A. If there is a tie, we use any tree topology that appears the most frequently.

- Step 1: For all sets A of four leaves, determine a dominant quartet tree on A and store it in a set $\mathcal{T}$.
- Step 2: Apply the All Quartets Method from Section 3.4.2 to the set $\mathcal{T}$.
  - If the set $\mathcal{T}$ of four-leaf trees is compatible, then the All Quartets Method will return a tree T, which we return.
  - Else the set $\mathcal{T}$ of four-leaf trees is not compatible, and the All Quartets Method returns Fail. In this case, we also return "Fail."

Theorem 10.4 *SUSTE is a statistically consistent method for estimating the unrooted species tree from unrooted true gene trees under the MSC model.*

Proof Let T be a model rooted species tree on $n \geq 4$ leaves under the MSC model. As the number of true gene trees increases, then for any set A of four species, with probability converging to 1 the dominant unrooted quartet tree on A will be topologically identical to the unrooted true species tree on A. When this is true for all sets A, the set $\mathcal{T}$ of unrooted four-taxon trees will be identical to $Q(T)$. Since the All Quartets Method applied to $Q(T)$ will return T, as the number of unrooted true gene trees increases, with probability converging to 1 SUSTE will return the unrooted true species tree. Therefore, SUSTE is a statistically consistent method for estimating the unrooted species tree from unrooted true gene trees under the MSC model. □

The problem with SUSTE is its reliance on the All Quartets Method, which means that it requires the correct estimation of every quartet tree. In other words, a single quartet tree error will result in SUSTE either failing to return anything (which is the most likely outcome) or returning some tree other than the true tree. Thus, SUSTE is the coalescent version of the Naive Quartet Method (which also uses the All Quartets Method) – statistically consistent, but best seen as a mathematical construct rather than as a method to use in practice.

In order to enable highly accurate species tree estimation using quartet trees, errors in selected quartet trees must be tolerated. There are several ways we could approach this. For example, in the SUSTE algorithm, we selected one quartet tree for every four species, and then sought a tree that agreed with all these quartet trees; alternatively, we could have sought to maximize the number of quartet trees that were satisfied in the output tree. Another approach would be to weight each quartet tree topology by the number of genes that induce the given quartet tree, and define the total support of the species tree by the total weight of its induced quartet trees. This is the Maximum Quartet Support Supertree problem. We can also reformulate the Maximum Quartet Support Supertree problem in terms of quartet-tree distances between gene trees and species trees, and so obtain the Quartet Median Tree problem. These two problems were originally introduced in the context of supertree estimation in Section 7.7.1. The two problems are equivalent, in that any optimal solution to one problem is an optimal solution to the other.

Theorem 10.5 *An exact solution to the Quartet Median Tree problem is a statistically consistent method for estimating the unrooted species tree from unrooted true gene trees under the MSC.*

Proof Let $\mathcal{T}$ be a set of k true gene trees and let T^* be the true species tree, and assume all trees have the same leafset. It is easy to see that the quartet median tree T of $\mathcal{T}$ maximizes $\sum_{t \in \mathcal{T}} Sim_q(T,t)$, where $Sim_q(T,t)$ is the number of four-taxon subsets of S on which T and t induce the same homeomorphic subtrees; thus, a quartet median tree is also an optimal solution to the Maximum Quartet Support Supertree problem.

Let $N(T,X)$ be the number of trees in $\mathcal{T}$ that agree with T on set X. It is easy to see that

$$\sum_{t \in \mathcal{T}} Sim_q(T,t) = \sum_X N(T,X),$$

where X ranges over the possible sets of four species of S. As the number k of genes increases, then for all sets X of four species, the dominant gene tree will be equal to the most probable gene tree with probability that converges to 1. Equivalently, for any $\varepsilon > 0$, there is a $K > 0$ so that given $k > K$ genes then for all sets X of four species, with probability at least $1 - \varepsilon$ the dominant gene tree on X will be equal to $T^*|X$ (the true species tree on X). Now suppose that for every set X of four species, the dominant gene tree on X is identical to $T^*|X$. To see that $\sum_X N(T,X)$ attains its maximum at $T = T^*$, consider a tree T' with a different topology. Then for some set of four taxa, $\{a,b,c,d\}$, T' and T^* induce different quartet trees. Since we assume that the dominant gene tree on a,b,c,d is topologically identical to the species tree on a,b,c,d, it follows that $N(T^*,\{a,b,c,d\}) > N(T',\{a,b,c,d\})$. Hence T' must have a worse score than T^*. In other words, with probability converging to 1 as the number of genes increases, the true species tree T^* will be the unique optimal solution to the Quartet Median Tree problem. Therefore, any algorithm that finds an exact solution to the Quartet Median Tree problem is a statistically consistent method for estimating the unrooted species tree from unrooted true gene trees under the MSC. □

Although the Quartet Median Tree optimization problem is NP-hard (Lafond and Scornavacca, 2016), if we constrain the species tree to be binary and draw its bipartitions from a set X, then an optimal tree can be found in $O(|X|^2 nk)$ time, where n is the total number of species and k is the total number of input gene trees (Mirarab and Warnow, 2015). Note that if we set X to be all possible bipartitions on S, then $|X|$ is exponential in $|S|$, and an exact solution to the constrained quartet median tree problem is an exact solution to the *unconstrained* problem. Thus, the benefit of this approach is only for those sets X that are proper subsets of the full set of all possible bipartitions. For example, if we choose $X = \cup_{t \in \mathcal{T}} C(t)$, then $|X|$ is $O(nk)$ where $|S| = n$ and $|\mathcal{T}| = k$. Interestingly, the following is also true:

Theorem 10.6 *(From Mirarab and Warnow (2015)) An exact solution to the constrained Quartet Median Tree problem, where the constraint set X contains the bipartitions from the input gene trees, is a statistically consistent method for estimating the unrooted species tree from unrooted true gene trees under the MSC model.*

Proof By Theorem 10.5, an exact solution to the unconstrained Quartet Median Tree problem is statistically consistent under the MSC model. The proof of Theorem 10.5 showed that when the set of gene trees is large enough then the unrooted species tree topology T will have an optimal score, from which the theorem followed. Thus, to prove that the constrained Quartet Median Tree problem remains statistically consistent when the constraint set X contains all the bipartitions from the input gene trees, we only need to prove that the bipartitions of T will be contained in the set X with probability converging to 1 as the number of genes increases. We have already noted that every gene tree has strictly positive probability under the MSC, and so the true species tree T has strictly probability of appearing among the gene trees; hence, with probability converging to 1, the set X will contain the bipartitions of the true species tree as the number of genes increases. This completes the proof. □

While this theorem establishes statistical consistency, it does not provide bounds on the number of true gene trees needed to recover all the bipartitions of the species tree with high probability; see Uricchio et al. (2016) for some initial work on this problem.

ASTRAL (Mirarab et al., 2014a) is a method for estimating species trees from gene trees that is based on this theorem. It sets X to be the bipartitions from the input gene trees, and then uses dynamic programming to find the species tree that has the minimum total quartet tree distance to the input gene trees, subject to drawing its bipartitions from X (i.e., it solves the constrained Quartet Median Tree problem). ASTRAL-2 (Mirarab and Warnow, 2015) is an improved version of ASTRAL (e.g., it achieves the $O(|X|^2 nk)$ running time) that modifies how X is set by better handling inputs whenever the gene trees are missing some of the species.

Corollary 10.7 *ASTRAL and ASTRAL-2 (Mirarab and Warnow, 2015) are statistically consistent methods under the MSC, even when run in default mode.*

Proof ASTRAL and ASTRAL-2 are exact algorithms for the constrained Quartet Median Tree problem. In their default mode, each defines the constraint set X so that it is guaranteed to contain all the bipartitions from the gene trees. Hence by Theorem 10.6, each is statistically consistent under the MSC. □

Other quartet-based summary methods. Another summary method that uses quartet trees is the population tree computed by BUCKy (Larget et al., 2010). However, BUCKy computes its set of quartet trees using a different technique: it first computes a distribution of trees for every gene from its sequence alignment using a Bayesian MCMC method (e.g., MrBayes (Ronquist and Huelsenbeck, 2003)), and then computes concordance factors for the quartet trees to determine the best quartet tree for each set of four leaves. The technique used to combine quartet trees differs between BUCKy and ASTRAL, but both use the same basic approach of computing quartet trees and then combining them into a tree on the full set of species. Note that any quartet amalgamation method could be used with quartet trees, whether those quartet trees are estimated by BUCKy or by selecting dominant quartet trees.

Distance-based estimation. Another type of summary method for estimating species trees from unrooted gene trees produces a dissimilarity matrix from the input set of unrooted gene trees, and then computes a species tree from the dissimilarity matrix using a distance-based tree estimation method.

Given a gene tree t on taxon set S, we define the "internode distance" between two leaves x, y to be the number of internal vertices on the path between x and y in t. Then, given an input set of k unrooted gene trees, the matrix of average internode distances between every pair of leaves is computed. We restate a theorem from Kreidl (2011) (see also Allman et al. (2016)):

Theorem 10.8 *Given a set of* k *true gene trees sampled randomly from the distribution defined by the model species tree* T, *the matrix of average internode distances converges, as* k → ∞, *to an additive matrix that corresponds to* (T, w), *for some edge-weighting function* w.

In other words, as we sample more gene trees, the matrix of average internode distances will converge to a matrix that is additive for the true species tree. Therefore, if we apply a distance-based tree estimation methods that has a positive safety radius (see Section 5.10) to the matrix of average internode distances, we can estimate the species tree in a statistically consistent manner. NJst (Liu and Yu, 2011) was the first method to use such an approach, and was based on neighbor joining. Since neighbor joining has a positive safety radius, NJst is statistically consistent under the MSC model. ASTRID (Vachaspati and Warnow, 2015) is a modification of NJst that uses the distance-based method FastME to compute the species tree from the matrix of average internode distances. Since FastME has a positive safety radius, ASTRID is statistically consistent under the MSC.

ASTRID also differs from NJst in that it is better able to handle inputs that result in matrices with missing entries (i.e., inputs where some pair of species is not in any gene tree together). For such inputs, ASTRID uses BioNJ* (Criscuolo and Gascuel, 2008), a distance-based method that is designed to deal with missing data, and that tends to produce better trees when the dissimilarity matrix has missing entries. ASTRID is much faster than NJst and typically more accurate, while the comparison to ASTRAL-2 shows it is much faster and often has comparable accuracy (Vachaspati and Warnow, 2015).

10.5.2 The Impact of Gene Tree Estimation Error

By definition, summary methods combine gene trees, and so their guarantees depend on the properties of the gene trees they combine. The proofs of statistical consistency we have presented assume that the input is a set of true gene trees, and so are error-free. Yet, estimated gene trees typically have some estimation error, and if gene sequence lengths are kept short in order to avoid recombination within a locus, then gene tree estimation error can be high. In addition, rooted gene trees are more likely to have error than unrooted gene trees, since in addition to estimating the tree topology they also estimate the location of the root. For example, using outgroup taxa to root gene trees is standard practice but problematic: when deep coalescence (ILS) occurs, the outgroup taxon may not be an outgroup within the gene tree, and so the use of outgroup taxa to root a gene tree may produce incorrectly rooted gene trees. Hence, estimation error in gene trees, and perhaps especially in rooted gene trees, is likely, and hence can potentially impact species tree estimation. Hence, we need to consider how gene tree estimation error impacts species tree estimation performed using summary methods.

Unfortunately, we have no theory yet that addresses this issue, at least not for any of the standard coalescent-based methods. However, from an empirical standpoint, it is clear that gene tree estimation error impacts the accuracy of species trees estimated using summary methods (Huang et al., 2010; Bayzid and Warnow, 2013; Patel et al., 2013; DeGiorgio and Degnan, 2014; Gatesy and Springer, 2014; Mirarab et al., 2014b; Bayzid et al., 2015; Gatesy et al., 2016; Meiklejohn et al., 2016; Springer and Gatesy, 2016). That is, as gene tree estimation error increases, species tree estimation error increases. Indeed, increases in gene tree estimation error can result in increased errors in species tree branch lengths and the incidence of strongly supported false positive branches (Mirarab et al., 2014b; Bayzid et al., 2015). The impact of gene tree estimation error on species tree estimation is clear, and in some cases can be substantial.

One approach that has been taken to ameliorate the impact of gene tree estimation error due to inadequate data for the different genes is to combine genes that seem to have similar phylogenetic trees together, and then recompute the gene trees based on these "supergenes." The goal of this type of analysis is to produce more accurate estimates of the gene trees by using the sequence data from other loci, so that species trees estimated on these re-estimated gene trees can be more accurate. An example of this kind of approach is treeCL (Gori et al., 2016), which seeks to produce groups of genes that not only have similar tree topologies but also similar branch lengths. However, the supergenes can also be defined by similarity in just the gene tree topologies without concern for branch lengths, as used in statistical binning (Mirarab et al., 2014b) and weighted statistical binning (Bayzid et al., 2015).

The input to statistical binning (weighted or unweighted) is a set of gene trees with branch support values (e.g., computed using bootstrapping), along with a user-selected threshold B. For every pair of gene trees, the branches with support below B are collapsed, thus producing incompletely resolved gene trees. Two gene trees that are incompatible after collapsing the low support branches are considered incompatible, and otherwise they are considered compatible. Statistical binning then computes a graph in which every node represents a gene, and the branches represent incompatibility between the trees on the two genes after collapsing the low support branches. The vertices of the graph are then assigned colors, so that no two vertices that are adjacent receive the same color, with the objective being to use a small number of colors, while keeping the number of vertices per color class close. This *balanced minimum vertex coloring* problem is likely NP-hard (the minimum vertex coloring problem is one of the classic NP-hard problems), and so a heuristic is used to find the coloring. Once the coloring is completed, the genes are partitioned into bins with one bin for each color. Then, a fully partitioned maximum likelihood analysis is performed for the genes in each bin, producing a new gene tree for the bin, which is referred to as a supergene tree. In the original formulation of the statistical binning method, these supergene trees are then given to a summary method, and a species tree is estimated; the weighted version of statistical binning makes as many copies of each supergene tree as there are genes in its bin, and then computes the species tree from this (larger) set of supergene trees.

Statistical binning was used with MP-EST to estimate a species tree for 48 birds and about 14,000 "genes" (Jarvis et al., 2014). As shown in Mirarab et al. (2014b) and Bayzid et al. (2015), both weighted and unweighted statistical binning tend to improve the accuracy of gene trees and species tree topologies, improve the accuracy of species tree branch lengths, and reduce the incidence of false positive branches. Unweighted statistical binning has the undesirable property of not being statistically consistent under the MSC, but weighted statistical binning is consistent (Bayzid et al., 2015). Thus, as the number of loci and the number of sites per locus increase, pipelines based on weighted statistical binning followed by statistically consistent summary methods will converge to the true species tree. Given the potential for improved gene tree estimation and species tree estimation, this is a fruitful direction for future research.

10.5.3 Site-based Methods

Another type of coalescent-based method uses individual sites within different unlinked loci, and estimates the species tree from the distribution it obtains on site patterns. Examples of this type of approach are SNAPP (Bryant et al., 2012), METAL (Dasarathy et al., 2015), SMRT-ML (DeGiorgio and Degnan, 2010), SVDquartets (Chifman and Kubatko, 2014, 2015), and some of the distance-based methods presented in Mossel and Roch (2015).

Although site-based methods are beginning to be used in phylogenomic analyses (Rheindt et al., 2014; Sun et al., 2014; Giarla and Esselstyn, 2015; Leaché et al., 2015; Leavitt et al., 2016; Meiklejohn et al., 2016; Moyle et al., 2016), much less is known about the accuracy of these methods compared to summary methods. The few studies evaluating site-based methods such as SVDquartets and METAL (Chou et al., 2015; Hosner et al., 2016; Rusinko and McParlon, 2017) have shown mixed performance; however, it is too early to draw any clear trends from these studies.

SVDquartets may be the most frequently used of the site-based methods, and has been integrated into PAUP* (Swofford, 2002). The input to SVDquartets is a set of sites sampled randomly from the genome, so that each site comes from a separate c-gene. SVDquartets uses the site patterns to compute quartet trees for all sets of four species. The specific approach to computing the quartet trees is based on the singular-value decomposition (SVD) of a matrix computed from the input data. SVDquartets can therefore be combined with quartet amalgamation methods to estimate a species tree.

To compute a quartet tree, SVDquartets calculates a cost (the SVD score, see Chifman and Kubatko (2014)) for each of the three possible unrooted trees on the four species, and the quartet tree with the lowest cost is selected as the best tree for the four species (Chifman and Kubatko, 2014). This approach can also be extended to provide estimates of the statistical support for each quartet tree, as shown in Gaither and Kubatko (2016). Under the assumption that the different c-genes evolve within the species tree under the MSC, that the sequences for each gene evolve under the GTR+G+I sequence evolution model (i.e., where site evolution is GTR, and sites are either variable or invariable, but if variable they draw rates from a gamma distribution) and obey a strict molecular clock, and that a single site is sampled from every c-gene, the unrooted quartet tree is identifiable from the distribution on site patterns (Chifman and Kubatko, 2014, 2015). Since any unrooted binary tree is defined by its set of quartet trees, this means that the distribution of site patterns defines the species tree, and also suggests methods for estimating the species tree from site patterns. We summarize the first point with the following theorem:

Theorem 10.9 *Let* T *be a model species tree so that gene trees evolve within* T *under the MSC, and assume that for all c-genes, sequences evolve under the GTR+G+I model with strict molecular clock, and that a single site is sampled from every c-gene. Then the tree* T *is identifiable from the distribution on the site patterns.*

Proof Under the assumptions of the theorem, Chifman and Kubatko (2015) showed that the quartet trees are identifiable from the site patterns. Since the unrooted species tree T is identifiable from $Q(T)$, this proves that the unrooted tree T is identifiable from the distribution on the site patterns. □

SVDquartets can be used to estimate species trees with more than four species. In the first phase, the quartet trees are computed using SVDquartets, and in the second phase they are combined into a species tree using a quartet amalgamation method. If the quartet amalgamation method is guaranteed to return T when given $Q(T)$, then, under the assumption that all the estimated quartet trees agree with the species tree, the full species tree can be constructed, using the quartet amalgamation method. An example of a quartet amalgamation method with this property is the All Quartets Method; other methods that fail to have this property can be modified to have this property by first running the All Quartets Method, and then only using the original approach if the input set is incompatible.

Therefore, to establish statistical consistency of a pipeline that uses SVDquartets to compute quartet trees and then uses a quartet amalgamation method to combine the quartet trees, the quartet tree estimation will need to be proven statistically consistent and the quartet amalgamation method will have to return tree T given $Q(T)$. The quartet amalgamation method used in PAUP*'s version of SVDquartets is a modification of QFM (Reaz et al., 2014), a quartet amalgamation method that seeks to maximize the number of input quartet trees it satisfies. It is not known if QFM will return T given $Q(T)$, but the modification of QFM in PAUP* may have this property.

However, Chifman and Kubatko (2014, 2015) only established identifiability of the model species tree in its unrooted form from the site patterns, and did not establish consistency of SVDquartets in estimating quartet trees. While it seems likely that SVDquartets is a statistically consistent technique for estimating quartet trees, no proof has yet been provided.

There are multiple advantages of site-based methods such as SVDquartets over standard summary methods. Perhaps the most important one is that there is no need to estimate gene trees, and hence site-based methods are likely to be particularly valuable for datasets where recombination may produce very short c-genes (i.e., genomic regions that are recombination-free), which reduces accuracy for summary methods. For the same reason, site-based methods have the potential to be faster than standard summary methods, since the most computationally intensive part of a phylogenomic analysis of many loci using a summary method is typically the calculation of gene trees.

There are some factors that may reduce the accuracy of SVDquartets in practice. First, although SVDquartets has performed well under conditions with relaxed clocks (Chou et al., 2015), its theoretical guarantees have been established only under the strict molecular clock. SVDquartets estimates quartet trees independently from each other; hence, like any quartet-based method, accuracy could suffer under conditions where many quartet trees are difficult to estimate with high accuracy. Finally, quartet amalgamation methods are generally attempts to solve NP-hard optimization problems (Jiang et al., 2001), and so may not have good accuracy on large datasets. Despite these issues, the potential for site-based methods such as SVDquartets to provide improved accuracy in the presence of ILS, especially when recombination-free loci are necessarily short, is high.

10.5.4 Co-estimation of Gene Trees and Species Trees under the MSC

Another type of coalescent-based method operates by co-estimating the gene trees and the species tree from the sequence alignments for the different loci; this approach has the benefit of not depending on an accurate gene tree for each gene.

The most well known of these methods, and the most computationally efficient, is the Bayesian method *BEAST (Heled and Drummond, 2010). *BEAST uses an MCMC technique to sample from the space of model species trees and gene trees; the result of the analysis is an estimate of the posterior distribution of the species tree and also of the posterior distribution for each gene tree. Point estimates of the species tree and gene trees can be obtained from these posterior distributions using standard techniques.

As shown in Bayzid and Warnow (2013), species trees computed by *BEAST can be more accurate than species trees estimated using summary methods, and the gene trees computed by *BEAST can also be more accurate than maximum likelihood trees estimated on individual gene sequence alignments. However, its running time can be excessively large (Bayzid and Warnow, 2013; McCormack et al., 2013; Leavitt et al., 2016), so that *BEAST analyses are typically limited to at most 30 or so species and perhaps 100 loci (and even analyses of datasets of these sizes can take weeks).

10.5.5 Fixed-Length Statistical Consistency

Statistical consistency guarantees have been established for coalescent-based methods under the assumption that the number of sites and the number of loci both increase to infinity. Yet, the more biologically relevant question is whether a method can be proven to converge to the true tree as the number of loci increases, but the number of sites per locus is a fixed constant. Methods that remain statistically consistent in this regime are said to be "fixed-length consistent."

The distinction between these regimes for statistical consistency (i.e., the usual version where the number of loci and the number of sites both go to infinity, and the more constrained version where only the number of loci goes to infinity) is discussed in Roch and Warnow (2015), Warnow (2015), and Mirarab et al. (2015b). Roch and Warnow (2015) presented summary methods that are fixed-length consistent even for the case where every gene has only a single site, as long as the sequences evolve under a strict molecular clock; however, they failed to establish any positive results for standard summary methods, such as MP-EST, ASTRAL, ASTRID, etc. Similarly, while unpartitioned maximum likelihood has been established to be inconsistent under the standard regime (where sequence lengths per locus increase to infinity) (Roch and Steel, 2015), it is unknown whether unpartitioned maximum likelihood is consistent or inconsistent in the fixed-length regime; see discussion in Warnow (2015) and Mirarab et al. (2015b).

A careful examination of the proofs of statistical consistency of other methods reveals that only a few methods have been proven to be fixed-length statistically consistent. For example, SMRT-ML (DeGiorgio and Degnan, 2010) is fixed-length statistically consistent under the assumption of a strict molecular clock and METAL (Dasarathy et al., 2015) is fixed-length statistically consistent even without the assumption of the strict molecular clock. It may well be that other methods are fixed-length consistent, but no proofs of this have yet been provided.

10.6 Improving Scalability of Coalescent-based Methods

Many of the most popular coalescent-based methods are computationally intensive. As an example, *BEAST is limited to small numbers of loci and species because it uses an MCMC analysis to co-estimate gene trees and species trees. Other coalescent-based methods, such as MP-EST and STELLS, are also computationally intensive, because they use heuristics to seek optimal trees with respect to maximum likelihood or maximum pseudo-likelihood. Thus, several techniques have been developed to improve the scalability of computationally intensive coalescent-based methods. Here we present two of these techniques, each using a divide-and-conquer strategy.

*Improving *BEAST's scalability to larger numbers of loci.* One of the major reasons that *BEAST is more accurate than summary methods is that it is able to produce estimated gene trees that are more accurate than maximum likelihood analyses of individual gene sequence alignments (Bayzid and Warnow, 2013). The BBCA method (Zimmermann et al., 2014) is a simple divide-and-conquer technique that takes advantage of this observation. BBCA randomly divides the set of loci into bins (e.g., of 25 genes per bin), runs *BEAST on each bin, and then combines the gene trees estimated by *BEAST into a species tree using a summary method. As shown in Zimmermann et al. (2014), using BBCA with MP-EST produced excellent results when using bins of size 25. For example, *BEAST was able to converge quickly to the stationary distribution on each of the 25-gene bins, but unable to converge on bins of 100 genes within reasonable time frames (e.g., one week).

As expected, gene trees computed using *BEAST for each bin were highly accurate, and more so than maximum likelihood (ML) trees on the individual alignments. Species trees computed using MP-EST on these *BEAST gene trees were highly accurate, and more so than species trees estimated using MP-EST on the original ML gene trees; in contrast, species trees obtained by letting *BEAST run for a week were not more accurate (and were sometimes less accurate) than species trees estimated using BBCA in which *BEAST was limited to 24 hours for each bin.

BBCA thus improves the scalability of *BEAST to larger number of loci, enabling highly accurate species trees to be obtained. The key to BBCA is that by restricting *BEAST to small sets of genes, it enables *BEAST to provide better gene trees, and then uses fast summary methods to combine the better gene trees into a species tree. However, the BBCA technique does not reduce the number of taxa that are analyzed, and so only addresses limitations in terms of the number of loci.

Improving MP-EST's scalability to larger numbers of loci. MP-EST uses a heuristic search strategy to seek optimal pseudo-likelihood trees; hence, its running time increases quickly with the number of taxa, as observed in Mirarab and Warnow (2015). An iterative divide-and-conquer strategy based on DACTAL (Nelesen et al., 2012), and hence referred to as "DACTAL-boosting," was developed to improve the scalability of MP-EST to large numbers of taxa in Bayzid et al. (2014).

Each iteration begins with the tree computed in the previous step, and then computes a new tree. In a given iteration, the taxon set is divided into smaller, overlapping subsets, using the tree from the previous step. The division into subsets is accomplished using the DACTAL decomposition (Definition 11.9), which uses recursion and the topology of the current tree to produce subsets of the desired size, and where each taxon subset is a set of species occupying a local part of the species tree. Then, for each taxon subset, it restricts the gene trees to the taxa in that set, and computes a species tree using the specified summary method. Finally, it uses the supertree method SuperFine+MRL (Nguyen et al., 2012; Swenson et al., 2012b) to combine the smaller estimated species trees together.

After iterating three or four times, the trees that are computed in each iteration are scored with respect to their quartet distance to the input gene trees (just as in the Quartet Median Tree problem), and the tree with the highest quartet support is returned (i.e., the same criterion as optimized by ASTRAL and ASTRAL-2). As shown in Bayzid et al. (2014), DACTAL-boosting improved the scalability of MP-EST by reducing the running time needed to analyze datasets. Interestingly, it also improved the topological accuracy of the resultant tree. DACTAL is an example of a disk-covering method (DCM), and is described further in Section 11.8.

10.7 Species Tree Estimation under Duplication and Loss Models

Gene family evolution. The duplication of genomic regions that contain genes is very common, and leads to **gene families**. These gene families can have different functions,

10.7 Species Tree Estimation under Duplication and Loss Models

with the new copies able to evolve new functions because mutations in the additional copies may not be deleterious to the organism (since the original copy is still present). Thus, when genes evolve with duplications, each chromosome in an organism can have multiple copies of the gene, each different at the sequence level and potentially having different functions.

Figure 10.2 shows an example of a rooted species tree $(a,(b,c))$, and shows how a gene (g) evolves down the tree with duplication and losses. Note that g is duplicated on the branch above the root (indicated with *), leading to two copies of the gene that enter the branches leading to a and to the MRCA of b and c. However, copies of g are lost on the various branches, so that only one copy of g ends up in the three species at the leaves of the tree. As a result of this gene duplication and loss process, the tree for gene g is $T = ((a,b),c)$, and so differs from the species tree, which is $(a,(b,c))$.

Orthology detection. One approach that is used is to pick one copy of the gene from each species, and then examine the tree that is computed for that set of gene copies. If the gene copies in the set have all evolved from the common ancestor via speciation events but without duplications (i.e., if the copies are all **orthologs**), then the tree on that set of copies will be topologically identical to the species tree. However, if the set of copies evolved with duplications (i.e., if some of them are **paralogs**), then the gene tree may not match the species tree. Also, as Figure 10.2 shows, it is possible for a gene to seem to be single copy even though it evolved with gene duplication and loss. Hence, what appears to be a set of orthologous genes may not be orthologous. The inference of orthology is one of the challenges in phylogenetic estimation. Despite the many methods that have been developed to detect orthology (surveyed in Sjölander et al. (2011); Lechner et al. (2014); Altenhoff et al. (2016)), even the best orthology detection methods will make mistakes, and in general the gene trees will not be perfect matches to the species tree.

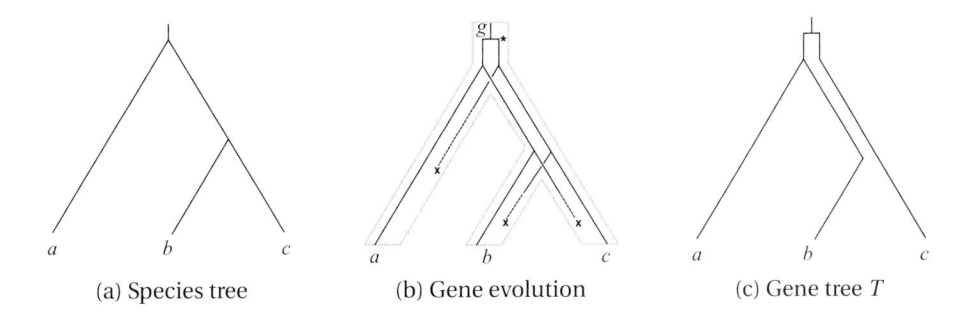

(a) Species tree (b) Gene evolution (c) Gene tree T

Figure 10.2 (Figure 11.17 in Huson et al. (2010)) Gene evolution within a species tree under a gene duplication and loss scenario. The gene g evolves down the species tree given in (a). First it duplicates on the branch above the root, and so two copies of the gene proceed down the tree towards the leaves. As a result of gene losses, only one copy of the gene appears at the leaves. Note that the tree describing how the gene evolves down the tree is different from the original species tree.

Improving gene trees using species trees. While species tree estimation and gene tree estimation are not always easy, in some cases the species tree is well established but the gene tree is more difficult to estimate with good accuracy. When the species tree is available, then it becomes possible to select between different estimated gene trees on the basis of the fit to the species tree. This observation has resulted in several "integrative methods" (surveyed in Noutahi et al. (2015)) that use both sequence data and the species tree to estimate gene trees or to correct estimated gene trees. While some of these methods only consider gene duplication and loss, others consider additional processes that can cause discord (such as ILS) and other sources of information, such as chromosomal location, to improve the accuracy of the estimated gene trees.

Species tree estimation via Gene Tree Parsimony. Species tree estimation from gene trees that evolve with duplication and loss has been approached in a parsimony framework, where the input is a set of rooted gene trees, and the objective is a rooted species tree that implies the minimum total number of duplications, or the minimum total number of duplications and losses. This is called "Gene Tree Parsimony."

Estimating the number of duplications and losses implied by a rooted gene tree with respect to a rooted species tree, and identifying the locations of the duplications and losses, is referred to as "gene tree reconciliation" (Hallett and Lagergren, 2000). Gene tree reconciliation is one of the major steps involved in correcting or estimating a gene tree given sequence data and the species tree (Vernot et al., 2008; Doyon et al., 2011; Stolzer et al., 2012; Swenson et al., 2012a; Jacox et al., 2016; Nakhleh, 2016).

The most common approach to estimating a species tree in the presence of gene duplication and loss is based on heuristic searches through treespace, where each candidate species tree is scored with respect to the input gene trees using one of the reconciliation techniques mentioned above, and the tree with the minimum reconciliation cost is returned.

Examples of methods for gene tree parsimony include DupTree (Wehe et al., 2008), iGTP (Chaudhary et al., 2010), and the methods described in Bansal and Eulenstein (2013). Several species trees have been constructed using these methods, including a "plant tree of life" for nearly 19,000 gene trees (Burleigh et al., 2011).

Since the optimization problems involved in Gene Tree Parsimony are NP-hard, globally optimal solutions are unlikely to be found. An alternative approach uses dynamic programming to find globally optimal solutions within a constrained search space (Hallett and Lagergren, 2000; Bayzid et al., 2013). For example, Hallett and Lagergren (2000) constrain the search space using clades that appear in the input gene trees, and Bayzid et al. (2013) constrain the search space using subtree-bipartitions that appear in the input gene trees, where a subtree-bipartition in a tree T is a pair of clades A, B in T where $A \cup B$ is also a clade in T. Given these constraints on the search space, Hallett and Lagergren (2000) and Bayzid et al. (2013) then use dynamic programming to find an optimal solution within that constrained space, and their algorithms run in polynomial time.

This approach of constraining the search space and then solving the optimization problem exactly within the search space using dynamic programming was introduced by Hallett and Lagergren (2000), but has subsequently been used in many other species tree estimation methods. For example, ASTRAL and ASTRAL-2 (Mirarab et al., 2014a; Mirarab and Warnow, 2015), the algorithm for the minimize deep coalescence (MDC) criterion (Than and Nakhleh, 2009; Yu et al., 2011a) in Phylonet (Nakhleh et al., 2003), FastRFS (Vachaspati and Warnow, 2016), and the method of Bryant and Steel (2001) for optimizing the quartet support, all use the same basic strategy: Constrain the search space, and then find a species tree that optimizes some criterion within that space.

Another method that uses this technique is ALE (Szöllősi et al., 2013), a method for computing rooted gene trees for multi-locus datasets given a rooted species tree and assuming a stochastic model of gene duplication and loss. ALE uses the sequence alignments for each gene and seeks the best gene tree for each gene using a cost function that depends on both the sequence evolution model and the stochastic model for gene duplication and loss. To make this computation feasible, ALE constrains the search space for each gene tree using a set of estimated ML trees for each gene; thus, each bipartition for each gene tree must be found in at least one ML gene tree. As with other methods of this type, ALE finds an optimal tree for each gene using dynamic programming.

Nearly all these methods use the clades, bipartitions, or subtree-bipartitions from the input gene trees to define the search space. For such methods, the representation of the search space will be polynomial in the input, but the search space will be larger (potentially exponential in the number of constraints). Dynamic programming allows this space to be searched efficiently, using at most a polynomial (in the input size) number of steps.

The specific way the search space is constrained and the specific details of the dynamic programming algorithm differ in ways that suggest opportunities for further advances. For example, most of these methods (though not ASTRAL-2 and not FastRFS) use the clades, bipartitions, or subtree bipartitions from the input gene trees to constrain the search. This approach is of limited value when the input gene trees are missing some species (i.e., are incomplete), and alternative approaches to constraining the search space are needed in those cases. ASTRAL-2 (Mir arabbaygi, 2015) and FastRFS (Vachaspati and Warnow, 2016) explicitly address this issue by expanding the set of allowed bipartitions, using different strategies. Obviously, the larger the constraint space, the better the criterion score, and so enlarging the constraint space can provide improved accuracy with respect to finding good solutions to the optimization problem; furthermore, if the constraint space is enlarged carefully, then it can also lead to improvements in topological accuracy (Mirarab and Warnow, 2015; Vachaspati and Warnow, 2016). Unfortunately, increasing the constraint space also increases the running time. Hence, one direction for future research is to develop strategies for defining the constraint space that provide improved accuracy without excessively increasing the running time.

Species tree estimation using statistical methods. Finding species trees from gene trees that minimize the total number of duplications and/or losses is a parsimony style reconstruction method. Alternative approaches based on statistical models of gene evolution that allow for duplications and losses have also been developed. Unsurprisingly, these are more computationally intensive than the parsimony-style methods. Nevertheless, because the techniques are grounded in statistical models, the potential for improved accuracy is substantial.

Phyldog (Boussau et al., 2013) is an example of such a statistical method. Specifically, Phyldog assumes an explicit stochastic model of evolution in which genes evolve within a species tree under duplication and loss, and then tries to estimate this species tree, as well as the gene trees for each gene, from the input set of sequence alignments. Phyldog uses a heuristic search strategy to find the gene trees and model species tree that maximizes the probability of the observed sequence data; thus, Phyldog is a maximum likelihood method that co-estimates gene trees and species trees under a gene duplication and loss model. As shown in Boussau et al. (2013), Phyldog produces highly accurate species trees.

Another statistical approach to estimate species trees in the presence of gene duplication and loss is MixTreEM (Ullah et al., 2015). The input is a set of gene families, each with a multiple sequence alignment, and the output is a species tree that is expected to have high likelihood under the gene duplication and loss model. MixTreEM has two phases: The first phase uses a variant of the expectation maximization (EM) algorithm to produce a set of candidate species trees that is expected to either contain the true species tree or to have a candidate tree that is close to the true species tree. The second phase then selects the best tree from the set, using both the gene tree input as well as the sequence alignments associated to the genes. MixTreEM and Phyldog showed comparable accuracy on a collection of simulated datasets, but MixTreEM was much faster, especially on the larger datasets (Ullah et al., 2015).

Finally, *guenomu*, a Bayesian supertree method (De Oliveira Martins et al., 2016), represents a third type of approach. There are multiple differences between Phyldog, MixTreEM, and *guenomu*, which make comparisons between them complicated. First, while Phyldog and MixTreEM only explicitly address duplication and loss, *guenomu* also models deep coalescence and HGT as possible sources of gene tree discord. Second, the input to Phyldog and MixTreEM is a set of gene sequence alignments, but the input to *guenomu* is a set of distributions of estimated gene family trees, with one distribution per gene. Third, Phyldog and MixTreEM output the species tree with the highest ML score it finds, while *guenomu* outputs a distribution on species tree topologies. Fourth, and perhaps most importantly, Phyldog and MixTreEM base their selection of the final tree using a likelihood calculation under an explicit stochastic model of gene evolution involving gene duplication and loss, while *guenomu*'s analysis uses the reconciliation distance to score trees.

A comparison between *guenomu*, iGTP (Chaudhary et al., 2010), and coalescent-based summary methods showed that *guenomu* was generally more accurate than the competing methods, with iGTP the closest competitor. No comparison has yet been made between *guenomu* and Phyldog or MixTrEM in terms of their accuracy on data, but *guenomu* is very likely to be the fastest of the three methods, as it was able to run on datasets with 80 species in reasonable time frames (De Oliveira Martins et al., 2016) and the other methods are much slower.

10.8 Constructing Trees in the Presence of Horizontal Gene Transfer

Horizontal gene transfer, also referred to as lateral gene transfer (LGT), is a process other than inheritance from a parent in which DNA sequences are transferred from one organism into another. HGT is considered to be a major contributor to diversity in prokaryotes. Because of the ubiquity of HGT, the inference of a species phylogeny, and even the definition of a species for prokaryotes, is difficult. In fact, Gogarten et al. (2002) and Gogarten and Townsend (2005) argued that only appropriate representation of the species history in the presence of massive HGT is a phylogenetic network, a topic we discuss in Section 10.9. However, others have argued that there are *core genes* that are not as highly impacted by HGT, and that it should be possible to infer an underlying tree down which the core genes evolve, even in the presence of otherwise massive HGT (Ge et al., 2005). In other words, there should be a well-defined underlying species tree, on top of which the HGT events occur.

Many techniques have been proposed to estimate the underlying species tree from a set of gene trees. An early approach, called the Median Tree Algorithm (Kim and Salisbury, 2001), is as follows: The matrix of median normalized pairwise distances between taxa across all the genes is computed, and then a neighbor joining tree is computed from the distance matrix. As shown in Kim and Salisbury (2001), the Median Tree Algorithm is more accurate than the greedy consensus at computing the underlying tree. This method was subsequently used in Ge et al. (2005) to construct a Tree of Life.

Recent theoretical results (Roch and Snir, 2013; Steel et al., 2013; Daskalakis and Roch, 2016) provide a theoretical justification for the hypothesis that the underlying species tree should be discernible from the set of gene trees, at least under bounded amounts of random HGT, and suggest statistically consistent methods for inferring the underlying tree. Steel et al. (2013) and Daskalakis and Roch (2016) address the inference of a rooted species tree from rooted gene trees, and Roch and Snir (2013) addresses the inference of an unrooted species tree from unrooted gene trees; all three papers are of interest, but we focus on Roch and Snir (2013), since they address the inference of the species tree from unrooted gene trees, which can be estimated from data without requiring a molecular clock.

Roch and Snir (2013) assume that random HGT events occur during the evolutionary history, under a continuous time Poisson process. Under this stochastic HGT model, if the rate of HGT is low enough, then for every four species, the most probable unrooted quartet tree will be topologically identical to the species tree on the four species. Therefore, under this bounded and random HGT model, the underlying species tree is identifiable from the distribution on gene trees. Furthermore, under this model, methods such as ASTRAL (Mirarab et al., 2014a; Mirarab and Warnow, 2015), which use appropriate quartet-based techniques to construct species trees, are consistent methods for estimating the species tree. The performance of species tree estimation methods under conditions in which both ILS and HGT occur was explored by Davidson et al. (2015), who observed that ASTRAL and weighted Quartets MaxCut (Avni et al., 2015) generally had better accuracy than both NJst (Liu and Yu, 2011) and unpartitioned concatenation using ML. Thus, quartet-based estimation provides both theoretical and empirical advantages in terms of estimating an underlying species tree in the presence of HGT and/or ILS.

These results only address the question of estimating an underlying species tree, and do not address how to identify the location of the HGT events on the species tree, nor how to determine whether a given gene has evolved with HGT. More generally, these methods do not provide a full description of the evolutionary history of the dataset, which – by necessity – would not be appropriately modeled by a tree.

10.9 Phylogenetic Networks

10.9.1 Reticulate Evolution

As we have seen, evolutionary processes operating on species trees, such as gene duplication and loss and ILS, can create conditions where true gene trees differ from the species tree, and hence also from each other. However, in these cases, the true evolutionary history for the species is still a tree, despite what may be extremely high levels of heterogeneity among the gene trees. Thus, heterogeneity among gene trees by itself does not indicate that a species phylogeny isn't a tree, and estimating a species tree in the presence of heterogeneous gene trees can still make sense.

On the other hand, some evolutionary processes, such as hybridization, recombination, and HGT, cannot be adequately modeled by a tree. These processes deviate from tree-like evolution, and are referred to as "reticulations." When reticulation is present, then graphical models that are more general than trees are necessary.

Phylogenetic networks are graphical models of evolution that can be used to represent evolutionary histories in the presence of reticulate evolutionary processes, and many methods have been developed to construct phylogenetic networks for datasets where reticulate evolution is believed to have occurred. However, there are many kinds of phylogenetic networks, and the construction and interpretation of these networks is much more complicated than the construction and interpretation of trees. As a result, phylogenetic network construction is a major research area in its own right. Huson et al. (2010) and Morrison (2011) are two books that cover species phylogenetic networks, but the focus of Gusfield (2014) is on ancestral recombination graphs (ARGs), a type of phylogenetic network that aims to model the evolution of a population rather than of a set of species. Since this book is focused on species phylogenies, the discussion that follows will not discuss ARGs, or other types of networks that are specifically focused on representing the evolution of a population.

10.9.2 Evolutionary Networks

There are fundamentally two different objectives in phylogenetic network construction. The first objective is to provide a graphical representation of the evolutionary history for the dataset; this type of network is referred to either as an explicit network (Huson, 2007) or as an evolutionary network (Morrison, 2010a). **Evolutionary networks** are rooted directed acyclic graphs, in which the edges either represent descent from an ancestor, or an event such as HGT. For example, an evolutionary network representing HGT could be obtained by taking a rooted phylogenetic tree, and adding directed edges to indicate where the HGT takes place. Similarly, to represent hybridization (i.e., where individuals from two different species mate and have viable offspring, termed "hybrids"), a phylogenetic network could be obtained by taking the rooted tree for all the taxa other than the hybrid, and then adding a node for the hybrid taxon with edges coming into it from its two parents. Note that in each of these cases, every site in the genomes within the species at the leaves evolves down some tree contained within the network. Hence, these networks provide explicit explanations for the evolutionary processes that underlie the dataset, and this is why they are referred to as explicit phylogenetic networks, or evolutionary networks.

By construction, evolutionary networks have some specific set of evolutionary processes they address; hence, a network might only model hybridization, another might model hybridization and also ILS, and a third might model recombination and ILS. The choice of processes to model is critical to the algorithm design and also to interpreting a constructed phylogenetic network.

Many methods for constructing evolutionary networks are based on optimization problems that seek to construct minimal networks. As an example, if hybridization is the expected cause for the discord between gene trees, then a rooted phylogenetic network might be sought that displays all the trees in the input, and that has a minimum number of hybridization events. This type of approach can be referred to more generally as a parsimony-style analysis. The first methods to construct an evolutionary network from a set of gene trees appear in Nakhleh et al. (2004, 2005a); see also Nakhleh et al. (2005b), Jin et al. (2007), Yu et al. (2011b), Wu (2013), and Wheeler (2015) for other methods for constructing evolutionary networks.

Figure 10.3 shows two evolutionary networks, N_1 and N_2, computed on two gene trees, T_1 and T_2. The reticulations in these networks represent hybridization events, and so these are hybridization networks, and each gene tree is displayed in each of the two networks. The networks each have three hybridizations, which is the minimum needed to display the two gene trees.

In addition to the parsimony-style approaches, methods for scoring or constructing networks have been developed that are based on parametric statistical models of evolution that include processes such as hybridization, HGT, or ILS. These statistical approaches (Strimmer and Moulton, 2000; Jin et al., 2006; Kubatko, 2009; Meng and Kubatko, 2009; Bloomquist and Suchard, 2010; Yu et al., 2012; Jones et al., 2013; Yu et al., 2013, 2014;

(a) Tree T_1 for *phyB*

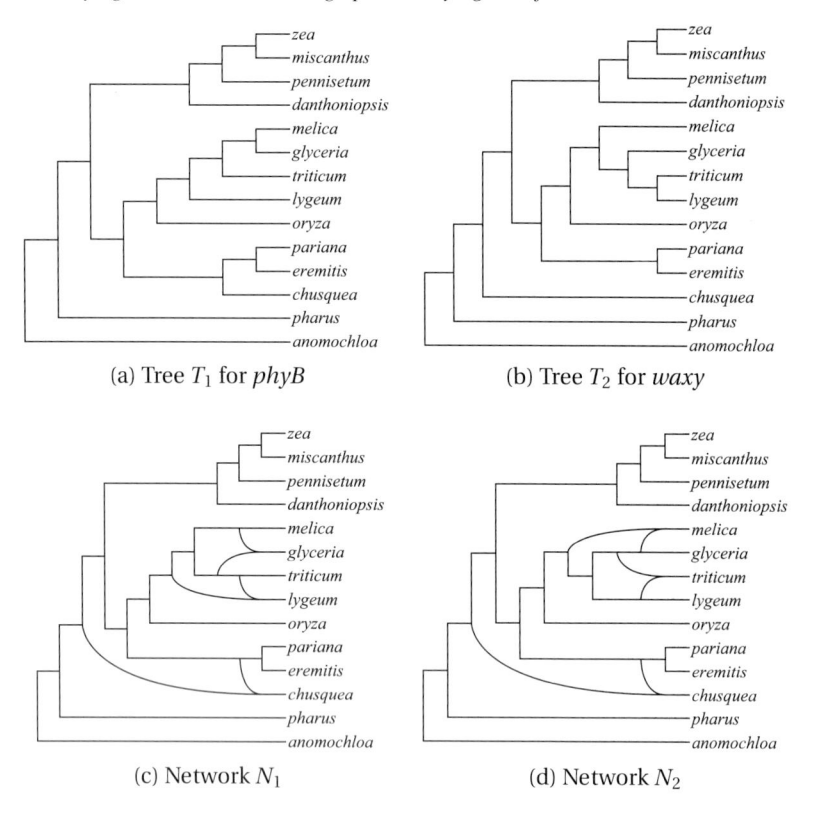

(b) Tree T_2 for *waxy*

(c) Network N_1

(d) Network N_2

Figure 10.3 (Figure 11.14 in Huson et al. (2010)) This figure presents two gene trees (a) T_1 and (b) T_2 (Grass Phylogeny Working Group, 2001) and two hybridization networks (c) N_1 and (d) N_2, each on 14 grasses. T_1 is a gene tree for the *phyB* gene, and T_2 is a gene tree for the *waxy* gene. The two hybridization networks N_1 and N_2 display both gene trees, and use a minimum number of hybridization events.

Kubatko and Chifman, 2015; Yu and Nakhleh, 2015b; Wen et al., 2016; Solís-Lemus and Ané, 2016) provide improved accuracy and interpretability compared to the parsimony-style evolutionary networks.

Unfortunately, statistical methods for estimating evolutionary networks tend to be very computationally intensive. For example, HyDe (Kubatko and Chifman, 2015) can construct a hybridization network on three species, and can be used to detect hybrid taxa in larger datasets, but is not otherwise applicable to constructing a larger evolutionary network. One of the challenges in building these networks is that even *scoring* a network with respect to the input (e.g., a set of trees) generally depends on the ability to compute all the trees that are contained in a network, which can be exponential in the number of reticulations in the network. Thus, as the network becomes more reticulate, the cost of scoring the network rises. Another challenge is selecting the appropriate graphical model, since the number of evolutionary networks on any fixed number of leaves is unbounded (unlike the case of trees). Thus, estimating evolutionary networks using statistical approaches tends to be very computationally intensive, and current methods are limited to very small numbers of species (e.g., under ten).

Distance-based estimation of evolutionary networks modeling hybridization is possible, using theoretical results relating average distances between species, given a network (Willson, 2012, 2013; Francis and Steel, 2015; Bordewich and Semple, 2016; Bordewich and Tokac, 2016). However, these assume that each gene tree is displayed by the network, which may not hold in the presence of ILS. To address this case, Yu and Nakhleh (2015a) presented a distance-based approach to construct an evolutionary network modeling both ILS and hybridization that is fast enough to run on large datasets. These distance-based methods are statistical in the sense that they are based on explicit stochastic models of gene evolution within evolutionary networks, and they can be quite fast. However, all these methods, including the method in Yu and Nakhleh (2015a), are impacted by errors in the estimated distances, and so are not expected to be as accurate as statistical methods that combine estimated gene trees or that are based on sequences.

Methods for computing evolutionary networks addressing hybridization and incomplete lineage sorting using both statistical and parsimony-style approaches are available in Phylonet (Nakhleh et al., 2003) and PhyloNetworks (Ané, 2016).

10.9.3 Data-Display Networks

The other type of phylogenetic network aims to provide a graphical representation of the input data, without providing an explicit evolutionary scenario. Because these networks do not provide any explicit explanation of the evolutionary events that led to the observed data, they are referred to either as **data-display networks** (Morrison, 2010a) or as implicit networks (Huson, 2007). The simplest type of data-display network is where the input is a distance matrix between the species (computed in some manner), and the output is a tree only when the input matrix is additive, and otherwise the output is a non-tree network. Data-display networks can be based also on sequence data, gene trees, or other types of characters. Examples of data-display networks are given in Figures 10.4–10.6. None of these networks provide an explicit description of an evolutionary scenario.

Methods to construct data-display networks have several advantages over methods that construct evolutionary networks. For example, they can be computed for single-locus datasets, while methods to construct evolutionary networks are usually restricted to multi-locus datasets. Data-display network methods are also generally much faster to compute than the statistically based evolutionary networks discussed above (and can be faster than the parsimony-style evolutionary networks as well).

The main advantage of evolutionary networks over data-display networks is that they provide an explicit hypothesis of *what happened*, and so can be interpreted easily. However, many users of phylogenetic networks are unaware that data-display networks are really designed for different purposes, as noted by David Morrison (Morrison, 2011: 47):

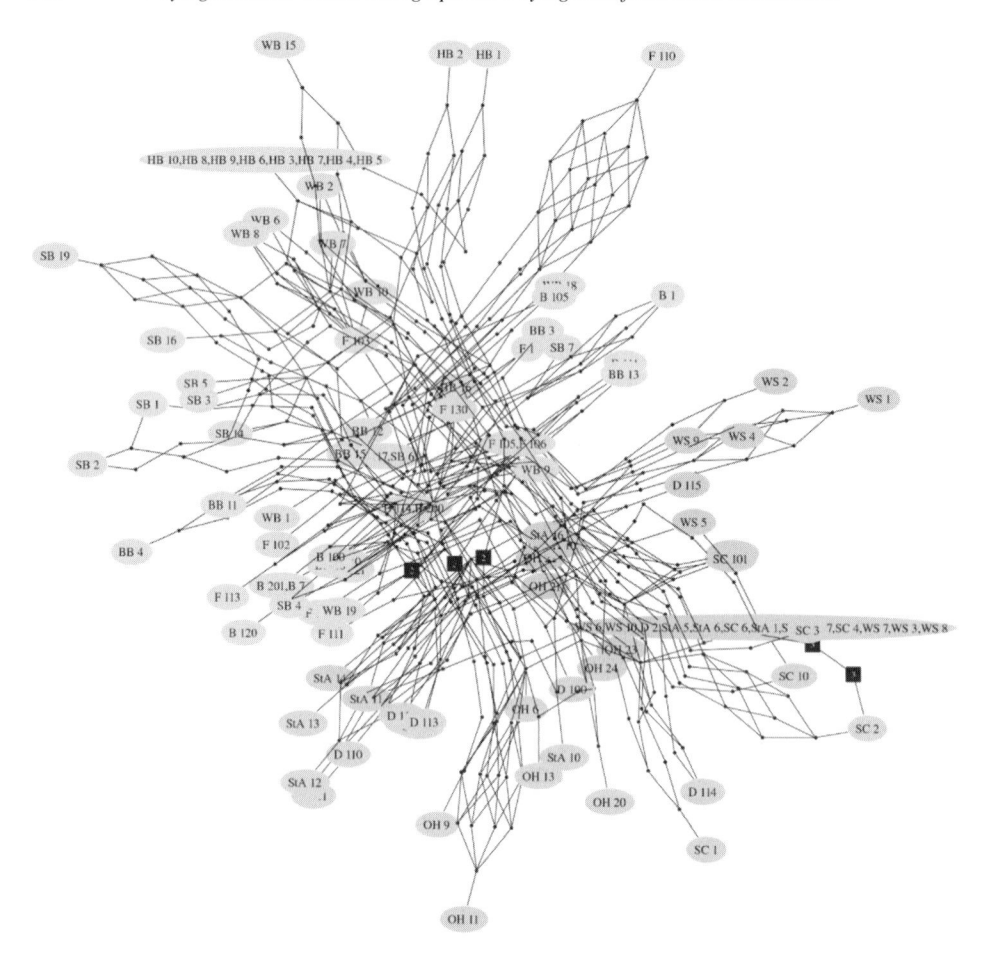

Figure 10.4 (From Ayling and Brown (2008)) A data-display network computed for 110 sequences from the spacer regions of 5S rDNA for sea beet (Ayling and Brown, 2008), obtained by pruning the quasi-median network. Reproduced from *BMC Bioinformatics* 9:115 (2008) under the Creative Commons Attribution License.

The basic issue, of course, is the simple fact that data-display networks and evolutionary networks can look the same. That is, they both contain reticulations even if they represent different things... Many people seem to have confused the two types of network, usually by trying to interpret a data-display network as an evolutionary network... The distinction between the two types of networks has frequently been noted in the literature, so it is hardly an original point for me to make here. Interestingly, a number of authors have explicitly noted the role of display networks in exploratory data analysis and then proceeded to treat them as genealogies anyway. It is perhaps not surprising, then, that non-experts repeatedly make the same mistake.

Although data-display networks do not provide an explicit description of the evolutionary history underlying a dataset, because they are fast and easy to use, they have the

10.9 Phylogenetic Networks 265

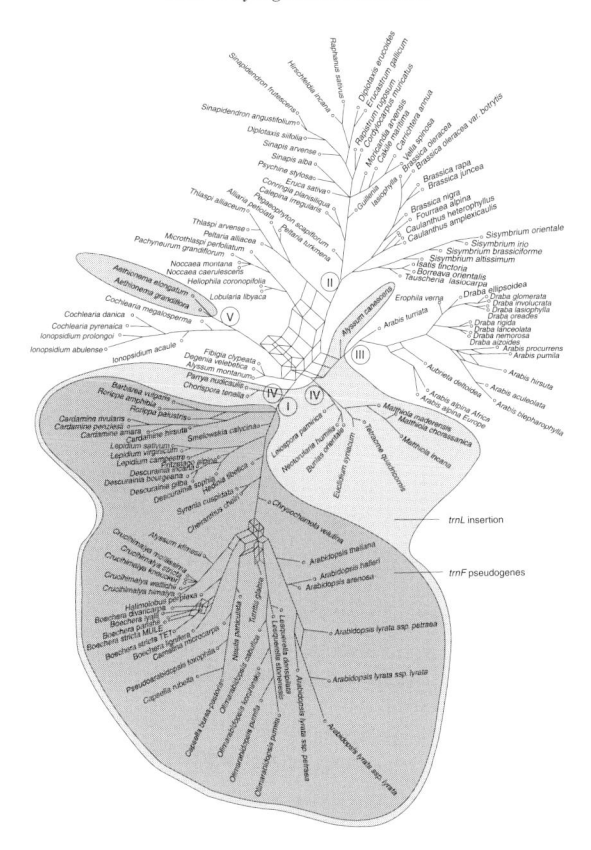

Figure 10.5 (From Koch et al. (2007) by permission of Oxford University Press) A data-display network computed for 71 plant taxa, constructed from five maximum parsimony gene trees, and then combined using the super-split method (Koch et al., 2007).

potential to be highly valuable for exploratory data analyses (Tukey, 1997) of multi-locus datasets (Morrison, 2010b).

Exploratory data analysis (EDA) of phylogenetic datasets is a standard part of systematics. For example, a systematist analyzing a dataset typically examines different choices for aligning sequences and computing trees on the alignments, to see whether the trees that are obtained by different methods are similar. Sometimes these analyses suggest that one or more taxa are causing instabilities in the analyses, and should be removed, or may indicate the need to gather more data in order to break up long branches. Thus, EDA for a single gene dataset can be valuable. The analysis of multi-locus datasets presents potential additional benefits for EDA, since species tree estimation in the presence of gene tree heterogeneity (whether due to ILS, gene duplication and loss, HGT, or hybridization) will require non-standard methods. Furthermore, if the EDA is able to indicate the cause for the gene tree heterogeneity, then it can be used to inform the next step of the analysis pipeline.

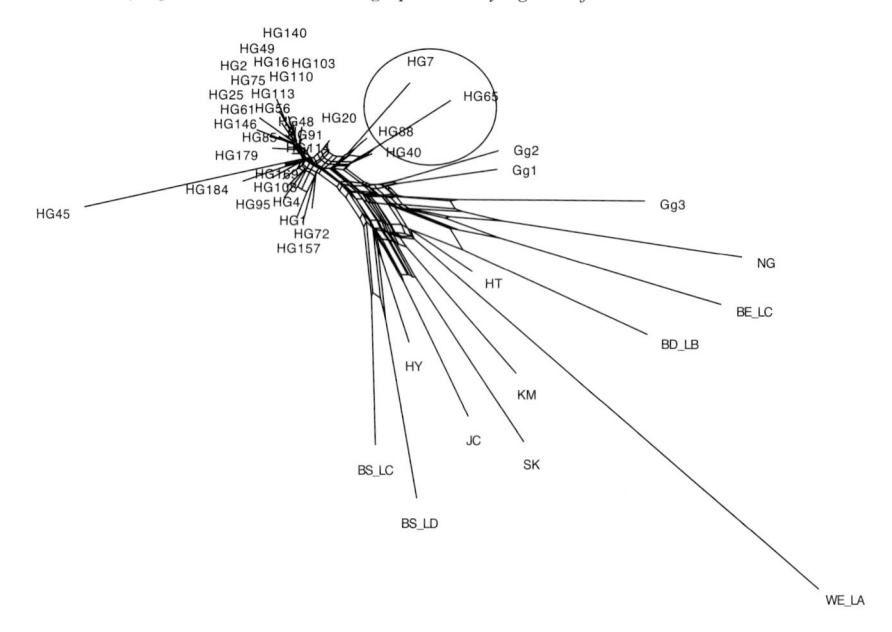

Figure 10.6 (From Berthouly et al. (2009)) A data-display network computed for 44 chicken populations (Berthouly et al., 2009) using the NeighborNet method. Reproduced from *BMC Genetics* 10:1 (2009) under the Creative Commons Attribution License.

For studies exploring the use of data-display networks for EDA in multi-locus datasets, see Holland et al. (2004, 2005), Wägele and Mayer (2007), and Morrison (2010b).

There are many types of data-display networks, including median networks (Bandelt, 1994; Bandelt et al., 2000), split decomposition networks (Bandelt and Dress, 1992), parsimony splits networks (Bandelt and Dress, 1993), NeighborNet (Bryant and Moulton, 2002), consensus networks (Holland and Moulton, 2003; Holland et al., 2004, 2005, 2006), and supernetworks (Huson et al., 2004; Holland et al., 2007). Another interesting approach is a method called Nye's Tree of Trees (Nye, 2008), which provides a visualization of a collection of trees. NeighborNet (Bryant and Moulton, 2002), a distance-based method for constructing a data-display network, is one of the most popular of these data-display networks, due to its speed, ability to analyze large datasets, and ease of use. In addition, Morrison (2010b) notes "if there is a large amount of conflicting character information in a dataset, then the median network will have a large complex set of cycles, whereas the split decomposition and parsimony splits networks will be unresolved, and the neighbor-net will be somewhere in between. These are all valid ways of representing the conflict, but in this case the neighbor-net will be the one that is clearest to interpret." NeighborNet and many of these data-display networks are available in the software package SplitsTree4 (Huson, 2016).

One of the challenges in using data-display networks for EDA in a multi-locus dataset is that interpreting the results is difficult. For example, suppose that the input is a set of

1000 estimated gene trees, and nearly all of them are different from each other. Data-display networks will demonstrate this heterogeneity. However, as we have previously noted, gene tree heterogeneity in itself, even if it is substantial, does not indicate that the species phylogeny is reticulate. Furthermore, unless the EDA is able to suggest a particular cause (or set of causes) for the gene tree heterogeneity, it will not help in determining how the species phylogeny should be computed.

10.9.4 Evaluating Phylogenetic Networks for Accuracy

As with trees, we need to have metrics for evaluating estimated phylogenetic networks for accuracy. Since data-display networks aim to represent the variance in the input data rather than construct an explicit hypothesis about the evolutionary history, this issue can realistically only be considered in the context of evolutionary networks. The determination of how to quantify error is more complicated for phylogenetic networks than for trees, since a phylogenetic network cannot be represented by its set of clades. See Nakhleh (2010) and Cardona et al. (2009) for metrics to quantify the difference between two networks.

10.9.5 Improving Evolutionary Network Estimation

Statistically based methods for constructing evolutionary networks are likely to provide the greatest accuracy, compared to parsimony-style or distance-based methods. Unfortunately, statistically based methods are limited to very small datasets, and enabling these methods to scale to larger datasets is an important direction. Approaches have been developed that aim to achieve this, by combining smaller evolutionary phylogenetic networks into a larger evolutionary phylogenetic network. An example of a recent such method is TriLoNet (Oldman et al., 2016), which provided improved accuracy compared to competing methods for constructing phylogenetic networks on simulated datasets.

Many methods to construct evolutionary networks operate by combining estimated gene trees together, and assume that each of the estimated gene trees evolves within the network under a stochastic process that includes some processes (e.g., ILS, HGT, and hybridization) that cause gene trees to deviate from a species tree. However, estimated gene trees nearly always have some estimation error, and the impact of gene tree estimation error on phylogenetic network estimation is likely to be considerable (Morrison, 2014b).

Finally, it is challenging to extend standard methods for producing statistical support from trees to networks. For example, the support on a branch in a tree is the frequency with which the bipartition appears in estimated trees, for example produced using non-parametric bootstrapping or within a Bayesian MCMC analysis. However, while trees can be naturally represented by their bipartitions, the same is not true for phylogenetic networks (indeed, it is not even clear what a bipartition means for a network). Thus, when the phylogenetic network is substantially non-treelike, the calculation of support becomes much more complicated.

10.10 Further Reading

Coalescent-based tree estimation. The main focus of this chapter has been on the estimation of species trees assuming that the gene sequences and/or trees have already been computed. For more information on the practical issues involved in designing such a phylogenomic study, see Knowles and Kubatko (2011). Many of the methods we have described estimate the tree topology but not the associated numeric parameters (e.g., branch lengths in coalescent units, times at speciation events, and population sizes). The estimation of these numeric parameters is of interest, especially in population genetics, and is discussed in greater detail elsewhere; see Wakeley (2009), Rannala and Yang (2003), and Helmkamp et al. (2012) for an entry into this literature.

The impact of new sequencing technologies. The newer sequencing technologies have made massive amounts of sequence data available at lower cost than the earlier sequencing technologies. As a result, there has been a huge increase in the amount of sequence data available for phylogenomic analysis, both in terms of the number of species for which we have sequence data and the number of whole genomes that have been sequenced and assembled (Lemmon and Lemmon, 2013; Olson et al., 2016). Many researchers consider these data to have very high potential to provide substantial novel insights into the evolutionary histories of major groups (Hallström and Janke, 2010; Jarvis et al., 2014; Wickett et al., 2014; Maddison, 2016). Phylogenomic analyses of these datasets have revealed the limitations of existing methods, and these realizations have led to the development of novel methods; for example, ASTRAL (Mirarab et al., 2014a; Mirarab and Warnow, 2015) was developed to enable a phylogenomic analysis of the 1KP plant transcriptome dataset (Wickett et al., 2014), and statistical binning (Mirarab et al., 2014b) was developed to enable a phylogenomic analysis of the avian project dataset (Jarvis et al., 2014).

Yet, the development of new phylogeny estimation methods is just one of the challenges in large-scale phylogenomic analysis. Another issue that arises and that we have not discussed at all is the impact of sequencing error on phylogeny estimation pipelines. Since the newer sequencing technologies (e.g., Illumina) have higher error rates than earlier sequencing technologies (e.g., Sanger), this is a potentially important problem. Thus, the newer sequencing technologies create massive datasets, but the datasets are not necessarily of uniformly high quality. In order to address the variable quality in the sequencing output, phylogenomic pipelines involve multiple steps in which the data are reduced to just the highest-quality sequences, and this creates an additional challenge. To quote Morrison (2017):

High-throughput genome sequencing involves many filtering steps in which parts of the genome are excluded from the final data set. These steps are intended to provide high-quality sequence data, but little attention has been paid to whether they simultaneously provide high-quality sampling... It is inconceivable that this filtering is unbiased, because the regions that are problematic to sequence are not randomly distributed in any genome... It would be nice if this potential bias was not problematic for systematics, or any other part of biology, but no one has checked and no one seems to be interested in checking.

This potential for bias was also commented on by Huang and Knowles (2016), who noted that the common practice of restricting phylogenomic data to loci without too much missing data has the unforseen consequence of disproportionally removing loci that evolve quickly. These issues may reduce as sequencing technologies improve, but it is also possible that the issues may become more serious as newer sequencing technologies are developed and used that are relatively inexpensive to use, but have higher error rates. Thus, another direction for further research is in evaluating the impact of sequencing error and genome assembly methods on phylogenomics, and the impact of the various pipelines for filtering genomic data for use in a phylogenomic analysis. In general, there are many practical issues involved in phylogenomic analysis, beyond the ones discussed here (Knowles and Kubatko, 2011).

Gene tree estimation using multi-locus datasets. One of the applications of phylogenomics is that it enables a more accurate estimation of individual gene trees. For example, *BEAST and BBCA (its heuristic version) co-estimate gene trees and species trees under the MSC, and produce more accurate gene trees as a result than ML applied to individual gene sequence alignments (Bayzid and Warnow, 2013). Phyldog, which co-estimates gene trees and species trees under a gene duplication and loss model, can produce more accurate gene trees than ML applied to individual alignments. Weighted and unweighted statistical binning has also led to improved estimates of gene trees, and hence to improved estimates of species trees. Thus, co-estimation of gene trees and the species tree is a powerful approach for multiple settings. Similarly, species phylogenies – whether trees or networks – can enable more accurate estimations of gene trees (Szöllősi et al., 2013). Future research into methods that can provide high accuracy and yet be computationally feasible is of great value.

Thus, one of the applications of phylogenomics is improving the inference of gene trees. Since traits are based on genes (sometimes collections of genes rather than a single gene), the most appropriate analyses should explore trait evolution using one or more gene trees, rather than a species tree (Hahn and Nakhleh, 2016). Hence, one of the benefits of phylogenomic analysis is that it can improve the understanding of trait evolution.

Constructing evolutionary histories in the presence of multiple sources of discord. An essential challenge is that methods for constructing phylogenies from multi-locus datasets are typically designed for just one, rather than two or more, causes for gene tree heterogeneity. The consequence is that such methods will interpret all heterogeneity as due to just the one process they are designed for, and can therefore have reduced accuracy when some other process is also operating (Zhu et al., 2016). As an example, suppose that the multi-locus dataset has substantial gene tree heterogeneity due to ILS and also has one hybridization. A method that constructs evolutionary networks where hybridization has to explain all the heterogeneity will interpret all the heterogeneity as due to hybridization and output a phylogenetic network with a lot of hybridization events, which will not be correct. Similarly, a method that produces trees and interprets all the heterogeneity as due to ILS will fail to detect the hybridization event. Thus, if a dataset has heterogeneity due to two or more causes, methods that are designed only for one cause may fail to produce a reasonable phylogeny for the dataset. The same kind of problem can occur if the dataset has only one source of heterogeneity but it doesn't match the assumptions of the method. Thus, methods that can address two or more causes for gene tree heterogeneity are needed.

Unfortunately, nearly all the methods described so far have focused on one source of discord between gene trees and species histories (whether species trees or phylogenetic networks). Yet in most (perhaps all) biological datasets in which gene tree heterogeneity is present, it is likely that multiple processes are producing the discord, and distinguishing between these different processes can be difficult (Mallet, 2005). In the last few years, some methods have been developed that specifically address species tree or network estimation in the presence of multiple sources of discord (e.g., the supertree method *guenomu* discussed earlier, and also the methods discussed in Kubatko (2009), Meng and Kubatko (2009), Joly et al. (2009), Yu et al. (2011b), Rasmussen and Kellis (2012), Szöllősi et al. (2012), Yu et al. (2013), and Wen et al. (2016)). While there are some methods that have performed well in the presence of multiple sources of discord (e.g., HGT and ILS, as shown in Davidson et al. (2015)), there is clearly a need for integrated statistical models that can represent these multiple processes and methods to estimate under these statistical models (Szöllősi et al., 2015). Because methods that are based explicitly on complex parametric models of evolution tend to be computationally intensive, scalable methods that can address multiple processes are very much needed.

Visualization of sets of trees. Related to data display networks is the use of visualization tools to explore sets of trees (potentially where each tree is a gene tree). For example, bipartition distances can be computed between trees, and then the metric space can be projected onto a lower dimensional space using multidimensional scaling (Amenta and Klingner, 2002; Hillis et al., 2005; Huang et al., 2016). Approaches like these can be used in multiple ways, including the exploration of whether a heuristic search for an optimal tree has converged. The visualization can assist in dividing the set of trees into distinct clusters, which need not be disjoint, so that each cluster of trees provides an alternative hypothesis for the evolutionary history underlying the set of sequences.

Genome rearrangement phylogeny estimation. Genomes evolve under processes beyond insertions, deletions, and substitutions. For example, rearrangement events, such as inversions, transpositions, and inverted transpositions, occur, which change the order and strandedness of genomic regions within genomes. Genomic regions are also duplicated, inserted, or deleted, which changes the copy number of the regions. Finally, fissions and fusions occur, which change the number of chromosomes and the distribution of genes among the chromosomes. These events are less common than substitutions and indels, and so create signatures in the genomes that can be used to infer evolutionary histories.

Perhaps the first use of genome rearrangements for phylogeny estimation appeared in the mid-1930s (Sturtevant and Dobzhansky, 1936). After this initial study, there was a lag until the mid-1980s, when several papers were published, including Watterson et al. (1982), Jansen and Palmer (1987), Raubeson and Jansen (1992), and Sankoff et al. (1992). These and other studies interested mathematicians, computer scientists, and statisticians, and led to new algorithmic problems, including the computation of breakpoint phylogenies (Blanchette et al., 1997, 1999; Sankoff and Blanchette, 1998) and the calculation of inversion distances between genomes (Kececioglu and Sankoff, 1995).

Genome rearrangements are more computationally challenging to analyze than sequence data, but substantial progress in analyzing these data has been made. Some edit distances between genomes can be computed efficiently, but computing the simplest of these distances (e.g., minimizing the number of inversions) requires sophisticated graph-theoretic algorithms (Hannenhalli and Pevzner, 1995; Bafna and Pevzner, 1996, 1998; Kaplan et al., 1997; Bader et al., 2001). However, other distances, such as the transposition distance (Bulteau et al., 2011), can be NP-hard to compute. Computing medians of three or more genomes to minimize the total edit distance is helpful in solving parsimony-style problems, but also tends to be NP-hard (Pe'er and Shamir, 1998).

Statistical models of genome evolution have also been developed. The Nadeau–Taylor Model (Nadeau and Taylor, 1984) of genome evolution only addressed inversions, and the Generalized Nadeau–Taylor model extended this to include transpositions and inverted transpositions (Wang and Warnow, 2001). The Double Cut-and-Join model (Yancopoulos et al., 2005) was also developed to enable a broader range of evolutionary events, including duplications, which change the number of copies of each gene within each genome. Polynomial time methods for estimating true evolutionary distances under these models have been developed, thus enabling statistically consistent distance-based tree estimation based on these genome rearrangements; see Wang and Warnow (2006) and Lin and Moret (2008) for examples of this kind of approach.

Parsimony-style methods that seek to minimize the total number of evolutionary events (e.g., minimize the number of inversions) have also been developed, and often have good accuracy, but are more computationally intensive than distance-based methods. Bayesian approaches have also been developed, but so far have not been able to scale to large datasets. In other words, all the usual types of methods have been developed, but they are more computationally expensive to run, and mathematically more involved. Nevertheless, although the estimation of phylogenies using gene content and order information for whole genomes is much more complex than using sequence data, the phylogenetic signal seems to be very strong, so that highly accurate trees can be computed. The current state of the art provides excellent accuracy even for large datasets, and some of the methods (especially the distance-based methods) are sufficiently fast that they can be used on large datasets.

Furthermore, because these models include gene duplications, methods that have been developed to compute distances between genomes or to infer ancestral genomes under these models typically provide mappings between gene copies that define orthology. As we have discussed earlier, orthology inference is a very important and unsolved problem. See Fu et al. (2007) for one of the early methods that inferred orthology taking genome rearrangements and duplications into account, and Shao and Moret (2016) for a recent method and a survey of the literature on this problem.

This is just a brief overview of the methods used to construct phylogenies based on chromosomal architectures; for an entry into this literature, see Moret et al. (2013) and El-Mabrouk and Sankoff (2012).

10.11 Review Questions

1. What does the term "HGT" refer to?
2. What does the term "MSC" refer to?
3. What is an anomalous gene tree?
4. Are there any anomalous rooted gene trees with three leaves?
5. Are there any anomalous rooted gene trees with four leaves?
6. Give an example of a method that is statistically consistent under the MSC for estimating the unrooted species tree, and explain why it works.
7. Is a concatenation analysis using unpartitioned maximum likelihood statistically consistent under the MSC?
8. What is the Quartet Median Tree problem?
9. What is the difference between orthology and paralogy?
10. What is the meaning of homology?
11. What is Gene Tree Parsimony?
12. What is the difference between an evolutionary network and a data-display network?
13. Is it the case that concatenation is always statistically consistent under the MSC, except in the anomaly zone?

10.12 Homework Problems

1. Let T be a rooted species tree with topology $(a,(b,(c,d)))$, and assume that the probability of coalescence on the edge above the most recent common ancestor of c and d is $\varepsilon > 0$.
   - Compute the probability of the unrooted gene tree $t_1 = (a,(b,(c,d)))$ under the multi-species coalescent model (this will be a function of ε). For what values of ε is this greater than 0.5?
   - Compute the probability of the unrooted gene tree $t_2 = (a,(c,(b,d)))$ under the multi-species coalescent model (this will be a function of ε). For what values of ε is this greater than 0.5?

2. Consider the model species tree from the previous problem. Prove or disprove: $\exists \varepsilon > 0$ such that the most probable unrooted gene tree is not the unrooted species tree.

3. Consider the rooted model species tree $((a,b),(c,d))$, and assume each of the two internal edges has the same coalescence probability ε. Prove or disprove: $\exists \varepsilon > 0$ such that the most probable rooted gene tree is not the rooted species tree.

4. Apply the SRSTE algorithm to the following input set of rooted gene trees:
   $\mathscr{T} = \{(a,(b,(c,d))),(a,(c,(b,d))),(b,(a,(c,d)))\}$.

5. Apply the SRSTE algorithm to the following input set of rooted gene trees:
   $\mathscr{T} = \{(a,(b,(c,d))),(a,(c,(b,d))),(b,(a,(c,d))),(b,(c,(a,d)))\}$.

6. Consider the inputs to SRSTE given in the previous two problems. For each of those inputs, interpret the gene trees as unrooted gene trees, and apply the SUSTE algorithm. What do you obtain?

7. Let T be an arbitrary model species tree, and consider a set $\mathscr{T}$ of rooted gene trees generated by T under the multi-species coalescent model. For the sake of this problem, assume that every gene tree is correctly computed. Suppose you were to compute the strict consensus tree for $\mathscr{T}$ (i.e., treating each gene tree as an unrooted tree). What would you expect to obtain, in the limit, as the number of gene trees in $\mathscr{T}$ increases?

8. Suppose you have a rooted species tree T, with branch lengths in coalescent units. Recall that every such species tree defines a distribution on rooted gene trees under the multi-species coalescent model. We define $p_{sib}(x,y)$ to be the probability that taxa x and y are siblings in a rooted gene tree that is sampled at random from the distribution. Is it the case that the pair x,y that maximizes $p_{sib}(x,y)$ must be siblings in the species tree?

9. Suppose you want to use ASTRAL to find a species tree for a set $\mathscr{T}$ of gene trees. ASTRAL requires a set X of allowed bipartitions, and so you set $X = C(T)$, where T is a supertree you have computed for the set of gene trees. What does ASTRAL return, and why?

10. Imagine you are a graduate student working in a bioinformatics laboratory, and your professor asks you to construct some trees for four different species, using different loci. For the sake of simplicity, we'll call the species H, C, G, and R; you can think of them as being human (H), chimp (C), gorilla (G), and rhesus monkey (R), but they could be any four species. You are fortunate that all the genomes have been assembled and aligned, and you have several thousand loci you can compare. You select five loci at random from these genomes, and use the best method you can to construct a tree for each of the five loci. Suppose four of these trees have topology $((H,C),(G,R))$, but in the fifth you get $((H,G),(C,R))$. You report your results to your professor, and he says "The first four gene trees are fine, but the last gene tree must be wrong. You must have made a mistake – you shouldn't get $((H,G),(C,R))$, because I'm sure H and C are siblings in the species tree." How would you respond to this?

11
Designing Methods for Large-Scale Phylogeny Estimation

11.1 Introduction

The construction of large phylogenies is of increasing interest, and the size of these datasets can be enormous. For example, the insect portion of the Tree of Life already contains "nearly a million described species," and the evolutionary relationships between these species is far from resolved (Maddison, 2016). Yet methods for constructing large trees are typically based on approaches that were designed for much smaller datasets, and very few available software packages have adequate accuracy on large datasets. Thus, there is a gap in terms of basic algorithmic strategies and in terms of software development for large-scale phylogeny estimation.

In this chapter, we investigate techniques for constructing phylogenetic trees for large datasets. We explore algorithm design, including standard heuristics used in many software packages, and also divide-and-conquer techniques used to scale computationally intensive methods to large datasets. Some of the initial parts of this chapter appear in earlier chapters, and are repeated here for the sake of completeness.

11.2 Standard Approaches

Many phylogeny estimation methods fall into the following categories: distance-based methods, subtree assembly-based methods, heuristics for NP-hard optimization methods, or Bayesian methods. Understanding each of these types of methods is helpful in developing methods for large datasets.

Distance-based methods. Generally the fastest phylogeny estimation methods operate by computing a matrix of pairwise "distances" between every pair of taxa, and then using that distance matrix to compute a tree. Distance-based methods (described in Chapter 5) typically run in $O(n^3)$ time, but some even run in $O(n^2)$ time, and many are fast enough to be used on even very large datasets.

Subtree assembly-based methods. Some methods operate by computing trees on a collection of small subsets of the taxon set, and then combine the subset trees together into a tree. We call these "subtree assembly-based methods" since they operate by estimating subtrees and then assembling them into a larger tree. Many subtree assembly-based methods compute quartet trees using maximum likelihood, and then combine the quartet trees using some quartet amalgamation method, such as Quartet Puzzling (Strimmer and von Haeseler, 1996), Weight Optimization (Ranwez and Gascuel, 2001), Quartet Joining (Xin et al., 2007), Quartets MaxCut (Snir and Rao, 2010), and Quartet FM (Reaz et al., 2014). Quartet amalgamation methods are generally not feasible for large numbers of sequences (e.g., even datasets with just a few hundred sequences can be very challenging), unless they restrict the set of computed quartet trees to a fairly small subset of the possible quartet trees; however, reducing the set of quartet trees can reduce accuracy compared to analyses of the full set of quartet trees (Swenson et al., 2010). While some quartet-based phylogeny estimation methods have been shown to be as accurate as neighbor joining under some conditions (Xin et al., 2007; Snir et al., 2008), to date, quartet-based estimation methods (and more generally subtree assembly-based methods) have not been shown to provide comparable accuracy to maximum likelihood heuristics or Bayesian methods. This disappointing performance of quartet-based methods have led some researchers to assume that quartet-based methods, where quartet trees are estimated independently, may have inherent limitations (Ranwez and Gascuel, 2001).

Heuristics for NP-hard optimization problems. Heuristics for maximum likelihood and other NP-hard optimization problems (such as maximum parsimony) typically combine hill-climbing to find local optima (trees that are better than any tree in the neighborhood they can search) with randomization to get out of local optima. Sometimes these steps alternate, and sometimes they are combined with search strategies (like simulated annealing) that accept poorer solutions with some probability.

All heuristics begin with starting trees that are typically computed using greedy techniques that add taxa, one by one, to a growing tree. Hill-climbing strategies explore the trees within a topologically defined neighborhood of the current tree until they find a better tree, at which point the search begins again at the new tree.

These neighborhoods are obtained by modifying the tree topology, typically using one of the following moves: NNI (nearest neighbor interchanges), SPR (subtree prune-and-regraft), and TBR (tree bisection and reconnection) (Felsenstein, 2004) (see Figures 11.1 to 11.3). By design, the neighbors of a tree under NNI are a subset of the neighbors under SPR, which are in turn a subset of the neighbors under TBR. Thus, searches using NNI will potentially get stuck in local optima with greater probability than searches using SPR, and similarly searches using SPR will get stuck in local optima with greater probability than searches using TBR.

Other strategies, including p-ECR (Ganapathy et al., 2003, 2004), where p edges are contracted and then the tree is refined optimally, or generalized NNI where some constant number of NNI moves are allowed (Sankoff et al., 1994), have also been explored. Both p-ECR and generalized NNI are generalizations of NNI moves, and so trees have a larger set of generalized NNI neighbors and p-ECR neighbors (for $p > 1$) than NNI neighbors.

However, otherwise the relationship between the sets of neighboring trees is more complicated (i.e., neither neighborhood contains the other). Other techniques for moving through treespace that have been used successfully on large datasets are described in Goloboff (1999).

One of the desirable properties of a local search strategy is that all possible trees should be reachable by a path that uses only the specified move. Therefore, we may ask whether every two trees are connected by a path of NNI moves. As shown in Gordon et al. (2013), not only is the space of trees connected via NNI moves (so that all two trees can be visited by a path of NNI moves), but there is a path that visits every tree exactly once (i.e., a Hamiltonian path).

When a local optimum is reached, then the search begins at a new tree, often obtained using a strategy that employs randomness. For example, the parsimony ratchet (Nixon, 1999) operates by (1) producing a bootstrap alignment (i.e., a modification of the original multiple sequence alignment produced by sampling with replacement from the sites of the alignment); (2) running the heuristic search strategy anew on this bootstrap alignment until a local optimum is obtained; and then (3) continuing the search from the new tree, but now based on scores computed using the original alignment. This alternation between searches based on the original alignment and the bootstrap replicate alignment can be repeated several times.

This process repeats until a desired stopping criterion is met (e.g., a time limit, or evidence that the search strategy is not likely to find a better scoring tree). At the end, the set of all best trees is returned; this set can be very large for maximum parsimony, but searches for maximum likelihood trees or for minimum evolution trees usually produce a single "best" tree.

The challenge in using heuristics to find good solutions for NP-hard problems is that even the best heuristics are only ensured to find good local optima, and local optima are not necessarily global optima. Furthermore, these heuristic search strategies can take a long time (and use large amounts of memory) on large datasets, meaning days, weeks, or even

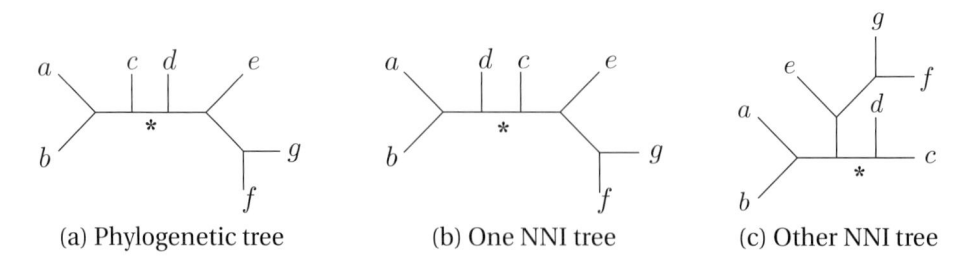

(a) Phylogenetic tree

(b) One NNI tree

(c) Other NNI tree

Figure 11.1 (Figure 3.14 in Huson et al. (2010)) Trees related by Nearest Neighbor Interchange (NNI) moves. In an NNI move, an internal edge in the tree is selected, and the four subtrees around the edge are identified. Then, two of these subtrees around the edge (one from each side) are swapped. Thus, trees (b) and (c) are each obtained by one NNI move applied to tree (a).

11.2 Standard Approaches

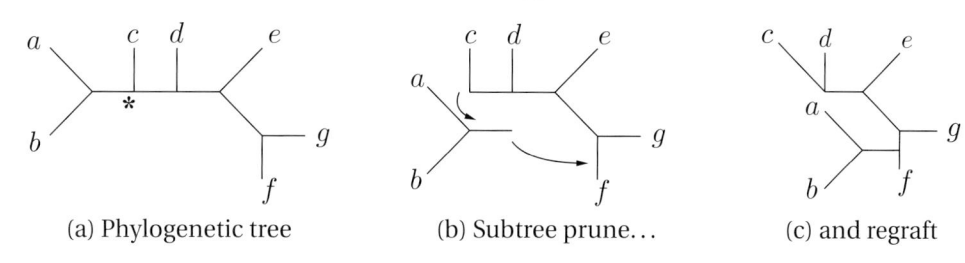

(a) Phylogenetic tree   (b) Subtree prune...   (c) and regraft

Figure 11.2 (Figure 3.15 in Huson et al. (2010)) Trees related by a Subtree Prune and Regraft (SPR) move. In an SPR move, an edge is deleted, creating two trees; these two trees are then re-attached by adding an edge between them, with the constraint that the new edge uses one of the endpoints of the edge that was deleted. Note that an NNI move is an SPR move. Tree (c) is obtained by applying one SPR move to tree (a).

(a) Tree bisection...   (b) reconnection choice...   (c) and reconnection

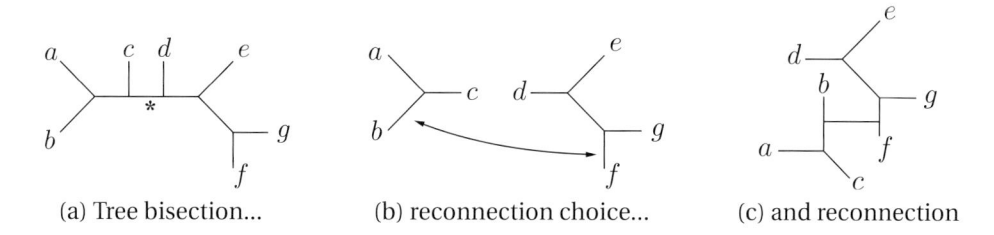

Figure 11.3 (Figure 3.16 in Huson et al. (2010)) Trees related by a Tree Bisection and Reconnection (TBR) move. In a TBR move, an edge is deleted from the tree, creating two trees; the two trees are then re-attached by the introduction of an edge between them. Note that every NNI or SPR move is a TBR move. Tree (c) is obtained by applying one TBR move to tree (a).

more, of CPU time. As an example, a maximum likelihood analysis for 48 avian species, using a concatenated alignment of about 41.8 million base pairs, took more than 200 CPU years and 1Tb of memory to find a *local optimum* (Jarvis et al., 2014). Thus, maximum likelihood tree estimation can be *very* computationally intensive, even on small numbers of sequences!

Bayesian MCMC methods. Bayesian analyses are based on the same statistical models as in maximum likelihood searches, but are not attempts at a "best" tree. Instead, they seek to sample from the distribution on model trees proportionally to the probability of the data under each model tree (Yang, 2009). This objective typically cannot be solved analytically, and so instead is often based on MCMC (Markov Chain Monte Carlo) techniques, which in essence are random walks through model treespace.

After the MCMC analysis is run long enough that statistical tests suggest that the MCMC walk may have reached the stationary distribution, this initial portion of the analysis is treated as "burn-in," and the trees that were visited are discarded. Then, the MCMC walk continues, and a sparse random sample of the visited model trees is retained. This sample is then used to estimate the posterior probability distribution on the parameters of the model tree. For example, to determine the posterior probability that a particular bipartition is present in the tree, the fraction of the sampled trees that have that bipartition is computed. In addition, the sample can be used to produce a point estimate of the model tree using various methods, such as computing a greedy consensus of the sampled trees or the maximum *a posteriori* (MAP) tree (i.e., the tree topology that appears the most frequently in the sampled trees). On large datasets, MCMC methods can take a very long time to converge. This convergence time grows with the number n of taxa, as the number of possible tree topologies grows exponentially in n. However, MCMC methods also have running times that grow with the number of characters. For example, the running time needed to reach convergence may increase with the sequence length when using MCMC methods that estimate gene trees under DNA sequence evolution models. Similarly, MCMC methods that co-estimate species trees and gene trees from multi-locus sequence data under the multi-species coalescent (MSC) model of gene tree evolution will take longer to converge when the number of genes is large rather than small (Zimmermann et al., 2014). Thus, both dimensions of the dataset – the number of taxa (whether species or individuals) and the number of characters (whether sites or genes) – impact the time needed to reach convergence.

Comparisons between methods. The choice of method to analyze a given dataset depends on many factors, including the preferences of the user (e.g., some biologists prefer maximum likelihood, others prefer Bayesian methods, others distance-based methods or maximum parsimony, etc.). However, there are practical considerations that can rule out certain approaches, and the choice of method will also depend on the dataset and the question that the biologist wishes to answer.

Although there are only a few studies examining the relative accuracy of phylogeny estimation methods on large datasets, the best-performing maximum likelihood and Bayesian MCMC methods have generally been found to be at least as accurate as the alternative approaches, and have the theoretical advantage over maximum parsimony of being statistically consistent under standard sequence evolution models. However, Bayesian MCMC methods are not used on very large datasets nearly as often as maximum likelihood heuristics, largely because of running time. Of the many methods for maximum likelihood, RAxML (Stamatakis, 2006), FastTree-2 (Price et al., 2010), PhyML (Guindon and Gascuel, 2003), and IQTree (Nguyen et al., 2015a) are perhaps the most effective for datasets where the number of sequences is large. Comparisons between methods on small to moderate-sized datasets sometimes favor one of IQTree, RAxML, or PhyML, but FastTree-2 generally outperforms the other methods on datasets with very large numbers of sequences in terms of the time vs. accuracy tradeoff (Liu et al., 2012a). However, FastTree-2 is not designed for very long sequences, and some other methods, such as ExaML (Kozlov et al., 2015) are more efficient under those conditions. Software packages also differ in terms of the sequence evolution models they allow. For example, RAxML allows more protein models than FastTree-2 does. For DNA sequence evolution models, nhPhyml (Boussau and Gouy, 2006) enables non-homogeneous models and IQTree enables a wide variety of models, including the PoMo model (De Maio et al., 2015; Schrempf et al., 2016) that allows for polymorphism.

11.3 Introduction to Disk-Covering Methods (DCMs)

11.3.1 Objectives of a DCM: Boosting a Base Method

As we have seen, standard phylogeny estimation methods typically have good performance (in terms of accuracy and speed) for small enough datasets that evolve under slow enough rates of evolution. However, most methods degrade in accuracy with increasing dataset size and heterogeneity, and running times also increase with dataset size. In this chapter, we describe divide-and-conquer strategies to improve phylogeny estimation methods. These methods, which we refer to as "disk-covering methods," or DCMs, have been used in a number of contexts with a variety of types of phylogeny estimation methods, which we will refer to as "base methods."

11.3.2 The Design of a DCM

Disk-covering methods vary in their design, with some of them being fairly simple and others being fairly elaborate. However, each DCM employs a divide-and-conquer strategy with the following basic format:

- Step 1: Given the set S of taxa and associated data (e.g., sequence alignments or gene trees), construct a triangulated graph G in which the vertices correspond to taxa in S.
- Step 2: Use properties about triangulated graphs to decompose the set S into overlapping subsets, and construct trees on the subsets using the base method.
- Step 3: Use a supertree method to merge the trees on the subsets of S into a tree on S.

Thus, DCMs are inherently closely connected to triangulated graphs, and designing and analyzing DCMs depends on understanding the properties of triangulated graphs. We provide some initial introduction to these graphs here, and details and proofs are provided at the end of the chapter.

Definition 11.1 A graph $G = (V, E)$ is **triangulated** (also referred to as **chordal**) if the largest induced simple cycle in G has length 3.

For an example of a triangulated graph, see Figure 11.4. The graph has several cycles that have more than three vertices; for example, $a - k - d - c - b - a$ is a five-cycle. However, it is not a simple cycle, since the edge (b, k) is a chord in the cycle. In fact, it can be verified by inspection that all the largest simple cycles have only three vertices, which makes it triangulated. Later in this chapter, we will prove that the graph is triangulated.

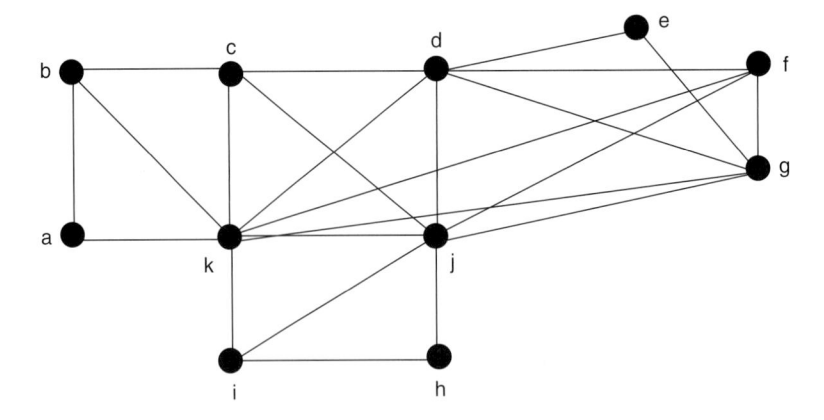

Figure 11.4 (Figure 4 from Berry et al. (2010)). This figure shows a triangulated graph. Reproduced from *Algorithms* 3(2):197–215 under the Creative Commons Attribution License.

Recall that Step 1 of a DCM involves the construction of a triangulated graph; as we will see, triangulated graphs arise naturally in the course of a phylogenetic analysis, and typically are fast to compute. Step 2 is where we use the decomposition of the graph to decompose the taxon set, and construct trees on the subsets using the base method. As we will see, the decomposition step can also be computed efficiently, using properties of triangulated graphs. To understand the decomposition strategies we use, we will need to define maximal cliques and minimal vertex separators in a graph.

Definition 11.2 A **maximal clique** in $G = (V, E)$ is a subset $V_0 \subseteq V$ that induces a clique (i.e., all elements of V_0 are pairwise adjacent), and where no superset V_1 of V_0 with $V_1 \neq V_0$ induces a clique. Now, let u and v be two vertices in a connected graph $G = (V, E)$. We will refer to any subset V_0 of V that separates u and v as a $u - v$ separator, and will say that V_0 is a minimal $u - v$ separator if no proper subset of V_0 is a $u - v$ separator. Finally, we will say that V_0 is a **minimal vertex separator** if it is a minimal $u - v$ separator for some pair of vertices u, v.

For example, the triangulated graph in Figure 11.4 has minimal vertex separators $\{b, k\}$, $\{c, k\}$, $\{d, j, k\}$, $\{j, k\}$, $\{i, j\}$, and $\{d, g\}$. It is easy to see that all of these sets are vertex separators, but it may be surprising that $\{d, j, k\}$ is a superset of $\{j, k\}$, and yet is considered a minimal vertex separator. The reason is that the removal of $\{d, j, k\}$ separates c from f, and the removal of $\{j, k\}$ does not.

- Every triangulated graph on n vertices has at most $n - 1$ maximal cliques, and these can be found in polynomial time. Hence, given a triangulated graph G, the **maximal clique decomposition** returns the set of maximal cliques of G.
- The minimal vertex separators in a triangulated graph on n vertices are cliques, and there are at most $n - 1$ of them; furthermore, these can be found in polynomial time. Hence,

given a triangulated graph G, a **separator-component** decomposition returns the sets $X \cup B_1, X \cup B_2, \ldots, X \cup B_k$, where X is one of the clique separators for G and B_i is a component of $G - X$ for each $i = 1, 2, \ldots, k$.

It is worth noting that the number of minimal vertex separators in a general graph can be exponential (Berry et al., 2010; Berry and Pogorelcnik, 2011), and so triangulated graphs have special properties.

The decomposition of the vertices of the triangulated graph G into overlapping subsets defines a decomposition of the taxon set S into overlapping subsets, since the taxon set S is the set of vertices for G. After we compute the decomposition of S into overlapping subsets $S_1, S_2, \ldots, S_k$, we can compute trees on each set S_i using the selected base method.

Step 3 involves applying a supertree method to the trees on the subsets computed in Step 2, using a selected supertree method. We will be able to prove guarantees about some DCMs we develop by taking advantage of the fact that the graphs we are working with are triangulated.

Since DCMs are designed to improve the accuracy or scalability of a base method, the design of a DCM depends on the given base method and the issues that primarily impact that method. For example, some base methods are very impacted by dataset size, and so can only feasibly analyze datasets that are fairly small. Other base methods are not so impacted by dataset size but have error rates that increase substantially with the heterogeneity within the dataset (e.g., the maximum evolutionary distance between any two sequences in the dataset). For a method that is primarily impacted by dataset size, the decomposition should produce small subsets, while decompositions that reduce heterogeneity are best for methods that degrade in the presence of high heterogeneity. Many methods are vulnerable to both issues (dataset size and heterogeneity), so approaches that reduce subset size and heterogeneity are often best.

Recall that the short quartet trees of a tree T are sufficient to define T (Corollary 3.9); hence, we want to design DCMs so that the triangulated graphs will have enough edges to contain all the short quartets as four-cliques, and then we want to design decompositions that create subsets on which we can compute (hopefully) accurate trees. Since accurate trees are generally easier to compute on smaller subsets than on larger subsets, these two objectives are – in a sense – pulling us in opposite directions.

There are other trends that impact the design strategies for DCMs. In particular, there is substantial evidence that dense taxonomic sampling improves phylogenetic accuracy; hence, phylogenetic analyses benefit by including intermediate taxa. This impacts the design of the triangulated graphs, and tends to increase the number of edges in the graph and hence the size of the subsets. Finally, the accuracy of the final tree that is returned depends on the ability of the supertree method to retain accuracy in the subset trees; hence smaller numbers of subsets may improve the final result, suggesting that larger subsets might have better results. On the other hand, it is well established that methods degrade in accuracy and/or running time with increases in dataset size, so smaller subsets will always provide at least some advantage. With these comments in mind, we introduce some simple DCMs.

11.4 DCMs that Use Distance Matrices
11.4.1 Matrix Threshold Graphs

Let $S = \{s_1, s_2, \ldots, s_n\}$ be a set of taxa, and let **d** be a distance matrix on S so that d_{ij} is the distance between i and j. Let $q \geq 0$ be any non-negative real number, which we will refer to as the "threshold." We now show how to compute a graph, called the matrix threshold graph, given **d** and q.

Definition 11.3 The **matrix threshold graph** $TG(d,q)$ has vertex set S and edge (s_i, s_j) for all i,j where $d_{ij} \leq q$.

We will generally refer to these as "threshold graphs," noting that the term has also been used for another type of graph. Various properties about threshold graphs are easily established. For example, $TG(d,q_1)$ is a subgraph of $TG(d,q_2)$ if $q_1 \leq q_2$. Also, if $q \geq max_{ij}\{d_{ij}\}$ then $TG(d,q)$ is a complete graph, while if $q < min_{i \neq j}\{d_{ij}\}$ then $TG(d,q)$ will be a graph without any edges. Threshold graphs based on intermediate values of q will contain edges, but may not be connected if q is small. Hence, the choice of q will impact the decomposition, and only a subset of the values (ones that are large enough to create a connected threshold graph, and not so large that they create a complete graph) will produce decompositions that could be useful.

The connection between threshold graphs and phylogeny estimation follows from Theorem 11.4, which shows that threshold graphs defined for additive distance matrices are always triangulated.

Theorem 11.4 *(From Huson et al. (1999a)) Let* **d** *be an additive matrix, and let* q *be a real number. Then* TG(d,q) *is triangulated.*

In prior chapters we have shown several conditions where it is possible to define distances between species so that as the amount of data (either sequence length or number of gene trees) increases, the distance matrix converges to an additive matrix defining the model tree. Therefore, this theorem implies that given enough data, with high probability the threshold graph computed on the matrix of estimated distances will be triangulated. For example,

Corollary 11.5 *Let* (T, θ) *be a Jukes–Cantor model tree with* n *leaves, and let* S *be a set of* n *sequences that evolve down* (T, θ). *Let* **d** *be the matrix of Jukes–Cantor distances computed on* S, *and let* q *be any real number. Then, as the sequence length for* S *increases, the threshold graph* TG(d,q) *converges to a triangulated graph.*

This corollary holds for any model of sequence evolution where the dissimilarity matrix of estimated distances can be corrected for the model of evolution, so that it converges to an additive matrix for the model tree (see Chapter 5). Furthermore, the corollary also holds if the input is a set of gene trees that differ from each other due to incomplete lineage sorting and the distances between species are computed in a way that converges to an additive matrix defining the species tree. For example, Kreidl (2011) and Allman et al.

(2016) proved that the matrix of average internode distances, as used in the coalescent-based species tree methods NJst (Liu and Yu, 2011) and ASTRID (Vachaspati and Warnow, 2015), converges to an additive matrix for the species tree (see Section 10.5.1).

Triangulating a threshold graph. Even though the threshold graph converges to a triangulated graph, in practice the amount of data (whether sequence length or number of gene trees) may not be sufficient for $TG(d,q)$ to be triangulated. Therefore, to use these decompositions, we may need to modify the threshold graph so that it becomes triangulated. One way to achieve this is by adding edges. The choice of edges to add impacts the accuracy and theoretical guarantees; in particular, selecting edges so as to minimize the largest weight of the added edges ensures good theoretical properties but is NP-hard (Huson et al., 1999a; Warnow et al., 2001). Lagergren (2002) developed a polynomial time distance-based algorithm that produces a triangulation with good theoretical properties, and a greedy technique (described in Huson et al. (1999a)) to triangulate the threshold graph can produce good empirical results.

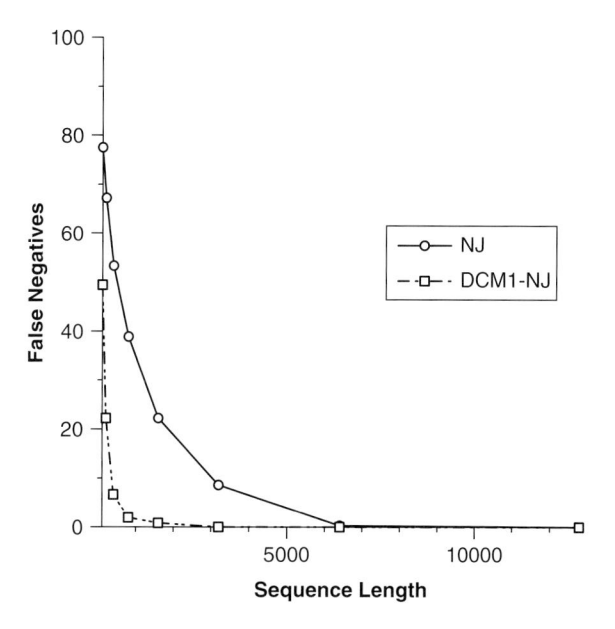

Figure 11.5 (Figure 7 from Huson et al. (1999a)) Comparison of neighbor joining and DCM1-NJ. We show average tree error (false negative rate, expressed as a percentage) of neighbor joining and DCM1-NJ methods as a function of the sequence length on a Jukes–Cantor model tree based on a phylogeny estimated for a biological dataset, and then scaled to a high rate of evolution. The decomposition is based on the matrix threshold graph for multiple thresholds. All thresholds that create connected graphs are used to compute trees. Finally, a Greedy Asymmetric Median Tree (Phillips and Warnow, 1996) is computed for these trees. Each point is the average of ten replicates.

11.4.2 DCM1-NJ, Followed by the Greedy Asymmetric Median Tree

Huson et al. (1999a) described several DCMs based on threshold graphs, including "DCM1" (i.e., Disk Covering Method 1), which has turned out to be very versatile. The input to DCM1 is a dissimilarity matrix **d** and a selected distance-based method. We describe DCM1-NJ, and how the method operates when the base method is neighbor joining.

Phase 1. In the first phase, the threshold graphs $TG(d,q)$ are computed for every q that appears in the matrix **d**. For each threshold graph that is connected, the threshold graph is triangulated using a greedy heuristic, and then decomposed into maximal cliques. Neighbor joining trees are computed on each of the maximal cliques, and then the subset trees are combined using the Strict Consensus Merger (see Section 7.8). This process typically produces an unresolved tree that depends on the threshold q, and which we denote t_q.

Phase 2. In the second phase, the trees created in the first phase are combined together using a greedy heuristic for the asymmetric median tree (see Section 6.2.5), which produces a fully resolved (i.e., binary) tree.

Figure 11.5 compares DCM1-NJ and neighbor joining (NJ) on simulated data. As the sequence length increases, the error for both methods decreases, but the error for NJ is much higher than that for DCM1-NJ, except for very long sequences. Huson et al. (1999a) also showed that this approach provided even bigger improvements for two other distance-based methods: the Buneman Tree (see Section 5.5.2) and the Agarwala *et al.* 3-approximation algorithm for the L_∞-nearest tree (see Section 5.8). It is clear that improvements obtained using this technique depend on the rate of evolution and number of leaves in the tree, and that with low enough rates of evolution or small trees, there may be no improvement. However, for high rates of evolution, such as in this study, the improvement obtained can be quite large.

11.4.3 DCM1-NJ, Followed by Tree Selection

A variation on this approach is to select a single tree from the set of trees computed in Phase 1. When NJ is used to construct the subset trees for each divide-and-conquer step and the short quartet support (SQS) criterion is used to select the best tree, the method is referred to as DCM1-NJ+SQS. The SQS score of a tree t is the largest value w such that $w = d_{ij}$ where **d** is the input dissimilarity matrix, and every four-taxon subset $A \subset S$ with all pairwise distances at most w induces the same tree in t as the Four Point Method applied to A.

As shown in Warnow et al. (2001), DCM1-NJ+SQS is an absolute fast converging (*afc*) method, provided that a minimum cost triangulation is used (see Chapter 8). The same result holds for DCM-Buneman+SQS and DCM-Agarwala+SQS (where Agarwala refers to the 3-approximation algorithm for the L_∞-nearest tree developed by Agarwala et al. (1998)), and for many other methods. Thus, any base method that is *exponentially converging* becomes *afc*, when used within the DCM1+SQS framework.

Other approaches for selecting the best tree from the set have also been examined, including using the maximum parsimony (MP) or maximum likelihood (ML) score of each tree with respect to the original sequence data (Nakhleh et al., 2001a). For example, DCM1-NJ+MP is the method that uses the MP score to select from the set of trees, and where NJ is used to construct the subset trees; similarly, DCM1-NJ+ML is the corresponding method where ML is used to select from the set of trees. Although DCM1-NJ+MP is not likely to have any particularly strong statistical properties (e.g., it probably isn't *afc* and may not even be statistically consistent), it produced results that were at least as accurate as DCM1-NJ+SQS on simulated datasets (Nakhleh et al., 2001b).

Many of these algorithmic steps can be modified. For example, instead of using NJ to construct trees on the subsets, other phylogeny estimation methods could be used (and these need not be limited to distance-based methods). Similarly, supertree methods other than the Strict Consensus Merger (SCM) can be used, typically to great advantage since the SCM tends to be a poor technique for computing a supertree. See Nakhleh et al. (2001b), Moret et al. (2002), and Nakhleh et al. (2001a, 2002) for studies of the impact of these DCMs on tree accuracy.

11.5 Tree-based DCMs

11.5.1 The Short Subtree Graph

In a tree-based DCM, the decomposition of the taxon set S is based on a tree T that is leaf-labeled by S. Thus, tree-based DCMs require either that the input includes a tree on S, or they will first compute a tree for S. Since these trees will normally be estimated from S, we may also comfortably assume that the edges of the tree T have positive lengths. Given a tree with edge weights, we also have an additive matrix **A** that we can use in developing the divide-and-conquer strategy.

The tree that begins a given iteration will generally not be the same as the tree at the end of the iteration; hence, a tree-based DCM can be combined with iteration, in which each iteration begins with the tree computed in the previous iteration. When iteration is used, then a stopping rule must be used (e.g., a time limit or the number of iterations).

This general algorithmic strategy is presented in Figure 11.6. Note the iterative technique, in which each iteration decomposes the dataset based on the tree from the previous iteration, re-estimates trees on each dataset, and then combines the subset trees into a tree on the full set of taxa.

The two decompositions we present each use a graph we can compute from the input tree T with its positive edge weighting, which we call the "Short Subtree Graph," which we

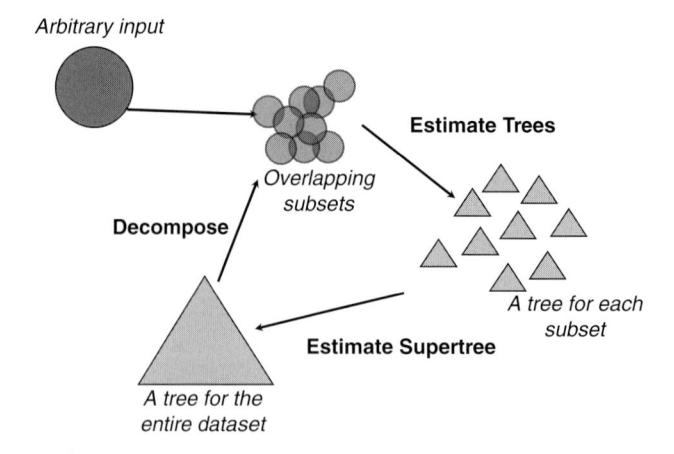

Figure 11.6 (Figure modified from Nelesen et al. (2012)) Combining iteration and tree-based DCMs to scale methods to large datasets. The input is an arbitrary set of taxa with associated data (e.g., sequences). In the first step, the dataset is decomposed into overlapping small subsets of a desired size. Two ways of decomposing a sequence dataset into subsets have been developed: One just uses the sequences, but the other begins by computing a tree (through a fast but approximate method). Then trees are computed on each subset, and the subset trees are combined together using a supertree method. If desired, the cycle can then begin again, using the current tree. Each subsequent iteration begins with the current tree, divides the dataset into subsets using the tree, computes trees on subsets, and combines the subset trees using the supertree method. (This is a modification of the DACTAL (Nelesen et al., 2012) algorithm, so it can be used generically.)

now describe. We define the distance of a leaf x to an edge e to be the sum of the lengths of the edges on the edges in the path from x to the nearest endpoint of e.

Definition 11.6 Let e be an edge of a binary tree T where every edge e has a positive weight $w(e)$. Let t_1, t_2, t_3, and t_4 be the four subtrees around the edge e (i.e., t_1 through t_4 are the components of $T - \{a,b\}$, where $e = (a,b)$). Let x_i denote those leaves in t_i that are closest to the edge e (using the path lengths defined by the edge-weighting on T). Let $X(e) = x_1 \cup x_2 \cup x_3 \cup x_4$. The **short subtree graph** (denoted by $SSG(T,w)$) is obtained by creating a graph with one vertex for every leaf in T, and making a clique out of every set $X(e)$. Thus, $SSG(T,w)$ has vertex set $S = \{s_1, s_2, \ldots, s_n\}$ and edge set $\{(s_i, s_j) : \exists e \in E(T), \{s_i, s_j\} \subset X(e)\}$.

Note the relationship between the sets $X(e)$ and the leaves that are involved in short quartets around edge e (see Definition 3.6). Hence, this decomposition ensures that $SSG(T,w)$ contains all the short quartets of (T,w).

Theorem 11.7 *(From Roshan et al. (2004)) Let* T *be any tree with positive edge-weighting* w. *Then the short subtree graph* SSG(T,w) *is triangulated.*

Since every short subtree graph is triangulated, we can decompose these graphs using either the maximal clique decomposition or a separator-component decomposition, as discussed above.

The padded short subtree graph. The padded short subtree graph is a variant of the short subtree graph where extra edges are added. The objective in adding edges is to increase the number of short quartets from the true species tree that induce a four-clique in the resultant graph, and hopefully cover all of the short quartets (see Theorem 11.7).

The padded short subtree graph takes an integer parameter p that is greater than 1 (but generally very small), and is denoted by $SSG_p(T,w)$. Instead of using just the single closest leaf in each subtree around an edge, the p closest leaves in each subtree (including all vertices that tie for p closest) are selected. The padding parameter has the impact of ensuring that the subsets have higher overlap, but also increases the size of the subsets. Given (T,w), the vertices of $SSG_p(T,w)$ are the leaves of T, and the edges of the padded short subtree graph are those (u,v) such that there is an edge e in T where u and v are among the p nearest leaves to e in a subtree (not necessarily the same subtree) off e.

11.5.2 Other Tree-based Decompositions

Suppose we have a tree that is leaf-labeled by the taxon set S. In addition to computing decompositions based on the short subtree graph or the padded short subtree graph, we can decompose the taxon set into overlapping subsets in additional ways. Here we describe two possible ways.

Decomposing into small subsets: centroid edge decomposition. Let e be an edge in T whose removal from T divides the leafset into two roughly equal sizes (i.e., e is a centroid edge). If we remove e and its endpoints a,b from T we decompose the leafset of T into four sets: A,B,C, and D. We compute $X(e)$, and then also $X(e) \cup A, X(e) \cup B, X(e) \cup C$, and $X(e) \cup D$. If this collection contains at least two distinct subsets, then this is the decomposition we will use. Otherwise, the collection does not contain at least two distinct subsets, and so we will continue the search. We first mark e as visited, and we then find a new edge e' incident to e that has not yet been visited, and repeat the search from e'. We repeat this process until we find an edge that creates at least two subsets; if no such edge exists, we return the full set (no acceptable decomposition is possible).

This process will substantially reduce the subset size, but may not substantially reduce the maximum evolutionary diameter (a proxy for the heterogeneity in the subset). However, finding a centroid edge is easily done in $O(n)$ time where n is the number of leaves in the tree.

Decomposing into low heterogeneity subsets: midpoint decomposition. Let P be a longest path (taking branch lengths into account) in the tree T, and let e be the edge containing the midpoint of this path. In other words, if you subdivide edge e by adding a new node v and split the length of e into two (not necessarily equal) parts, then you would have the longest path $P_{a,b}$ in the tree going through v, and the distance from a to v would be identical to the distance from v to b. Let $A, B, C,$ and D be the components of T after e and its endpoints are deleted. Compute the sets $X(e) \cup A, X(e) \cup B, X(e) \cup C$, and $X(e) \cup D$. As with the centroid edge decomposition, we check to see if this collection contains at least two distinct subsets, and if so we accept it. Otherwise, we mark e as visited, and replace e by an edge incident to e that has not yet been visited, and repeat this process until you find an edge that creates at least two subsets. If no such edge exists, return the full set (no acceptable decomposition is possible). This decomposition produces subsets that will tend to have substantially reduced evolutionary diameters, but may not be that small. Finding the midpoint edge is easily done in $O(n)$ time.

Decompositions based on short subtree graphs of this form (find a clique set $X(e)$, and then form the subsets $A, B, C,$ and D) are particularly beneficial, because of the following theorem. Recall Definition 3.6, which defined the set of short quartets in an edge-weighted tree), and so also in designing DCMs:

Theorem 11.8 *(From Nelesen et al. (2012)) Let $t_1, t_2, \ldots, t_k$ be unrooted binary trees and let S_i be the leafset of t_i. Let T be a tree with leafset $\cup_i S_i$. Assume that $S_i = A_i \cup X$, $A_i \cap A_j = \emptyset$ for all $i \neq j$, that $t_i = T|S_i$, and that every short quartet of T is in some S_i. Then SCM applied to $\{t_1, t_2, \ldots, t_k\}$ returns T, independent of the order in which the trees are merged.*

11.6 Recursive Decompositions of Triangulated Graphs

Once we have a triangulated graph, we can decompose it using the maximal clique technique, or using the separator–component technique. While the maximal clique produces a unique decomposition, the separator–component technique produces multiple decompositions, depending on the choice of separator. This suggests that examining several separator–component decompositions and picking one that optimizes some criterion could be helpful. Alternatively, the short subtree graph (or some other triangulated graph) can be decomposed recursively using separator–component decompositions, until each subset is small enough or has low evolutionary diameter.

11.7 Creating Multiple Trees

There are several ways in which DCMs can be used to create multiple trees. One of them is iteration, where each iteration uses the tree from the previous iteration to perform a tree-based decomposition, constructs trees on the subsets, and then combines them; DACTAL (Nelesen et al., 2012) is an example of such a DCM. The matrix threshold graph approach also can be used to create multiple trees, since each choice of a threshold leads to a different graph and hence a potentially different decomposition and final tree.

Once we have a set of trees, we can return a tree using several different techiques; for example, by selecting a tree that optimizes some criterion (as in Warnow et al. (2001)), or by computing a consensus of the trees (as in Huson et al. (1999a)).

11.8 DACTAL: A General Purpose DCM

11.8.1 DACTAL's Algorithmic Design

DACTAL (Nelesen et al., 2012) is one of the most versatile DCMs we have developed, and combines iteration with divide-and-conquer. In the general case, DACTAL uses a fast but not necessarily highly accurate method to compute a starting tree. Each iteration of DACTAL begins with a DACTAL decomposition of the taxon set, then has subset trees constructed using the selected base method, and then a supertree is constructed on the subset trees using a preferred supertree method. The process can then iterate. Thus, DACTAL is a combination of divide-and-conquer with iteration, and depends on how each step is completed. The base method for computing subset trees depends on the data and optimization problem, and new and improved supertree methods have continued to be developed. However, the DACTAL decomposition has been stable, and so is stated here:

Definition 11.9 The **DACTAL decomposition** is based on an input tree T that is leaf-labeled by the set S of taxa. Using this tree, a padded Short Subtree Graph with $p = 2$ or $p = 3$ is computed. DACTAL then computes a separator–component decomposition based on a centroid edge, and recurses until each subset is small enough (where "small enough" depends on the limitations of the base method, and is a parameter that the user can specify). This produces a decomposition of the set S into overlapping subsets.

Any desired supertree method can be used to combine the subset trees into a tree on the full taxon set. However, our studies have examined the use of SuperFine with MRP or MRL to resolve polytomies in the SCM tree. Since the decomposition strategy is recursive, we use the decomposition tree to define the order in which we apply SuperFine. We begin by merging the subset trees formed by the last DACTAL decomposition into a tree, and merge the trees formed in this way until we reach the root of the decomposition tree. This recursive approach to merging subset trees enables us to prove the following theorem about DACTAL:

Theorem 11.10 *Suppose that the supertree method used in DACTAL is SuperFine, using some base supertree method to refine polytomies in the strict consensus merger (SCM) tree. Let* S *be the input set of taxa. Suppose that the starting tree in some iteration is the true tree* T *and let* $S_1, S_2, \ldots, S_k$ *be the DACTAL decomposition of* S *into overlapping subsets. If* *the base method for computing subset trees produces* $T|S_i$ *for each* $i = 1, 2, \ldots, k$, *then the supertree computed using SuperFine at the end of the iteration is* T.

Proof Under the assumptions of this theorem, every short quartet in T is in some subset S_i and all the subset trees are correct. The order in which the subset trees are merged into a tree on the full dataset follows the recursive decomposition. Since the supertree method that is used is SuperFine, it begins by computing the SCM. By Theorem 11.8, the merger of subset trees performed in this recursive fashion using SCM has no collisions and so returns the unique compatibility tree in every merger. Hence, the final tree at the end of the iteration will be T. □

Thus, although DACTAL is iterative, if it reaches T during one of the iterations and the subset trees are correctly computed, it will return T in the next iteration as well.

11.8.2 Using DACTAL to Compute Trees from Unaligned Sequences

DACTAL was originally designed to enable tree estimation for very large datasets, for which multiple sequence alignment was too difficult to do with high accuracy or was too time consuming. The name DACTAL stands for "Divide-And-Conquer Trees (almost) without ALignments," with "almost" indicating that although no full multiple sequence alignment is ever computed, some subset alignments are computed (Nelesen, 2009; Nelesen et al., 2012).

Two different strategies can be used to obtain the initial decomposition of the sequence dataset into subsets. The most common approach is to compute a fast sequence alignment using methods such as PartTree (Katoh and Toh, 2007), and then compute a tree on the sequence alignment using methods such as FastTree-2 (Price et al., 2010); then a DACTAL decomposition can be computed and can be applied to the tree. An alternative approach uses BLAST (Altschul et al., 1990) to compute clusters around each sequence in S, and hence divide S into small overlapping subsets containing sequences that are highly similar to each other.

After the dataset is divided into subsets, multiple sequence alignments and trees are computed on each subset using the preferred alignment and tree estimation methods, thus producing a collection of overlapping subset trees. Finally, the subset trees are combined using a preferred supertree method. DACTAL can then iterate, using the tree returned in the previous iteration to produce new decomposition of the sequence set into overlapping subsets.

Note that this algorithm design *does* involve multiple sequence alignment but only on small subsets of the leafset. Since the subsets overlap and the different multiple sequence alignments are not constrained to be identical on the sequences they share, these alignments cannot be merged into an alignment on the full sequence set. Hence, DACTAL outputs a tree but not a multiple sequence alignment of the input sequences.

Figure 11.7 compares default DACTAL to two-phase methods (i.e., first align, then compute an ML tree) on three large biological datasets from Cannone et al. (2002) with

11.8 DACTAL: A General Purpose DCM

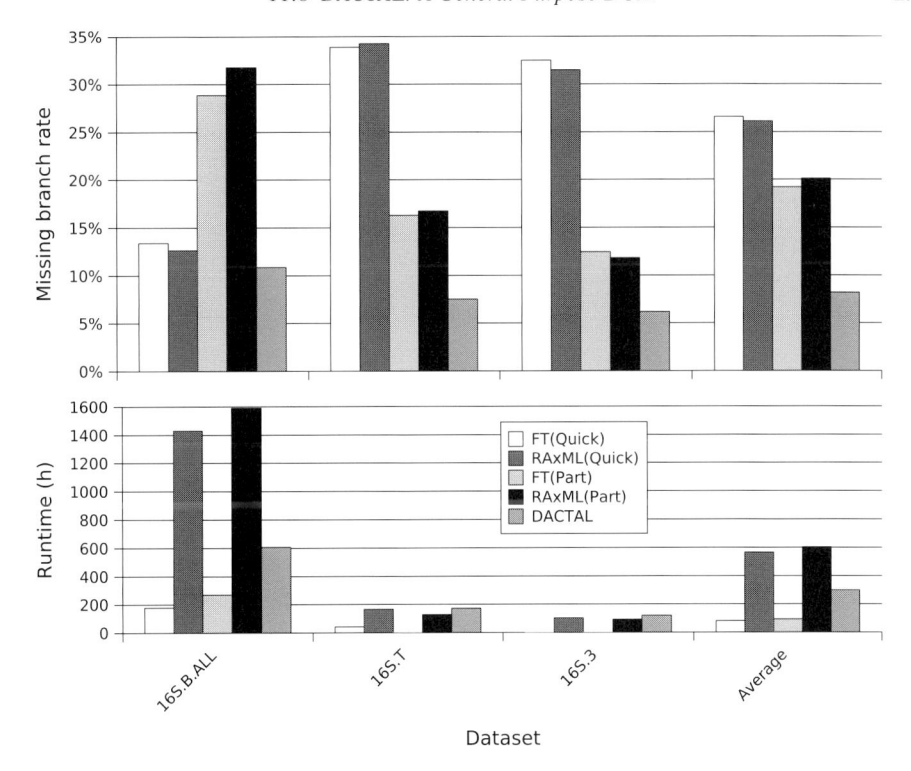

Figure 11.7 (From Nelesen et al. (2012)) DACTAL (based upon five iterations) compared to ML trees computed on alignments of three large biological datasets with 6323 to 27,643 sequences. We used FastTree-2 (FT) and RAxML to estimate ML trees on the PartTree (Part) and Quicktree (Quick) alignments. The starting tree for DACTAL on each dataset is FT(Part).

curated reference alignments. The reference trees for these alignments are the RAxML trees computed on the reference alignments, restricted to the highly supported branches. Default DACTAL decomposes into subsets of 200 sequences each, computes RAxML(MAFFT) trees on each subset, and then combines subset trees using SuperFine+MRP (see Section 7.9). DACTAL achieves substantially lower error rates than the two-phase methods, even though it never computes any alignment with more than 200 sequences.

Figure 11.8 shows the impact of iteration on DACTAL's accuracy in comparison to SATé (Liu et al., 2009a) (see Section 9.16.3) on the 16S.T dataset from Cannone et al. (2002), with 7350 sequences. SATé is a precursor to the PASTA method for co-estimating multiple sequence alignments and trees, and also has an iterative divide-and-conquer strategy. The first iteration provides the biggest improvement in accuracy, but the next few iterations also provide some improvement. On a per-iteration basis, DACTAL and SATé provide about the same accuracy, but each DACTAL iteration is much faster than each SATé iteration.

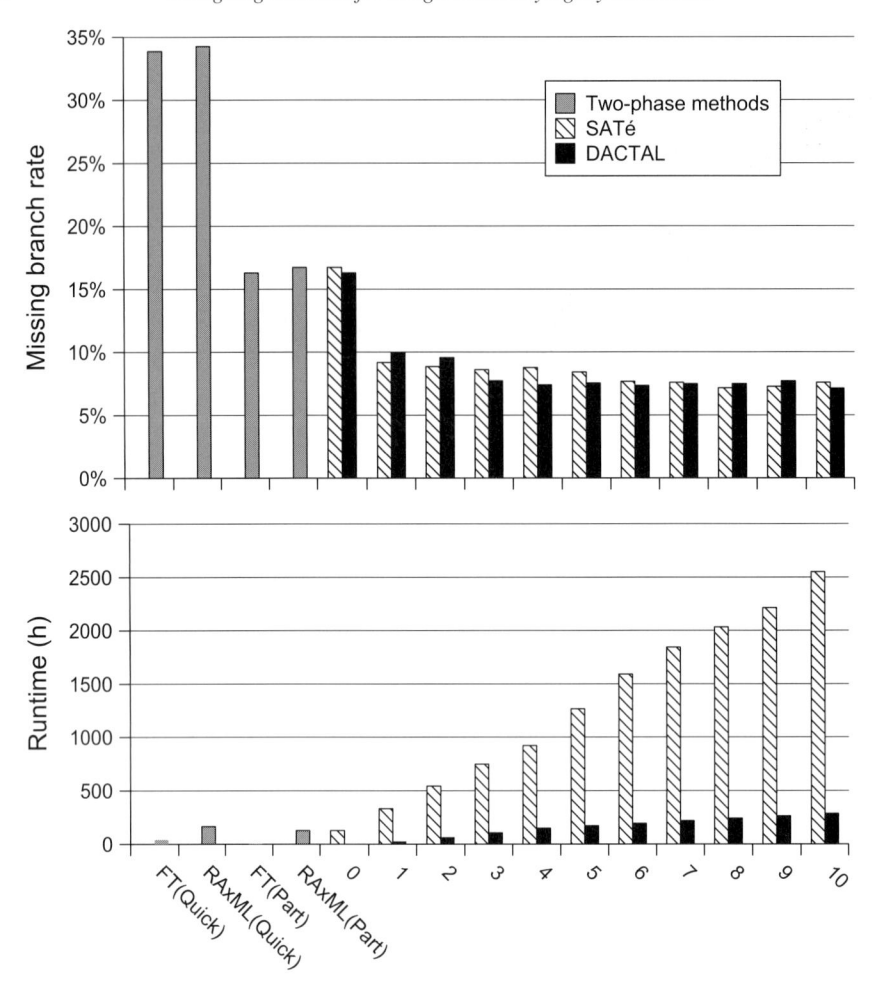

Figure 11.8 (From Nelesen et al. (2012)) Comparisons of DACTAL and SATé iterations with two-phase methods on the 16S.T dataset with 7350 sequences. The starting trees were RAxML(Part) for SATé and FT(Part) for DACTAL (computed in iteration 0). We show missing branch rates (top) and cumulative runtimes in hours (bottom); $n = 1$ for each reported value.

11.8.3 Using DACTAL for Coalescent-based Species Tree Estimation

DACTAL has also been used to improve the accuracy and scalability of MP-EST, one of the methods for coalescent-based species tree estimation (see Section 10.6). In this case, the input is a set of gene trees for a set S of taxa. In the first iteration, a simple method (e.g., supertree or consensus method) is used to compute the starting species tree T, and the species set S is decomposed into overlapping subsets based on T. Each subset A of species then defines a reduced set of gene trees, obtained by restricting the leafset of each gene tree to just those species in A. Then, for each subset A, MP-EST is used to compute a new species tree on A, and these species trees are combined together using a supertree method. The result of this iteration is a new species tree, which can then be used in the next iteration. As shown in Bayzid et al. (2014), using three iterations of DACTAL-boosting and selecting the tree that had the best quartet support score with respect to the gene trees (i.e., the same criterion as used in ASTRAL, another species tree estimation method) improved the topological accuracy of the species trees compared to MP-EST.

11.9 Triangulated Graphs

Here we provide the definitions and theorems about triangulated graphs that are relevant to phylogeny estimation and DCM design in particular, but without proofs. Triangulated graphs were introduced in Berge (1967), and their theoretical properties were established in a sequence of papers, including Dirac (1974), Rose (1970), Gavril (1972, 1974), and Buneman (1974a). For an overview of triangulated graphs, see Golumbic (2004), which provides proofs and a much richer set of results about triangulated graphs, and Tarjan (1985), Leimer (1993), Blair and Peyton (1993), Berry et al. (2010), and Berry and Pogorelcnik (2011).

Definition 11.11 A graph that has no induced simple cycles of length greater than three is a **triangulated graph**.

Here we present an equivalent definition, which will be useful for establishing theorems about the dataset decompositions we employ. Let T be a tree with vertex set V. Thus, T is just a connected acyclic graph, and there is no constraint on the degrees of the vertices in V; also, T is not rooted. A *subtree* of T is just a connected subgraph of T, and so is defined by a subset of the vertices of V. Two subtrees are said to intersect if they share any vertices. Thus, given an arbitrary set X of subtrees of an arbitrary tree T, the intersection graph of X has node set X and edges (t_1, t_2) where t_1 and t_2 share at least one node in common. It is not hard to see that any intersection graph of a set of subtrees of a tree T is triangulated, but it is also true that every triangulated graph can be written as such as intersection graph! We summarize this with the following theorem:

Theorem 11.12 *(From Gavril (1974); Buneman (1974a)) A graph* G *is triangulated if and only if* G *is the intersection graph of a set of subtrees of a tree.*

Definition 11.13 A **perfect elimination ordering** (also called a perfect elimination scheme) for a graph $G = (V, E)$ is an ordering of the vertices $\sigma = v_1, v_2, \ldots, v_n$, so that for each $i = 1, 2, \ldots, n-1$, $\Gamma(v_i) \cap \{v_{i+1}, v_{i+2}, \ldots, v_n\}$ is a clique (recall that $\Gamma(v_i)$ indicates the neighbor set of v_i). We denote $\Gamma(v_i) \cap \{v_{i+1}, v_{i+2}, \ldots, v_n\}$ by X_i and $X_i \cup \{v_i\}$ by Y_i.

Theorem 11.14 *(From Rose (1970)) A graph* $G = (V, E)$ *is triangulated if and only if it has a perfect elimination scheme; furthermore, given a triangulated graph, a perfect elimination scheme can be found in polynomial time.*

It is easy to see that a graph that has a perfect elimination scheme is triangulated: otherwise there is a chordless cycle of at least four vertices but the vertex that appears earliest in the perfect elimination scheme is simplicial, from which we derive a contradiction. The proof of the other direction is given in Rose (1970), along an $O(n^2d^2)$ algorithm to find a perfect elimination scheme where n is the number of vertices and d is the maximum degree of any vertex in G; Rose et al. (1976) improved this and gave an $O(n+m)$ algorithm where m is the number of edges in G. In addition, a slower but very simple polynomial time algorithm that finds a perfect elimination scheme or determines that no perfect elimination scheme exists is as follows. Initialize the perfect elimination scheme to the empty set. Then, check each vertex in the graph until you find a simplicial vertex (i.e., a vertex whose neighbors form a clique). If no vertex is simplicial, return *Fail* – the graph is not triangulated. Otherwise, append the vertex to the perfect elimination scheme, remove this vertex, and recurse on the remaining graph, adding simplicial vertices as you find them to the growing perfect elimination scheme.

Recall the graph in Figure 11.4 that we claimed was triangulated. The ordering of the vertex set given by h,a,b,i,c,e,d,f,g,j,k is a perfect elimination scheme; hence, by Theorem 11.10, the graph is triangulated. (There are many other perfect elimination schemes for this graph.)

Theorem 11.15 *(From Gavril (1972)) Every triangulated graph* $G = (V,E)$ *has at most* $n = |V|$ *maximal cliques, and these can be found in polynomial time.*

Proof Given a triangulated graph, we can construct a perfect elimination ordering $\sigma = v_1, v_2, \ldots, v_n$. Hence, Y_i (see Definition 11.13) is a clique for every i. Let X be a maximal clique in G, and let $j = min\{i : v_i \in X\}$. Then $X \subseteq Y_j$ (since otherwise $\exists y \in X \setminus Y_j$ with $(v_j,y) \in E$, contradicting how j and Y_j are defined). Furthermore, since X is a maximal clique, $X = Y_j$. A polynomial time algorithm to find the set of maximal cliques can therefore proceed by computing all Y_j, and returning those sets Y_j that are not subsets of any Y_k with $k \neq j$. □

Therefore,

Theorem 11.16 *(From Gavril, 1972) For any triangulated graph, a maximum clique can also be found in polynomial time.*

Proof Remember that a maximum clique is one that has the largest number of vertices, and that finding a maximum clique is in general NP-hard. However, every maximum clique is a maximal clique, so if we can compute all the maximal cliques we can then determine the largest clique in the graph. Therefore, given a triangulated graph, we compute a perfect elimination ordering and then the sets X_i and Y_i based on the perfect elimination ordering; this can be computed in polynomial time. Finding the largest Y_i after the sets are computed is clearly polynomial time. Hence the maximum clique problem can be solved in polynomial time for triangulated graphs. □

11.9 Triangulated Graphs

Recall Definition 11.2 of a minimal vertex separator in a graph, which is a subset A of the vertices such that the removal of A from the graph separates a pair u, v of vertices, and where no proper subset of A also separates u, v. We now show how to compute minimal vertex separators in a triangulated graph.

Theorem 11.17 *(From Dirac (1974)) Let $G = (V, E)$ be a triangulated graph, let σ be a perfect elimination ordering for G, and let the sets $X_1, X_2, \ldots, X_n$ be defined as in Definition 11.13. Then every minimal vertex separator in G is equal to some X_i.*

Example 11.18 We will use the algorithm described in the proof of Theorem 11.16 to compute the set of the maximal cliques for the triangulated graph from Figure 11.4, and from the set we will compute the maximum clique. We will also use Theorem 11.17 to enumerate all the minimal vertex separators in the graph.

Recall that we obtained the perfect elimination ordering $h, a, b, i, c, e, d, f, g, j, k$ for this graph. Instead of writing the sets $X_1, X_2, X_3, \ldots$, we will write the sets X_i and Y_i in terms of the vertex in the ith position within the perfect elimination ordering; hence instead of X_1 and Y_1 we will write X_h and Y_h, since h is the first vertex in the perfect elimination ordering.

- $Y_h = \{h, i, j\}$ and $X_h = \{i, j\}$
- $Y_a = \{a, b, k\}$ and $X_a = \{b, k\}$
- $Y_b = \{b, c, k\}$ and $X_b = \{c, k\}$
- $Y_i = \{i, j, k\}$ and $X_i = \{j, k\}$
- $Y_c = \{c, d, j, k\}$ and $X_c = \{d, j, k\}$
- $Y_e = \{d, e, g\}$ and $X_e = \{d, g\}$
- $Y_d = \{d, f, g, j, k\}$ and $X_d = \{f, g, j, k\}$
- $Y_f = \{f, g, j, k\}$ and $X_f = \{g, j, k\}$
- $Y_g = \{g, j, k\}$ and $X_g = \{j, k\}$
- $Y_j = \{j, k\}$ and $X_j = \{k\}$
- $Y_k = \{k\}$ and $X_k = \emptyset$

Note that every Y_α and X_α is a clique in this graph, where α ranges across the vertex set of the graph. The largest set of the form Y_α is Y_d, which has five vertices, $\{d, f, g, j, k\}$. Thus, the largest clique in this graph has size five.

To find all the minimal vertex separators, we examine the sets X_α, and retain the ones that are minimal vertex separators. The minimal vertex separators we find are $X_a = \{b, k\}$, $X_b = \{c, k\}$, $X_c = \{d, j, k\}$, $X_g = \{j, k\}$, $X_h = \{i, j\}$, and $X_e = \{d, g\}$. Note that not all of the X_α are minimal vertex separators. For example, $X_d = \{f, g, j, k\}$ is a vertex separator, in that its removal produces two components $\{h, i\}$ and $\{a, b, c, d, e\}$. However, $X_g = \{j, k\}$ is also a vertex separator whose removal produces two components $\{h, i\}$ and $\{a, b, c, d, e, f, g\}$; hence, any pair u, v of vertices separated by X_d is also separated by X_g, which means that X_d is *not* a minimal vertex separator. It is also worth noting that $X_c = \{d, j, k\}$ and $X_i = \{j, k\}$ are both listed as minimal vertex separators, yet X_c is a superset of X_i. The reason is that the removal of $X_c = \{d, j, k\}$ separates c from f, and the removal of $X_i = \{j, k\}$ does not separate that pair of vertices. In fact, no proper subset of X_c separates c from f, which makes X_c a minimal vertex separator.

11.10 Further Reading

Numerical issues. The best accuracy tends to be obtained using likelihood-based methods, such as maximum likelihood and Bayesian MCMC. However, likelihood calculations on very large model trees are challenging to obtain with high precision, which is necessary in order to compare different model trees. See Izquierdo-Carrasco et al. (2011) for an entry into this literature.

Exploring treespace. Many of the approaches we have described for computing trees are based on heuristics for NP-hard optimization problems. These searches are based on different ways of moving through treespace, such as NNI, SPR, TBR, etc., and the choice of topological move has implications for both running time and likelihood of getting stuck in local optima. Once the search ends, the set of all the best of the explored trees (and perhaps the near-optimal trees) can be explored.

One of the basic challenges in exploring the space involves computing distances between trees, which can be based on the topological moves used to explore treespace or more sophisticated techniques that have mathematically desirable properties, and that can be used to compare trees with branch lengths. For example, the geodesic distance between trees has been explored in Billera et al. (2001), and the branch score distance was explored in Kuhner and Felsenstein (1994).

The properties of treespace and the challenges in exploring it are discussed in St. John (2016), a survey article that includes topics such as the computational complexity of computing distances between trees and the sizes of the neighborhoods of trees using different operations (e.g., NNI, uSPR, rSPR, TBR, etc). SPR-based moves (and hence also TBR) are generally superior to NNI moves in searching treespace, and the choice of technique for exploring treespace can have a larger impact on topological accuracy than the sequence evolution model that is selected for the dataset (Money and Whelan, 2012).

Disk-covering methods. The DCM described in this chapter have their origins in the development of absolute fast converging (*afc*) methods (see Section 8.12). One of the earliest *afc* methods was DCM-Buneman (Huson et al., 1999a), which used a divide-and-conquer strategy to improve the Buneman Tree accuracy, while also providing guarantees about the sequence lengths that would suffice for accuracy with arbitrarily high probability. This divide-and-conquer technique was extended in Warnow et al. (2001), where it was shown to improve neighbor joining's accuracy. These *afc* methods relied heavily on the properties of triangulated graphs to provide theoretical guarantees for the methods they developed.

Other early DCMs were developed for use within heuristic searches for maximum parsimony (Huson et al., 1999b; Roshan et al., 2004) or maximum likelihood (Du et al., 2005a,b; Dotsenko et al., 2006). Another use of DCMs has been to improve the scalability of methods for genome rearrangement phylogeny estimation (Tang and Moret, 2003). As we have noted, DACTAL is one of the most versatile DCMs, and can be used in many different contexts. However, the design and use of DCMs in phylogenetics has only been partially explored. For example, DCMs could potentially be used to improve the scalability of evolutionary phylogenetic network estimation methods, which are extremely computationally expensive. Further work in this area is likely to be very helpful and interesting.

Adding sequences to a tree. Another approach to large-scale phylogeny estimation is to build the tree greedily, adding sequences into a growing tree. For example, given a dataset with 10,000 sequences, an initial tree could be built on a subset of just 1000 sequences, and then each subsequent sequence could be inserted into the tree, optimizing some criterion. The problem of adding a sequence into a tree, given a multiple sequence alignment on all the sequences, is called phylogenetic placement. Several methods for phylogenetic placement have been developed that try to optimize maximum likelihood, including pplacer (Matsen et al., 2010) and EPA (Berger et al., 2011). A related problem is adding sequences into an existing alignment, and then placing the sequence into an existing tree for the alignment; SEPP (Mirarab et al., 2012), PAGAN (Löytynoja et al., 2012), PaPaRa (Berger and Stamatakis, 2011), and UPP (Nguyen et al., 2015b) are some of the methods designed for this problem.

11.11 Review Questions

1. Draw a tree T with seven leaves labeled $1, 2, \ldots, 7$, and show a tree T' that is a TBR-neighbor of T but not an NNI-neighbor.
2. What is the objective of a Bayesian method? How is it different from maximum likelihood?
3. What is the definition of a triangulated graph?
4. What is the definition of a minimal vertex separator in a graph?
5. What is a perfect elimination scheme?
6. How many maximal cliques can there be in a triangulated graph on n vertices?
7. Give an example of a graph G on four vertices that is not triangulated.
8. Give an example of a connected graph G' on six vertices that is triangulated but is not a clique.
    a. List all the maximal cliques in the triangulated graph G'.
    b. Find at least one minimal vertex separator A for the triangulated graph G', and specify the pair of vertices for which it is a minimal vertex separator.
    c. What are the components of $G' - A$?
9. What is a matrix threshold graph?

11.12 Homework Problems

1. Find three perfect elimination schemes for the graph given in Figure 11.4.
2. Prove that every triangulated graph on two or more vertices is either a clique or has at least two non-adjacent simplicial vertices.
3. Give an example of a connected triangulated graph on n vertices that has exactly $n-1$ maximal cliques.
4. Consider the graph G on vertex set $\{a,b,c,d,e\}$ where e is adjacent to every other vertex and $a-b-c-d-a$ is a cycle (i.e., G is a wheel graph). Draw G. Is G triangulated?
5. Describe an algorithm to find a centroid edge in a binary tree T with n leaves, and analyze its running time.
6. Give an example of a tree T in which there are two or more centroid edges. Prove or disprove: The set of centroid edges is connected for all trees T.
7. What is the largest number of centroid edges that is possible in a binary tree T on n leaves?
8. Describe an algorithm to find the edge containing the midpoint of the longest (taking edge weights into account) leaf-to-leaf path in an edge-weighted tree (T,w) on n leaves, and analyze its running time. Is this edge guaranteed to be unique?
9. Let T_n be the caterpillar tree $(1,(2,(3,\ldots,(n-1,n)\ldots)$ and suppose all edges have the unit weight (i.e., $w(e)=1$ for all edges e). How many short quartets are there in this edge-weighted tree, with $n \geq 7$ leaves? List them all for $n=7$.
10. Write down the additive matrix $\mathbf{D}$ for T_8 with unit edge weights (see the previous problem for how T_n is defined). Find the largest entry q in $\mathbf{D}$ for which the threshold graph $TG(D,q)$ is not a clique. For this threshold graph:
    - find a perfect elimination ordering;
    - list all the maximal cliques;
    - list all the minimal vertex separators; and
    - find a maximum clique in the graph.

Appendix A
Primer on Biological Data and Evolution

This chapter will provide a basic introduction to evolution, biological data, and standard approaches to biological data analysis.

A.1 Phylogeny Estimation Pipeline

Phylogeny estimation is a multi-step process, and understanding the process is helpful to developing and evaluating methods.

- The biologist identifies a question they wish to answer, and based on the question they select the species and genes they will analyze.
- Sequence data are collected for the species and genes. This may be done by accessing public databases, or by going into the field and gathering specimens. Typically, at the end of this step, for each selected gene and each species, at most one sequence is obtained. Often this is a DNA sequence, but in some cases it may be an RNA or amino acid sequence (also called a "protein sequence").
- For each gene, a **multiple sequence alignment** of the different sequences is obtained. This process puts the sequences into a matrix so that the rows correspond to the different species, and the columns represent **homologies** (nucleotides having a common evolutionary history).
- For each gene, a phylogenetic tree is estimated based on the multiple sequence alignment computed in the previous step. This analysis is almost always based on a statistical model of sequence evolution, and most methods combine graph-theoretic methods with statistical estimation techniques. Furthermore, many methods attempt to solve NP-hard optimization problems, so heuristic techniques are often used.
- Statistical support for the individual branches of each gene tree is computed, typically using methods such as non-parametric bootstrapping. These support values let the biologist assess which aspects of the estimated evolutionary history are considered highly reliable, and which ones are not as reliable.
- Now we have a collection of trees and multiple sequence alignments, one for each gene. When the gene trees are either identical or very similar to each other, then the

species tree is often estimated by concatenating the sequence alignments together, and then using standard techniques on the large "super-alignment" to compute a species tree. However, some biological phenomena cause gene trees to be different from each other, so that estimating the species tree can require different techniques, although the choice of technique depends on the cause for the discordance. When the discordance is due to horizontal gene transfer or hybrid speciation, then a phylogenetic network is needed. However, sometimes the species evolution is truly treelike but gene trees can still be different from the species tree (e.g., in the presence of "incomplete lineage sorting"). Thus, estimating species phylogenies (whether phylogenetic trees or phylogenetic networks) requires **phylogenomic** methods.

- Statistical support for the individual branches of the species tree is computed using a variety of techniques, including non-parametric bootstrapping.
- After the species tree or phylogenetic network is computed, other aspects of the history (e.g., dates at internal nodes, whether selection has occurred, and how some specific trait evolved within the species phylogeny) are estimated. Phylogenies and the associated alignments are also used to predict protein structure and function (Sjölander, 2004), to understand gene regulation and construct gene regulatory networks, and for other systems biology questions. These are called **post-tree** analyses, and are often the main goal of the study.

The pipeline described above is often varied somewhat. For example, instead of computing a single multiple sequence alignment, sometimes several alignments are computed, and a tree is constructed for each alignment. Furthermore, instead of computing a single tree for a single alignment, sometimes multiple trees are computed (perhaps based on different tree estimation methods, or sometimes even just one tree estimation method). The set of trees is then explored to determine the features that are consistent across the different techniques, and a **consensus tree** is computed. In addition, sometimes the alignment and tree are co-estimated together, rather than having the alignment estimated first and then the tree based on that alignment.

Note that the final species tree or phylogenetic network depends on the individual gene trees and multiple sequence alignments, and that the gene trees themselves depend on the multiple sequence alignments. This dependency suggests that errors in these initial analyses could result in errors in the downstream analyses, and hence lead to errors in the conclusions of the scientific study.

Most of the steps we have described in this pipeline involve statistical tests to evaluate uncertainty and test for the fit of the model to the data. See Holmes (1999) for a discussion of some of the issues that arise.

A.2 Taxonomies and Phylogenies

While the true evolutionary history of life on earth is by no means fully understood, some of its features are generally accepted. For example, **eukaryote**, **bacteria**, and **archaea** are three groups (also called **domains**) of organisms, and it is hypothesized that these form monophyletic groups (i.e., clades within the Tree of Life). Bacteria and archaea together constitute the **prokaryotes**, and animals, plants, and fungi are examples of eukaryotes.

Taxonomies are attempts to classify all living organisms into a hierarchical structure. The main ranks of a taxonomy (from the top down), are: **Domain, Kingdom, Phylum, Class, Order, Family, Genus,** and **Species**.

A.3 Genes and Alleles

Genes are stretches of genomic DNA that form the basic units of heredity. For **diploid** organisms (i.e., organisms with two copies of each chromosome), there are two copies of each gene in each organism, one inherited from each parent. These two copies may differ slightly, and are called **alleles**. In a given species, there can be multiple different alleles for a given gene, each potentially producing a different phenotype.

The exact definition of "gene" has changed over the years, because no simple definition has turned out to quite suffice (Gerstein et al., 2007); biology, after all, is always complicated. One recent definition is that a gene is a "union of genomic sequences encoding a coherent set of potentially overlapping functional products" (Gerstein et al., 2007).

A.4 Types of Molecular Sequence Data

In essence there are three different types of molecular sequence data: DNA, RNA, and amino acid sequences. DNA and RNA sequences have four possible states (A, C, T, G for DNA, and A, C, U, G for RNA, with U taking the place of T). Amino acid sequences have 20 possible states, one for each of the 20 standard amino acids, and can create proteins.

While the truth is not quite as simple as this, the central dogma of molecular biology is that DNA makes RNA, and RNA makes proteins. The process of changing a DNA sequence for a gene into messenger RNA (mRNA) is called **transcription**. The first step creates the pre-mRNA, and the second step splices out portions of the pre-mRNA sequence to create the **mature mRNA** sequence. The portions that are spliced out are called **introns** and the portions that remain are called **exons**; **alternative splicing** can also occur, and results in different mature mRNAs being created from a given pre-mRNA. The mature mRNA sequence is then used by the **ribosome** to synthesize an amino acid sequence, in a process that operates three nucleotides at a time (the **codons**). With a few exceptions, each codon is used to synthesize an amino acid, and the output is a protein (as described by its amino acid sequence). However, some codons do not create amino acids (i.e., AUG is the **start codon**, and UAA, UGA, and UAG are **stop codons**).

There are 64 possible codons, and after removing the four codons that do not create amino acids, there are 60 codons that will create amino acids; however, there are only 20 amino acids. There is a many-to-one mapping from codons to amino acids, and this is called a **genetic code**. There are multiple genetic codes, although in general the individuals in each species will share the same genetic code.

Because of the **redundancy** of the genetic code, multiple codons can code for the same amino acid. For example, CAU and CAC both create the same amino acid, and similarly for GCU, GCC, GCA, and GCG. Thus, a nucleotide sequence can change without changing the amino acid sequence that it codes for, and these changes are called **synonymous changes**; substitutions that change the amino acid are called **non-synonymous changes**.

Thus, DNA sequences are transcribed into mRNA sequences, and mRNA sequences are translated into amino acid sequences. However, not everything in the DNA sequence ends up in the corresponding mRNA sequence (e.g., introns, but also other genomic regions). Furthermore, the intronic component of a DNA gene sequence can be quite long, often much longer than the exonic component. Thus, DNA sequences can be much longer than their corresponding mature mRNA sequences.

Mutations in DNA sequences include substitutions, insertions, and deletions (among others), and any of these can change the protein encoded by a DNA gene sequence. Because of the redundancy in the genetic code, some mutations will have no impact on the protein sequence; furthermore, some amino acids are sufficiently similar in terms of their biochemistry that substitutions of amino acids in certain positions may not change the protein in any important way. Indels (insertions and deletions), however, can cause **frame shifts**, and will result in very different proteins being created. Furthermore, since the frame shift can move the position of the first stop codon substantially, the length of the protein that is created can be quite different. Some frameshift mutations can cause severe genetic diseases (e.g., Tay Sachs), others can simply result in non-functional proteins. Some mutations occur during DNA replication, but DNA repair processes can correct many of these errors (Pray, 2004).

DNA sequences are double stranded; this enables the DNA repair processes mentioned above. RNA sequences and amino acid sequences are single stranded, and fold into structures that enable them to interact with other molecules (Jones and Thornton, 1996).

A.5 Models of Sequence Evolution

Each type of molecule has properties that are reflected in the models that are used to describe them, and in particular in the sequence evolution models that are used to infer phylogenetic trees. Nearly all models for DNA sequence evolution are **neutral models** reflecting only genetic drift, and so ignore the impact of substitutions and indels on fitness. However, because different mutations can have quite different impacts on the proteins that the DNA sequences might code for, these models may be better suited for non-coding regions (e.g., introns) than for exons. Similarly, since RNAs form secondary structures, the most appropriate models might be those that allow for correlations or dependencies between sites, which aren't generally allowed in standard DNA sequence evolution models. Finally, changes in amino acid sequences can have large impacts on phenotype and hence on fitness, making the neutral models of evolution less appealing.

A.6 Detecting Selection

Phylogenies and multiple sequence alignments are used for many purposes, including understanding how a particular group of species evolved, how humans migrated across the globe, etc.; however, the detection of natural selection (i.e., changes in genomic DNA that result in differential distributions of alleles in populations due to their effect on the organismal phenotype) is one of the major applications. Mutations can be roughly partitioned into three categories: neutral (having no effect on phenotype), deleterious (reducing fitness), or advantageous (improving fitness), with deleterious mutations more likely than advantageous mutations.

Deleterious mutations may have no effect on the population because the individuals with these mutations may not be successful in reproducing, which is an example of **negative selection**. Conversely, advantageous mutations may increase the number of offspring produced by the individual, and so may lead to increased frequency of the resultant alleles in the population; this is referred to as **positive selection**. Other types of selection also occur, which are based on the impact on the population; examples include **balancing selection**, **diversifying selection**, and **stabilizing selection**. Of these different types of selection, positive selection is considered to be the major mechanism of **adaptation** (i.e., the origins of phenotypes in response to changes in the environment) (Vitti et al., 2013), and so methods to detect positive selection have received the greatest attention.

The detection of selection is often performed using multiple sequence alignments and phylogenetic trees for coding regions within genomic DNA. Then, given the coding sequence for a given gene in different organisms, the different mutations that have occurred can be inferred using the alignment and the tree. This approach makes it possible to determine the relative proportion of rates of synonymous to non-synonymous substitutions in coding regions. Since synonymous substitutions have no impact on the amino acids they code for, if the two rates are substantially different from their background rates, then some kind of selective pressure is presumed to be operating.

Because this type of analysis depends on the alignment, errors in the alignment can result in errors in the detection of selection. A common alignment error is *over-alignment*, where substitutions are favored over indels in the alignment process, resulting in estimated alignments that are substantially shorter than the true alignment. Since substitutions in the coding sequence are more likely to be non-synonymous than synonymous, over-alignment tends to result in false positive detection of selection. The impact of alignment error on methods for detecting selection has been documented in several studies; see Fletcher and Yang (2010) for one such study.

There are other types of techniques to detect and quantify selection, and different types of techniques are needed for different timescales. Vitti et al. (2013) provides an excellent survey of these techniques and the issues involved in using them, as well as a discussion of some of the open problems in the field.

Appendix B
Algorithm Design and Analysis

This appendix provides necessary background in algorithm design and analysis, and in proving algorithms correct. It covers techniques for computing the running time of an algorithm, the standard "Big-O" notation, what it means for a problem to be NP-hard or NP-complete, and for a problem to be polynomial time. However, the material in this appendix is not meant as a substitute for undergraduate courses in discrete mathematics and algorithm design and analysis, but rather to provide some of the material that the reader needs to know. We recommend that readers without sufficient background in this area consult other textbooks for background material.

B.1 Discrete Mathematics

B.1.1 Graph Theory

A **graph** G consists of a set V of vertices and a set E of edges, where the edges are unordered pairs of vertices; we often write graphs using notation $G = (V, E)$. The edges of a **simple graph** are always distinct (thus, there are no self-loops, which are edges between a vertex and itself), and no parallel edges (two edges with the same endpoints). In this text, we will only discuss simple graphs.

Two vertices that are connected by an edge are said to be **adjacent** and the edge that connects them is said to be **incident** to its vertex endpoints. If a vertex a is adjacent to vertex b, it is said to be a **neighbor** of b. The **degree** of a vertex is the number of edges that are incident with the vertex, which is the same as the number of vertices that are neighbors of the vertex.

A **path** in a graph is a sequence of vertices $v_1, v_2, \ldots, v_k$ so that v_i is adjacent to v_{i+1} for each $i = 1, 2, \ldots, k-1$, with the assumption that no two vertices are the same except perhaps v_1 and v_k. A path in which any vertex can be repeated is called a **walk**. A graph for which every two vertices are connected by a path is said to be **connected**. The maximal connected subgraphs of a graph are called the **components** of the graph. (Saying that "X is a maximal connected subgraph" means that there is no subgraph Y of G that strictly contains X and is also connected. Thus, "maximal" is not the same as "maximum.")

A **cycle** in a graph is a path in which the first and last vertices are the same. A graph that has no cycles is said to be **acyclic**. A **tree** is a graph that is connected and acyclic. Note that this definition of a tree may differ from what you are used to seeing; in particular, this definition of a tree does not provide a "root" for the tree, and makes no constraints on the degrees of the nodes in a tree. Later, we will distinguish between "rooted" and "unrooted" trees, but graph-theoretically the definition of a tree is quite simple: it's just an acyclic connected graph.

Some concepts you should know include:

- **Clique**: A clique X in a graph $G = (V, E)$ is a subset of the vertex set V, such that all pairs of vertices in X are adjacent.
- **Independent set**: An independent set X in a graph $G = (V, E)$ is a subset of the vertex set V, such that no pair of vertices in X are adjacent.
- **Hamiltonian graph**: A graph G is Hamiltonian if there is a cycle in the graph that covers every vertex exactly once.
- **Eulerian graph**: A graph G is Eulerian if there is a cycle in the graph that covers every edge exactly once.

Finally, what we have described so far is the usual kind of graph, in which edges are simply pairs of vertices. Sometimes, it is useful to direct the edges from one vertex to another, and graphs in which all the edges are directed are called **directed graphs** or **digraphs**. We use $\langle a, b \rangle$ or $a \to b$ to indicate the directed edge (also called an *arc*) from a to b in a digraph. If the digraph has no directed cycles, then it is referred to as a **directed acyclic graph** or **DAG**.

B.1.2 Binary Relations

Binary relations are used to represent many real-world situations. Mathematically, a **binary relation** on a set S is a set R of ordered pairs of elements of S; thus, $R \subseteq S \times S$. For example, consider the binary relation R on integers where $\langle a, b \rangle \in R$ means that a divides b without remainder. Hence, $\langle 3, 6 \rangle \in R$ but $\langle 3, 5 \rangle \notin R$, and $\langle 6, 3 \rangle \notin R$. However, binary relations do not have to *mean* anything in particular; you could for example take an arbitrary set X of ordered pairs and consider it a binary relation. Two types of binary relations are frequently used – partial orders and equivalence relations. We will use both in this text, and so discuss these further, below.

B.1.3 Hasse Diagrams and Partially Ordered Sets

A **partial order** is a binary relation R on a set S satisfying:

- $\langle A, B \rangle \in R$ and $\langle B, C \rangle \in R$ implies that $\langle A, C \rangle \in R$;
- $\langle A, A \rangle \in R$ for all $A \in S$;
- $\langle A, B \rangle \in R$ and $\langle B, A \rangle \in R$ implies that $A = B$.

We say that two elements A and B are comparable if $\langle A,B \rangle \in R$ or $\langle B,A \rangle \in R$, and when all pairs of elements are comparable then the partial order is called a "total order." However, for most partial orders, not all pairs of elements of the set S are comparable. Suppose we define the partial order R_{div} on positive integers so that $\langle A,B \rangle \in R_{div}$ if and only if A divides B evenly (without remainder). It is easy to see that R_{div} is a partial order, but not a total order, since 3 and 5 are not comparable.

A **partially ordered set** (or **poset**) is a set S with a partial order. Thus, the set of positive integers with the relation R_{div} constitutes a partially ordered set. Another example of a partially ordered set is the set of all subsets of the integers, with partial order R_{subset} defined by $\langle A,B \rangle \in R_{subset}$ if and only if $A \subseteq B$.

A **Hasse Diagram** is a drawing of the transitive reduction of a partially ordered set in the plane.[1] To construct the Hasse Diagram, create one vertex for each element in S and a directed edge $x \to y$ if $\langle x,y \rangle \in R$ and $x \neq y$. Order them from bottom to top on your page so that all the directed edges go upward. Then, repeatedly remove directed edges $x \to y$ if there is a third vertex z such that $\langle x,z \rangle \in R$ and $\langle z,y \rangle \in R$.

B.1.4 Equivalence Relations

A binary relation R on the set S is said to be an **equivalence relation** if it satisfies the following properties:

- $\langle a,a \rangle \in R$ for all $a \in S$; this is called the **reflexive property**.
- If $\langle a,b \rangle \in R$, then $\langle b,a \rangle \in R$; this is called the **symmetric property**.
- If $\langle a,b \rangle \in R$ and $\langle b,c \rangle \in R$, then also $\langle a,c \rangle \in R$; this is called the **transitivity property**.

When two elements x and y in the set S are in the relation (i.e., $\langle x,y \rangle \in R$), then we say that x and y are equivalent. The equivalence relation R thus partitions the set S into **equivalence classes**.

For example, you can define an equivalence relation R on a set S of people by saying x and y are equivalent if they earn the same salary (in dollars, not counting the amount past the decimal point). You can also define an equivalence relation R on the set of positive integers by saying $\langle x,y \rangle \in R$ if they have the same set of distinct prime factors. For example, under this relation, $\langle 6,18 \rangle \in R$, since 6 and 18 have the same prime factors (2 and 3).

B.1.5 Transitive Closure

Given a binary relation R on a set S, the **transitive closure** of R is the binary relation R^+ that is obtained by repeatedly adding $<A,B>$ to R whenever R contains both $\langle A,C \rangle$ and $\langle C,B \rangle$ for some $C \in S$, until no additional ordered pairs can be added. Equivalently, the transitive closure of R is the smallest binary relation containing R that is transitive.

---

[1] Hasse Diagrams are named after Helmut Hasse, a mathematician who used Hasse Diagrams in his research in algebraic number theory, but were introduced earlier in Vogt (1895).

B.1.6 Combinatorial Counting

The running time analysis of algorithms depends on being able to count the number of operations the algorithm executes as a function of its input size. This analysis then depends on a kind of discrete mathematics called "combinatorial counting" (or "counting," for short).

Suppose S is a set of n distinct objects, $s_1, s_2, \ldots, s_n$. Consider the following questions:

1. How many possible subsets of S are there, including the empty set and the set S?
2. How many non-empty subsets are there?
3. How many subsets are there that contain s_1?
4. How many subsets are there that do not contain s_1?
5. How many subsets are there that contain exactly one of s_1 and s_2?
6. How many ways can you partition this set into two non-empty sets?
7. How many functions are there from $\{1, 2, \ldots, k\}$ to S?
8. How many functions are there from S to $\{1, 2, \ldots, k\}$?
9. How many ways can you order the elements of S?

Techniques for combinatorial counting vary from very easy (enumerate all the objects algorithmically) to somewhat complicated (use Inclusion–Exclusion).

B.2 Proof Techniques

You will often need to prove theoretical results, and different techniques can be used to prove these results. Here we describe a few basic techniques.

B.2.1 Proof by Induction

Suppose that the sequence $a_1, a_2, \ldots,$ is defined recursively by $a_1 = 2$ and $a_i = 3 \times a_{i-1}$ for $i \geq 2$. Thus, $a_2 = 6, a_3 = 18, a_4 = 54$, etc. We will prove that $a_i = 2 \times 3^{i-1}$ for all $i \geq 1$ by induction on i.

In a proof by induction, you have to establish that a statement is true for the smallest value of some parameter (here, i); this is called "proving the base case." The next step is the "inductive hypothesis": You *assume* it is true for some arbitrary setting of the parameter. If you can then show that it will be true for the *next* value of the parameter, then it will be true for all settings of the parameter, starting with the base case.

Theorem B.1 *Let* $a_1 = 2$ *and* a_i *be defined to be* $3 \times a_{i-1}$ *for integers* $i \geq 2$. *Then* $a_i = 2 \times 3^{i-1}$ *for all integers* $i \geq 1$.

Proof We will prove this statement by induction on i. Hence, we can think of this as proving that the statement $S(i) = $ "$a_i = 2 \times 3^{i-1}$" is true for all $i \geq 1$. We begin by showing that the statement is true for the base case (the smallest value for i). We then assume the statement is true for some arbitrary value of i, and infer from this that it is true for the next value of i. Equivalently, we assume that the statement $S(I)$ is true, and we use that to infer that $S(I+1)$ is also true, where I is an arbitrarily chosen value for i.

The base case is $i = 1$. We know that $a_1 = 2$, by definition. We then check that $2 \times 3^{i-1} = 2$ when $i = 1$. We note that $S(1)$ is true, and so the base case holds.

We then assume that $S(I)$ is true; hence, $a_I = 2 \times 3^{I-1}$. This is called the *inductive hypothesis*, and is a statement about what happens when $i = I$ and not about any other value for i. Now, by definition, $a_{I+1} = 3 \times a_I$. By the inductive hypothesis, $a_I = 2 \times 3^{I-1}$. Hence, $a_{I+1} = 3 \times (2 \times 3^{I-1}) = 2 \times 3^I$. This is what we wanted to prove, so we are done. □

B.2.2 Proof by Contradiction

In a proof by contradiction, to prove that a statement is true, you assume it is not and then derive a contradiction. For example, here's a simple proof of a relatively obvious fact.

Theorem B.2 *There is an infinite number of prime numbers.*

Proof To prove this, we assume there is a finite number of prime numbers, and try to derive a contradiction. So let $\{p_1, p_2, \ldots, p_k\}$ be the set of all primes, and let

$$Y = 1 + \prod_{i=1}^{k} p_i.$$

By construction Y is strictly bigger than every p_i. Note also that each prime is at least 2. There are two cases to consider: Y is prime or Y is not prime.

If Y is prime, then Y must be in the set of primes, $\{p_1, p_2, \ldots, p_k\}$. However, by construction, Y is bigger than every prime, and so we derive a contradiction.

If Y is not prime, it can be written as a product of its prime divisors, which are all in the set $\{p_1, p_2, \ldots, p_k\}$. Hence, there must be some i between 1 and k such that p_i is one of Y's prime factors. Recall that the primes are always at least 2, so $p_i \geq 2$. Because $Y = 1 + \prod_j p_j$ we can rewrite Y as $1 + p_i Z$, where $Z = \prod_{j \neq i} p_j$. But since Z is a product of integers, Z is also an integer. Let $M = \frac{Y}{p_i}$. Since p_i is a prime factor of Y, M must be an integer. By definition, $M = \frac{1}{p_i} + Z$. Since $\frac{1}{p_i}$ is not an integer, M is not an integer. Hence we have derived a contradiction.

Thus when Y is prime we derive a contradiction, and when Y is not prime we derive a contradiction. Hence it is not possible for the set of primes to be finite, and so it must be infinite. □

B.3 Running Time Analysis

B.3.1 Pseudocode

Before we analyze the running time of a method, we need to be able to describe what the method does. For the purpose of describing methods, pseudocode is better than using a real programming language. The objective is to make it as easy as possible for your reader to understand the algorithm, so the description should be simple, provide all the information necessary to understand it, and not require knowledge of any particular programming language. Your pseudocode can certainly include English, but when you use English make sure that you aren't omitting necessary information. The important thing is that pseudocode should be very easy to understand.

You will need to have symbols that express assignment of values to variables (I use x := y or x ← y). You will want to have symbols that express comparisons of variables, but be careful to use a different symbol to test equality than to do assignment (i.e., "X=Y" should not mean that X and Y have the same value and also be used to assign value Y to variable X). Otherwise, you can use all the usual things (arithmetic operations, logical expressions such as IF/THEN/ELSE, and WHILE and FOR loops). You can use subscripts to refer to variables (i.e., a_i) or use elements in an array (i.e., $a[i]$), as you prefer.

To analyze the running time of the method, we need to count the number of operations the algorithm uses on an input of size n; note that this means we need to be able to quantify, in some way, what we mean by the "size" of an input. For graphs, the size depends on the representation. We can represent a graph G on n vertices with an $n \times n$ **adjacency matrix M** where $M_{ij} = 1$ if and only if the edge $(v_i, v_j) \in E$, or with an **adjacency list** $L = (L_1, L_2, \ldots, L_n)$, where L_i is the list of indices that are adjacent to v_i. If we don't worry about the space needed to represent the indices $1, 2, \ldots, n$, the space usage for the first representation is quadratic, in the number of vertices and the space usage for the second representation is linear in the sum of the number of vertices and edges. When the graph is sparse (i.e., has relatively few edges), then an adjacency list can require less space than an adjacency matrix, but the space usage for the two data structures are about the same for dense graphs.

To provide a proper running time analysis, you need to specify the data structures you use to represent your input, and the choice of data structure can impact the running time. For algorithms on graphs, the choice is generally between adjacency lists and adjacency matrices. Despite the fact that adjacency lists are more efficient representations of graphs, some algorithms are more efficiently implemented using adjacency matrices instead of adjacency lists. However, if the only issue is whether the running time is polynomial or not, then either representation can be used.

Now consider algorithms that are not based on graphs; here, too, the representation of the input can impact the running time. For example, what if your algorithm is attempting to determine if the input K is a prime number? A simple algorithm would look at every integer i between 2 and $\sqrt{K}$ and see if i divides K. This would take $O(\sqrt{K})$ time, if every division operation takes $O(1)$ time. However, is this polynomial in the input size? The question comes down to how we can efficiently represent integers. If we represent K using base 10 (the usual representation), this will use $O(log_{10}(K))$ digits. If we switch to a binary representation, this will use $O(log_2(K))$ bits. These two representations differ only by a constant factor. Therefore, for representations of integers, we say that the "size of n" is

$\log n$ (where $\log$ can be any base greater than 1). And so the question becomes whether $\sqrt{K}$ is bounded from above by a polynomial in $\log_{10}(K)$. (What do you think the answer is?)

We also have to say which operations are allowed and how much they cost. Running time analyses normally just consider every operation to have the same (unit) cost, and allow standard arithmetic operations, I/O operations, and logical operations. Thus, assigning values to variables, adding or multiplying numbers, comparing two numbers, and IF/THEN/ELSE operations all have the same cost.

B.3.2 Big-O Analysis

Computational methods that are designed to solve problems should be highly accurate (preferably completely accurate) and also fast. In this section, we discuss how to characterize the *asymptotic running time* of a method.

We are normally concerned with obtaining an *upper bound* on the running time, and when we talk about "Big-O" running times, we are providing an upper bound on the running time. This upper bound essentially hides all the constants that are involved. Thus, if we say that an algorithm has $O(n^4)$ ("Big-O of n to the fourth") running time, then we are saying that the running time on inputs of size n will never be larger than Cn^4, for some constant $C > 0$ and large enough values of n.

Note that we can use Big-O analysis to compare two functions, not just to characterize the running time of a method. For example, the following statements are all true:

1. $5n^4$ is $O(n^4)$
2. $5n^4 + 500n^3 + 300,000$ is $O(n^4)$
3. $5n^2 + \log n$ is $O(n^2)$
4. $500n^2 - 3n$ is $O(n^2)$
5. $5n^2$ is $O(n^3)$

To verify these statements are true, you'd need to be able to find the constants C that make the statements true for large enough n. For example, statement (1) is easily seen as true, by letting $C = 5$, since then the statement becomes $5n^4 \leq 5n^4$, which is always true. For the second statement, if you set $C = 5$ the statement will not be true because $5n^4 + 500n^3 + 300,000$ is greater than $5n^4$ when $n > 0$. However, if you set $C = 6$, then the inequality becomes $5n^4 + 500n^3 + 300,000 \leq Cn^4 = 6n^4$ for $n > N_0$. Note that if you let $N_0 = 1$, the inequality doesn't hold, but it should hold for some larger N_0. (For example, see if $N_0 = 500$ makes the statement true.) Hence, the second statement can be proven true as well. Similarly, every one of these statements can be proven.

The most confusing of these statements is probably (5), which says that $5n^2$ is $O(n^3)$. To understand why this is true you have to remember that the "Big-O" analysis is just an upper bound on the growth of the function, and doesn't need to be a tight upper bound.

Another way of describing running time is using $\Theta(f(n))$. To say that $g(n)$ is $\Theta(f(n))$ means that $f(n)$ and $g(n)$ are *each* Big-O of each other. Thus, it's equivalent to saying

$\exists N_0 > 0, C_1 > 0, C_2 > 0$ such that for all $n > N_0$, $C_1 f(n) \leq g(n) \leq C_2 f(n)$. Thus, saying that $f(n)$ is $O(g(n))$ just provides an upper bound on $f(n)$, but saying that $f(n)$ is $\Theta(g(n))$ provides an upper and a lower bound!

Thus, saying $f(n)$ is $O(g(n))$ or $\Theta(g(n))$ are really quite different. In particular, while it is true that $5n^2$ is $O(n^3)$, it isn't true that $5n^2$ is $\Theta(n^3)$! However, in general, running times are typically given with the Big-O description, rather than with Θ, even though the person writing this may be thinking in terms of Θ.

B.4 Different Types of Problems: Decision, Optimization, and Construction Problems

Decision problems. Decision problems simply ask for the answer to a *Yes/No* question. A simple example of this would be "Does this array have the value 5 in it?" The answer to the problem is just *Yes* or *No*. Determining the answer to this particular problem is easy: Just scan the array looking for an entry with value 5, then return *Yes* if you find it, and otherwise return *No*.

Other decision problems aren't always as easy to answer. For example, you might want to know if you can assign three colors (red, yellow, and blue) to the vertices of a graph so that no two adjacent vertices have the same color. Such a vertex coloring is called a "proper three-coloring," where "proper" means that there are no adjacent vertices having the same color. How efficiently can we determine if a graph has a proper three-coloring?

Exhaustive search techniques evaluate all possible solutions, and so can solve problems exactly when the number of possible solutions is finite. Since the number of possible three-colorings of a graph with n vertices is finite (i.e., only 3^n), the exhaustive search solution will certainly work; however, checking all 3^n different colorings is not efficient (and certainly not polynomial time). More efficient strategies exist, but none of them have been shown to have polynomial worst case running times.

On the other hand, determining whether a graph has a proper two-coloring (i.e., whether you can assign just two colors to the graph so that no vertices with the same color are adjacent) is easy to solve. Yes, you could still use exhaustive search (which would require checking 2^n different colorings), but there are more efficient ways to solve the problem. In fact, a simple algorithm will correctly solve this problem: Color one vertex red, then color its neighbors blue, then color vertices adjacent to those blue vertices red, and continue until all reachable vertices have been colored. If there are any vertices not yet colored, then start again with some uncolored vertex. At the end, all vertices will be colored, and you just need to check that no two vertices with the same color are adjacent. This type of algorithm is called "greedy" because it makes decisions and doesn't revisit its decision.

Many of these problems can be described as examining a search space, where the search space is potentially very large (exponential in the input size, for example). Sometimes the problem can be solved in polynomial time, even with an exponentially large search space, and sometimes it seems that no polynomial time solution can be found. Thus, some decision problems seem to be easy to solve, and others seem harder.

Optimization problems. Optimization problems are another type of problem, where instead of finding the answer to a Yes/No question, you want to find the score of the best possible solution to some problem. For example, you might want to find the largest value k so that a graph has a vertex v of degree k (i.e., a vertex that has k neighbors). Or you might want to find the largest k so that the graph has a clique of size k (i.e., the *Maximum Clique* problem). Or you might want to find the smallest value k so that the graph can be properly vertex-colored using k colors (the *Minimum Vertex Coloring* problem).

Again, exhaustive search will provide correct solutions to optimization problems, but sometimes these approaches have exponential running times. However, some optimization problems can be solved in polynomial time, even when exhaustive search is exponential. For example, suppose we are given an array of integers, and we want to find an ordering of the integers $x[1], x[2], \ldots, x[n]$ so that we minimize $\sum_{i=1}^{n-1} |x[i] - x[i+1]|$. If we use exhaustive search, we can evaluate the result of using every possible ordering, but there are $n!$ orderings. However, it is not hard to see that the best ordering is obtained by sorting the elements, which can be solved in polynomial time.

Nevertheless, many optimization problems are hard to solve efficiently, in that despite many efforts, no polynomial time algorithms have been found for them. Examples of such problems include the Minimum Vertex Coloring and Maximum Clique problems, defined above.

Construction problems. Finally, construction problems are ones where you want to find an object (if it exists). For example, you might want to find a maximum sized clique in a graph, or a proper vertex coloring of a graph using a minimum number of colors; these would be the construction problems for a given optimization problem. We can also define the construction problem version of a decision problem. For example, the construction problem version of the decision problem "Does the graph have a proper three-coloring?" is "If the graph has a proper three-coloring, then find one."

Thus, decision problems, optimization problems, and construction problems are different types of problems. However, there are connections between the different types of problems. For example, suppose you have an algorithm to find a three-coloring in a graph, when it exists; obviously, you can use that algorithm to determine if the graph has a three-coloring. Thus, algorithms for construction problems can be used to solve decision problems. What is less obvious is that if you can solve the decision problem "Does the graph have a proper three-coloring?" then you can use the answer to that decision problem to construct a proper three-coloring.

Here's an algorithm that shows how you can do this. Assume that you have an algorithm A that takes as input a graph G and returns *Yes* if the graph G can be properly three-colored, and else returns *No*. Suppose you want to construct a proper three-coloring for graph G_0, if it exists. You first apply algorithm A to G_0. If the answer is that there is no proper three-coloring, you immediately give up (no point in trying to construct something that doesn't exist). However, if algorithm A says there is a proper three-coloring, you continue, as follows.

- If G_0 has three or fewer vertices, then give a different color to every vertex, and you are done.
- Else, make a list L of all pairs of vertices that are not adjacent to each other. Maintain an equivalence relation on the vertices of G_0 in which initially all vertices are in their own equivalence class. Let G' be a copy of G_0. Then, for every pair v, w in the list L, DO:
  - Make a new graph G_1 in which v and w are collapsed into a new node x, which is adjacent to all nodes that v and w were adjacent to in G'. Note that G_1 has one fewer vertex, and the number of edges in G_1 is at most the number of edges in G_0.
  - Use A to find out if G_1 has a proper three-coloring. If the answer is *Yes*, then replace G' by G_1, merge the equivalence classes for v and w, and rename the equivalence class vw.

After processing the list, the graph G' will have at most three vertices, and each vertex will represent an equivalence class of vertices in the original graph G_0. Assign the three colors arbitrarily to the (at most three) equivalence classes; all the vertices in the same equivalence class get the same color. This will be a proper three-coloring of the graph G_0.

B.5 The Classes P and NP of Decision Problems

Some decision problems can be solved exactly in polynomial time. As we have seen, the two-colorability problem, which asks "Does the graph have a proper two-coloring?", can be solved in polynomial time. Other decision problems can also be solved in polynomial time, although the polynomial time algorithms that solve them may be less obvious. For example, a graph G is said to be Eulerian if there is a walk in the graph that covers every edge exactly once. The decision "Eulerian Graph" problem, which asks "Is this graph Eulerian?" can also be solved in polynomial time using a very simple algorithm, but proving the algorithm correct is more challenging (and is given as a homework problem).

The class P is the set of decision problems that can be solved exactly in polynomial time. Thus, the two-colorability and the Eulerian Graph problems are both in P. Determining whether a given decision problem is in P is equivalent to asking whether it can be solved in polynomial time.

The class NP is also a set of decision problems, but is defined differently. Instead of asking whether the problem can be solved in polynomial time, we are interested in whether we can *prove* that the answer is *Yes* in polynomial time. The trick here, however, is that we don't consider the time to come up with the proof. For example, suppose that the problem is three-colorability, so we want to know if the vertices of the input graph can be assigned three colors, so that no two adjacent vertices have the same color. Some graphs can be three-colored, and some cannot. For example, a four-clique cannot be three-colored, but every cycle, no matter how big, can be three-colored. We are not concerned with the graphs that cannot be three-colored, because these are not ones for which the answer is *Yes*. However, suppose that the graph G can be three-colored. How can we prove this, and make the proof take only polynomial time?

Don't consider the time it takes to figure out that the graph can be three-colored (e.g., we might go ahead and examine every possible three-coloring, to find one that is proper, but we won't count the time to do this). Once we have found the proper three-coloring, how can we use this to prove that the graph can be three-colored?

The answer is surprisingly simple: just write down the three-coloring, and then verify, edge by edge, that no two adjacent vertices have the same color. The running time is easily seen to be polynomial: writing down the three-coloring takes $O(n)$ time, and verifying that no edge connects vertices of the same color takes $O(n^2)$ time (since there are fewer than n^2 edges in a graph with n vertices).

Thus, while we do not know how to solve three-colorability in polynomial time, when a graph G *can* be three-colored then we can prove that G has a three-coloring in polynomial time. This means that three-colorability is in the class *NP*.

B.6 The NP-Complete Problems
B.6.1 Introduction

Informally, the NP-complete problems are the hardest problems in NP. As a result, NP-complete problems are decision problems, since they are problems in NP, and NP is a set of decision problems. But what do we mean by "the hardest problems in NP"?

Saying that a problem X is at least as hard as any problem in NP means that if X could be solved in polynomial time, then *every problem* in NP could be solved in polynomial time. For example, three-colorability (i.e., the decision problem that asks whether the vertices of the input graph can be properly colored using three colors) has been proven to be one of the NP-complete problems (Karp, 1972). What this means is that if anyone ever manages to develop a polynomial time algorithm for three-colorability, then every other problem in NP could be solved in polynomial time.

To prove that a problem X is NP-complete you therefore need to prove two things: (1) that it is in NP (this is generally the easy part), and (2) that X is at least as hard as any other problem in NP (this is the hard part), i.e., that X is NP-hard. Proving that X is NP-hard is often challenging, but is also sometimes easy. To prove that X is NP-hard, we find some other NP-hard problem Y, and we then prove that if we can solve X in polynomial time, then we can solve Y in polynomial time. Once we do this, the result follows because if X can be solved in polynomial time, then so can Y, and hence so can every other problem in NP (because Y is NP-hard).

We will show how to do this with a very simple example: proving that four-colorability is NP-complete. It is trivial to see that four-colorability is in NP, since we can verify that an assignment of four colors to the vertices of a graph is proper (i.e., does not assign the same color to the endpoints of any edge) in time that is linear in the number of edges in the graph. To complete the proof, we need to show that four-colorability is NP-hard.

We will use the fact that three-colorability is NP-hard to prove that four-colorability is NP-hard. What we will do is give a technique that will take any input I to three-colorability and turn it into an input $f(I)$ to four-colorability.

B.6.2 Karp Reductions

The technique we describe is an example of a Karp reduction (named after Richard Karp, who developed this technique for establishing that problems are NP-complete). Karp reductions are functions that map inputs to problem Y (already established to be NP-complete) to inputs to problem X (the problem you wish to prove NP-complete).

Before we define what a Karp reduction is, we need to introduce the concept of "yes-instances" and "no-instances." An instance to a decision problem is an input to the problem. Thus, suppose π is a decision problem and that algorithm A is guaranteed to answer the decision problem correctly. If I is an input to a decision problem π, then we say that I is a yes-instance for π if $A(I) = Yes$ and otherwise I is a no-instance. For example, if π is the three-colorability problem then the yes-instances are the graphs that can be three-colored and the no-instances are all the other graphs. Hence, if G is a cycle of length 10 then G is a yes-instance for three-colorability, while if G' is a clique with four vertices then G' is a no-instance for three-colorability. A Karp reduction from problem Y to problem X is a function f with the following properties:

- f maps inputs (instances) to problem Y to inputs to problem X, and the time $f(I)$ takes is polynomial in the size of I.
- For inputs I to problem Y, the size of $f(I)$ is bounded by a polynomial in the size of I.
- I is a yes-instance for Y if and only if $f(I)$ is a yes-instance for problem X.

We use the notation

$$f : Y \propto X$$

to indicate that f is a Karp reduction from Y to X.

The size of an input: Note that the size of an input I is an important quantity that determines whether f is a Karp reduction or not. In essence, the size is what is needed to represent I. For example, to represent an integer x in binary, we use $log_2(x) + 1$ space, and to represent it in base 10 (the way we write it) uses $log_{10}(x) + 1$ space. Although $log_{10}(x) < log_2(x)$, they are related by a constant C (i.e., $log_2(x) = C log_{10}(x)$). Hence, a polynomial in $log_{10}(x)$ is a polynomial in $log_2(x)$. Therefore, for the purposes of defining a Karp reduction, the specific choice of base does not matter. In fact, for Big-O analyses, you will often see *log* without any base provided, since the base only changes the constant and has no effect on the Big-O analysis.

As another example, how much space do we need to represent a simple graph with n vertices, $v_1, v_2, \ldots, v_n$? Recall that we can represent a graph $G = (V, E)$ with an adjacency matrix or with an adjacency list, and both use a polynomial in n amount of space (but different polynomials). Hence, although the space needed to represent a graph using an adjacency list can be less than the space needed to represent it using an adjacency matrix, both are equivalent in terms of the implications for Karp reductions.

Suppose four-colorability can be solved in polynomial time, and that A is a polynomial time algorithm that solves four-colorability exactly. Now suppose also that we have a Karp reduction f from three-colorability to four-colorability. We will show how we will use f and A to solve three-colorability in polynomial time. Given an input G to three-colorability (i.e., G is a graph), we compute $f(G) = G'$. Because f is a Karp reduction, the calculation of G' uses polynomial time, and the graph G' has at most $poly(n)$ vertices, where n is the number of vertices in G and $poly(n)$ is a polynomial in n. Furthermore, because f is a Karp reduction, G can be properly three-colored if and only if G' can be properly four-colored.

If we apply algorithm A on G', the output is *Yes* or *No*, depending on whether G' can be four-colored. If the answer is *Yes*, then we know G' has a proper four-coloring, and so G has a proper three-coloring; similarly if the answer is *No*, then we know G does not have a proper three-coloring. Running algorithm A on G' uses polynomial time in the number of vertices in G', which is bounded by a polynomial in the number of vertices in G. Hence, the running time of A on G' is bounded by the composition of two polynomials, which is itself polynomial in the number of vertices in G. Therefore, the entire process (computing G' given G, and running the algorithm A on G') completes in time that is bounded from above by a polynomial in the size of G. Thus, this is a polynomial time algorithm for three-colorability.

Hence, if we can find a Karp reduction from three-colorability to four-colorability, then if four-colorability can be solved in polynomial time then so can three-colorability. This analysis had nothing to do with the details of the two problems – all we needed was the Karp reduction. And, since three-colorability is already established to be NP-hard, then if four-colorability can be solved in polynomial time, then every problem in NP can be solved in polynomial time.

We summarize this discussion.

Theorem B.3 *Suppose* X *is a problem that is established to be* NP*-hard, and* Y *is some other problem. If* $\exists f : X \propto Y$ *(i.e., if there is a Karp reduction from* X *to* Y*), then* Y *is* NP*-hard. Furthermore, if any* NP*-hard problem can be solved in polynomial time, then all problems in* NP *can be solved in polynomial time.*

To prove that four-colorability is NP-complete, we need to show that it is in NP, and then present the Karp reduction from some NP-complete problem Y to four-colorability. It is easy to see that four-colorability is in *NP* (just present a proper four-coloring, and verify that all vertices are colored one of four colors, and no edge connects vertices of different colors). Now we need to come up with the Karp reduction f : three-colorability $\propto$ four-colorability.

A Karp reduction from three-colorability to four-colorability. Given graph $G = (V, E)$, the graph $f(G) = G'$ is formed by adding one vertex v^* to G, and making v^* adjacent to every vertex in G. We need to show that f is a Karp reduction.

We begin by analyzing the size of G', and showing that it is bounded by a polynomial in the size of G. Note that G' has $n+1$ vertices (where $n = |V|$) and $m+n$ edges (where $m = |E|$). Thus, the size of G' is bounded by a polynomial in the size of G. It is trivial to see that $f(G)$ takes polynomial time to compute. The next part is to show that f maps yes-instances to yes-instances, and no-instances to no-instances. So suppose G is a yes-instance of three-colorability, and so G has a proper three-coloring; if we use the same coloring on G' and then add a new color for vertex v^*, we have a proper four-coloring on G'. Conversely, suppose G' has a proper four-coloring, using colors red, blue, yellow, and green. Without loss of generality, suppose v^* is colored red. Since v^* is adjacent to every vertex in G, then the proper four-coloring does not assign red to any other vertex in G', and thus defines a proper three-coloring of G. In other words, we have shown that G has a proper three-coloring if and only if G' has a proper four-coloring. Thus, we have proven that the transformation is a Karp reduction, and so four-colorability is NP-hard. Since we already established that four-colorability is in NP, it follows that four-colorability is NP-complete.

B.7 The NP-Hard Problems

The difference between *NP*-hard and *NP*-complete is only that a problem that is NP-complete must be in the class NP. Thus, NP-hard problems are computational problems (maybe not decision problems, however) that are *at least* as hard as any problem in NP. Hence, to say that a problem X is NP-hard means that if it could be solved in polynomial time, then every problem in NP could be solved in polynomial time.

As we said earlier, algorithms for decision problems can be used to solve optimization and construction problems, and vice versa. For example, the Minimum Vertex Coloring problem takes as input a graph G and returns the smallest value k such that G has a proper k-coloring but does not have a proper $(k-1)$-coloring. Since Minimum Vertex Coloring is not a decision problem, it is not in the set NP. However, we will show it is NP-hard, which means that if we can solve Minimum Vertex Coloring in polynomial time, then we can solve any problem in *NP* in polynomial time. We will do this by showing that if we can solve Minimum Vertex Coloring in polynomial time, then we can solve three-colorability in polynomial time.

So suppose we have a polynomial time algorithm A for Minimum Vertex Coloring, and G is an input to three-colorability. We run algorithm A on G. If the answer is three or less, then we know that G can be properly three-colored, and otherwise we know G cannot be properly three-colored. Hence, if Minimum Vertex Coloring can be solved in polynomial time, then so can three-colorability. Since three-colorability is NP-hard, this means that Minimum Vertex Coloring is NP-hard.

B.8 General Algorithm Design Techniques

B.8.1 Introduction

Algorithms researchers design methods that solve problems, whether decision, optimization, or construction problems. The usual objective is an algorithm that is guaranteed to solve the problem exactly, and that does so efficiently. Thus, we would like not only a polynomial time algorithm, but one that is as fast as possible. Thus, we distinguish between methods that run in $\Theta(n^3)$ time and $\Theta(n^2)$ time, and prefer the quadratic running times to the cubic running times. (Similarly, we would prefer a $\Theta(n)$ algorithm to a $\Theta(n^2)$ algorithm.)

As we have seen, some problems are NP-hard, and the only exact solutions found for NP-hard problems have required more than polynomial time. However, for problems that can be solved in polynomial time, algorithm design techniques can make a difference between an exponential time and a polynomial time algorithm, or between a $\Theta(n^3)$ algorithm and a $\Theta(n^2)$ algorithm. Here we describe a few of these techniques.

B.8.2 Dynamic Programming

In a **dynamic programming** algorithm, the idea is to decide in advance all the subproblems you need to solve, the order in which you'll solve them, and how solving all the subproblems allows you to solve the entire problem. As long as the number of subproblems is at most polynomial in the input size and solving a subproblem uses at most polynomial time once the earlier subproblems are solved, the entire approach uses polynomial time. An important point about dynamic programming is that when you solve the subproblem, you have to save its solution in memory so that you can easily access it later (without having to recalculate it).

A very simple example of a dynamic programming algorithm is one that computes the nth Fibonacci number. Recall that the Fibonacci numbers are formed by having the first two numbers equal to 1, and then each successive number is the sum of the previous two numbers. Thus, letting $F(i)$ denote the ith Fibonacci number, then $F(1)$ and $F(2)$ are both set to 1 and $F(i)$ is the sum of $F(i-1)$ and $F(i-2)$; in other words, the Fibonacci numbers are *defined recursively*. To compute the nth Fibonacci number, we design a dynamic programming algorithm, as follows.

The input to the problem is n, and the output is the nth Fibonacci number $F(n)$. We let the jth subproblem be the computation of $F(j)$ and we note that we need to compute $F(j)$ for all j between 3 and n (because $F(1)$ and $F(2)$ are already known to be 1). The algorithm computes these values in increasing order for j, and stores the results in an array $Fib[1...n]$. We first compute $Fib[3]$, then $Fib[4]$, etc., until we obtain $Fib[n]$. To compute $Fib[j]$ given all the previously computed values, we set $Fib[j]$ to the sum of the previous two values. Thus, computing $Fib[j]$ takes constant time, as long as we calculate $Fib[j-1]$ and $Fib[j-2]$ before we compute $Fib[j]$. In other words, the algorithm has the following form:

Computing F(n), *the* n*th Fibonacci number:* Note that we will compute values and store them in an array, from the smallest index to the largest. We then return the value in the last element of the array.

If n is not a positive integer, return Null.
Fib[1] ← 1, Fib[2] ← 1

For j=3 up to n DO:

    Fib[j] ← Fib[j-1] + Fib[j-2]

End(For)

Return Fib[n]

A running time analysis shows that the initialization (before the loop is entered) uses three operations (one to check that n is a positive integer, and the other two to set the values of Fib[1] and Fib[2]). Then, every time the loop is entered, we use four operations (two to look at the values of Fib[j − 1] and Fib[j − 2], one operation to add those values, and then one operation to set the value of Fib[j]). The loop is entered $n - 2$ times. Hence, the total time to compute the nth Fibonacci number using this algorithm is $O(n)$.

B.8.3 Recursive Algorithms

Recursive algorithms can look a lot like dynamic programming algorithms, because the solution to a problem depends on solving smaller subproblems. However, unlike the dynamic programming approach, the number of *possible* subproblems need not be small, as long as only a proper subset of the possible subproblems is solved during the course of the algorithm.

Analyzing the running time for a recursive algorithm typically amounts to analyzing a recursively defined function for the running time $t(n)$, where n is the input (or the size of the input). For example, consider the simple sorting algorithm that scans an array, finds the largest value, swaps that largest value with the last element of the array, and then recurses on the first $n-1$ elements. In this case, the running time $t(n)$ for arrays of size n satisfies

$$t(n) = Cn + t(n-1)$$

and

$$t(1) = C',$$

where C and C' are two positive constants. This recursively defined function can be solved exactly, and yields $t(n) \leq Cn^2 + C'$. Hence, $t(n)$ is $O(n^2)$, and the recursive algorithm runs in $O(n^2)$ time.

Note that using dynamic programming to sort the array would not have been so pleasing; if we decided to sort all subarrays, the number of possible subarrays would have been exponential, and the running time would have been exponential as well. Thus, recursion trumps dynamic programming in this case, because even though there are an exponential number of possible subproblems, the recursive algorithm only explores a linear number of subproblems, and each one can be solved in polynomial time.

On the other hand, sometimes recursion is less efficient than dynamic programming. For example, we saw above that we could use dynamic programming to compute the nth Fibonacci number $F(n)$. If we try to use recursion to compute $F(n)$, the algorithm would look like this:

If n is not a positive integer, return Null
Else, if $n \leq 2$ return 1
Else, return $F(n-1) + F(n-2)$

Note that in this pseudocode, when $n \geq 3$, we are calling the algorithm recursively on $n-1$ and $n-2$; therefore, the running time $t(n)$ for the algorithm on input n satisfies $t(n) = t(n-1) + t(n-2) + C$ (for some constant C) and $t(1) = t(2) = C'$ (for some constant C'). Finding a closed form for this function is not easy, but it is not hard to show that $t(n) > g^n$ for some constant $g > 1$ (see homework!). Hence, the recursive algorithm for computing $F(n)$ is much more computationally intensive than the dynamic programming algorithm, which used linear time.

The problem with using recursion here is that although the number of subproblems is polynomial, unlike with dynamic programming, we may compute each subproblem more than once. For example, if $n = 15$, then when we compute $F(15)$ we recursively compute $F(14)$ and $F(13)$. But when we compute $F(14)$ we also recursively compute $F(13)$. So we computed $F(13)$ twice during this analysis. It is not hard to see that $F(12)$ is computed even more often.

B.8.4 Divide-and-Conquer

Divide-and-conquer algorithms are a type of recursive algorithm, but typically have a more elaborate design. For example, *Merge Sort* is a divide-and-conquer approach to sorting an array of n integers: The array is divided into two approximately equal sets, each subset is recursively sorted, and then the two sorted arrays are merged together. Since the merger technique takes only $O(n)$ time, the running time $t(n)$ can be shown to be $O(n \log n)$.

B.9 Designing Algorithms for NP-Hard Problems

If you learn that a problem is NP-hard, what does this mean in terms of practice? As we have discussed, by definition, if any NP-hard problem can be solved in polynomial time, then they can all be solved in polynomial time. Equivalently, this is the same as saying that if some NP-hard problem is solvable in polynomial time, then P = NP. Whether "P = NP" or not is one of the most fundamental questions in computer science, and while it is still open (i.e., unsolved), most researchers assume that P $\neq$ NP. In other words, most computer scientists assume that no NP-hard problem can be solved in polynomial time.

Therefore, if you know a problem is NP-hard, then don't try to solve it exactly, unless you are willing to use more than polynomial time on some inputs. In other words, you need to sacrifice something – either running time (take a lot of time on some inputs) or guarantees of accuracy. Similarly, if you know a problem is NP-hard and you have a dataset that is too large for an exact solution, do not be too confident in the result you obtain by using some software package to analyze your dataset. No fast method is currently guaranteed to find the correct solution to NP-hard problems on large inputs!

B.10 Method Evaluation

When we evaluate methods – such as tree estimation methods, or multiple sequence alignment methods – we want to be able to quantify their error and compare them to other methods. To do this, we use a reference tree or reference alignment, and quantify the error (or accuracy) with respect to the reference. Fundamentally, this kind of evaluation comes down to evaluating the method in terms of false positives, false negatives, true positives, and true negatives.

Given this, we can relate these error and/or accuracy evaluations to basic statistical concepts, such as *sensitivity* (also called the true positive rate, or recall rate), *specificity* (also called the true negative rate), and *precision*.

Each of these concepts has a mathematical definition in terms of false positives (FP), true positives (FP), false negatives (FN), and true negatives (TN), which are also important to understand. FP, TP, FN, and TN arise in the context of a test (or binary classifier), which can be described as a function that maps objects to "Positive" (has a trait) or "Negative" (does not have the trait). The "positives" are the objects that are mapped to "Positive" and the "negatives" are the objects that are mapped to "Negative." Some of these classifications are correct and some are not correct; the classifications that are correct are either "True Positives" or "True Negatives," and the classifications that are incorrect are either "False Positives" or "False Negatives." For example, a "False Negative" is an object the classifier mapped to "Negative" but should have been mapped to "Positive."

To make this concrete, consider a test to predict whether a patient has influenza. Assume that all test results are definitive – positive or negative – and so everyone you test is either characterized as having the flu or not having the flu. Thus, the test is a binary classifier. The people for whom the test comes out positive and who do have the flu are the TP, while the people for whom the test comes out positive but don't have the flu are the FP. The people for whom the test comes out negative but do have the flu are the FN, and the people for whom the test comes out negative and who don't have the flu are the TN.

The *precision* of a binary classifier is the fraction of the "positives" that are true positives: i.e., it is the ratio of TP to TP+FP (i.e., all the ones you classified as positives). So when you say that a binary classifier has 80 percent precision, this means that 80 percent of the objects classified as having the trait actually *do* have the trait. Precision is also referred to as *positive predictive value* (PPV).

The *recall* of a binary classifier is the fraction of the objects that truly do have the trait that you correctly detect as having the trait. In other words, it is the ratio of TP to TP+FN. *Sensitivity* is another term for recall.

The *false discovery rate* (FDR) is the fraction of the objects you identify as having the trait that do *not* have the trait; this is the same as $1 - PPV$. Think of this as the fraction of the people you think have the flu but actually don't have the flu.

- Precision/Positive Predictive Value (PPV) = $\frac{TP}{TP+FP}$

- Sensitivity/Recall = $\frac{TP}{TP+FN}$
- Specificity = $\frac{TN}{FP+TN}$

B.11 Homework Problems

For all these problems involving graphs, assume that the graph has no self-loops or parallel edges. If the problem requires an algorithm (but not code), describe your algorithm using pseudocode. You should specify the data structure representations for your input and provide a running time analysis based on those data structures (e.g., for algorithms on graphs, the running time will depend on whether the graph is represented by an adjacency list or an adjacency matrix). Be careful to explain what each of your variables mean.

1. Prove that the number of vertices with odd degree in a simple graph must be even.
2. Consider the binary relation R on integers containing exactly those ordered pairs $\langle x, y \rangle$ for which x divides y. Thus, $\langle 3, 15 \rangle \in R$ but $\langle 3, 8 \rangle \notin R$. Draw the Hasse Diagram for the partially ordered set defined by R on the set of integers between 1 and 15 (i.e., $\{1, 2, 3, 4, \ldots, 15\}$).
3. Do all the counting problems in Section B.1.6. (Explain your answers.)
4. Consider the binary relation R defined on the vertices of a given graph G: $\langle x, y \rangle \in R$ if there is a path from x to y with at most two edges.

    a. Give an example of a graph G_1 for which this binary relation is an equivalence relation, and graph G_2 for which the binary relation is not an equivalence relation.
    b. What properties does this relation always hold, independent of the graph G?
    c. What happens if you allow G to be a directed graph, instead of an undirected graph?

5. Let G be a graph with $n > 1$ vertices, and let v_0 be one of its vertices. Consider the binary relation R defined by: $\langle x, y \rangle \in R$ if the distance from v_0 to x is at most the distance from v_0 to y. (Recall that the distance from a to b is the number of edges in the shortest path from a to b.) What properties does this relation always hold, independent of the graph G?
6. Give a real-world example of a partial order that is not a total order.
7. Let S be the set $\{a, b, c, d\}$. Give an example of a relation $R \neq S \times S$ on S whose transitive closure is R.
8. Let S be the set $\{a, b, c, d\}$. Give an example of a relation $R \neq S \times S$ on S whose transitive closure is $S \times S$.
9. Consider a simple graph $G = (V, E)$ and the binary relation R on V containing those pairs $\langle x, y \rangle$ such that there is a path in G from x to y.

    a. Prove that R is an equivalence relation.
    b. Given an example of a graph G for which this equivalence relation has exactly three equivalence classes.

10. Consider a recursively defined function $t(n)$ defined by $t(1) = 1$ and $t(n) = 3 + t(n-1)$ when $n \geq 2$. Find a closed form solution for $t(n)$ and prove it correct using induction.

B.11 Homework Problems

11. Consider a recursively defined function $t(n)$ defined by $t(1) = t(2) = 1$ and $t(n) = t(n-1) + t(n-2)$. Prove, using induction, that $t(n)$ is $O(2^n)$.
12. Prove by contradiction that $\sqrt{2}$ is not rational.
13. Prove by contradiction that $\sqrt{3}$ is not rational.
14. Let $(0,1)$ denote the open interval between 0 and 1 (i.e., the set $\{x : 0 < x < 1\}$). Prove by contradiction that $(0,1)$ does not contain a smallest element.
15. Prove by contradiction that the number of functions from the set of positive integers to the set $\{0,1\}$ is infinite.
16. Consider a simple undirected graph $G = (V, E)$ in which every vertex has even degree, and assume that G is connected. Prove by induction on the number of edges in G that there is a walk in G that covers every edge exactly once where the first and last vertices are the same.
17. Prove the five statements about Big-O running times in Section B.3.2, by finding the positive constants C, C'.
18. Suppose you have an oracle that correctly answers Yes/No questions of the form "Does graph G have a clique of size 5?" (where you can specify G). Show how to use the oracle to find a five-clique (if it exists) in an input graph G on n vertices, without calling the oracle more than $O(n)$ times.
19. Suppose you have an oracle that correctly answers Yes/No questions of the form "Does graph G have a clique of size k?" (where you can specify G and k). Show how to use the oracle to find a maximum clique in an input graph G on n vertices, without calling the oracle more than $O(n+k)$ times.
20. Use the fact that maximum clique is NP-hard to prove that maximum independent set is NP-hard.
21. Consider the function f that maps inputs for two-colorability to inputs for three-colorability, that operates as follows: $f(G)$ is the graph G' formed by adding a vertex v^* to G and making it adjacent to every other vertex in G.

    - Prove that this function is a Karp reduction.
    - Since three-colorability is NP-complete, does this mean that two-colorability is NP-complete?
    - Since two-colorability can be solved in polynomial time, does this mean that three-colorability can be solved in polynomial time?

22. Design a dynamic programming algorithm, and provide a running time analysis, that computes the length of a longest increasing subsequence in an input array of n integers.
23. Design a dynamic programming algorithm, and provide its running time analysis, that computes the length of a longest common subsequence given two arrays of integers.
24. Consider the recursively defined algorithm for computing the nth Fibonacci number $F(n)$. Find and prove (using induction) lower and upper bounds on its running time. (Full points where both bounds are exponential.)

25. Consider the recursively defined algorithm for computing the nth Fibonacci number $F(n)$. Let $n = 7$. How many times is $F(4)$ computed? How many times is $F(3)$ computed?

26. Find and prove (using induction) upper and lower bounds for $t(n)$ (as a function of n), for $t(n)$ defined as follows:
    - $t(1) = t(2) = 2$
    - $t(n) = t(n-1) + t(n-2) + 1$ for $n > 2$

27. Find and prove (using induction) a closed form formula for $t(n)$ (as a function of n), for $t(n)$ defined as follows:
    - $t(1) = 5$
    - $t(n) = t(n-1) + 3$ for $n > 1$

28. Find and prove (using induction) a closed form formula for $t(n)$ (as a function of n), for $t(n)$ defined as follows:
    - $t(1) = 2$
    - $t(n) = 5t(n-1)$ for $n > 1$

29. Find and prove (using induction) lower and upper bounds for $t(n)$ (as a function of n), for $t(n)$ defined as follows:
    - $t(1) = 2$
    - $t(n) = 5t(n-1) + n$ for $n > 1$

30. Design a dynamic programming algorithm, and provide its running time analysis, for the following problem. The input is a graph G with vertices $v_1, v_2, \ldots, v_n$, and with positive weights on the edges. The output is a symmetric $n \times n$ matrix $\mathbf{M}$ where $M[i,j]$ is the length of the shortest path from vertex v_i to v_j. (Hint: consider solving subproblems that give the length of the shortest path from v_i to v_j, using at most k edges. Alternatively, consider solving subproblems that give the length of the shortest path from v_i to v_j in which the only additional vertices that are permitted are $v_1, v_2, \ldots, v_k$. Both of these approaches yield polynomial time algorithms, but have different running times.)

31. Design an $O(n)$ dynamic programming algorithm and provide the running time analysis for the following problem. The input is a rooted binary tree T with n leaves, $s_1, s_2, \ldots, s_n$, and with internal nodes labeled $s_{n+1}, s_{n+2}, \ldots, s_{2n-1}$. The output is an array $w[1\ldots 2n-1]$ where the ith element $w[i]$ is the number of leaves in the subtree of T rooted at s_i. (Note that $w[i] = 1$ for $1 \leq i \leq n$.)

32. Design an $O(n^2)$ dynamic programming algorithm and provide the running time analysis for the following problem. The input is a rooted binary tree with n leaves $s_1, s_2, \ldots, s_n$, and with internal nodes labeled by $s_{n+1}, s_{n+2}, \ldots, s_{2n-1}$. The output is an $n \times n$ matrix $\mathbf{MRCA}$ where $MRCA(i,j) = k$ means that the most recent common ancestor of s_i and s_j is s_k.

33. Design an $O(n)$ dynamic programming algorithm and provide the running time analysis for the following problem. The input is a rooted binary tree T with n leaves $s_1, s_2, \ldots, s_n$, and with every internal node also labeled by $s_{n+1}, s_{n+2}, \ldots, s_{2n-1}$.

The edges of the rooted binary tree have positive lengths. The output is an array $Longest[1\ldots 2n-1]$, where $Longest[i]$ is the length of the longest path from s_i to a leaf in the subtree of T rooted at s_i. Thus, $Longest[i] = 0$ for $1 \leq n$, but $Longest[i] > 0$ for all i such that $n+1 \leq i \leq 2n-1$

34. Design an efficient algorithm to find the edge that contains the midpoint of a longest leaf-to-leaf path in a binary tree T on n leaves where every edge has positive length. Analyze its running time.

35. Consider the following two-player game. At the start of the game there are two piles of stones, and at least one pile has at least one stone. Thus, you can consider the starting point to be a pair (p,q), where p is the number of stones in the first pile, q is the number of stones in the second pile, $p+q \geq 1$, and both are non-negative. The first player starts, and then they take turns, until the game ends. In each turn, the player must take a stone off of at least one pile, but cannot take more than one stone off of any pile; thus, the choice is between taking a stone off of each pile or a stone off of one pile. The game ends when the last stone is removed, and the player who took the last stone wins the game. For this problem, do the following:

   a. Write a dynamic programming algorithm to determine which player wins, given input values for (p,q). Have your dynamic programming algorithm output a $(p+1) \times (q+1)$ matrix **Winner**, where $Winner[i,j]$ is T if the first player wins on input i,j, and is F if the first player does not win on input i,j. You should assume both players play optimally.
   b. Analyze the running time.
   c. Give the matrix for **Winner** where $p = q = 5$.

36. Consider the following two-player game. At the start of the game there are three piles of stones and at least one pile has at least one stone. A player must take off at least one stone, and can take as many as two, but can only take off one stone from any one pile. Again, you need to determine who wins. The input is the triplet of the number of stones on each pile, (p,q,r), where p is the number of stones on the first pile, q is the number of stones on the second pile, and r is the number of stones on the third pile, and $p+q+r \geq 1$.

   a. Write a dynamic programming algorithm to determine which player wins, given input values for (p,q,r). Have your dynamic programming algorithm output a $(p+1) \times (q+1) \times (r+1)$ matrix **Winner**, where $Winner[i,j,k]$ is T if the first player wins on input i,j,k, and is F if the first player does not win on input i,j,k. You should assume both players play optimally.
   b. Analyze the running time.
   c. Give the matrix **Winner** where $p = 3, q = 2, r = 1$.

37. Design an exact algorithm for maximum clique and analyze its running time.
38. Design an exact algorithm for maximum independent set and analyze its running time.
39. Design an exact algorithm for three-colorability and analyze its running time.
40. Design an exact algorithm for Hamiltonian path and analyze its running time.

41. We know that the optimization problem Max Clique of finding the size of the largest clique in a graph is NP-hard. Suppose that your uncle is a software developer, and has created an algorithm for Max Clique that runs in polynomial time and claims to solve the problem optimally. What do you think is going on?

42. Suppose that you have designed a test for a disease that comes up either positive (indicating that the person has the disease) or negative (indicating that the person does not have the disease). Suppose in your population there are 1000 people, ten of them have the disease, and 990 people do not have the disease. You use your test on these 1000 people.

- Suppose all the tests come back positive. What is the true positive rate? What is the true negative rate? What are the positive predictive value (PPV), sensitivity, and specificity?
- Suppose all tests come back negative. What is the true positive rate? What is the true negative rate? What are the PPV, sensitivity, and specificity?
- Suppose nine of the ten people with the disease come back positive, ten people without the disease come back positive, and the rest come back negative. What is the true positive rate? What is the true negative rate? What are the PPV, sensitivity, and specificity?
- Suppose that nine of the ten people with the disease come back positive, 100 other people come back positive, and every one else comes back negative. Now suppose that you learn that Sarah has a positive test, but you don't know whether she has the disease. What is the probability that she has the disease, based upon the information you have?
- Suppose that nine of the ten people with the disease come back positive, 500 other people come back positive, and every one else comes back negative. Now suppose that you learn that Sarah has a positive test, but you don't know whether she has the disease. What is the probability that she has the disease, based upon the information you have?

43. Implement the algorithm for problem 30. You should assume that the input to describe the graph on n vertices is a pair of $n \times n$ matrices $\mathbf{D}$ and $\mathbf{W}$, where $\mathbf{D}$ is the adjacency matrix (so that $D_{ij} = 1$ if there is an edge between v_i and v_j and otherwise $D_{ij} = 0$) and $\mathbf{W}$ indicates the weight of the edges (so W_{ij} is the weight of the edge between v_i and v_j if it is present, and otherwise W_{ij} is -1 to indicate the edge is not present). Set the output matrix $\mathbf{M}$ with $M[i,j] = -1$ if there is no path between v_i and v_j. Submit well-documented code. Show the output of your algorithm on a graph with ten vertices that has two components.

44. Implement the algorithm for problem 35, and submit your well-documented code and its output on the case $p = q = 5$. Report the running time as you change the value of p from 0 to 100 but leave $q = 5$.

45. Implement the algorithm for problem 36, and submit your well-documented code and its output for the case $p = 3, q = 2, r = 1$. Report the running time as you change the value of p from 0 to 100, but leave $q = r = 3$.

Appendix C
Guidelines for Writing Papers About Computational Methods

Becoming a good writer is tremendously important, for many reasons. The most obvious one is that your papers will be easier to understand and this will help you get published and also contribute to the research field. However, in addition, being a good writer will also help you find mistakes in your work, which is perhaps even more important!

There are many ways to write well, and you should find your own style. However, all good scientific writing should have the following properties:

- *clarity of exposition*, so that both you and the reader understand what you've done and can draw correct inferences from the data;
- *reproducibility*, so that the experiment can be performed by someone else, using the exact same methods and data;
- *rigor*, so that what you infer makes sense; and
- *scientific relevance*, so that what you generate is relevant to some real data analysis.

To become a good writer, you should read the scientific literature as much as you can, and note what you like and don't like about each paper, and why. Extensive reading helps in terms of developing a good writing style, and also developing skills in designing and doing experiments that are convincing and appropriate. It's also very good practice to read the supplementary materials of all the papers you like, because often it's only in the supplement that you will find out the details that are the most important. For that matter, some journals make it quite difficult to provide sufficient detail, due to space limitations, and may not even make it feasible to provide supplementary materials on the journal's website. So, as you read, develop a sense of how the different journals enable or discourage reproducibility. It may be that this will end up informing your thoughts on where *you* wish to publish.

It should be obvious that *any* paper you want to write should be written well, with a clear introduction that provides a context for the study and engages the reader, a discussion of what was done and why, the observations that you made from your work, detailed discussion of what the observations suggest and how they relate to the rest of the scientific literature, and – of course – conclusions. Yet in the context of a study evaluating computational methods, especially one introducing a new method, some additional guidelines may be of benefit.

Rule 1: Clearly describe your new method, and make the code available so that it can be run by others. It's important to do both – be clear about the method so that the user understands what it is doing, but also to provide the code; one without the other is better than nothing, but doing both enables the user to both understand the work and redo your experiments (and so potentially confirm your results). In particular, avoid using terms that have multiple meanings to describe your method, and if your method relies on other software, then specify exactly how you use the other software. To fully explain your new method, you may need to provide the details in a supplementary document (especially since some journals have page limits).

Rule 2: Evaluate your method on appropriate datasets. The datasets you pick to evaluate your method are essential, but you need to justify the choices. If there are established benchmarks, use them (or explain why you don't use them). If you are using simulations, make sure the simulations produce datasets with properties that are biologically relevant. You should think about what your objective is: speed, memory usage, accuracy, or something else. If your concern is speed, then make sure you include datasets that are challenging for running time, and similarly for memory usage, accuracy, or whatever criterion you are focusing on.

Try to ensure that your datasets match the empirical properties of the real datasets that are of current interest (or future interest). For example, if the objective is to enable highly accurate analyses of datasets of a certain size, make sure that your simulated datasets have this size. Also, it doesn't make a lot of sense to focus on datasets that are either too easy (so all methods can be extremely accurate) or too hard (so no methods do well); at a minimum, you will want to make sure that your collection of datasets includes some where the best current methods have accuracy levels that are worth improving, but are not terrible.

One way to decide what level of accuracy is reasonable for the simulation is to think about what happens on real data – if the typical error rates on real data seem to be within some range (say, 5 percent to 20 percent), then reducing error from 1 percent to 0.5 percent may not be exciting to the practitioner (because the datasets are too easy), and similarly reducing error from 90 percent to 80 percent may not be exciting (because the datasets are too difficult). Always think about whether your datasets match the real datasets in ways that are convincing to the practitioner. Remember that relative performance under one model condition may not hold under another model condition!

In general, it's best to explore many model conditions – but the trick is to not do so many model conditions that the results cannot be comprehended. Also, to fully understand the impact of the model conditions, a good practice is to divide the study into a few experiments, each changing one variable at a time; that way, you will better understand how each variable impacts accuracy.

Rule 3: Consider statistical significance. In a simulation study, you can generate enough datasets that you can test for the statistical significance of a difference in performance between two methods. This is important, since sometimes differences are really due to random fluctuations in performance, and you won't know if an improvement you see for your method compared to another isn't just the result of randomness. But, make sure you correct for multiple tests, so that you reduce the false discovery rate.

Rule 4: Don't make a big deal out of a small difference in performance. Avoid getting excited about small differences, whether they are in favor of your method or against your method. For example, if you reduce error from 0.01 percent to 0.008 percent, it may not matter to anyone. And don't confuse statistical significance with importance – even if the result is statistically significant, it may not matter in terms of practice.

Rule 5: Don't avoid the cases where your method doesn't perform well. Most methods will have some weaknesses, and being able to find out what those weaknesses are is important. Explore many datasets, varying the model conditions for your simulations or the empirical properties of your real datasets, to find out where your method performs well and where it doesn't. Then, report all of this. You may not be comfortable doing this (no one likes to reveal weaknesses), but there are many benefits to doing this. First, you will learn essential things about your method that you won't otherwise. Second, you will earn the respect and trust of your reviewers and readers, because (unlike some other authors), you will gain a reputation of not over-hyping your results. Third, it's really much better if you criticize your own method, rather than having a follow-up study by your competition criticizing your method!

Rule 6: Don't test on your training data. Many methods have algorithmic design parameters that can be modified, for example to suit different datasets. If you use datasets to set these parameters, don't then report results based on the same datasets. Instead, use a sample of the data to set the parameters for the algorithm, and then test on other datasets!

Rule 7: Compare your methods to the best alternative methods. Comparing your method to other methods is important, but which other methods you select is critical. If your comparison is to methods that are no longer considered leading methods, then the comparison is not helpful. Again, consider your objective (speed, memory usage, or accuracy), and pick methods that are best for your objective. If you cannot use a leading method for some reason, then state why you didn't use it, and modify your conclusions appropriately. Also, make sure you are using the current version for the method, and the best way of running the method (and, of course, provide full details for how you ran the method).

Rule 8: Make all your data available. Reproducibility is an important objective, and so making all the data you use available is key to this. Since simulated datasets can be large, many authors may prefer to simply provide commands for regenerating the data; however, software can change, and simulated datasets are not always exactly reproduced, even using the same commands. If random number generators are used, then make sure you provide the seeds you used, since otherwise the same datasets will not be generated. If you can, put your data into some public depository (with a DOI), rather than hosting them on your own machines, as too often datasets you think you have stored securely end up moved to another location, or deleted, and you won't be able to find them. In other words, do what you can to ensure that your data will be available in a semi-permanent way, and make the data easy to find.

Related to this, make sure that *how* you generated your data is completely described in your paper. This may be best accomplished by providing the commands and the software you used to generate the data.

Rule 9: Show your results visually in a way that is most helpful for understanding trends. How you display your results is also important. Sometimes, tables can make small differences seem big, so consider using figures instead. Show error bars, since overlapping error bars can suggest that differences are not statistically significant. Make your figures easy to understand (with informative x-axes and y-axes, and enough detail in the caption that the reader quickly understands the trends). Be careful with how you set the ranges for your x-axes and y-axes of different figures for the same questions, so that they can be compared to each other.

Rule 10: Compare to other studies. If you are working in an area where there is other literature, make sure you discuss the most important related papers. That comparison may include early work, but should also include the recent work on the topic. If you observe the same trends, say so; but if you find differently, then indicate this, and try to understand why there are differences. Sometimes the differences are due to different datasets with different properties, sometimes due to the choice of method, or sometimes due to how the method was run. It's also possible that the other study made a big deal out of something small, and so it's not that your data suggest something different from their data, but perhaps only that your conclusions are different! So, don't just read the conclusions in the other papers – look at their experiments carefully, and decide if you agree with their conclusions. Learning to be a careful reader is important, and essential to being a good researcher.

Discussion. These rules are very basic, but point to the difficulties in doing rigorous work in method evaluation that is also relevant to a real application, and which can be understood. Other rules might be even more important than these, so please don't think of these as exhaustive or more important than others you might think of.

Appendix D
Projects

D.1 Introduction

There are three types of projects in this collection: short projects, long projects, and projects that involve the development of novel methods. Each project requires data analysis, either on real or simulated data, and also writing. Therefore, even the short projects will require about a week for completion.

The main purpose of the short projects is to familiarize the student with the process of computing and interpreting alignments and trees on datasets. Because the data analysis part of these projects should be fast to complete, they are focused on relatively small nucleotide datasets. If the student has access to sufficient computational resources, then analyses of larger datasets or amino acid datasets are possible. Each short project also asks the student to explore the impact of method choice (i.e., alignment method or tree estimation method) or dataset on the resultant tree, typically using visualization tools.

The long projects build on the short projects, but do more exploration of the impact of method choice (for alignment estimation or tree estimation) or dataset on phylogeny estimation. Some of these projects examine scalability of methods to large datasets, and so will require substantial computational resources. As the student will learn, the degree to which the method selection impacts the final phylogeny can depend on the properties of the data, such as number of sequences, number of sites (i.e., sequence length), rate of evolution, percentage of missing data, etc. The use of both biological and simulated data will help the students evaluate the impact of the different factors on the final outcomes.

The projects aimed at novel method development are likely to be the most difficult, and success in these projects will probably require substantial effort beyond the period of the course. However, a student who wishes to do a novel method development project is usually best served by starting with a long project to identify the competing methods and select datasets that are best able to differentiate between methods.

Final projects for the course are typically long projects rather than novel method development projects, and are focused on comparisons of leading computational methods on simulated or biological datasets, with an eye toward assessing the relative performance of these methods, and gaining insight into the conditions that impact each method. Studies that provide such insights can be published in bioinformatics conferences and journals, as well as in biology journals focusing on phylogenetics and systematics.

Each of these projects, including the short projects, requires the use of external software for computing alignments, computing trees, and visualizing alignments and/or trees. The long projects also require external software for bootstrapping, computing error rates of estimated trees and alignments, and comparing trees to each other. These external tools are under rapid development, and the projects should be based on the current best methods for each part of the analysis. Therefore, this list of projects does not suggest specific software to try to "beat." Instead, the choice of method for each step should be based on the current research in the field.

Note: Before starting any project, whether long or short, read Appendix C. It is likely that not all the advice will be specifically relevant to your study, but much of the advice is generally relevant to any scientific study concerned with methods and their performance on data.

D.2 Short Projects

For each of these short projects, find a small sequence dataset, all for the same gene. Make it small enough that you can run your analyses without too much effort; 20 sequences is probably a good number (more is fine, but check the computational requirements for whatever you try to do).

1. *Objective: Learn how to read a computational paper carefully.* Find a paper that performs a phylogenetic analysis, and try to figure out exactly what they did. What program did they use to align the sequences? What version number, and what command was used? Did they perform any site masking (also called "trimming")? What program did they use to compute a tree? What version number, and what command was used? Are you able to find the data they analyzed? What about the alignments they computed? To what extent is the analysis repeatable?

2. *Objective: Learn how to run alignment and phylogenetic estimation software, and interpret the output.* Compute a multiple sequence alignment and phylogenetic tree on your dataset using any standard methods. Visualize the tree. If your dataset has an outgroup, you can root the tree at that outgroup (though this may not be an accurate way of rooting the tree in some conditions); otherwise be careful not to interpret the tree as rooted. If the tree has numbers on the branches, what do they mean? What does the tree suggest about your dataset?

3. *Objective: Compare different tree construction methods on the same dataset.* Compute a multiple sequence alignment on the dataset using any standard method. Now compute a UPGMA tree, a maximum parsimony tree, and a maximum likelihood tree on the alignment (any software package you like). Visualize the trees. What differences, if any, do you see?

4. *Objective: Compare different alignment estimation methods on the same dataset.* Compute two different multiple sequence alignments on the dataset using any standard methods. Now compute a tree on each alignment, using any approach you like. Visualize the trees. What differences, if any, do you see?
5. *Objective: Evaluate the impact of changing the sequence evolution model.* Compute a multiple sequence alignment on your dataset using any standard method. Now compute maximum likelihood trees on the alignment under the Jukes–Cantor model and also under the GTRGAMMA model. Visualize the trees. What differences, if any, do you see?
6. *Objective: Explore gene tree heterogeneity.* Find a multi-locus dataset (all for the same set of species), and compute trees on each locus with bootstrapping. Compare the trees to each other. Are they different? If so, are the edges in which they are different highly supported?
7. *Objective: Explore the impact of alignment error on site-based coalescent species tree methods.* Some methods for calculating species trees in the presence of incomplete lineage sorting are based on site patterns; examples include SVDquartets and SNAPP. For an input set of unaligned sequences for a collection of different genes, compute alignments using at least two methods, thus producing two different super-alignments. Now apply your site-based species tree method to the resultant two super-alignments. Do you get the same trees?

D.3 Longer Projects

Each of these longer projects has a stated purpose that explores the impact of method choice or dataset property on the final alignment and/or tree.

1. *Objective: Explore the choice of methods for estimating gene trees from unaligned sequences on the resultant gene trees.* Compare gene trees computed on a biological dataset with at least 50 unaligned sequences using at least two different techniques (i.e., you can vary the alignment method, or you can vary the tree estimation method, or you can vary both techniques).
    - Get branch support on the trees you compute.
    - Compare the gene trees, taking branch support into account. Where are they different? Are these differences interesting or important? What is your interpretation of these differences? If one method did particularly poorly, was there something about the data that was difficult for the method? What did you learn about the methods you used?
2. *Objective: Evaluate the impact of the species tree estimation method on the estimated species tree for multi-locus datasets with gene tree heterogeneity.* Compare species trees computed on a biological dataset with at least five genes and 10–100 species. It would be most interesting if you pick a dataset where gene tree heterogeneity has been observed or where it is expected.

- Compute gene sequence alignments and gene trees using reasonable methods. (If you are using a dataset from a published study, these may already be computed for you!)
- Compute species trees using at least two coalescent-based methods and one concatenation analysis. Unless you have access to substantial computational resources, try to select reasonably fast methods so that they each complete on your dataset within 24 hours and do not have high memory requirements.
- Compare the species trees that you obtain using different species tree estimation methods. Where are they different? Are these differences interesting or important? What is your interpretation of these differences? What does this tell you about the methods you used?

3. *Objective: Explore how modifying your input data affects species tree estimation.* Take a multi-locus dataset, and compute gene trees with bootstrap support. Now make versions of the dataset as follows. Suppose you only want to consider gene trees that have average bootstrap support on the branches above some threshold B (e.g., $B = 75$ percent or $B = 90$ percent). Delete all genes that have average bootstrap support below B. Now re-estimate the species trees from these new gene trees. What differences do you see? Are these differences interesting or important? What is your interpretation of these differences?

4. *Objective: Explore the impact of missing data on gene tree estimation using two-phase methods.* Find or create a small nucleotide sequence dataset for a single gene; call this M. Compute a multiple sequence alignment A on M using any standard method; then pick an arbitrary sequence x in the dataset, and delete the first 50 percent of the nucleotides in x. Now re-compute the alignment on this new dataset using the same method you used to produce A, and call this alignment A'. Compute trees on both A and A' using the same method (e.g., neighbor joining, maximum likelihood, parsimony, etc). What changes do you see? Vary the experiment to explore the impact of method choice in the presence of missing data, for example:

- Vary the alignment estimation method.
- Vary the tree estimation method.
- Vary the experiment by removing more nucleotides (varying from 50 percent to 90 percent) from x.
- Vary the experiment by modifying more sequences (up to 50 percent of the original dataset).

5. *Objective: Evaluate the impact of including random sequences in a dataset.* Find a small (at most 20 sequences) nucleotide sequence dataset for a single gene; call this dataset M. Pick a sequence x in M, and replace the DNA sequence by a random nucleotide sequence of the same length; call this new dataset M'. Construct trees on M and M' using the same protocol. What changes do you see? Remove x from both trees; do the trees (without x) look the same? Modify the experiment by changing the method you use for constructing a tree; does anything change?

6. *Objective: Evaluate techniques for distance-correction when datasets are "saturated."* When analyzing sequence datasets using distance-based methods such as neighbor joining, a corrected distance-matrix must be computed. Recall that the calculation for Jukes–Cantor distances implicitly assumes that the normalized Hamming distance between every two sequences is strictly less than 75 percent. Yet, for fast-evolving sequences, or for sequences spanning large evolutionary distances, this may not be the case. In fact, a dataset is said to be *saturated* when at least one pair of sequences has normalized Hamming distance that matches or exceeds the expected value for a random pair of sequences (which is 75 percent for DNA sequences). The question here is how to handle such datasets, and in particular how to correct distances so that phylogenies computed on these corrected distance matrices are as accurate as they can be. To do this project, you should find out how this situation is treated in general, and then think about whether you can handle it better. Evaluate multiple ways of correcting the distances, and also evaluate multiple ways of computing trees on the resultant distance matrices. Compare the resultant phylogenies to each other.

7. *Objective: Evaluate the impact of "masking" sites within multiple sequence alignments on phylogeny estimation.* Several techniques have been developed to identify sites that are noisy, and perhaps have substantial error, within multiple sequence alignments. After these sites are identified, they can be deleted from the alignment, thus producing a new alignment that contains a subset of the sites from the original alignment; this is called "masking." Early studies suggested that masking alignments would lead to improved phylogeny estimation, but were limited to very long alignments and small numbers of sequences. Evaluate the impact of masking multiple sequence alignments on phylogeny estimation, using larger numbers of taxa and/or single gene datasets.

8. *Objective: Compare heuristics for maximum likelihood estimation.* Maximum likelihood (ML) phylogeny estimation is an NP-hard problem for all the standard models of sequence evolution, such as Jukes–Cantor and GTR. How well the various heuristics for ML solve the optimization problem can impact the accuracy of the parameters they are estimating (e.g., gene tree topology, branch lengths, and substitution matrix) but also impacts the running time. Liu et al. (2012a) is an example of a study from 2012 comparing two ML heuristics, but the current best ML heuristics have probably changed since then. Using both simulated and biological datasets, explore these questions on the current leading ML methods for large datasets.

9. *Objective: Compare Bayesian methods to ML methods for phylogeny estimation in terms of the accuracy of point estimates of the tree.* Bayesian and ML phylogeny estimation methods are both likelihood-based and highly popular, yet little is known about the relative performance of these methods with respect to accuracy and computational requirements. Evaluate this on a collection of biological and simulated datasets. Note that since Bayesian methods produce a distribution on treespace rather than a point estimate of the tree, to use Bayesian methods to produce a point estimate you would need to summarize the distribution in some way.

10. *Objective: Explore the impact of model misspecification on gene tree estimation.* Create simulated data in which the sequences have evolved down a tree under a complex model (e.g., GTR+Gamma+I), and then estimate the tree under the correct model as well as under a simpler model (e.g., Jukes–Cantor). Compare the two estimations of the tree you obtain. How are they different? Is one tree more accurate than the other? Explore this under various simulation conditions, varying rates of evolution and numbers of sequences.

11. *Objective: Explore the impact of using inputs from Bayesian gene tree methods instead of ML on coalescent-based species trees estimation using summary methods.* Many methods have been developed to estimate the species tree from collections of gene trees; these are called "summary methods." The population tree from BUCKy was designed to work with inputs computed using Bayesian gene tree estimators, so that each gene was represented by a collection of gene trees produced by a Bayesian MCMC analysis; most other summary methods are designed for use with a single tree for each gene, typically computed using ML heuristics. Examine the impact of using a distribution of gene trees computed by a Bayesian MCMC analysis instead of a single ML tree on the point estimate produced by these summary methods for species tree estimation.

12. *Objective: Explore statistical coalescent-based species tree estimation methods on datasets with very few loci.* Statistical methods for species tree estimation from multiple loci can have good accuracy when there is a large number of loci, but little is known about how well they perform under conditions where the number of loci is small and there is substantial gene tree heterogeneity. Explore this problem using a combination of simulated and biological datasets. Also, compare results obtained by sampling a small number of loci from a larger set; does the species tree estimation error change as you sample fewer loci? Can you design good techniques for determining which loci to include?

13. *Objective: Explore statistical coalescent-based species tree estimation methods on datasets where each locus has very few sites.* Most studies of coalescent-based methods have been performed on datasets where each locus has a sufficient number of sites that the gene trees have reasonable, even if imperfect, accuracy. Yet some researchers have argued that analyses of coalescent-based methods should be based on very short genomic regions, in order to avoid intra-locus recombination, which violates the assumptions of the MSC (multi-species coalescent) model. Explore the impact of short loci on species tree estimation produced using different coalescent-based methods. See Chou et al. (2015) for an example of an early study of this issue.

14. *Objective: Revisit a controversial evolutionary biology question, using new methods.* Many questions in evolutionary biology are controversial; even how mammals evolved is still debated. Find a recent paper addressing one of these questions where they perform a phylogenetic analysis; then, redo the analysis using the best current methods for each step, and see if your conclusions are the same as the published paper.

D.4 Projects Involving Novel Method Development

The work you do for this project could lead to a publication; document everything you do so that it is reproducible, and save your data so that you can enable others to verify your results. Have fun!

D.4.1 Multiple Sequence Alignment and/or Gene Tree Estimation

1. *Objective: Design a new heuristic for ML gene tree estimation so it can run more effectively than current methods on datasets with thousands of sequences.* The estimation of ML gene trees from multiple sequence alignments containing thousands of sequences is computationally very intensive. Develop a new heuristic, and compare it to the current heuristics methods for ML. Note the computational effort (e.g., running time and peak memory usage) and likelihood scores. If you use simulated data, then also record the topological accuracy of the trees each method produces.
2. *Objective: Design methods to detect foreign (i.e., non-homologous) sequences in a dataset.* Multiple sequence alignment and phylogeny estimation methods assume that all the sequences in the input are homologous, which means that they share a common ancestor. Little is understood about the impact of the inclusion of non-homologous sequences in phylogenetic datasets. Evaluate the consequence of including non-homologous sequences in input datasets on the resultant phylogenies, and also develop a method for detecting the non-homologs so that they can be removed from the dataset.
3. *Objective: Design better divide-and-conquer methods for large-scale multiple sequence alignment.* PASTA and SATé use divide-and-conquer strategies in which a dataset of unaligned sequences is divided into subsets, the sequence subsets are then aligned using external alignment methods (e.g., MAFFT), and then the subset alignments are merged together using profile–profile alignment methods. Design your own such strategy, but vary the algorithmic design. For example, use HMM–HMM alignment instead of profile–profile alignment to merge subset alignments.

D.4.2 Species Tree and Supertree Estimation

1. *Objective: Design better methods for quartet tree amalgamation.* Many methods (e.g., BUCKy-pop, ASTRAL, and SVDquartets) for coalescent-based species tree estimation operate by estimating quartet trees (perhaps with weights on each quartet tree) and then combining the estimated quartet trees together. A standard optimization problem for this purpose is to find a tree whose total (weighted) quartet support is maximal. Develop a heuristic for this optimization problem and test it within a coalescent-based species tree estimation pipeline.
2. *Objective: Design supertree methods that can provide branch length and/or branch support on the supertree edges.* Most supertree methods do not provide branch lengths

or support values on the supertree edges, even though in many cases the input source trees have branch lengths or branch support values. Design a supertree method that can provide these values, and explore its performance on a collection of biological and simulated datasets. (See Binet et al. (2016) for an example of one method that *does* compute branch lengths.)

3. *Objective: Develop a new method to construct a tree from a distance matrix with missing values.* Many methods for estimating species trees operate by combining gene trees or smaller species trees estimated on subsets of the taxon set; some of these methods have two steps where the first step computes a dissimilarity matrix of estimated distances between the species, and the second step computes a tree from the dissimilarity matrix. The challenge is how to compute a tree from dissimilarity matrices that are incomplete, in that they are missing entries. Standard methods (e.g., neighbor joining and FastME) can be excellent at computing trees from complete matrices, but either do not run when the matrices are incomplete or have poor accuracy. There are some methods (Criscuolo and Gascuel, 2008) that can analyze incomplete matrices, but clearly new methods are needed. Develop a new approach, or modify existing methods, and test your method under various ways of calculating distance matrices that naturally create incomplete matrices. See, for example, the coalescent-based methods ASTRID (Vachaspati and Warnow, 2015) and NJst (Liu and Yu, 2011), and the ways these methods were evaluated.

References

Afrasiabi, C., Samad, B., Dineen, D., Meacham, C., and Sjölander, K. 2013. The PhyloFacts FAT-CAT web server: ortholog identification and function prediction using fast approximate tree classification. *Nucleic Acids Research*, **41**(W1), W242–W248.

Agarwala, R., and Fernández-Baca, D. 1994. A polynomial-time algorithm for the perfect phylogeny problem when the number of character states is fixed. *SIAM Journal on Computing*, **23**, 1216–1224.

Agarwala, R., and Fernández-Baca, D. 1996. Simple algorithms for perfect phylogeny and triangulating colored graphs. *International Journal of Foundations of Computer Science*, **7**, 11–21.

Agarwala, R., Bafna, V., Farach, M., Paterson, M., and Thorup, M. 1998. On the approximability of numerical taxonomy (fitting distances by tree metrics). *SIAM Journal on Computing*, **28**(3), 1073–1085.

Aho, A., Sagiv, Y., Szymanski, T., and Ullman, J. 1978. Inferring a tree from lowest common ancestors with an application to the optimization of relational expressions. Pages 54–63 of: *Proceedings of the 16th Annual Allerton Conference on Communication, Control, and Computing*.

Ailon, N., and Charikar, M. 2005. Fitting tree metrics: hierarchical clustering and phylogeny. Pages 73–82 of: *46th Annual IEEE Symposium on Foundations of Computer Science (FOCS'05)*. doi:10.1109/SFCS.2005.36.

Akaike, H. 1974. A new look at the statistical model identification. *IEEE Transactions on Automatic Control*, **19**, 716–723.

Aldous, D.J. 1991. The continuum random tree II: an overview. Pages 23–70 of: Barlow, M.T., and Bingham, N.H. (eds.), *Stochastic Analysis*. Cambridge: Cambridge University Press.

Aldous, D.J. 2001. Stochastic models and descriptive statistics for phylogenetic trees, from Yule to today. *Statistical Science*, **16**, 23–34.

Alfaro, M.E., and Holder, M.T. 2006. The posterior and the prior in Bayesian phylogenetics. *Annual Review of Ecology, Evolution, and Systematics*, 19–42.

Allison, L., Wallace, C.S., and Yee, C.N. 1992a. Finite-state models in the alignment of macromolecules. *Journal of Molecular Evolution*, **35**, 77–89.

Allison, L., Wallace, C.S., and Yee, C.N. 1992b. Minimum message length encoding, evolutionary trees and multiple alignment. Pages 663–674 of: *Hawaii International Conference Systems Science*, vol. 1.

Allman, E.S., Ané, C., and Rhodes, J. 2008. Identifiability of a Markovian model of molecular evolution with Gamma-distributed rates. *Advances in Applied Probability*, **40**(1), 229–249.

Allman, E.S., Degnan, J.H., and Rhodes, J.A. 2011. Identifying the rooted species tree from the distribution of unrooted gene trees under the coalescent. *Journal of Mathematical Biology*, **62**(6), 833–862.

Allman, E., Degnan, J.H., and Rhodes, J. 2016. Species tree inference from gene splits by unrooted STAR methods. *IEEE/ACM Transactions on Computational Biology and Bioinformatics*. doi: 10.1109/TCBB.2016.2604812.

Altenhoff, A.M., Boeckmann, B., Capella-Gutierrez, S., Dalquen, D.A., et al. 2016. Standardized benchmarking in the quest for orthologs. *Nature Methods*, **13**, 425–430.

Altschul, S.F. 1998. Generalized affine gap costs for protein sequence alignment. *Proteins: Structure, Function and Genomics*, **32**, 88–96.

Altschul, S.F., Gish, W., Miller, W., Myers, E.W., and Lipman, D.J. 1990. Basic local alignment search tool. *Journal of Molecular Biology*, **215**(3), 403–410.

Altschul, S.F., Madden, T.L., Schäffer, A.A., Zhang, J., Zhang, Z., Miller, W., and Lipman, D.J. 1997. Gapped BLAST and PSI-BLAST: a new generation of protein database search programs. *Nucleic Acids Research*, **25**, 3389–3402.

Amenta, N., and Klingner, J. 2002. Case study: Visualizing sets of evolutionary trees. Pages 71–74 of: *Proceedings of Information Visualization (INFOVIS) 2002*.

Amenta, N., Clarke, F., and St. John, K. 2003. A linear-time majority tree algorithm. Pages 216–227 of: Benson, G., and Page, R. (eds.), *Proceedings of the Workshop on Algorithms in Bioinformatics (WABI) 2003*. Berlin Heidelberg: Springer.

Amir, A., and Keselman, D. 1994. Maximum agreement subtree in a set of evolutionary trees: metrics and efficient algorithms. In: *Proceedings of the 35th IEEE Symposium on the Foundations of Computer Science (FOCS)*.

Ané, C. 2016. *PhyloNetworks: analysis for phylogenetic networks in Julia*. Software available online at *https://github.com/crsl4/PhyloNetworks.jl*.

Angiuoli, S.V., and Salzberg, S.L. 2011. Mugsy: fast multiple alignment of closely related whole genomes. *Bioinformatics*, 27(3): 334–342.

Anisimova, M., and Kosiol, C. 2009. Investigating protein-coding sequence evolution with probabilistic codon substitution models. *Molecular Biology and Evolution*, **26**(2), 255–271.

Atteson, K. 1999. The performance of neighbor-joining methods of phylogenetic reconstruction. *Algorithmica*, **25**, 251–278.

Avni, E., Cohen, R., and Snir, S. 2015. Weighted quartets phylogenetics. *Systematic Biology*, **64**(2), 233–242.

Ayling, S.C., and Brown, T.A. 2008. Novel methodology for construction and pruning of quasi-median networks. *BMC Bioinformatics*, **9**, 115.

Bader, D.A., Moret, B.M.E., and Yan, M. 2001. A linear-time algorithm for computing inversion distances between signed permutations with an experimental study. *Journal of Computational Biology*, **8**(5), 483–491.

Bafna, V., and Pevzner, P.A. 1996. Genome rearrangements and sorting by reversals. *SIAM Journal on Computing*, **25**(2), 272–289.

Bafna, V., and Pevzner, P.A. 1998. Sorting by transpositions. *SIAM Journal on Discrete Mathematics*, **11**(2), 224–240.

Bandelt, H.-J. 1994. Phylogenetic networks. *Verhandl Naturwiss Vereins Hamburg*, **34**, 51–71.

Bandelt, H.-J., and Dress, A.W.M. 1992. Split decomposition: a new and useful approach to phylogenetic analysis of distance data. *Molecular Phylogenetics and Evolution*, **1**, 242–252.

Bandelt, H.-J., and Dress, A.W.M. 1993. A relational approach to split decomposition. Pages 123–131 of: Optiz, O., Lausen, B., and Klar, R. (eds.), *Information and Classification*. Berlin: Springer.

Bandelt, H.-J., Macauley, V., and Richards, M. 2000. Median networks: speedy construction and greedy reduction, one simulation, and two case studies from human mtDNA. *Molecular Phylogenetics and Evolution*, **16**, 8–28.

Bansal, M.S., and Eulenstein, O. 2013. Algorithms for genome-scale phylogenetics using gene tree parsimony. *IEEE/ACM Transactions on Computational Biology and Bioinformatics*, **10**(4), 939–956.

Bansal, M., Burleigh, J.G., Eulenstein, O., and Fernández-Baca, D. 2010. Robinson–Foulds Supertrees. *Algorithms for Molecular Biology*, **5**, 18.

Barthélemy, J.-P., and McMorris, F.R. 1986. The median procedure for n-trees. *Journal of Classification*, **3**, 329–334.

Bastkowski, S., Moulton, V., Spillner, A., and Wu, T. 2016. The minimum evolution problem is hard: a link between tree inference and graph clustering problems. *Bioinformatics*, **32**(4), 518–522.

Bateman, A., Birney, E., Cerruti, L., Durbin, R., Etwiller, L., Eddy, S.R., Griffiths-Jones, S., Howe, K.L., Marshall, M., and Sonnhammer, E.L.L. 2002. The Pfam protein families database. *Nucleic Acids Research*, **30**, 276–280.

Baum, B.R., and Ragan, M.A. 2004. The MRP method. Pages 17–34 of: Bininda-Emonds, Olaf R.P. (ed.), *Phylogenetic Supertrees: Combining Information to Reveal the Tree of Life*. Dordrecht: Kluwer Academic.

Baum, L.E. 1972. An equality and associated maximization technique in statistical estimation for probabilistic functions of Markov processes. *Inequalities*, **3**, 1–8.

Bayzid, Md. S., and Warnow, T. 2013. Naive binning improves phylogenomic analyses. *Bioinformatics*, **29**(18), 2277–2284.

Bayzid, Md. S., Hunt, T., and Warnow, T. 2014. Disk covering methods improve phylogenomic analyses. *BMC Genomics*, **15**(Suppl 6), S7.

Bayzid, Md. S., Mirarab, S., and Warnow, T. 2013. Inferring optimal species trees under gene duplication and loss. Pages 250–261 of: *Pacific Symposium on Biocomputing*, vol. 18.

Bayzid, Md. S., Mirarab, S., Boussau, B., and Warnow, T. 2015. Weighted statistical binning: enabling statistically consistent genome-scale phylogenetic analyses. *PLoS One*, **10**(6), 30129183.

Berge, C. 1967. Some classes of perfect graphs. Pages 155–166 of: Harary, F. (ed.), *Graph Theory and Theoretical Physics*. New York: Academic Press.

Berger, M.P., and Munson, P.J. 1991. A novel randomized iterative strategy for aligning multiple protein sequences. *Computer Applications in the Biosciences: CABIOS*, **7**(4), 479–484.

Berger, S.A., and Stamatakis, A. 2011. Aligning short reads to reference alignments and trees. *Bioinformatics*, **27**(15), 2068–2075.

Berger, S.A., Krompass, D., and Stamatakis, A. 2011. Performance, accuracy, and web server for evolutionary placement of short sequence reads under maximum likelihood. *Systematic Biology*, **60**(3), 291–302.

Bernardes, J., Zaverucha, G., Vaquero, C., and Carbone, A. 2016. Improvement in protein domain identification is reached by breaking consensus, with the agreement of many profiles and domain co-occurrence. *PLoS Computational Biology*, **12**(7), e1005038.

Berry, A., and Pogorelcnik, R. 2011. A simple algorithm to generate the minimal separators and the maximal cliques of a chordal graph. *Information Processing Letters*, **111**(11), 508–511.

Berry, A., Pogorelcnik, R., and Simonet, G. 2010. An introduction to clique minimal separator decomposition. *Algorithms*, **3**(2), 197–215.

Berry, V., and Bryant, D. 1999. Faster reliable phylogenetic analysis. Page 59–68 of: *Proceedings 3rd Annual International Conference on Computational Molecular Biology (RECOMB)*.

Berry, V., and Gascuel, O. 1997. Inferring evolutionary trees with strong combinatorial evidence. Pages 111–123 of: *Proceedings 3rd Annual International Conference on Computing and Combinatorics (COCOON97)*. Berlin: Springer.

Berry, V., Jiang, T., Kearney, P., Li, M., and Wareham, T. 1999. Quartet cleaning: improved algorithms and simulations. Pages 313–324 of: *Proceedings European Symposium on Algorithms (ESA99)*. Berlin: Springer.

Berry, V., Bryant, D., Kearney, P., Li, M., Jiang, T., Wareham, T., and Zhang, H. 2000. A practical algorithm for recovering the best supported edges in an evolutionary tree. Pages 287–296 of *Proceedings ACM/SIAM Symposium on Discrete Algorithms (SODA)*.

Berthouly, C., Leroy, G., Van, T.N., et al. 2009. Genetic analysis of local Vietnamese chickens provides evidence of gene flow from wild to domestic populations. *BMC Genetics*, **10**(1), 1.

Billera, L.J., Holmes, S.P., and Vogtmann, K. 2001. Geometry of the space of phylogenetic trees. *Advances in Applied Mathematics*, **27**, 733–767.

Binet, M., Gascuel, O., Scornavacca, C., Douzery, E.J.P., and Pardi, F. 2016. Fast and accurate branch lengths estimation for phylogenomic trees. *BMC Bioinformatics*, **17**(1), 23.

Bishop, M.J., and Thompson, E.A. 1986. Maximum likelihood alignment of DNA sequences. *Journal of Molecular Biology*, **190**, 156–165.

Blackburne, B.P., and Whelan, S. 2013. Class of multiple sequence alignment algorithm affect genomic analysis. *Molecular Biology and Evolution*, **30**(3), 642–653.

Blair, J.R.S., and Peyton, B.W. 1993. An introduction to chordal graphs and clique trees. Pages 1–29 of: George A., et al. (eds.), *Graph Theory and Sparse Matrix Computation*. New York: Springer.

Blanchette, M., Bourque, G., and Sankoff, D. 1997. Breakpoint phylogenies. *Genome Informatics*, **8**, 25–34.

Blanchette, M., Kunisawa, T., and Sankoff, D. 1999. Gene order breakpoint evidence in animal mitochondrial phylogeny. *Journal of Molecular Evolution*, **49**(2), 193–203.

Blanchette, M., Kent, W.J., Riemer, C., Elnitski, L., Smit, A.F.A., Roskin, K.M., Baertsch, R., Rosenbloom, K., Clawson, H., Green, E.D., Haussler, D., and Miller, W. 2004. Aligning multiple genomic sequences with the threaded blockset aligner. *Genome Research*, **14**, 708–715.

Bloomquist, E.W., and Suchard, M.A. 2010. Unifying vertical and nonvertical evolution: a stochastic ARG-based framework. *Systematic Biology*, **59**(1), 27–41.

Blum, M.G.B., and Francois, O. 2006. Which random processes describe the Tree of Life? A large-scale study of phylogenetic tree imbalance. *Systematic Biology*, **55**, 685–691.

Bodlaender, H., Fellows, M., and Warnow, T. 1992. Two strikes against perfect phylogeny. Pages 273–283 of: *Proceedings 19th International Colloquium on Automata, Languages, and Programming (ICALP)*. Berlin: Springer.

Bogusz, M., and Whelan, S. 2017. Phylogenetic tree estimation with and without alignment: new distance methods and benchmarking. *Systematic Biology*, **66**(2), 218–231.

Bordewich, M., and Mihaescu, R. 2010. Accuracy guarantees for phylogeny reconstruction algorithms based on balanced minimum evolution. Pages 250–261 of: Moulton, V., and Singh, M. (eds.), *Proceedings of the 2010 Workshop on Algorithms for Bioinformatics*. Berlin Heidelberg: Springer.

Bordewich, M., and Semple, C. 2016. Determining phylogenetic networks from inter-taxa distances. *Journal of Mathematical Biology*, **73**(2), 283–303. doi:10.1007/s00285-015-0950-8.

Bordewich, M., and Tokac, N. 2016. An algorithm for reconstructing ultrametric tree-child networks from inter-taxa distances. *Discrete Applied Mathematics*, **213**, 47–59.

Bordewich, M., Gascuel, O., Huber, K.T., and Moulton, V. 2009. Consistency of topological moves based on the balanced minimum evolution principle of phylogenetic inference. *IEEE/ACM Transactions on Computational Biology and Bioinformatics (TCBB)*, **6**, 110–117.

Bouchard-Côté, A., and Jordan, M.I. 2013. Evolutionary inference via the Poisson Indel Process. *Proceedings of the National Academy of Sciences (USA)*, **110**(4), 1160–1166.

Boussau, B., and Gouy, M. 2006. Efficient likelihood computations with non-reversible models of evolution. *Systematic Biology*, **5**(55), 756–768.

Boussau, B., Szöllősi, G.J., Duret, L., Gouy, M., Tannier, E., and Daubin, V. 2013. Genome-scale co-estimation of species and gene trees. *Genome Research*, **23**, 323–330.

Bradley, R.K., Roberts, A., Smoot, M., Juvekar, S., Do, J., Dewey, C., Holmes, I., and Pachter, L. 2009. Fast statistical alignment. *PLoS Computational Biology*, **5**(5), e1000392.

Brandstetter, M.G., Danforth, B.N., Pitts, J.P., Faircloth, B.C., Ward, P.S., Buffington, M.L., Gates, M.W., Kula, R.R., and Brady, S.G. 2016. Phylogenomic insights into the evolution of stinging wasps and the origins of ants and bees. *Current Biology*, **27**(7), 1019–1025.

Bray, N., and Pachter, L. 2004. MAVID: constrained ancestral alignment of multiple sequences. *Genome Research*, **14**, 693–699.

Brinkmeyer, M., Griebel, T., and Böcker, S. 2011. Polynomial supertree methods revisited. *Advances in Bioinformatics*, **2011**. Article ID 524182.

Brown, D.G., and Truszkowski, J. 2012. Fast phylogenetic tree reconstruction using locality-sensitive hashing. Pages 14–29 of: *International Workshop on Algorithms in Bioinformatics*. Berlin Heidelberg: Springer.

Brown, D.P., Krishnamurthy, N., and Sjölander, K. 2007. Automated protein subfamily identification and classification. *PLoS Computational Biology*, **3**(8), 31–60.

Brown, M., Hughey, R., Krogh, A., Mian, I.S., Sjölander, K., and Haussler, D. 1993. Using Dirichlet mixture priors to derive hidden Markov models for protein families. Pages 47–55 of: *Proceedings of the First International Conference on Intelligent Systems for Molecular Biology (ISMB)*.

Brudno, M., Chapman, M., Gottgens, B., Batzoglou, S., and Morgenstern, B. 2003a. Fast and sensitive multiple alignment of long genomic sequences. *BMC Bioinformatics*, **4**:66.

Brudno, M., Do, C., Cooper, G., Kim, M., Davydov, E., NISC Comparative Sequencing Program, Green, E.D., Sidow, A., and Batzoglou, S. 2003b. LAGAN and Multi-LAGAN: efficient tools for large-scale multiple alignment of genomic DNA. *Genome Research*, **13**, 721–731.

Bruno, W.J., Socci, N.D., and Hapern, A.L. 2000. Weighted neighbor joining: a likelihood-based approach to distance-based phylogeny reconstruction. *Molecular Biology and Evolution*, **17**(1), 189–197.

Bryant, D. 1997. *Hunting for trees, building trees and comparing trees: theory and method in phylogenetic analysis*. Ph.D. thesis, University of Canterbury, Christchurch, New Zealand.

Bryant, D. 2003. A classification of consensus methods for phylogenetics. Pages 163–184 of: Janowitz, M.F., Lapointe, F.-J., McMorris, F.R., Mirkin, B., and Roberts, F.S. (eds.), *Bioconsensus*. Providence, RI: American Mathematical Society.

Bryant, D. 2005. On the uniqueness of the selection criterion of neighbor-joining. *Journal of Classification*, **22**, 3–15.

Bryant, D., and Moulton, V. 1999. A polynomial time algorithm for constructing the refined Buneman tree. *Applied Mathematics Letters*, **12**, 51–56.

Bryant, D., and Moulton, V. 2002. NeighborNet: an agglomerative method for the construction of planar phylogenetic networks. Pages 375–391 of: *Proceeding of Workshop on Algorithms for Bioinformatics*. Berlin/Heidelberg: Springer.

Bryant, D., and Steel, M.A. 2001. Constructing optimal trees from quartets. *Journal of Algorithms*, **38**, 237–259.

Bryant, D., and Steel, M.A. 2009. Computing the distribution of a tree metric. *IEEE/ACM Transactions on Computational Biology and Bioinformatics*, **6**(3), 420–426.

Bryant, D., and Waddell, P. 1998. Rapid evaluation of least-squares and minimum-evolution criteria on phylogenetic trees. *Molecular Biology and Evolution*, **15**, 1346–1359.

Bryant, D., Bouckaert, R., Felsenstein, J., Rosenberg, N.A., and RoyChoudhury, A. 2012. Inferring species trees directly from biallelic genetic markers: bypassing gene trees in a full coalescent analysis. *Molecular Biology and Evolution*, **29**(8), 1917–1932.

Bulteau, L., Fertin, G., and Rusu, I. 2011. Sorting by transpositions is difficult. In: *Proceedings of the 38th International Colloquium on Automata, Languages, and Programming (ICALP 2011)*. Berlin: Springer.

Buneman, P. 1971. The recovery of trees from measures of dissimilarity. Pages 387–395 of: Hodson, F., Kendall, D., and Tautu, P. (eds.), *Mathematics in the Archaeological and Historical Sciences*. Edinburgh: Edinburgh University Press.

Buneman, P. 1974a. A characterization of rigid circuit graphs. *Discrete Mathematics*, **9**(3), 205–212.

Buneman, P. 1974b. A note on the metric properties of trees. *Journal of Combinatorial Theory (B)*, **17**, 48–50.

Burleigh, J.G., Eulenstein, O., Fernández-Baca, D., and Sanderson, M.J. 2004. MRF supertrees. Pages 65–86 of: Bininda-Emonds, O.R.P. (ed.), *Phylogenetic Supertrees: Combining Information to Reveal the Tree of Life*. Dordrecht: Kluwer Academic.

Burleigh, J.G., Bansal, M.S., Eulenstein, O., Hartmann, S., Wehe, A., and Vision, T.J. 2011. Genome-scale phylogenetics: inferring the plant tree of life from 18,896 gene trees. *Systematic Biology*, **60**(2), 117–125.

Cannone, J.J., Subramanian, S., Schnare, M.N., Collett, J.R., D'Souza, L.M., Du, Y., Feng, B., Lin, N., Madabusi, L.V., Muller, K.M., Pande, N., Shang, Z., Yu, N., and Gutell, R.R. 2002. The Comparative RNA Web (CRW) site: an online database of comparative sequence and structure information for ribosomal, intron and other RNAs. *BMC Bioinformatics*, **3**(15).

Capella-Gutiérrez, S., and Galbadón, T. 2013. Measuring guide-tree dependency of inferred gaps for progressive aligners. *Bioinformatics*, **29**(8), 1011–1017.

Cardona, G., Llabrés, M., Roselló, F., and Valiente, G. 2009. On Nakhleh's metric for reduced phylogenetic networks. *IEEE/ACM Transactions on Computational Biology and Bioinformatics*, **6**(4), 629–638.

Cartwright, Reed A. 2006. Logarithmic gap costs decrease alignment accuracy. *BMC Bioinformatics*, **7**(527).

Cavender, J.A. 1978. Taxonomy with confidence. *Mathematical Biosciences*, **40**, 271–280.

Chan, C.X., and Ragan, M.A. 2013. Next-generation phylogenomics. *Biology Direct*, **8**(3).

Chan, C.X., Bernard, G., Poirion, O., Hogan, J.M., and Ragan, M.A. 2014. Inferring phylogenies of evolving sequences without multiple sequence alignment. *Scientific Reports*, **4**: article number 6504.

Chang, J.T. 1996. Full reconstruction of Markov models on evolutionary trees: identifiability and consistency. *Mathematical Biosciences*, **137**, 51–73.

Chang, M., and Benner, S. 2004. Empirical analysis of protein insertions and deletions: determining parameters for the correct placement of gaps in protein sequence alignments. *Journal of Molecular Biology*, **341**, 671–631.

Chaudhary, R. 2015. MulRF: a software package for phylogenetic analysis using multi-copy gene trees. *Bioinformatics*, **31**, 432–433.

Chaudhary, R., Bansal, M.S., Wehe, A., Fernández-Baca, D., and Eulenstein, O. 2010. iGTP: a software package for large-scale gene tree parsimony analysis. *BMC Bioinformatics*, **11**, 574.

Chaudhary, R., Burleigh, J.G., and Fernández-Baca, D. 2012. Fast local search for unrooted Robinson–Foulds supertrees. *IEEE/ACM Transactions on Computational Biology and Bioinformatics*, **9**, 1004–1013.

Chen, D., Diao, L., Eulenstein, O., Fernández-Baca, D., and Sanderson, M.J. 2003. Flipping: a supertree construction method. Pages 135–160 of: *Bioconsensus*. Providence, RI: American Mathematical Society.

Chen, D., Eulenstein, O., Fernández-Baca, D., and Burleigh, J.G. 2006. Improved heuristics for minimum-flip supertree construction. *Evolutionary Bioinformatics*, **2**, 401–410.

Chen, M.-H., Kuo, L., and Lewis, P.O. (eds.). 2014. *Bayesian Phylogenetics: Methods, Algorithms, and Applications*. Boca Raton, FL CRC Press.

Chifman, J., and Kubatko, L. 2014. Quartet inference from SNP data under the coalescent. *Bioinformatics*, **30**(23), 3317–3324.

Chifman, J., and Kubatko, L. 2015. Identifiability of the unrooted species tree topology under the coalescent model with time-reversible substitution processes, site-specific rate variation, and invariable sites. *Journal of Theoretical Biology*, **374**, 35–47.

Chou, J., Gupta, A., Yaduvanshi, S., Davidson, R., Nute, M., Mirarab, S., and Warnow, T. 2015. A comparative study of SVDquartets and other coalescent-based species tree estimation methods. *BMC Genomics*, **16**(Suppl 10), S2.

Collingridge, P., and Kelly, S. 2012. MergeAlign: improving multiple sequence alignment performance by dynamic reconstruction of consensus multiple sequence alignments. *BMC Bioinformatics*, **13**: 117.

Cotton, J.A., and Wilkinson, M. 2007. Majority rule supertrees. *Systematic Biology*, **56**(3), 445–452.

Cotton, J.A., and Wilkinson, M. 2009. Supertrees join the mainstream of phylogenetics. *Trends in Ecology & Evolution*, **24**, 1–3.

Criscuolo, A., and Gascuel, O. 2008. Fast NJ-like algorithms to deal with incomplete distance matrices. *BMC Bioinformatics*, **9**(166).

Criscuolo, A., Berry, V., Douzery, E., and Gascuel, O. 2006. SDM: a fast distance-based approach for (super) tree building in phylogenomics. *Systematic Biology*, **55**, 740–755.

Cummings, M.P., Otto, S.P., and Wakeley, J. 1995. Sampling properties of DNA sequence data in phylogenetic analysis. *Molecular Biology and Evolution*, **12**, 814–822.

Darling, A.C.E., Mau, B., Blatter, F.R., and Perna, N.T. 2004. Mauve: multiple alignment of conserved genomic sequence with rearrangements. *Genome Research*, **14**, 1394–1403.

Darling, A.C.E., Mau, B, and Perna, N.T. 2010. progressiveMauve: multiple genome alignment with gene gain, loss and rearrangement. *PLoS One*, **5**(6), e11147.

Darriba, D., Taboada, G.L., Doallo, R., and Posada, D. 2011. ProtTest 3: fast selection of best-fit models of protein evolution. *Bioinformatics*, **27**(8), 1164–1165.

Darriba, D., Taboada, G.L., Doallo, R., and Posada, D. 2012. jModelTest 2: more models, new heuristics and parallel computing. *Nature Methods*, **9**(8), 772.

Dasarathy, G., Nowak, R., and Roch, S. 2015. Data requirement for phylogenetic inference from multiple loci: a new distance method. *IEEE/ACM Transactions on Computational Biology and Bioinformatics*, **12**(2), 422–432.

Daskalakis, C., and Roch, S. 2010. Alignment-free phylogenetic reconstruction: sample complexity via a branching process analysis. *The Annals of Applied Probability*, **23**(2), 693–721.

Daskalakis, C., and Roch, S. 2016. Species trees from gene trees despite a high rate of lateral genetic transfer: a tight bound. Pages 1621–1630: *Proceedings of the 27th Annual ACM-SIAM Symposium on Discrete Algorithms (SODA)*.

Daskalakis, C., Mossel, E., and Roch, S. 2006. Optimal phylogenetic reconstruction. Pages 159–168 of: *STOC06: Proceedings of the 38th Annual ACM Symposium on Theory of Computing*.

Davidson, R., Vachaspati, P., Mirarab, S., and Warnow, T. 2015. Phylogenomic species tree estimation in the presence of incomplete lineage sorting and horizontal gene transfer. *BMC Genomics*, **16**(Suppl 10), S1.

Day, W.H.E. 1987. Computational complexity of inferring phylogenies from dissimilarity matrices. *Bulletin of Mathematical Biology*, **49**, 461–467.

Day, W.H.E., and Sankoff, D. 1986. Computational complexity of inferring phylogenies by compatibility. *Systematic Biology*, **35**(2), 224–229.

Dayhoff, M.O., Schwartz, R.M., and Orcutt, B.C. 1978. A model of evolutionary change in proteins. Pages 345–352 of: *Atlas of Protein Sequence and Structure*, vol. 5. Washington, DC: National Biomedical Research Foundation.

De Maio, N., Holmes, I., Schlötterer, C., and Kosiol, C. 2013. Estimating empirical codon hidden Markov models. *Molecular Biology and Evolution*, **30**(3), 725–736.

De Maio, N., Schrempf, D., and Kosiol, C. 2015. PoMo: an allele frequency-based approach for species tree estimation. *Systematic Biology*, **64**(6), 1018–1031.

De Oliveira Martins, L., Mallo, D., and Posada, D. 2016. A Bayesian supertree model for genome-wide species tree reconstruction. *Systematic Biology*, **65**(3), 397–416.

DeBlasio, D., and Kececioglu, J. 2015. Learning parameter-advising sets for multiple sequence alignment. *IEEE/ACM Transactions on Computational Biology and Bioinformatics*, **PP**(99), 1.

DeGiorgio, M.I., and Degnan, J.H. 2010. Fast and consistent estimation of species trees using supermatrix rooted triples. *Molecular Biology and Evolution*, **27**(3), 552–569.

DeGiorgio, M., and Degnan, J. H. 2014. Robustness to divergence time underestimation when inferring species trees from estimated gene trees. *Systematic Biology*, **63**(1), 66–82.

Degnan, J.H. 2013. Anomalous unrooted gene trees. *Systematic Biology*, **62**(4), 574–590.

Degnan, J.H., and Rosenberg, N.A. 2006. Discordance of species trees with their most likely gene trees. *PLoS Genetics*, **2**, 762–768.

Degnan, J.H., DeGiorgio, M., Bryant, D., and Rosenberg, N.A. 2009. Properties of consensus methods for inferring species trees from gene trees. *Systematic Biology*, **58**, 35–54.

Della Vedova, G., Jiang, T., Li, J., and Wen, J. 2002. Approximating minimum quartet inconsistency. Pages 894–895 of: *Proceedings ACM/SIAM Symposium on Discrete Algorithms (SODA) 2002*.

Deng, X., and Cheng, J. 2011. MSACompro: protein multiple sequence alignment using predicted secondary structure, solvent accessibility, and residue–residue contacts. *BMC Bioinformatics*, **12**, 472.

DeSantis, T.Z., Hugenholtz, P., Keller, K., Brodie, E.L., Larsen, N., Piceno, Y.M., Phan, R., and Anderson, G.L. 2006. NAST: a multiple sequence alignment server for comparative analysis of 16S rRNA genes. *Nucleic Acids Research*, **34**, W394–399.

Desper, R., and Gascuel, O. 2002. Fast and accurate phylogeny reconstruction algorithm based on the minimum-evolution principle. *Journal of Computational Biology*, **9**, 687–705.

Desper, R., and Gascuel, O. 2004. Theoretical foundations of the balanced minimum evolution method of phylogenetic inference and its relationship to weighted least-squares tree fitting. *Molecular Biology and Evolution*, **21**(3), 587–598.

Desper, R., and Gascuel, O. 2005. The minimum-evolution distance-based approach to phylogeny inference. Pages 1–32 of: Gascuel, O. (ed.), *Mathematics of Evolution and Phylogeny*. Oxford: Oxford University Press.

Devroye, L., and Györfi, L. 1990. No empirical probability measure can converge in the total variation sense for all distributions. *The Annals of Statistics*, **18**(3), 1496–1499.

Dewey, C.N. 2007. Aligning multiple whole genomes with Mercator and MAVID. *Methods in Molecular Biology*, **395**, 221–236.

Dirac, G.A. 1974. On rigid circuit graphs. *Abhandlungen aus dem Mathematischen Seminar der Univiversität Hamburg*, **25**, 71–76.

Do, C.B., Mahabhashyam, M.S.P., Brudno, M., and Batzoglou, S. 2005. ProbCons: probabilistic consistency-based multiple sequence alignment. *Genome Research*, **15**(2), 330–340.

Dobzhansky, T. 1973. Nothing in biology makes sense except in the light of evolution. *American Biology Teacher*, **35**, 125–129.

Domelevo-Entfellner, J.-B., and Gascuel, O. 2008. Une approche phylo-HMM pour la recherche de séquences. Pages 133–139 of: van Helden, J., and Moreau, Y. (eds.), *JOBIM'08: Journées Ouvertes Biologie, Informatique, Mathématiques*, vol. 408.

Dotsenko, Y., Coarfa, C., Mellor-Crummey, J., Nakhleh, L., and Roshan, U. 2006. PRec-I-DCM3: a parallel framework for fast and accurate large scale phylogeny reconstruction. *International Journal on Bioinformatics Research and Applications (IJBRA)*, **2**, 407–419.

Douady, C.J., Delsuc, F., Boucher, Y., Doolittle, W.F., and Douzery, E.J.P. 2003. Comparison of Bayesian and maximum likelihood bootstrap measures of phylogenetic reliability. *Molecular Biology and Evolution*, **20**(2), 248–254.

Doyon, J.-P., Ranwez, V., Daubin, V., and Berry, V. 2011. Models, algorithms and programs for phylogeny reconciliation. *Briefings in Bioinformatics*, **12**(5), 392–400.

Dress, A., and Steel, M.A. 1992. Convex tree realizations of partitions. *Applied Mathematics Letters*, **5**(3), 3–6.

Du, S.Z., Zhang, Y., and Feng, Q. 1991. On better heuristic for Euclidian Steiner minimum trees. Pages 431–439 of: *Proceedings of the 32nd Symposium on the Foundations of Computer Science (FOCS)*.

Du, Z., Stamatakis, A., Lin, F., Roshan, U., and Nakhleh, L. 2005a. Parallel divide-and-conquer phylogeny reconstruction by maximum likelihood. Pages 776–785 of: *International Conference on High Performance Computing and Communications*. Berlin: Springer.

Du, Z., Lin, F., and Roshan, U. 2005b. Reconstruction of large phylogenetic trees: a parallel approach. *Computational Biology and Chemistry*, **29**(4), 273–280.

Dubchak, I., Poliakov, A., Kislyuk, A., and Brudno, M. 2009. Multiple whole-genome alignments without a reference organism. *Genome Research*, **19**, 682–689.

Durbin, R., Eddy, S., Krogh, A., and Mitchison, G. 1998. *Biological Sequence Analysis: Probabilistic Models of Proteins and Nucleic Acids*. Cambridge: Cambridge University Press.

Durrett, R. 2008. *Probability Models for DNA Sequence Evolution*. Second edition. New York: Springer.

Earl, D., Nguyen, N., Hickey, G., Harris, R.S., Fitzgerald, S., Beal, K., Seledtsov, I., Molodtsov, V., Raney, B.J., Clawson, H., Kim, J., Kemena, C., Chang, J.-M., Erb, I., Poliakov, A., Hou, M., Herrero, J., Kent, W. James, Solovyev, V., Darling, A.E., Ma, J., Notredame, C., Brudno, M., Dubchak, I., Haussler, D., and Paten, B. 2014. Alignathon: a competitive assessment of whole-genome alignment methods. *Genome Research*, **24**, 2077–2089.

Eddy, S.R. 2009. A new generation of homology search tools based on probabilistic inference. *Genome Informatics*, **23**, 205–211.

Edgar, R.C. 2004a. MUSCLE: a multiple sequence alignment method with reduced time and space complexity. *BMC Bioinformatics*, **5**(113), 113.

Edgar, R.C. 2004b. MUSCLE: a multiple sequence alignment with high accuracy and high throughput. *Nucleic Acids Research*, **32**(5), 1792–1797.

Edgar, R.C., and Sjölander, K. 2003. COACH: profile–profile alignment of protein families using hidden Markov models. *Bioinformatics*, **20**(8), 1309–1318.

Edgar, R.C., and Sjölander, K. 2004. A comparison of scoring functions for protein sequence profile alignment. *Bioinformatics*, **20**(8), 1301–1308.

Efron, B., Halloran, E., and Holmes, S. 1996. Bootstrap confidence levels for phylogenetic trees. *Proceedings of the National Academy of Science (USA)*, **93**, 13429–13434.

El-Mabrouk, N., and Sankoff, D. 2012. Analysis of gene order evolution beyond single-copy genes. Pages 397–429 of: Anisimova, M. (ed.), *Evolutionary Genomics: Statistical and Computational Methods*, vol. 1. Berlin: Springer.

Erdös, P.L., Steel, M.A., Székely, L., and Warnow, T. 1997. Local quartet splits of a binary tree infer all quartet splits via one dyadic inference rule. *Computers and Artificial Intelligence*, **16**(2), 217–227.

Erdös, P.L., Steel, M.A., Székely, L., and Warnow, T. 1999a. A few logs suffice to build (almost) all trees (I). *Random Structures and Algorithms*, **14**, 153–184.

Erdös, P.L., Steel, M.A., Székely, L., and Warnow, T. 1999b. A few logs suffice to build (almost) all trees (II). *Theoretical Computer Science*, **221**, 77–118.

Estabrook, G., Johnson, C., and McMorris, F. 1975. An idealized concept of the true cladistic character. *Mathematical Biosciences*, **23**, 263–272.

Evans, S.N., and Warnow, T. 2005. Unidentifiable divergence times in rates-across-sites models. *IEEE/ACM Transactions on Computational Biology and Bioinformatics*, **1**, 130–134.

Ewens, W.J. 2000. *Mathematical Population Genetics: I. Theoretical Introduction*. Second edition. Berlin: Springer.

Ewens, W.J., and Grant, G.R. 2001. *Statistical Methods in Bioinformatics: An Introduction*. New York: Springer.

Ezawa, K. 2016. General continuous-time Markov model of sequence evolution via insertions/deletions: are alignment probabilities factorable? *BMC Bioinformatics*, **17**(1), 304. erratur, 17, 457.

Faghih, F., and Brown, D.G. 2010. *Answer set programming or hypercleaning: where does the magic lie in solving maximum quartet consistency*. Technical Report, University of Waterloo, CS-2010-20.

Fakcharoenphol, J., Rao, S., and Talwar, K. 2003. A tight bound on approximating arbitrary metrics by tree metrics. Pages 448–455 of: *Proceedings 35th Annual ACM Symposium on the Theory of Computing (STOC)*. New York ACM Press.

Farach, M., and Thorup, M. 1994. Fast comparison of evolutionary trees. Pages 481–488 of: *Proceedings 5th Annual ACM-SIAM Symposium on Discrete Algorithms*.

Farris, J.S. 1973. A probability model for inferring evolutionary trees. *Systematic Zoology*, **22**, 250–256.

Felsenstein, J. 1978. Cases in which parsimony and compatibility methods will be positively misleading. *Systematic Zoology*, **27**, 401–410.

Felsenstein, J. 1981. Evolutionary trees from DNA sequences: a maximum likelihood approach. *Journal of Molecular Evolution*, **17**(6), 368–376.

Felsenstein, J. 1985. Confidence limits on phylogenies: an approach using the bootstrap. *Evolution*, **39**(4), 783–791.

Felsenstein, J. 2004. *Inferring Phylogenies*. Sunderland, MA: Sinauer Associates.

Felsenstein, J., and Churchill, G.A. 1996. A hidden Markov model approach to variation among sites in rates of evolution. *Molecular Biology and Evolution*, **13**, 93–104.

Feng, D.-F., and Doolittle, R. 1987. Progressive sequence alignment as a prerequisite to correct phylogenetic trees. *Journal of Molecular Evolution*, **25**, 351–360.

Fernández-Baca, D. 2000. The perfect phylogeny problem. Pages 203–234 of: Du, D.-Z., and Cheng, X. (eds.), *Steiner Trees in Industries*. Dordrecht: Kluwer Academic.

Finden, C.R., and Gordon, A.D. 1995. Obtaining common pruned trees. *Journal of Classification*, **2**, 255–276.

Finn, R.D., Clements, J., and Eddy, S.R. 2011. HMMER web server: interactive sequence similarity searching. *Nucleic Acids Research*, **39**, W29–W37.

Fiorini, S., and Joret, G. 2012. Approximating the balanced minimum evolution problem. *Operations Research Letters*, **40**, 31–55.

Fisher, R.A. 1922. On the dominance ratio. *Proceedings of the Royal Society of Edinburgh*, **42**, 321–341.

Fitch, W.M. 1971. Toward defining the course of evolution: minimal change for a specific tree topology. *Systematic Zoology*, **20**, 406–416.

Fitch, W.M., and Margoliash, E. 1967. Construction of phylogenetic trees. *Science*, **155**, 279–284.

Fletcher, W., and Yang, Z. 2009. Indelible: a flexible simulator of biological sequence evolution. *Molecular Biology and Evolution*, **26**(8), 1879–1888.

Fletcher, W., and Yang, Z. 2010. The effect of insertions, deletions, and alignment errors on the branch-site test of positive selection. *Molecular Biology and Evolution*, **27**(10), 2257–2267.

Foulds, L.R., and Graham, R.L. 1982. The Steiner problem in phylogeny is NP-complete. *Advances in Applied Mathematics*, **3**, 43–49.

Francis, A.R., and Steel, M.A. 2015. Tree-like reticulation networks: when do tree-like distances also support reticulation evolution? *Mathematical Biosciences*, **259**, 12–19.

Frati, F., Simon, C., Sullivan, J., and Swofford, D.L. 1997. Evolution of the mitochondrial COII gene in Collembola. *Journal of Molecular Evolution*, **44**, 145–158.

Frith, M.C., and Kawaguchi, R. 2015. Split-alignment of genomes finds orthologies more accurately. *Genome Biology*, **16**, 106.

Fu, Z., Chen, X., Vacic, V., Nan, P., Zhong, Y., and Jiang, T. 2007. MSOAR: a high-throughput ortholog assignment system based on genome rearrangement. *Journal of Computational Biology*, **14**(9), 1160–1175.

Gaither, J., and Kubatko, L. 2016. Hypothesis tests for phylogenetic quartets, with applications to coalescent-based species tree inference. *Journal of Theoretical Biology*, **408**, 179–186.

Ganapathy, G., and Warnow, T. 2001. Finding the maximum compatible tree for a bounded number of trees with bounded degree is solvable in polynomial time. Pages 156–163 of: *Proceedings of the First International Workshop on Algorithms and Bioinformatics (WABI)*. Berlin: Springer.

Ganapathy, G., and Warnow, T. 2002. Approximating the complement of the maximum compatible subset of leaves of k trees. Pages 122–134 of: *Proceedings of the Fifth International Workshop on Approximation Algorithms for Combinatorial Optimization*.

Ganapathy, G., Ramachandran, V., and Warnow, T. 2003. Better hill-climbing seaches for parsimony. Pages 245–258 of: *Proceedings of the Third International Workshop on Algorithms in Bioinformatics (WABI)*.

Ganapathy, G., Ramachandran, V., and Warnow, T. 2004. On contract-and-refine-transformations between phylogenetic trees. Pages 893–902 of: *ACM/SIAM Symposium on Discrete Algorithms (SODA'04)*.

Gardner, D.P., Xu, W., Miranker, D.P., Ozer, S., Cannonne, J., and Gutell, R. 2012. An accurate scalable template-based alignment algorithm. Pages 237–243 of: *Proceedings International Conference on Bioinformatics and Biomedicine, 2012*.

Gascuel, O. 1997. Concerning the NJ algorithm and its unweighted version, UNJ. Pages 149–170 of: Roberts, F.S., and Rzhetsky, A. (eds.), *Mathematical Hierarchies and Biology*. Providence, RI American Mathematical Society.

Gascuel, O. 2000. On the optimization principle in phylogenetic analysis and the minimum-evolution criterion. *Molecular Biology and Evolution*, **17**(3), 401–405.

Gascuel, O., and Steel, M.A. 2006. Neighbor-joining revealed. *Molecular Biology and Evolution*, **23**(11), 1997–2000.

Gascuel, O., and Steel, M.A. 2016. A "stochastic safety radius" for distance-based tree reconstruction. *Algorithmica*, **74**(4), 1386–1403.

Gascuel, O., Bryant, D., and Denis, F. 2001. Strengths and limitations of the minimum evolution principle. *Systematic Biology*, **50**(5), 621–627.

Gatesy, J.P., and Springer, M.S. 2014. Phylogenetic analysis at deep timescales: unreliable gene trees, bypassed hidden support, and the coalescence/concatalescence conundrum. *Molecular Phylogenetics and Evolution*, **80**, 231–266.

Gatesy, J., Meredith, R.W., Janecka, J.E., Simmons, M.P., Murphy, W.J., and Springer, M.S. 2016. Resolution of a concatenation/coalescence kertuffle: partitioned coalescence support and a robust family-level tree for Mammalia. *Cladistics*. DOI: 10.1111/cla.12170.

Gavril, F. 1972. Algorithms for minimum coloring, maximum clique, minimum covering by cliques, and maximum independent set of a chordal graph. *SIAM Journal on Computing*, **1**(2), 180–187.

Gavril, F. 1974. The intersection graphs of subtrees in trees are exactly the chordal graphs. *Journal of Combinatorial Theory (B)*, **16**, 47–56.

Ge, F., Wang, L.-S., and Kim, J. 2005. The cobweb of life revealed by genome-scale estimates of horizontal gene transfer. *PLoS Biology*, **3**(10), e316.

Gerstein, M.B., Bruce, C., Rozowsky, J.S., Zheng, D., Du, J., Korbel, J.O., Emanuelsson, O., Zhang, Z.D., Weissman, S., and Snyder, M. 2007. What is a gene, post-ENCODE? History and updated definition. *Genome Research*, **17**(6), 669–681.

Giarla, T.C., and Esselstyn, J.A. 2015. The challenges of resolving a rapid, recent radiation: empirical and simulated phylogenomics of Philippine shrews. *Systematic Biology*, **64**(5), 727–740.

Gogarten, J.P., and Townsend, J.P. 2005. Horizontal gene transfer, genome innovation and evolution. *Nature Reviews Microbiology*, **3**(9), 679–687.

Gogarten, J.P., Doolittle, W.F., and Lawrence, J.G. 2002. Prokaryotic evolution in light of gene transfer. *Molecular Biology and Evolution*, **19**(12), 2226–2238.

Goldman, N., and Yang, Z. 1994. A codon-based model of nucleotide substitution for protein-coding DNA sequences. *Molecular Biology and Evolution*, **11**, 725–736.

Goloboff, P.A. 1999. Analyzing large data sets in reasonable times: solution for composite optima. *Cladistics*, **15**, 415–428.

Golumbic, M. 2004. *Algorithmic Graph Theory and Perfect Graphs*. Second edition. New York: Elsevier.

Gordon, A.D. 1986. Consensus supertrees: the synthesis of rooted trees containing overlapping sets of labeled leaves. *Journal of Classification*, **3**, 335–348.

Gordon, K., Ford, E., and St. John, K. 2013. Hamiltonian walks of phylogenetic treespaces. *IEEE/ACM Transactions on Computational Biology and Bioinformatics*, **10**(4), 1076–1079.

Gori, K., Suchan, T., Alvarez, N., Goldman, N., and Dessimoz, C. 2016. Clustering genes of common evolutionary history. *Molecular Biology and Evolution*, **33**(6): 1590–1605.

Gotoh, O. 1990. Consistency of optimal sequence alignments. *Bulletin of Mathematical Biology*, **52**, 509–525.

Gotoh, O. 1994. Further improvement in methods of group-to-group sequence alignment with generalized profile operations. *Computer Applications in the Biosciences: CABIOS*, **10**, 379–387.

Gould, S.J., Raup, D.M., Spekowski, J.J., and Schopf, T.J.M. 1997. The shape of evolution: a comparison of real and random clades. *Paleobiology*, **3**, 23–40.

Grass Phylogeny Working Group. 2001. Phylogeny and subfamilial classification of the grasses (Poacea). *Annals of the Missouri Botanical Garden*, **88**(3), 373–457.

Grauer, D. 2016. *Molecular and Genome Evolution*. Sunderland, MA: Sinauer Associates.

Grauer, D., and Li, W.-H. 2000. *Fundamentals of Molecular Evolution*. Sunderland, MA: Sinauer Associates.

Graybeal, A. 1998. Is it better to add taxa or characters to a difficult phylogenetic problem? *Systematic Biology*, **47**(1), 9–17.

Gribskov, M., McLachlan, A.D., and Eisenberg, D. 1987. Profile analysis: detection of distantly related proteins. *Proceedings of the National Academy of Sciences (USA)*, **84**(13), 4355–4358.

Gu, X., and Li, W.H. 1995. The size distribution of insertions and deletions in human and rodent pseudogenes suggests the logarithmic gap penalty for sequence alignment. *Journal of Molecular Evolution*, **40**, 464–473.

Guindon, S., and Gascuel, O. 2003. A simple, fast, and accurate algorithm to estimate large phylogenies by maximum likelihood. *Systematic Biology*, **52**(5), 696–704.

Guo, S., Wang, L.-S., and Kim, J. 2009. Large-scale simulation of RNA macroevolution by an energy-dependent fitness model. arXiv:0912.2326.

Gusfield, D. 1991. Efficient algorithms for inferring evolutionary trees. *Networks*, **21**, 19–28.

Gusfield, D. 1993. Efficient methods for multiple sequence alignment with guaranteed error bounds. *Bulletin of Mathematical Biology*, **55**(1), 141–154.

Gusfield, D. 2014. *ReCombinatorics: The Algorithmics of Ancestral Recombination Graphs and Explicit Phylogenetic Networks*. Cambridge, MA: MIT Press.

Guyer, C., and Slowinski, J.B. 1991. Comparisons of observed phylogenetic topologies with null expectations among three monophyletic lineages. *Evolution*, **45**, 340–350.

Hagopian, R., Davidson, J.D., Datta, R.S., Jarvis, G., and Sjölander, K. 2010. SATCHMO-JS: a webserver for simultaneous protein multiple sequence alignment and phylogenetic tree construction. *Nucleic Acids Research*, **38** (Web Server Issue), W29–W34.

Hahn, M.W., and Nakhleh, L. 2016. Irrational exuberance for resolved species trees. *Evolution*, **70**(1), 7–17.

Hallett, M.T., and Lagergren, J. 2000. New algorithms for the duplication-loss model. Pages 138–146 of: *Proceedings of the Annual International Conference on Research in Computututional Molecular Biology (RECOMB) 2000*. New York: ACM Press.

Hallström, B.M., and Janke, A. 2010. Mammalian evolution may not be strictly bifurcating. *Molecular Biology and Evolution*, **27**, 2804–2806.

Hamada, M., and Asai, K. 2012. A classification of bioinformatics algorithms from the viewpoint of Maximizing Expected Accuracy (MEA). *Journal of Computational Biology*, **19**(5), 532–549.

Hannenhalli, S., and Pevzner, P.A. 1995. Transforming men into mice (polynomial algorithm for genomic distance problem). Pages 581–592 of: *Proceedings of the 36th Annual Symposium on the Foundations of Computer Science (FOCS)*.

Hartigan, J.A. 1973. Minimum mutation fits to a given tree. *Biometrics*, **29**(1), 53–65.

Hastings, W.K. 1970. Monte Carlo sampling methods using Markov chains and their applications. *Biometrika*, **57**(1), 97–109.

Haussler, D., Krogh, A., Mian, I.S., and Sjölander, K. 1993. Protein modelling using hidden Markov models: analysis of globins. Pages 792–802 of: *Proceedings Twenty-Sixth Annual Hawaii International Conference on System Sciences*, vol. 1. Washington, DC: IEEE Computer Society Press.

Haws, D., Hodge, T.L., and Yoshida, R. 2011. Optimality of the neighbor joining algorithm and faces of the balanced minimum evolution polytope. *Bulletin of Mathematical Biology*, **73**(11).

Heard, S.B., and Mooers, A.O. 2002. Signatures of random and selective mass extinctions in phylogenetic tree balance. *Systematic Biology*, **51**, 889–897.

Heath, T.A., Hedtke, S.M., and Hillis, D.M. 2008. Taxon sampling and the accuracy of phylogenetic analyses. *Journal of Systematics and Evolution*, **46**(3), 239–257.

Hein, J., Jiang, T., Wang, L., and Zhang, K. 1996. On the complexity of comparing evolutionary trees. *Discrete Applied Mathematics*, **71**, 153–169.

Hein, J., Wiuf, C., Knudsen, B., Møller, M.B., and Wibling, G. 2000. Statistical alignment: computational properties, homology testing and goodness-of-fit. *Journal of Molecular Biology*, **302**, 265–279.

Heled, J., and Drummond, A.J. 2010. Bayesian inference of species trees from multilocus data. *Molecular Biology and Evolution*, **27**, 570–580.

Helmkamp, L.J., Jewett, E.M., and Rosenberg, N.A. 2012. Improvements to a class of distance matrix methods for inferring species trees from gene trees. *Journal of Computational Biology*, **19**(6), 632–649.

Hendy, M., and Penny, D. 1989. A framework for the quantitative study of evolutionary trees. *Systematic Zoology*, **38**, 297–309.

Herman, J.L., Challis, C.J., Novák, A., Hein, J., and Schmidler, S.C. 2014. Simultaneous Bayesian estimation of alignment and phylogeny under a joint model of protein sequence and structure. *Molecular Biology and Evolution*, **31**(9), 2251–2266.

Herman, J.L., Novák, A., Lyngso, R., Szabó, A., Miklós, I., and Hein, J. 2015. Efficient representation of uncertainty in multiple sequence alignments using directed acyclic graphs. *BMC Bioinformatics*, **16**(1), 1–26.

Higgins, D.G., and Sharp, P.M. 1988. CLUSTAL: a package for performing multiple sequence alignment on a microcomputer. *GENE*, **73**(1), 237–244.

Hillis, D.M. 1996. Inferring complex phylogenies. *Nature*, **383**, 130–131.

Hillis, D. 1997. Response from D.M. Hillis. *Trends in Ecology & Evolution*, **12**(9), 358.

Hillis, D.M., Huelsenbeck, J.P., and Swofford, D.L. 1994. Hobgoblin of phylogenetics. *Nature*, **369**, 363–364.

Hillis, D.M., Moritz, C., and Mable, B.K. (eds.). 1996. *Molecular Systematics* Second edition. Sunderland, MA: Sinauer Associates.

Hillis, D.M., Pollock, D.D., McGuire, J.A., and Zwickl, D.J. 2003. Is sparse taxon sampling a problem for phylogenetic inference? *Systematic Biology*, **52**(1), 124–126.

Hillis, D.M., Heath, T.A., and St. John, K. 2005. Analysis and visualization of tree space. *Systematic Biology*, **54**(3), 471–482.

Hirosawa, M., Totoki, Y., Hoshida, M., and Ishikawa, M. 1995. Comprehensive study on iterative algorithms of multiple sequence alignment. *Computer Applications in the Biosciences: CABIOS*, **11**(1), 13–18.

Hoff, M., Orf, S., Riehm, B., Darriba, D., and Stamatakis, A. 2016. Does the choice of nucleotide substitution models matter topologically? *BMC Bioinformatics*, **17**, 143.

Hogeweg, P., and Hesper, B. 1984. The alignment of sets of sequences and the construction of phyletic trees: an integrated method. *Journal of Molecular Evolution*, **20**, 175–186.

Hohl, M., Kurtz, S., and Ohlebusch, E. 2002. Efficient multiple genome alignment. *Bioinformatics*, **18**, S312–S320.

Holder, M.T., and Lewis, P.O. 2003. Phylogeny estimation: traditional and Bayesian approaches. *Nature Reviews Genetics*, **43**, 275–284.

Holland, B., and Moulton, V. 2003. Consensus networks: a method for visualizing incompatibilities in collections of trees. Pages 165–176 of: *Proceedings of the Workshop on Algorithms in Bioinformatics*. Berlin: Springer.

Holland, B.R., Huber, K.T., Moulton, V., and Lockhart, P.J. 2004. Using consensus networks to visualize contradictory evidence for species phylogeny. *Molecular Biology and Evolution*, **21**, 1459–1461.

Holland, B.R., Delsuc, F., and Moulton, V. 2005. Visualizing conflicting evolutionary hypotheses in large collections of trees: using consensus networks to study the origins of placentrals and hexapods. *Systematic Biology*, **54**, 66–76.

Holland, B.R., Jermiin, L.S., and Moulton, V. 2006. Improved consensus network techniques for genome-scale phylogeny. *Molecular Biology and Evolution*, **23**, 848–855.

Holland, B., Conner, G., Huber, K., and Moulton, V. 2007. Imputing supertrees and supernetworks from quartets. *Systematic Biology*, **56**(1), 57–67.

Holmes, I., and Bruno, W.J. 2001. Evolutionary HMMs: a Bayesian approach to multiple alignment. *Bioinformatics*, **17**, 803–820.

Holmes, I., and Durbin, R. 1988. Dynamic programming alignment accuracy. *Journal of Computational Biology*, **5**, 493–504.

Holmes, S. 1999. Phylogenies: an overview. Pages 81–118 of: Halloran, M., and Geisser, S. (eds.), *Statistics and Genetics* New York: Springer-Verlag.

Holmes, S. 2003. Bootstrapping phylogenetic trees: theory and methods. *Statistical Science*, **18**(2), 241–255.

Holmes, S. 2005. Statistical approach to tests involving phylogenies. Pages 91–120 of: Gascuel, O. (ed.), *Mathematics of Evolution and Phylogeny*. Oxford: Oxford University Press.

Hosner, P.A., Faircloth, B.C., Glenn, T.C., Braun, E.L., and Kimball, R.T. 2016. Avoiding missing data biases in phylogenomic inference: an empirical study in the landfowl (Aves: Galliformes). *Molecular Biology and Evolution*, **33**(4), 1110–1125.

Huang, H., and Knowles, L.L. 2016. Unforeseen consequences of excluding missing data from next-generation sequences: simulation study of RAD sequences. *Systematic Biology*, **65**(3), 357–365.

Huang, H., He, Q., Kubatako, L.S., and Knowles, L.L. 2010. Sources of error inherent in species-tree estimation: impact of mutational and coalescent effects on accuracy and implications for choosing among different methods. *Systematic Biology*, **59**, 573–583.

Huang, W., Zhou, G., Marchand, M., Ash, J.R., Morris, D., Dooren, P. Van, Brown, J.M., Gallivan, K.A., and Wilgenbusch, J.C. 2016. TreeScaper: visualizing and extracting phylogenetic signal from sets of trees. *Molecular Biology and Evolution*, **33**(12), 3314–3316.

Hudson, R.R. 1983. Testing the constant-rate neutral allele model with protein sequence data. *Evolution*, **37**, 203–217.

Hudson, R.R. 1991. Gene genealogies and the coalescent process. Pages 1–44 of: Futuyma, D., and Antonovics, J. (eds.), *Oxford Surveys in Evolutionary Biology*, vol. 7. Oxford: Oxford University Press.

Huelsenbeck, J.P., and Crandall, K.A. 1997. Phylogeny estimation and hypothesis testing using maximum likelihood. *Annual Review of Ecology and Systematics*, **28**, 437–466.

Huelsenbeck, J.P., and Hillis, D.M. 1993. Success of phylogenetic methods in the four-taxon case. *Systematic Biology*, **42**(3), 247–265.

Huelsenbeck, J.P., Ronquist, F., Nielsen, R., and Bollback, J.P. 2001. Bayesian inference of phylogeny and its impact on evolutionary biology. *Science*, **294**(5550), 2310–2314.

Huson, D.H. 2007. Split networks and reticulate networks. Pages 247–276 of: Gascuel, O., and Steel, M.A. (eds.), *Reconstructing Evolution: New Mathematical and Computational Advances*. Oxford: Oxford University Press.

Huson, D.H. 2016. *SplitsTree4* Software for computing phylogenetic networks. Available online at *www.splitstree.org*.

Huson, D., Nettles, S., Parida, L., Warnow, T., and Yooseph, S. 1998. A divide-and-conquer approach to tree reconstruction. In: *Proceedings of the Workshop on Algorithms and Experiments (ALEX98)*. Trento.

Huson, D., Nettles, S., and Warnow, T. 1999a. Disk-covering, a fast converging method for phylogenetic tree reconstruction. *Journal of Computational Biology*, **6**(3), 369–386.

Huson, D., Vawter, L., and Warnow, T. 1999b. Solving large scale phylogenetic problems using DCM2. Pages 118–129 of: *Proceedings of the 7th International Conference on Intelligent Systems for Molecular Biology (ISMB'99)*. Cambridge, MA: AAAI Press.

Huson, D.H., Dezulian, T., Klöpper, T., and Steel, M.A. 2004. Phylognetic super-networks from partial trees. *IEEE/ACM Transactions on Computational Biology and Bioinformatics*, **1**, 151–158.

Huson, D.H., Rupp, R., and Scornavacca, C. 2010. *Phylogenetic Networks: Concepts, Algorithms and Applications*. Cambridge: Cambridge University Press.

Iantomo, S., Gori, K., Goldman, N., Gil, M., and Dessimoz, C. 2013. Who watches the watchmen? An appraisal of benchmarks for multiple sequence alignment. Pages 59–73 of: Russell, D. (ed.), *Multiple Sequence Alignment Methods*. New York: Springer.

Izquierdo-Carrasco, F., Smith, S., and Stamatakis, A. 2011. Algorithms, data structures, and numerics for likelihood-based phylogenetic inference of huge trees. *BMC Bioinformatics*, **12**(1), 470.

Jacox, E., Chauve, C., Szöllősi, G.J., Ponty, Y., and Scornavacca, C. 2016. ecceTERA: comprehensive gene tree-species tree reconciliation using parsimony. *Bioinformatics*, **32**(13), 2056–2058.

Janowitz, M.F., Lapointe, F.-J., McMorris, F.R., Mirkin, B., and Roberts, F.S. (eds.). 2003. *Bioconsensus*. Providence, RI: American Mathematical Society.

Jansen, R., and Palmer, J. 1987. A chloroplast DNA inversion marks an ancient evolutionary split in the sunflower family (Asteraceae). *Proceedings of the National Academy of Sciences (USA)*, **84**, 5818–5822.

Jarvis, E., Mirarab, S., Aberer, A., et al. 2014. Whole-genome analyses resolve early branches in the tree of life of modern birds. *Science*, **346**(6215), 1320–1331.

Jewett, E.M., and Rosenberg, N.A. 2012. iGLASS: an improvement to the GLASS method for estimating species trees from gene trees. *Journal of Computational Biology*, **19**(3), 293–315.

Jiang, T., Kearney, P., and Li, M. 2001. A polynomial time approximation scheme for inferring evolutionary trees from quartet topologies and its application. *SIAM Journal on Computing*, **30**, 1942–1961.

Jin, G., Nakhleh, L., Snir, S., and Tuller, T. 2006. Maximum likelihood of phylogenetic networks. *Bioinformatics*, **22**, 2604–2611.

Jin, G., Nakhleh, L., Snir, S., and Tuller, T. 2007. Inferring phylogenetic networks by the maximum parsimony criterion: a case study. *Molecular Biology and Evolution*, **24**(1), 324–337.

Joly, S., McLenachan, P.A., and Lockhart, P.J. 2009. A statistical approach for distinguishing hybridization and incomplete lineage sorting. *The American Naturalist*, **174**(2), E54–E70.

Jones, D.T., Taylor, W.R., and Thornton, J.M. 1992. The rapid generation of mutation data matrices from protein sequences. *Computing Applications in the Biosciences (CABIOS)*, **8**, 275282. doi:10.1093/bioinformatics/8.3.275.

Jones, G., Sagitov, S., and Oxelman, B. 2013. Statistical inference of allopolyploid species networks in the presence of incomplete lineage sorting. *Systematic Biology*, **62**(3), 476–478.

Jones, S., and Thornton, J.M. 1996. Principles of protein–protein interactions. *Proceedings of the National Academy of Sciences (USA)*, **93**(1), 13–20.

Jukes, T.H., and Cantor, C.R. 1997. Evolution of protein molecules. Pages 21–132 in Munro, H.N. (ed.), *Mammalian Protein Metabolism*. New York: Academic Press.

Kannan, S., and Warnow, T. 1997. A fast algorithm for the computation and enumeration of perfect phylogenies when the number of character states is fixed. *SIAM Journal on Computing*, **26**(6), 1749–1763.

Kannan, S., Warnow, T., and Yooseph, S. 1995. Computing the local consensus of trees. *SIAM Journal on Computing*, **27**(6), 1695–1724.

Kaplan, H., Shamir, R., and Tarjan, R.E. 1997. Faster and simpler algorithm for sorting signed permutations by reversals. Pages 344–351 of: *Proceedings of the ACM/SIAM Symposium on Discrete Algorithms (SODA)*.

Karp, R.M. 1972. Reducibility among combinatorial problems. Pages 85–103 of: *Complexity of Computer Computations*. New York: Plenum.

Karpinski, M., and Zelikovsky, A. 1997. New approximation algorithms for the Steiner Tree problems. *Journal of Combinatorial Optimization*, **1**(1), 47–65.

Katoh, K., and Toh, H. 2007. PartTree: an algorithm to build an approximate tree from a large number of unaligned sequences. *Bioinformatics*, **23**(3), 372–374.

Katoh, K., Kuma, K., Miyata, T., and Toh, H. 2005. Improvement in the acccuracy of multiple sequence alignment MAFFT. *Genome Informatics*, **16**, 22–33.

Kececioglu, J.D. 1993. The maximum weight trace problem in multiple sequence alignment. Pages 106–119 of: *Annual Symposium on Combinatorial Pattern Matching*. Berlin: Springer.

Kececioglu, J., and Sankoff, D. 1995. Exact and approximation algorithms for sorting by reversals, with application to genome rearrangement. *Algorithmica*, **13**(1–2), 180–210.

Kececioglu, J., and Starrett, D. 2004. Aligning alignments exactly. Pages 85–96 of: *Proceedings 8th ACM Conference on Research in Computational Molecular Biology (RECOMB)*.

Kececioglu, J., and Zhang, W. 1998. Aligning alignments. Pages 189–208 of: *Combinatorial Pattern Matching* Berlin: Springer.

Kececioglu, J., Kim, E., and Wheeler, T. 2010. Aligning protein sequences with predicted secondary structure. *Journal of Computational Biology*, **17**(3), 561–580.

Kemena, C., and Notredame, C. 2009. Upcoming challenges for multiple sequence alignment methods in the high-throughput era. *Bioinformatics*, **25**, 2455–2465.

Kendall, D.G. 1948. On the generalized "birth–death" process. *Annual Review Mathematical Statistics*, **19**, 1–15.

Kent, W.J., Baertsch, R., Hinrichs, A., Miller, W., and Haussler, D. 2003. Evolution's cauldron: duplication, deletion, and rearrangement in the mouse and human genomes. *Proceedings National Academy of Sciences (USA)*, **100**, 11484–11489.

Kidd, K.K., and Sgaramella-Zonta, L.A. 1971. Phylogenetic analysis: concepts and methods. *American Journal of Human Genetics*, **23**, 235–252.

Kim, J. 1996. General inconsistency conditions for maximum parsimony: effects of branch lengths and increasing numbers of taxa. *Systematic Biology*, **45**, 363–374.

Kim, J., and Ma, J. 2014. PSAR-Align: improving multiple sequence alignment through probabilistic sampling. *Bioinformatics*, **30**(7), 1010–1012.

Kim, J., and Salisbury, B.A. 2001. A tree obscured by vines: horizontal gene transfer and the median tree method of estimating species phylogeny. Pages 571–582 of: *Proceedings of the Pacific Symposium of Computing*, vol. 6.

Kimura, M. 1980. A simple method for estimating evolutionary rates of base substitutions through comparative studies of nucleotide sequences. *Journal of Molecular Evolution*, **16**, 111–120.

Kingman, J.F.C. 1982. The coalescent. *Stochastic Processes and Applications*, **13**, 235–248.

Knowles, L.L., and Kubatko, L.S. (eds.). 2011. *Estimating Species Trees: Practical and Theoretical Aspects*. Hoboken, NJ: John Wiley and Sons.

Koch, M.A., Dobes, C., Kiefer, C., Schmickl, R., Klimes, L., and Lysak, M.A. 2007. Supernetwork identifies multiple events of plastid *trn*F(GAA) pseudogene evolution in the Brassicaceae. *Molecular Biology and Evolution*, **24**(1), 63–73.

Kolaczkowski, B., and Thornton, J.W. 2004. Performance of maximum parsimony and likelihood phylogenetics when evolution is heterogeneous. *Nature*, **431**, 980–984.

Kosiol, C., Bofkin, L., and Whelan, S. 2006. Phylogenetics by likelihood: evolutionary modeling as a tool for understanding the genome. *Journal Biomedical Informatics*, **39**, 51–61.

Kosiol, C., Holmes, I., and Goldman, N. 2007. An empirical codon model for protein sequence evolution. *Molecular Biology and Evolution*, **24**(7), 1464–1479.

Kozlov, A.M., Aberer, A.J., and Stamatakis, A. 2015. ExaML version 3: a tool for phylogenomic analyses on supercomputers. *Bioinformatics*, **31**(15), 2577.

Kreidl, M. 2011. *Note on expected internode distances for gene trees in species trees*. arXiv:1108.5154v1.

Krishnamurthy, N., Brown, D., and Sjölander, K. 2007. FlowerPower: clustering proteins into domain architecture classes for phylogenomic inference of protein function. *BMC Evolutionary Biology*, **7**(1), 1–11.

Krogh, A., Brown, M., Mian, I., Sjölander, K., and Haussler, D. 1994. Hidden Markov models in computational biology: applications to protein modeling. *Journal of Molecular Biology*, **235**, 1501–1531.

Kruskal, J.B. 1956. On the shortest spanning subtree of a graph and the travelling salesman problem. *Proceedings of the American Mathematical Society*, **7**, 48–50.

Kubatko, L.S. 2009. Identifying hybridization events in the presence of coalescence via model selection. *Systematic Biology*, **58**(5), 478–488.

Kubatko, L., and Chifman, J. 2015. An invariants-based method for efficient identification of hybrid species from large-scale genomic data. bioRxiv: 034348.

Kubatko, L.S., and Degnan, J.H. 2007. Inconsistency of phylogenetic estimates from concatenated data under coalescence. *Systematic Biology*, **56**, 17.

Kubatko, L.S., Carstens, B.C., and Knowles, L.L. 2009. STEM: species tree estimation using maximum likelihood for gene trees under coalescence. *Bioinformatics*, **25**(7), 971–973.

Kuhner, M.K., and Felsenstein, J. 1994. A simulation comparison of phylogeny algorithms under equal and unequal evolutionary rates. *Molecular Biology and Evolution*, **11**(3), 459–468.

Kullback, S. 1987. Letter to the editor: The Kullback–Leibler distance. *The American Statistician*, **41**(4), 340–341.

Kupczok, A. 2011. Split-based computation of majority rule supertrees. *BMC Evolutionary Biology*, **11**.

Lacey, M.R., and Chang, J.T. 2006. A signal-to-noise analysis of phylogeny estimation by neighbor-joining: insufficiency of polynomial length sequences. *Mathematical Biosciences*, **199**(2), 188–215.

Lafond, M., and Scornavacca, C. 2016. On the Weighted Quartet Consensus problem. arXiv preprint arXiv:1610.00505.

Lagergren, J. 2002. Combining polynomial running time and fast convergence for the Disk-Covering Method. *Journal of Computer and System Science*, **65**(3), 481–493.

Lanfear, R., Calcott, B., Ho, S.Y.W., and Guindon, S. 2012. PartitionFinder: combined selection of partitioning schemes and substitution models for phylogenetic analyses. *Molecular Biology and Evolution*, **29**(6), 1695–1701.

Lapointe, F.-J., and Cucumel, G. 1997. The average consensus procedure: combination of weighted trees containing identical or overlapping sets of taxa. *Systematic Biology*, **46**(2), 306–312.

Lapointe, F.-J., Wilkinson, M., and Bryant, D. 2003. Matrix representations with parsimony or with distances: two sides of the same coin? *Systematic Biology*, **52**, 865–868.

Larget, B., Kotha, S.K., Dewey, C.N., and Ané, C. 2010. BUCKy: Gene tree/species tree reconciliation with the Bayesian concordance analysis. *Bioinformatics*, **26**(22), 2910–2911.

Leaché, A.D., Chavez, A.S., Jones, L.N., Grummer, J.A., Gottscho, A.D., and Linkem, C.W. 2015. Phylogenomics of phrynosomatid lizards: conflicting signals from sequence capture versus restriction site associated DNA sequencing. *Genome Biology and Evolution*, **7**(3), 706–719.

Leavitt, S.D., Grewe, F., Widhelm, T., Muggia, L., Wray, B., and Lumbsch, H.T. 2016. Resolving evolutionary relationships in lichen-forming fungi using diverse phylogenomic datasets and analytical approaches. *Scientific Reports*, **6**(22262). doi: 10.1038/srep22262.

Lechner, M., Hernandez-Rosales, M., Doerr, D., Wieseke, N., Thévenin, A., Stoye, J., Hartmann, R.K., Prohaska, S.J., and Stadler, P.F. 2014. Orthology detection combining clustering and synteny for very large datasets. *PLoS ONE*, **9**(8), e105015. doi:10.1371/journal.pone.0105015.

Lecointre, G.H., Philippe, H., Van Le, H.L., and Le Guyader, H. 1994. How many nucleotides are required to resolve a phylogenetic problem? The use of a new statistical method applicable to available sequences. *Molecular Phylogenetics and Evolution*, **3**, 292–309.

Leebens-Mack, J., Raubeson, L.A., Cui, L., Kuehl, J.V., Fourcade, M.H., Chumley, T.W., Boore, J.L., Jansen, R.K., and dePamphilis, C.W. 2005. Identifying the basal angiosperm node in chloroplast genome phylogenies: sampling one's way out of the Felsenstein zone. *Molecular Biology and Evolution*, **22**(10), 1948–1963.

Lefort, V., Desper, R., and Gascuel, O. 2015. FastME 2.0: a comprehensive, accurate, and fast distance-based phylogeny inference program. *Molecular Biology and Evolution*, **32**(10), 2798–2800.

Leimer, H.-G. 1993. Optimal decomposition by clique separators. *Discrete Mathematics*, **113**, 99–123.

Lemmon, E.M., and Lemmon, A.R. 2013. High-throughput genomic data in systematics and phylogenetics. *Annual Review of Ecology, Evolution, and Systematics*, **44**, 99–121.

LeQuesne, W.J. 1969. A method of selection of characters in numerical taxonomy. *Systematic Zoology*, **18**, 201–205.

Li, C., Medlar, A., and Löytynoja, A. 2016. Co-estimation of phylogeny-aware alignment and phylogenetic tree. *bioRxiv*, 077503. doi: *https://doi.org*/10.1101.077503

Li, W.-H. 1997. *Molecular Evolution*. Sunderland, MA: Sinauer Associates, Inc.

Li, W.-H., and Tanimura, M. 1987. The molecular clock runs more slowly in man than in apes and monkeys. *Nature*, **326**(6108), 93–96.

Lin, Y., and Moret, B. 2008. Estimating true evolutionary distances under the DCJ model. *Bioinformatics*, **24**(13), i114–i122.

Lin, Y., Rajan, V., and Moret, B. 2012a. A metric for phylogenetic trees based on matching. *IEEE/ACM Transactions on Computational Biology and Bioinformatics*, **9**(4), 1014–1022.

Lin, Y., Rajan, V., and Moret, B. 2012b. TIBA: a tool for phylogeny inference from rearrangement data with bootstrap analysis. *Bioinformatics*, **28**(24), 3324–3325.

Liu, K., and Warnow, T. 2012. Treelength optimization for phylogeny estimation. *PLoS One*, **7**(3), e33104.

Liu, K., Raghavan, S., Nelesen, S., Linder, C.R., and Warnow, T. 2009a. Rapid and accurate large-scale coestimation of sequence alignments and phylogenetic trees. *Science*, **324**(5934), 1561–1564.

Liu, K., Linder, C.R., and Warnow, T. 2012a. RAxML and FastTree: comparing two methods for large-scale maximum likelihood phylogeny estimation. *PLoS ONE*, **6**(11), e27731.

Liu, K., Warnow, T., Holder, M.T., Nelesen, S.M., Yu, J., Stamatakis, A.P., and Linder, C.R. 2012b. SATé-II: very fast and accurate simultaneous estimation of multiple sequence alignments and phylogenetic trees. *Systematic Biology*, **61**(1), 90–106.

Liu, L., and Yu, L. 2011. Estimating species trees from unrooted gene trees. *Systematic Biology*, **60**(5), 661–667.

Liu, L., Yu, L., Pearl, D.K., and Edwards, S.V. 2009b. Estimating species phylogenies using coalescence times among sequences. *Systematic Biology*, **58**(5), 468–477.

Liu, L., Yu, L., and Edwards, S.V. 2010. A maximum pseudo-likelihood approach for estimating species trees under the coalescent model. *BMC Evolutionary Biology*, **10**(1), 302.

Lockhart, P., Novis, P., Milligan, B.G., Riden, J., Rambaut, A., and Larkum, T. 2006. Heterotachy and tree building: a case study with Plastids and Eubacteria. *Molecular Biology and Evolution*, **23**(1), 40–45.

Lopez, P., Casane, D., and Philippe, H. 2002. Heterotachy, an important process of protein evolution. *Molecular Biology and Evolution*, **19**(1), 1–7.

Löytynoja, A., and Goldman, N. 2005. An algorithm for progressive multiple alignment of sequences with insertions. *Proceedings of the National Academy of Sciences (USA)*, **102**, 10557–10562.

Löytynoja, A., and Goldman, N. 2008a. A model of evolution and structure for sequence alignment. *Philosophical Transactions of the Royal Society (B)*, **363**, 3913–3919.

Löytynoja, A., and Goldman, N. 2008b. Phylogeny-aware gap placement prevents errors in sequence alignment and evolutionary analysis. *Science*, **320**(5883), 1632–1635.

Löytynoja, A., Vilella, A.J., and Goldman, N. 2012. Accurate extension of multiple sequence alignments using a phylogeny-aware graph algorithm. *Bioinformatics*, **28**, 1685–1691.

Lunter, G.A., Miklós, I., Drummond, A., Jensen, J.L., and Hein, J. 2003a. Bayesian phylogenetic inference under a statistical indel model. *Lecture Notes in Bioinformatics, Proceedings of the Workshop on Algorithms for Bioinformatics (WABI) 2003*, **2812**, 228–244.

Lunter, G.A., Miklos, I., Song, Y.S., and Hein, J. 2003b. An efficient algorithm for statistical multiple alignment on arbitrary phylogenetic trees. *Journal of Computational Biology*, **10**, 869–889.

Lunter, G.A., Miklós, I., Drummond, A., Jensen, J.L., and Hein, J. 2005a. Bayesian coestimation of phylogeny and sequence alignment. *BMC Bioinformatics*, **6**, 83.

Lunter, G.A., Drummond, A.J., Miklós, I., and Hein, J. 2005b. Statistical alignment: recent progress, new applications, and challenges. Pages 375–406 of: Nielsen, R. (ed.), *Statistical Methods in Molecular Evolution (Statistics for Biology and Health)*. Berlin: Springer.

Ma, B., Wang, Z., and Zhang, K. 2003. Alignment between two multiple alignments. Pages 254–265 of: *Proceedings of the 14th Annual Conference on Combinatorial Pattern Matching*. Berlin, Heidelberg: Springer.

Ma, J., Zhang, L., Suh, B.B., Raney, B.J., Burhans, R.C., Kent, W.J., Blanchette, M., Haussler, D., and Miller, W. 2006. Reconstructing contiguous regions of an ancestral genome. *Genome Research*, **16**(12), 1557–1565.

Ma, J., Wang, S., Wang, Z., and Xu, J. 2014. MRFalign: protein homology detection through alignment of Markov random fields. *PLoS Computational Biology*, **10**(3), e1003500. doi:10.1371/journal.pcbi.1003500.

Maddison, D. 1991. The discovery and importance of multiple islands of most parsimonious trees. *Systematic Zoology*, **40**, 315–328.

Maddison, D. 2016. The rapidly changing landscape of insect phylogenetics. *Current Opinion in Insect Science*, **18**, 77–82.

Maddison, W.P. 1997. Gene trees in species trees. *Systematic Biology*, **46**, 523–536.

Mallet, J. 2005. Hybridization as an invasion of the genome. *Trends in Ecology & Evolution*, **20**, 229–237.

Mallo, D., and Posada, D. 2016. Multilocus inference of species trees and DNA barcoding. *Philosophical Transactions of the Royal Society (B)*, **371**, 20150335. http://dx.doi.org/10.1098/rstb.2015.0335.

Matsen, F.A., Kodner, R.B., and Armbrust, E.V. 2010. pplacer: linear time maximum-likelihood and Bayesian phylogenetic placement of sequences onto a fixed reference tree. *BMC Bioinformatics*, **11**, 538.

McCormack, J., Harvey, M., Faircloth, B., Crawford, N., Glenn, T., and Brumfield, R. 2013. A phylogeny of birds based on over 1,500 loci collected by target enrichment and high-throughput sequencing. *PLoS One*, **8**, 54848.

McKenzie, A., and Steel, M.A. 2001. Properties of phylogenetic trees generated by Yule-type speciation models. *Mathematical Biosciences*, **170**, 91–112.

McMorris, F.R., Warnow, T.J., and Wimer, T. 1994. Triangulating vertex-colored graphs. *SIAM Journal on Discrete Mathematics*, **7**(2), 296–306.

Meiklejohn, K.A., Faircloth, B.C., Glenn, T.C., Kimball, R.T., and Braun, E.L. 2016. Analysis of a rapid evolutionary radiation using ultraconserved elements: evidence for a bias in some multispecies coalescent methods. *Systematic Biology*, **65**(4), 612–627.

Meng, C., and Kubatko, L.S. 2009. Detecting hybrid speciation in the presence of incomplete lineage sorting using gene tree incongruence: a model. *Theoretical Population Biology*, **75**, 35–45.

Metzler, D. 2003. Statistical alignment based on fragment insertion and deletion models. *Bioinformatics*, **19**(4), 490–499.

Mihaescu, R., and Pachter, L. 2008. Combinatorics of least-squares trees. *Proceedings of the National Academy of Sciences (USA)*, **105**(36), 13206–13211.

Mihaescu, R., Levy, D., and Pachter, L. 2009. Why neighbor-joining works. *Algorithmica*, **54**(1), 1–24.

Miklós, I. 2003. Algorithm for statistical alignment of sequences derived from a Poisson sequence length distribution. *Discrete Applied Mathematics*, **127**(1), 79–84.

Miklós, I., Lunter, G.A., and Holmes, I. 2004. A "long indel model" for evolutionary sequence alignment. *Molecular Biology and Evolution*, **21**(3), 529–540.

Miller, W., and Myers, E. 1988. Sequence comparison with concave weighting functions. *Bulletin of Mathematical Biology*, **50**, 97–120.

Mindell, D.P. 2013. The Tree of Life: metaphor, model, and heuristic device. *Systematic Biology* **62**(3), 478–489.

Mir arabbaygi (Mirarab), S. 2015. *Novel scalable approaches for multiple sequence alignment and phylogenomic reconstruction*. Ph.D. thesis, University of Texas at Austin.

Mirarab, S., and Warnow, T. 2011. FastSP: linear-time calculation of alignment accuracy. *Bioinformatics*, **27**(23), 3250–3258.

Mirarab, S., and Warnow, T. 2015. ASTRAL-II: coalescent-based species tree estimation with many hundreds of taxa and thousands of genes. *Bioinformatics*, **31**(12), i44–i52.

Mirarab, S., Nguyen, N., and Warnow, T. 2012. SEPP: SATé-enabled phylogenetic placement. Pages 247–58 of: *Pacific Symposium on Biocomputing*.

Mirarab, S., Reaz, R., Bayzid, Md. S., Zimmermann, T., Swenson, M.S., and Warnow, T. 2014a. ASTRAL: Accurate Species TRee ALgorithm. *Bioinformatics*, **30**(17), i541–i548.

Mirarab, S., Bayzid, Md. S., Boussau, B., and Warnow, T. 2014b. Statistical binning improves phylogenomic analysis. *Science*, **346**(6215), 1250463.

Mirarab, S., Nguyen, N., Wang, L.-S., Guo, S., Kim, J., and Warnow, T. 2015a. PASTA: ultra-large multiple sequence alignment of nucleotide and amino acid sequences. *Journal of Computational Biology*, **22**, 377–386.

Mirarab, S., Bayzid, M.S., Boussau, B., and Warnow, T. 2015b. Response to comment on "Statistical binning enables an accurate coalescent-based estimation of the avian tree." *Science*, **350**(6257), 171.

Mitchison, G.J. 1999. A probabilistic treatment of phylogeny and sequence alignment. *Journal of Molecular Evolution*, **49**, 11–22.

Mitchison, G.J., and Durbin, R.M. 1995. Tree-based maximum likelihood substitution matrices and hidden Markov models. *Journal of Molecular Evolution*, **41**, 1139–1151.

Money, D., and Whelan, S. 2012. Characterizing the phylogenetic tree-search problem. *Systematic Biology*, **61**(2), 228–239.

Mooers, A.O. 2004. Effects of tree shape on the accuracy of maximum likelihood-based ancestor reconstructions. *Systematic Biology*, **53**(5), 809–814.

Mooers, A., and Heard, S.B. 1997. Inferring evolutionary process from phylogenetic tree shape. *Quantitative Review of Biology*, **72**(1), 31–54.

Moret, B.M.E., Roshan, U., and Warnow, T. 2002. Sequence length requirements for phylogenetic methods. Pages 343–356 of: *Proceedings of the 2nd International Workshop on Algorithms in Bioinformatics (WABI'02)*. Berlin: Springer.

Moret, B.M.E., Lin, Y., and Tang, J. 2013. Rearrangements in phylogenetic inference: compare, model, or encode? Pages 147–171 of: Chauve, C., El-Mabrouk, N., and Tannier, E. (eds.), *Models and Algorithms for Genome Evolution*. London: Springer.

Morgenstern, B., Dress, A., and Wener, T. 1996. Multiple DNA and protein sequence based on segment-to-segment comparison. *Proceedings of the National Academy of Sciences (USA)*, **93**, 12098–12103.

Morgenstern, B., Frech, K., Dress, A., and Werner, T. 1998. DIALIGN: finding local similarities by multiple sequence alignment. *Bioinformatics*, **14**(3), 290–294.

Morlon, H., Potts, M.D., and Plotkin, J.B. 2010. Inferring the dynamics of diversification: a coalescent approach. *PLoS Biology*, **8**(9), e1000493.

Morrison, D.A. 2006. Multiple sequence alignment for phylogenetic purposes. *Australian Systematic Botany*, **19**, 479–539.

Morrison, D. 2010a. Phylogenetic networks in systematic biology (and elsewhere). Pages 1–48 of: Mohan, R. (ed.), *Research Advances in Systematic Biology*. Trivandrum: Global Research Network.

Morrison, D.A. 2010b. Using data-display networks for exploratory data analysis in phylogenetic studies. *Molecular Biology and Evolution*, **27**(5), 1044–1057.

Morrison, D.A. 2011. *Introduction to Phylogenetic Networks*. Uppsala: RJR Productions.

Morrison, D.A. 2014a. Is the Tree of Life the best mataphor, model, or heuristic for phylogenetics? *Systematic Biology*, **63**(4), 628–638.

Morrison, D.A. 2014b. Next generation sequencing and phylogenetic networks. *EMBnet.journal*, **20**, 3760.

Morrison, D.A. 2017. Next Generation Systematics. – Peter D. Olson, Joseph Hughe and James A. Cotton. *Systematic Biology*, **66**(1), 121–123. doi: *https://doi.org/10.1093/sysbio/syw081*.

Morrison, D.A., Morgan, M.J., and Kelchner, S.A. 2015. Molecular homology and multiple-sequence alignment: an analysis of concepts and practice. *Australian Systematic Biology*, **28**, 46–62.

Mossel, E., and Roch, S. 2011. Incomplete lineage sorting: consistent phylogeny estimation from multiple loci. *IEEE/ACM Transactions on Computational Biology and Bioinformatics*, **7**(1), 166–171.

Mossel, E., and Roch, S. 2013. Identifiability and inference of non-parametric rates-across-sites models on large-scale phylogenies. *Journal of Mathematical Biology*, **67**, 767–797.

Mossel, E., and Roch, S. 2015. Distance-based species tree estimation: information-theoretic trade-off between number of loci and sequence length under the coalescent. arXiv preprint. arXiv:1504.05289v1. To appear, *The Annals of Applied Probability*

Moyle, R.G., Oliveros, C.H., Andersen, M.J., Hosner, P.A., Benz, B.W., Manthey, J.D., Travers, S.L., Brown, R.M., and Faircloth, B.C. 2016. Tectonic collision and uplift of Wallacea triggered the global songbird radiation. *Nature Communications*, **7**.

Nadeau, J.H., and Taylor, B.A. 1984. Lengths of chromosome segments conserved since divergence of man and mouse. *Proceedings of the National Academy of Sciences (USA)*, **81**, 814–818.

Nakhleh, L. 2010. A metric on the space of reduced phylogenetic networks. *IEEE/ACM Transactions on Computational Biology and Bioinformatics*, **7**(2), 218–222.

Nakhleh, L. 2016. Computational approaches to species phylogeny inference and gene tree reconciliation. *Trends in Ecology & Evolution*, **28**, 719–728.

Nakhleh, L., Roshan, U., St. John, K., Sun, J., and Warnow, T. 2001a. Designing fast converging phylogenetic methods. *Bioinformatics*, **17**, 190–198.

Nakhleh, L., Roshan, U., St. John, K., Sun, J., and Warnow, T. 2001b. The performance of phylogenetic methods on trees of bounded diameter. Pages 189–203 of: *Proceedings 1st Workshop on Algorithms in BioInformatics (WABI01)*. Berlin: Springer.

Nakhleh, L., Moret, B.M.E., Roshan, U., St. John, K., Sun, J., and Warnow, T. 2002. The accuracy of fast phylogenetic methods for large datasets. Pages 211–222 of: *Proceedings of the 7th Pacific Symposium on Biocomputing (PSB02)*.

Nakhleh, L., Sun, J., Warnow, T., Linder, C.R., Moret, B.M.E., and Tholse, A. 2003. Towards the development of computational tools for evaluating phylogenetic network reconstruction methods. Pages 315–326 of: *Proceedings of the 8th Pacific Symposium on Biocomputing (PSB 2003)*.

Nakhleh, L., Warnow, T., and Linder, C.R. 2004. Reconstructing reticulate evolution in species: theory and practice. Pages 337–346 of: *Proceedings of the 8th International Conference on Computational Molecular Biology (RECOMB'04)*. New York: ACM Press.

Nakhleh, L., Warnow, T., Linder, C.R., and St. John, K. 2005a. Reconstructing reticulate evolution in species: theory and practice. *Journal of Computational Biology*, **12**(6), 796–811.

Nakhleh, L., Ruths, D., and Wang, L.-S. 2005b. RIATA-HGT: A fast and accurate heuristic for reconstructing horizontal gene transfer. Pages 84–93 of: *The Eleventh International Computing and Combinatorics Conference (COCOON'05)*. Berlin: Springer.

Nawrocki, E.P. 2009. *Structural RNA homology search and alignment using covariance models*. Ph.D. thesis, Washington University in Saint Louis, School of Medicine.

Nawrocki, E.P., Kolbe, D.L., and Eddy, S.R. 2009. Infernal 1.0: inference of RNA alignments. *Bioinformatics*, **25**, 1335–1337.

Needleman, S.B., and Wunsch, C.D. 1970. A general method applicable to the search for similarities in the amino acid sequence of two proteins. *Journal of Molecular Biology*, **48**, 443–453.

Nei, M. 1986. Stochastic errors in DNA evolution and molecular phylogeny. Pages 515–534 of: Karlin, S., and Nevo, E. (eds.), *Evolutionary Processes and Theory*. New York: Academic Press.

Nei, M., Kumar, S., and Kumar, S. 2003. *Molecular Evolution and Phylogenetics*. Oxford: Oxford University Press.

Nelesen, S. 2009. *Improved methods for phylogenetics*. Ph.D. thesis, University of Texas at Austin.

Nelesen, S., Liu, K., Zhao, D., Linder, C.R., and Warnow, T. 2008. The effect of the guide tree on multiple sequence alignments and subsequent phylogenetic analyses. Pages 15–24 of: *Pacific Symposium on Biocomputing*, vol. 13.

Nelesen, S., Liu, K., Wang, L.-S., Linder, C.R., and Warnow, T. 2012. DACTAL: divide-and-conquer trees (almost) without alignments. *Bioinformatics*, **28**, i274–i282.

Neuwald, A.F. 2009. Rapid detection, classification, and accurate alignment of up to a million or more related protein sequences. *Bioinformatics*, **25**, 1869–1875.

Neves, D.T., and Sobral, J.L. 2017. Parallel SuperFine: a tool for fast and accurate supertree estimation – features and limitations. *Future Generation Computer Systems*, **67**, 441–454.

Neves, D.T., Warnow, T., Sobral, J.L., and Pingali, K. 2012. Parallelizing SuperFine. Pages 1361–1367 of: *27th Symposium on Applied Computing (ACM-SAC), Bioinformatics*. ACM. doi: 10.1145/2231936.2231992.

Neyman, J. 1971. Molecular studies of evolution: a source of novel statistical problems. Page 127 of: Gupta, S.S., and Yackel, J. (eds.), *Statistical Decision Theory and Related Topics*. New York: Academic Press.

Nguyen, L.-T., Schmidt, H.A., von Haeseler, A., and Minh, B.Q. 2015a. IQ-TREE: a fast and effective stochastic algorithm for estimating maximum-likelihood phylogenies. *Molecular Biology and Evolution*, **32**(1), 268–274.

Nguyen, N., Mirarab, S., and Warnow, T. 2012. MRL and SuperFine+MRL: new supertree methods. *Algorithms for Molecular Biology*, **7**(3).

Nguyen, N., Mirarab, S., Liu, B., Pop, M., and Warnow, T. 2014. TIPP: taxonomic identification and phylogenetic profiling. *Bioinformatics*, **30**(24), 3548–3555.

Nguyen, N., Mirarab, S., Kumar, K., and Warnow, T. 2015b. Ultra-large alignments using phylogeny aware profiles. *Genome Biology*, **16**(124).

Nguyen, N., Nute, M., Mirarab, S., and Warnow, T. 2016. HIPPI: highly accurate protein family classification with ensembles of hidden Markov models. *BMC Bioinformatics*, **17** (Suppl 10), 765.

Nixon, K.C. 1999. The parsimony ratchet, a new method for rapid parsimony analysis. *Cladistics*, **15**, 407–414.

Notredame, C., Holm, L., and Higgins, D.G. 1998. COFFEE: an objective function for multiple sequence alignments. *Bioinformatics*, **14**(5), 407–422.

Notredame, C., Higgins, D.G., and Heringa, J. 2000. T-Coffee: a novel method for fast and accurate multiple sequence alignment. *Journal of Molecular Biology*, **302**, 205–217.

Noutahi, E., Semeria, M., Lafond, M., Seguin, J., Boussau, B., Guégen, L., and El-Mabrouk, N. 2015. Efficient gene tree correction guided by species and synteny evolution. HAL archive, *https://hal.archives-ouvertes.fr/hal*-01162963v2.

Novák, Á., Miklós, I., Lyngsoe, R., and Hein, J. 2008. StatAlign: an extendable software package for joint Bayesian estimation of alignments and evolutionary trees. *Bioinformatics*, **24**, 2403–2404.

Nute, M., and Warnow, T. 2016. Scaling statistical multiple sequence alignment to large datasets. *BMC Bioinformatics*, **17**(Suppl 10), 764.

Nye, T.M.W. 2008. Tree of trees: an approach to comparing multiple alternative phylogenies. *Systematic Biology*, **57**, 785–794.

Ogden, T.H., and Rosenberg, M.S. 2006. Multiple sequence alignment accuracy and phylogenetic inference. *Systematic Biology*, **55**(2), 314–328.

Ogden, T.H., and Rosenberg, M. 2007. Alignment and topological accuracy of the direct optimization approach via POY and traditional phylogenetics via ClustalW + PAUP*. *Systematic Biology*, **56**(2), 182–193.

Ohno, S. 1970. *Evolution by Gene Duplication*. New York: Springer.

Oldman, J., Wu, T., and van Iersel, L. 2016. TriLoNet: piecing together small networks to reconstruct reticulate evolutionary histories. *Molecular Biology and Evolution*, **33**(8), 2151–2162.

Olson, P.D., Hughes, J., and Cotton, J.A. (eds.). 2016. *Next Generation Systematics*. Cambridge: Cambridge University Press.

Ortuno, F.M., Valenzuela, O., Pomares, H., Rojas, F., Florido, J.P., Urquiza, J.M., and Rojas, I. 2013. Predicting the accuracy of multiple sequence alignment algorithms by using computational intelligent techniques. *Nucleic Acids Research*, **41**(1).

O'Sullivan, O., Suhre, K., Abergel, C., Higgins, D.G., and Notredame, C. 2004. 3DCoffee: combining protein sequences and structure within multiple sequence alignments. *Journal of Molecular Biology*, **340**, 385–395.

Page, R. 2002. Modified Mincut Supertrees. Pages 537–551 of: Guigó, R., and Gusfield, D. (eds.), *Algorithms in Bioinformatics*. Berlin: Springer.

Page, R., and Holmes, E. 1998. *Molecular Evolution: A Phylogenetic Approach*. Oxford: Blackwell Publishers.

Pais, F.S.-M., Ruy, P.d.C., Oliveira, G., and Coimbra, R.S. 2014. Assessing the efficiency of multiple sequence alignment programs. *Algorithms for Molecular Biology*, **9**, 4.

Pamilo, P., and Nei, M. 1998. Relationship between gene trees and species trees. *Molecular Biology and Evolution*, **5**, 568–583.

Papadopoulos, J.S., and Agarwala, R. 2007. COBALT: constraint-based alignment tool for multiple protein sequences. *Bioinformatics*, **23**(9), 1073–1079.

Pardi, F., Guillemot, S., and Gascuel, O. 2012. Combinatorics of distance-based tree inference. *Proceedings of the National Academy of Sciences (USA)*, **109**(41), 16443–16448.

Patel, S., Kimball, R.T., and Braun, E.L. 2013. Error in phylogenetic estimation for bushes in the Tree of Life. *Journal of Phylogenetics and Evolutionary Biology*, **1**, 110. doi:10.4172/2329-9002.1000110.

Paten, B., Herrero, J., Beal, K., Fitzgerald, S., and Birney, E. 2008. Enredo and Pecan: genome-wide mammalian consistency-based multiple alignment with paralogs. *Genome Research*, **18**, 1814–1828.

Paten, B., Herrero, J., Beal, K., and Birney, E. 2009. Sequence progressive alignment, a framework for practical large-scale probabilistic consistency alignment. *Bioinformatics*, **25**(3), 295–301.

Paten, B., Earl, D., Nguyen, N., Diekhans, M., Zerbino, D., and Haussler, D. 2011. Cactus: algorithms for genome multiple sequence alignment. *Genome Research*, **21**, 1512–1528.

Pauplin, Y. 2000. Direct calculation of a tree length using a distance matrix. *Molecular Biology and Evolution*, **51**, 41–47.

Pe'er, I., and Shamir, R. 1998. The median problems for breakpoints are NP-complete. In: *Electronic Colloquium on Computational Complexity*, vol. 71.

Pei, J., and Grishin, N.V. 2006. MUMMALS: multiple sequence alignment improved by using hidden Markov models with local structural information. *Nucleic Acids Research*, **34**, 4364–4374.

Pei, J., and Grishin, N.V. 2007. PROMALS: towards accurate multiple sequence alignments of distantly related proteins. *Bioinformatics*, **23**, 802–808.

Pei, J., Sadreyev, R., and Grishin, N.V. 2003. PCMA: fast and accurate multiple sequence alignment based on profile consistency. *Bioinformatics*, **19**, 427–428.

Penn, O., Privman, E., Landan, G., Graur, D., and Pupko, T. 2010. An alignment confidence score capturing robustness to guide tree uncertainty. *Molecular Biology and Evolution*, **27**(8), 1759–1767.

Pennisi, E. 2016. Shaking up the Tree of Life. *Science*, **354**, 817–821.

Phillimore, A.B., and Price, T.D. 2008. Density-dependent cladogenesis in birds. *PLoS Biology*, **6**(3), e71.

Phillips, C.A., and Warnow, T. 1996. The asymmetric median tree: a new model for building consensus trees. *Discrete Applied Mathematics*, **71**, 311–335.

Phuong, T.M., Do, C.B., Edgar, R.C., and Batzoglou, S. 2006. Multiple alignment of protein sequences with repeats and rearrangements. *Nucleic Acids Research*, **34**(20), 5932–5942.

Poe, S. 2003. Evaluation of the strategy of long-branch subdivision to improve the accuracy of phylogenetic methods. *Systematic Biology*, **52**, 423–428.

Pollock, D.D., Zwickl, D.J., McGuire, J.A., and Hillis, D.M. 2002. Increased taxon sampling is advantageous for phylogenetic inference. *Systematic Biology*, **51**, 664–671.

Posada, D. 2016. Phylogenomics for systematic biology. *Systematic Biology*, **65**, 353–356.

Posada, D., and Buckley, T.R. 2004. Model selection and model averaging in phylogenetics: advantages of Akaike information criterion and Bayesian approaches over likelihood ratio tests. *Systematic Biology*, **53**(5), 793–808.

Posada, D., and Crandall, K.A. 1998. Modeltest: testing the model of DNA substitution. *Bioinformatics*, **14**(9), 817–818.

Posada, D., and Crandall, K.A. 2001. Selecting the best-fit model of nucleotide substitution. *Systematic Biology*, **50**(4), 580–601.

Pray, L.A. 2004. Epigenetics: genome, meet your environment – as the evidence for epigenetics, researchers reacquire a taste for Lamarckism. *The Scientist*, 14.

Preusse, E., Quast, C., Knittel, K., Fuchs, B.M., Ludwig, W., Peplies, J., and Glockner, F.O. 2007. SILVA: a comprehensive online resource for quality checked and aligned ribosomal RNA sequence data compatible with ARB. *Nucleic Acids Research*, **35**, 718–796.

Price, M.N., Dehal, P.S., and Arkin, A.P. 2010. FastTree 2: approximately maximum-likelihood trees for large alignments. *PLoS ONE*, **5**(3), e9490. doi:10.1371/journal.pone.0009490.

Purvis, A., and Quicke, D.L.J. 1997a. Are big trees indeed easy? Reply from A. Purvis and D.L.J. Quicke. *Trends in Ecology & Evolution*, **12**(9), 357–358.

Purvis, A., and Quicke, D.L.J. 1997b. Building phylogenies: are the big easy? *Trends in Ecology & Evolution*, **12**(2), 49–50.

Qian, B., and Goldstein, R.A. 2001. Distribution of indel lengths. *Proteins*, **45**, 102–104.

Qian, B., and Goldstein, R.A. 2003. Detecting distant homologs using phylogenetic tree-based HMMs. *PROTEINS: Structure, Function, and Genetics*, **52**, 446–453.

Rannala, B., and Yang, Z. 1996. Probability distribution of molecular evolutionary trees: a new method of phylogenetic inference. *Journal of Molecular Evolution*, **43**, 304–311.

Rannala, B., and Yang, Z. 2003. Bayes estimation of species divergence times and ancestral population sizes using DNA sequences from multiple loci. *Genetics*, **164**, 1645–1656.

Rannala, B., Huelsenbeck, J.P., Yang, Z., and Nielsen, R. 1998. Taxon sampling and the accuracy of large phylogenies. *Systematic Biology*, **47**, 702–710.

Ranwez, V., and Gascuel, O. 2001. Quartet-based phylogenetic inference: improvements and limits. *Molecular Biology and Evolution*, **18**(6), 1103–1116.

Ranwez, V., Criscuolo, A., and Douzery, E.J. 2010. SuperTriplets: a triplet-based supertree approach to phylogenomics. *Bioinformatics*, **26**(12), i115–i123.

Raphael, B., Zhi, D., Tang, H., and Pevzner, P. 2004. A novel method for multiple alignment of sequences with repeated and shuffled elements. *Genome Research*, **14**, 2336–2346.

Rasmussen, M.D., and Kellis, M. 2012. Unified modelling of gene duplication, loss, and coalescence using a locus tree. *Genome Research*, **22**, 755–765.

Raubeson, L.A., and Jansen, R.K. 1992. Chloroplast DNA evidence on the ancient evolutionary split in vascular land plants. *Science*, **255**, 1697–1699.

Raup, D.M. 1985. Mathematical models of cladogenesis. *Paleobiology*, **11**(1), 42–52.

Raup, D.M., Gould, S.J., Schopf, T.J.M., and Simberloff, D.S. 1973. Stochastic-models of phylogeny and the evolution of diversity. *Journal of Geology*, **81**, 525–542.

Reaz, R., Bayzid, M.S., and Rahman, M.S. 2014. Accurate phylogenetic tree reconstruction from quartets: a heuristic approach. *PLoS One*. doi: 10.1371/journal.pone.0104008.

Redelings, B.D., and Suchard, M.A. 2005. Joint Bayesian estimation of alignment and phylogeny. *Systematic Biology*, **54**(3), 401–418.

Reeck, G.R., de Haen, C., Teller, D.C., Doolitte, R., Fitch, W., Dickerson, R.E., Chambon, P., McLachlan, A.D., Margoliash, E., Jukes, T.H., and Zuckerkandl, E. 1987. "Homology" in proteins and nucleic acids: a terminology muddle and a way out of it. *Cell*, **50**, 667.

Rheindt, F.E., Fujita, M.K., Wilton, P.R., and Edwards, S.V. 2014. Introgression and phenotypic assimilation in Zimmerius flycatchers (Tyrannidae): population genetic and phylogenetic inferences from genome-wide SNPs. *Systematic Biology*, **63**(2), 134–152.

Rieseberg, L.H. 1997. Hybrid origins of plant species. *Annual Review of Ecology and Systematics*, **28**, 359–389.

Rivas, E., and Eddy, S.R. 2008. Probabilistic phylogenetic inference with insertions and deletions. *PLoS Computational Biology*, **4**(9), e1000172.

Rivas, E., and Eddy, S.R. 2015. Parameterizing sequence alignment with an explicit evolutionary model. *BMC Bioinformatics*, **16**(1).

Roch, S. 2006. A short proof that phylogenetic tree reconstruction by maximum likelihood is hard. *IEEE/ACM Transactions on Computational Biology and Bioinformatics*, **3**(1), 92–94.

Roch, S. 2008. Sequence-length requirement for distance-based phylogeny reconstruction: breaking the polynomial barrier. Pages 729–738 of: *Proceedings of the Symposium on Foundations of Computer Science (FOCS)*.

Roch, S. 2010. Towards extracting all phylogenetic information from matrices of evolutionary distances. *Science*, **327**(5971), 1376–1379.

Roch, S., and Sly, A. 2016. Phase transition in the sample complexity of likelihood-based phylogeny inference. arXiv:1508.01964.

Roch, S., and Snir, S. 2013. Recovering the tree-like trend of evolution despite extensive lateral genetic transfer: a probabilistic analysis. *Journal of Computational Biology*, **20**, 93–112.

Roch, S., and Steel, M.A. 2015. Likelihood-based tree reconstruction on a concatenation of aligned sequence data sets can be statistically inconsistent. *Theoretical Population Biology*, **100**, 56–62.

Roch, S., and Warnow, T. 2015. On the robustness to gene tree estimation error (or lack thereof) of coalescent-based species tree methods. *Systematic Biology*, **64**(4), 663–676.

Ronquist, F., and Huelsenbeck, J.P. 2003. MrBayes 3: Bayesian phylogenetic inference under mixed models. *Bioinformatics*, **19**, 1572–1574.

Ronquist, F., Huelsenbeck, J.P., and Britton, T. 2004. Bayesian supertrees. Pages 193–224 of: Bininda-Emonds, O.R.P. (ed.), *Phylogenetic Supertrees*. New York: Springer.

Rose, D.J. 1970. Triangulated graphs and the elimination process. *Journal of Mathematical Analysis and Applications*, **32**, 597–609.

Rose, D.J., Tarjan, R.E., and Lueker, G.S. 1976. Algorithmic aspects of vertex elimination on graphs. *SIAM Journal on Computing*, **5**(2), 266–283.

Rosenberg, N.A. 2002. The probability of topological concordance of gene trees and species trees. *Theoretical Population Biology*, **61**(2), 225–247.

Rosenberg, N.A. 2013. Discordance of species trees with their most likely gene trees: a unifying principle. *Molecular Biology and Evolution*, **30**(12), 2709–2013.

Roshan, U., and Livesay, D.R. 2006. Probalign: multiple sequence alignment using partition function posterior probabilities. *Bioinformatics*, **22**, 2715–2721.

Roshan, U., Moret, B.M.E., Warnow, T., and Williams, T.L. 2004. Rec-I-DCM3: a fast algorithmic technique for reconstructing large phylogenetic trees. In: *Proceedings of the IEEE Computational Systems Bioinformatics Conference (CSB)*.

Rusinko, J., and McParlon, M. 2017. Species tree estimation using neighbor joining. *Journal of Theoretical Biology*, **414**, 5–7.

Rzhetsky, A., and Nei, M. 1993. Theoretical foundation of the minimum evolution method of phylogenetic inference. *Molecular Biology and Evolution*, **10**(5), 1073–1095.

Sadreyev, R., and Grishin, N. 2003. COMPASS: a tool for comparison of multiple protein alignments with assessment of statistical significance. *Journal of Molecular Biology*, **326**, 317–336.

Saitou, N., and Nei, M. 1987. The neighbor-joining method: a new method for reconstructing phylogenetic trees. *Molecular Biology and Evolution*, **4**, 406–425.

Sanderson, M.J., and Shaffer, H.B. 2002. Troubleshooting molecular phylogenetic analyses. *Annual Review of Ecology and Systematics*, **33**(1), 49–72.

Sankoff, D. 1975. Minimal mutation trees of sequences. *SIAM Journal of Applied Mathematics*, **28**(1), 35–42.

Sankoff, D., and Blanchette, M. 1998. Multiple genome rearrangement and breakpoint phylogeny. *Journal of Computational Biology*, **5**(3), 555–570.

Sankoff, D., and Cedergren, R.J. 1983. Simultaneous comparison of three or more sequences related by a tree. Pages 253–264 of: *Time Warps, String Edits, and Macromolecules: The Theory and Practice of Sequence Comparison*. Boston, MA.

Sankoff, D., and Rousseau, P. 1975. Locating the vertices of a Steiner tree in an arbitrary metric space. *Mathematical Programming*, **9**, 240–246.

Sankoff, D., Leduc, G., Antoine, N., Paquin, B., Lang, B.F., and Cedergren, R. 1992. Gene order comparisons for phylogenetic inference: evolution of the mitochondrial genome. *Proceedings of the National Academy of Sciences (USA)*, **89**(14), 6575–6579.

Sankoff, D., Abel, Y., and Hein, J. 1994. A tree- a window- a hill; generalization of nearest-neighbor interchange in phylogenetic optimization. *Journal of Classification*, **11**(2), 209–232.

Sattath, S., and Tversky, A. 1977. Additive similarity trees. *Psychometrika*, **42**(3), 319–345.

Schrempf, D., Minh, B.Q., De Maio, N., von Haeseler, A., and Kosiol, C. 2016. Reversible polymorphism-aware phylogenetic models and their application to tree inference. *Journal of Theoretical Biology*, **407**, 362–370.

Schwartz, S., Kent, W.J., Smit, A., Zhang, Z., Baertsch, R., Hardison, R.C., Haussler, D., and Miller, W. 2003. Human–mouse alignments with BLASTZ. *Genome Research*, **13**(1), 103–107.

Schwarz, G. 1978. Estimating the dimension of a model. *Annals of Statistics*, **6**, 461–464.

Schwikowski, B., and Vingron, M. 1997. The deferred path heuristic for the generalized tree alignment problem. *Journal of Computational Biology*, **4**(3), 415–431.

Schwikowski, B., and Vingron, M. 2003. Weighted sequence graphs: boosting iterated dynamic programming using locally suboptimal solutions. *Discrete Applied Mathematics*, **127**(1), 95–117.

Semple, C., and Steel, M.A. 2000. A supertree method for rooted trees. *Discrete Applied Mathematics*, **105**(13), 147–158.

Sevillya, G., Frenkel, Z., and Snir, S. 2016. Triplet MaxCut: a new toolkit for rooted supertree. *Methods in Ecology and Evolution*, **7**(11), 1359–1365.

Shao, M., and Moret, B.M.E. 2016. On computing breakpoint distances for genomes with duplicate genes. Pages 189–203 of: Singh, M. (ed.), *Proceedings of the 28th Annual Conference for Research in Computational Molecular Biology (RECOMB 2016)*. New York: Springer.

Shigezumi, T. 2006. Robustness of greedy type minimum evolution algorithms. Pages 815–821 of: *Proceedings of the International Conference on Computational Science*. Berlin: Springer.

Siepel, A., and Haussler, D. 2004. Combining phylogenetic and hidden Markov models in biosequence analysis. *Journal of Computational Biology*, **11**(2–3), 413–428.

Sievers, F., Wilm, A., Dineen, D., Gibson, T.J., Karplus, K., Li, W., Lopez, R., McWilliams, H., Remmert, M., Soding, J., Thompson, J.D., and Higgins, D. 2011. Fast, scalable generation of high-quality protein multiple sequence alignments using Clustal Omega. *Molecular Systems Biology*, **7**.

Sievers, F., Dineen, D., Wilm, A., and Higgins, D.G. 2013. Making automated multiple alignments of very large numbers of protein sequences. *Bioinformatics*, **29**(8), 989–995.

Sjölander, K. 2004. Phylogenomic inference of protein molecular function: advances and challenges. *Bioinformatics*, **20**(2), 170–179.

Sjölander, K., Karplus, K., Brown, M.P., Hughey, R., Krogh, A., Mian, I.S., and Haussler, D. 1996. Dirichlet mixtures: a method for improved detection of weak but significant protein sequence homology. *Computing Applications in the Biosciences (CABIOS)*, **12**, 327–45.

Sjölander, K., Datta, R.S., Shen, Y., and Shoffner, G.M. 2011. Ortholog identification in the presence of domain architecture rearrangement. *Briefings in Bioinformatics*, **12**(5), 413–422.

Smith, J.V., Braun, E.L., and Kimball, R.T. 2013. Ratite non-monophyly: independent evidence from 40 novel loci. *Systematic Biology*, **62**(1), 35–49.

Smith, S.A., Beaulieu, J.M., and Donoghue, M.J. 2009. Mega-phylogeny approach for comparative biology: an alternative to supertree and supermatrix approaches. *BMC Evolutionary Biology*, **9**, 37.

Smith, T.F., and Waterman, M.S. 1981. Identification of common molecular subsequences. *Journal of Molecular Biology*, **147**, 195–197.

Snir, S., and Rao, S. 2006. Using Max Cut to enhance rooted trees consistency. *IEEE/ACM Transactions on Computational Biology and Bioinformatics*, **3**(4), 323–333.

Snir, S., and Rao, S. 2010. Quartets MaxCut: a divide and conquer quartets algorithm. *IEEE/ACM Transactions on Computational Biology and Bioinformatics*, **7**(4), 704–718.

Snir, S., Warnow, T., and Rao, S. 2008. Short quartet puzzling: a new quartet-based phylogeny reconstruction algorithm. *Journal of Computational Biology*, **15**(1), 91–103.

Söding, J. 2005. Protein homology detection by HMM–HMM comparison. *Bioinformatics*, **21**(7), 951–960.

Sokal, R., and Michener, C. 1958. A statistical method for evaluating systematic relationships. *University of Kansas Science Bulletin*, **38**, 1409–1438.

Solís-Lemus, C., and Ané, C. 2016. Inferring phylogenetic networks with maximum pseudolikelihood under incomplete lineage sorting. *PLoS Genetics*, **12**(3), e1005896.

Spencer, M., Bordalejo, B., Wang, L.-S., Barbrook, A.C., Mooney, L.R., Robinson, P., Warnow, T., and Howe, C.J. 2003. Gene order analysis reveals the history of *The Canterbury Tales* manuscripts. *Computers and the Humanities*, **37**(1), 97–109.

Springer, M.S., and Gatesy, J. 2016. The gene tree delusion. *Molecular Phylogenetics and Evolution*, **94**, 1–33.

St. John, K. 2016. Review paper: the shape of phylogenetic treespace. *Systematic Biology*. doi:10.1093/sysbio/syw025.

St. John, K., Warnow, T., Moret, B.M.E., and Vawter, L. 2003. Performance study of phylogenetic methods: (unweighted) quartet methods and neighbor-joining. *Journal of Algorithms*, **48**, 173–193.

Stadler, T., and Degnan, J.H. 2012. A polynomial time algorithm for calculating the probability of a ranked gene tree given a species tree. *Algorithms for Molecular Biology*, **7**(7).

Stamatakis, A. 2006. RAxML-VI-HPC: Maximum likelihood-based phylogenetic analyses with thousands of taxa and mixed models. *Bioinformatics*, **22**, 2688–2690.

Steel, M.A. 1992. The complexity of reconstructing trees from qualitative characters and subtrees. *Journal of Classification*, **9**, 91–116.

Steel, M.A. 1994a. The maximum likelihood point for a phylogenetic tree is not unique. *Systematic Biology*, **43**(4), 560–564.

Steel, M.A. 1994b. Recovering a tree from the leaf colourations it generates under a Markov model. *Applied Mathematics Letters*, **7**, 19–24.

Steel, M.A. 2013. Consistency of Bayesian inference of resolved phylogenetic trees. *Journal of Theoretical Biology*, **336**, 246–249.

Steel, M.A., and Gascuel, O. 2006. Neighbor-joining revealed. *Molecular Biology and Evolution*, **23**(11), 1997–2000.

Steel, M.A., and Rodrigo, A. 2008. Maximum likelihood supertrees. *Systematic Biology*, **57**(2), 243–250.

Steel, M.A., and Warnow, T. 1993. Kaikoura tree theorems: the maximum agreement subtree problem. *Information Processing Letters*, **48**, 77–82.

Steel, M.A., Linz, S., Huson, D.H., and Sanderson, M.J. 2013. Identifying a species tree subject to random lateral gene transfer. *Journal of Theoretical Biology*, **322**, 81–93.

Stockham, C., Wang, L.-S., and Warnow, T. 2002. Postprocessing of phylogenetic analysis using clustering. *Bioinformatics* **18**(Supp.1), i285–i293.

Stolzer, M., Lai, H., Xu, M., Sathaye, D., Vernot, B., and Durand, D. 2012. Inferring duplications, losses, transfers and incomplete lineage sorting with nonbinary species trees. *Bioinformatics*, **28**(18), i409–i415.

Stoye, J., Evers, D., and Meyer, F. 1998. Rose: generating sequence families. *Bioinformatics*, **14**(2), 157–163.

Strimmer, K., and Moulton, V. 2000. Likelihood analysis of phylogenetic networks using directed graphical models. *Molecular Biology and Evolution*, **17**(6), 875–881.

Strimmer, K, and von Haeseler, A. 1996. Quartet puzzling: a quartet maximim-likelihood method for reconstructing tree topologies. *Molecular Biology and Evolution*, **13**(7), 964–969.

Studier, J.A., and Keppler, K.L. 1988. A note on the neighbor-joining algorithm of Saitou and Nei. *Molecular Biology and Evolution*, **5**(6), 729–731.

Sturtevant, A., and Dobzhansky, T. 1936. Inversions in the third chromosome of wild races of *Drosophila pseudoobscura* and their use in the study of the history of the species. *Proceedings of the National Academy of Sciences (USA)*, **22**, 448–450.

Subramanian, A.R., Kaufmann, M., and Morgenstern, B. 2008. DIALIGN-TX: greedy and progressive approaches for segment-based multiple sequence alignment. *Algorithms for Molecular Biology*, **3**, 6.

Suchard, M.A., and Redelings, B.D. 2006. BAli-Phy: simultaneous Bayesian inference of alignment and phylogeny. *Bioinformatics*, **22**, 2047–2048.

Sugiura, N. 1978. Further analysis of the data by Akaike's information criterion and the finite corrections. *Communications in Statistics: Theory and Methods*, **7**(1), 13–26.

Sullivan, J., and Joyce, P. 2005. Model selection in phylogenetics. *Annual Review of Ecology, Evolution, and Systematics*, **36**, 445–466.

Sullivan, J., and Swofford, D.L. 1997. Are guinea pigs rodents? The importance of adequate models in molecular phylogenetics. *Journal of Mammalian Evolution*, **4**, 77–86.

Sun, K., Meiklejohn, K.A., Faircloth, B.C., Glenn, T.C., Braun, E.L., and Kimball, R.T. 2014. The evolution of peafowl and other taxa with ocelli (eyespots): a phylogenomic approach. *Proceedings of the Royal Society of London (B): Biological Sciences*, **281**(1790), 20140823.

Swenson, K.M., Doroftei, A., and El-Mabrouk, N. 2012a. Gene tree correction for reconciliation and species tree inference. *Algorithms for Molecular Biology*, **7**(1), 1–11.

Swenson, M.S., Barbançon, F., Linder, C.R., and Warnow, T. 2010. A simulation study comparing supertree and combined analysis methods using SMIDGen. *Algorithms for Molecular Biology*, **5**, 8.

Swenson, M.S., Suri, R., Linder, C.R., and Warnow, T. 2011. An experimental study of Quartets MaxCut and other supertree methods. *Algorithms for Molecular Biology*, **6**, 7.

Swenson, M.S., Suri, R., Linder, C.R., and Warnow, T. 2012b. SuperFine: fast and accurate supertree estimation. *Systematic Biology*, **61**(2), 214–227.

Swofford, D.L. 2002. *PAUP*: Phylogenetic Analysis Using Parsimony (* And Other Methods) Ver. 4*. Sunderland, MA: Sinauer Associates.

Swofford, D.L., Olson, G.J., Waddell, P.J., and Hillis, D.M. 1996. *Phylogenetic Inference*. Second edition. Sunderland, MA: Sinauer Associates.

Syvanen, M. 1985. Cross-species gene transfer; implications for a new theory of evolution. *Journal of Theoretical Biology*, **112**, 333–343.

Szöllősi, G.J., Boussau, B., Abby, S.S., Tannier, E., and Daubin, V. 2012. Phylogenetic modeling of lateral gene transfer reconstructs the pattern and relative timing of speciations. *Proceedings of the National Academy of Sciences (USA)*, **109**, 17513–17518.

Szöllősi, G.J., Rosikiewicz, W., Boussau, B., Tannier, E., and Daubin, V. 2013. Efficient exploration of the space of reconciled gene trees. *Systematic Biology*, **62**(6), 901–912.

Szöllősi, G.J., Tannier, E., Daubin, V., and Boussau, B. 2015. The inference of gene trees with species trees. *Systematic Biology*, **64**, E42–E62.

Takahashi, H., and Matsuyama, A. 1980. An approximate solution for the Steiner problem in graphs. *Mathematica Japonica*, **24**, 463–470.

Talavera, G., and Castresana, J. 2007. Improvement of phylogenies after removing divergent and ambiguously aligned blocks from protein sequence alignments. *Systematic Biology*, **56**, 564–577.

Taly, J.-F., Magis, C., Bussotti, G., Chang, J.-M., Tommaso, P.D., Erb, I., Espinosa-Carrasco, J., Kemena, C., and Notredame, C. 2011. Using the T-Coffee package to build multiple sequence alignments of protein, RNA, DNA sequences and 3D structures. *Nature Protocols*, **6**, 1669–1682.

Tan, G., Muffato, M., Ledergerber, C., Herrero, J., Goldman, N., Gil, M., and Dessimoz, C. 2015. Current methods for automated filtering of multiple sequence alignments frequently worsen single-gene phylogenetic inference. *Systematic Biology*, **64**(5), 778–791.

Tang, J., and Moret, B.M.E. 2003. Scaling up accurate phylogenetic reconstruction from gene-order data. *Bioinformatics*, 19 (Suppl. 1), i305–i312.

Tarjan, R.E. 1985. Decomposition by clique separators. *Discrete Mathematics*, **55**, 221–232.

Tavaré, S. 1986. Some probabilistic and statistical problems in the analysis of DNA sequences. Pages 57–86 of: *Lectures on Mathematics in the Life Sciences*, vol. 17. Providence, RI: American Mathematical Society.

Taylor, M.S., Kai, C., Kawai, J., Carninci, P., Hayashizaki, Y., and Semple, C.A.M. 2006. Heterotachy in mammalian promoter evolution. *PLoS Genetics*, **2**(4), e30.

Than, C., and Nakhleh, L.. 2009. Species tree inference by minimizing deep coalescences. *PLoS Computational Biology*, **5**, 31000501.

Thompson, J., Plewniak, F., and Poch, O. 1999. BAliBASE: a benchmark alignments database for the evaluation of multiple sequence alignment programs. *Bioinformatics*, **15**, 87–88. Extended collection of benchmarks is available at *www-bio3d-igbmc.u-strasb.fr/balibase/*.

Thompson, J.D., Higgins, D.G., and Gibson, T.J. 1994. CLUSTAL W: improving the sensitivity of progressive multiple sequence alignment through sequence weighting, position-specific gap penalties and weight matrix choice. *Nucleic Acids Research*, **22**, 4673–4680.

Thorley, J.L., and Wilkinson, M. 2003. A view of supertree methods. *DIMACS Series in Discrete Mathematics and Theoretical Computer Science*, **61**, 185–194.

Thorne, J.L., Kishino, H., and Felsenstein, J. 1991. An evolutionary model for maximum likelihood alignment of DNA sequences. *Journal of Molecular Evolution*, **33**(2), 114–124.

Thorne, J.L., Kishino, H., and Felsenstein, J. 1992. Inching toward reality: an improved likelihood model of sequence evolution. *Journal of Molecular Evolution*, **34**(1), 3–16.

Toth, A., Hausknecht, A., Krisai-Greilhuber, I., Papp, T., Vagvolgyi, C., and Nagy, L.G. 2013. Iteratively refined guide trees help improving alignment and phylogenetic inference in the mushroom family *Bolbitiaceae*. *PLoS One*, **8**(2), e56143.

Tuffley, C., and Steel, M.A. 1997. Links between maximum likelihood and maximum parsimony under a simple model of site substitution. *Bulletin of Mathematical Biology*, **59**, 581–607.

Tukey, J.W. 1997. *Exploratory Data Analysis*. Reading, MA: Addison-Wesley.

Ullah, I., Parviainen, P., and Lagergren, J. 2015. Species tree inference using a mixture model. *Molecular Biology and Evolution*, **32**(9), 2469–2482.

Uricchio, L.H., Warnow, T., and Rosenberg, N.A. 2016. An analytical upper bound on the number of loci required for all splits of a species tree to appear in a set of gene trees. *BMC Bioinformatics*, **17**(14), 241.

Vach, W. 1989. Least squares approximation of additive trees. Pages 230–238 of: Opitz, O. (ed.), *Conceptual and Numerical Analysis of Data*. Berlin: Springer.

Vachaspati, P., and Warnow, T. 2015. ASTRID: Accurate Species TRees from Internode Distances. *BMC Genomics*, **16**(Suppl 10), S3.

Vachaspati, P., and Warnow, T. 2016. FastRFS: fast and accurate Robinson–Foulds Supertrees using constrained exact optimization. *Bioinformatics*. doi: 10.1093/bioinformatics/btw600.

Varón, A., Vinh, L.S., Bomash, I., and Wheeler, W.C. 2007. *POY Software*. Documentation by A. Varon, L.S. Vinh, I. Bomash, W. Wheeler, K. Pickett, I. Temkin, J. Faivovich, T. Grant, and W.L. Smith. Available for download at *www.amnh.org/our-research/ computational-sciences/research/projects/systematic-biology/poy*.

Vernot, B., Stolzer, M., Goldman, A., and Durand, D. 2008. Reconciliation with non-binary species trees. *Journal of Computational Biology*, **15**(8), 981–1006.

Vinga, S., and Almeida, J. 2003. Alignment-free sequence comparison: a review. *Bioinformatics*, **19**(4), 513–523.

Vingron, M., and Argos, P. 1991. Motif recognition and alignment for many sequences by comparison of dot-matrices. *Journal of Molecular Biology*, **218**, 33–43.

Vingron, M., and von Haeseler, A. 1997. Towards integration of multiple alignment and phylogenetic tree construction. *Journal of Computational Biology*, **4**(1), 22–34.

Vinh, L.S., and von Haeseler, A. 2005. Shortest triplet clustering: reconstructing large phylogenies using representative sets. *BMC Bioinformatics*, **6**(1), 1–14.

Vitti, J.J., Grossman, S.R., and Sabeti, P.C. 2013. Detecting natural selection in genomic data. *Annual Review of Genetics*, **47**, 97–120.

Vogt, H.G. 1895. *Leçons sur la résolution algébrique des équations*. Nony.

Wägele, J.W., and Mayer, C. 2007. Visualizing differences in phylogenetic information content of alignments and distinction of three classes of long-branch effects. *BMC Evolutionary Biology*, **7**, 147.

Wakeley, J. 2009. *Coalescent Theory: An Introduction*. Greenwood Village, CO: Roberts & Company.

Wallace, I.M., O'Sullivan, O., Higgins, D.G., and Notredame, C. 2006. M-Coffee: combining multiple sequence alignment methods with T-Coffee. *Nucleic Acids Research*, **34**, 1692–1699.

Wang, L., and Gusfield, D. 1997. Improved approximation algorithms for tree alignment. *Journal of Algorithms*, **25**, 255–273.

Wang, L., and Jiang, T. 1994. On the complexity of multiple sequence alignment. *Journal of Computational Biology*, **1**(4), 337–348.

Wang, L., Jiang, T., and Lawler, E.L. 1996. Approximation algorithms for tree alignment with a given phylogeny. *Algorithmica*, **16**, 302–315.

Wang, L., Jiang, T., and Gusfield, D. 2000. A more efficient approximation scheme for tree alignment. *SIAM Journal of Computing*, **30**(1), 283–299.

Wang, L.-S., and Warnow, T. 2001. Estimating true evolutionary distances between genomes. In: *Proceedings of the 33rd Symposium on the Theory of Computing (STOC01)*. New York: ACM Press.

Wang, L.-S., and Warnow, T. 2006. Reconstructing chromosomal evolution. *SIAM Journal on Computing*, **36**(1), 99–131.

Wang, L.-S., Warnow, T., Moret, B.M.E., Jansen, R.K., and Raubeson, L.A. 2006. Distance-based genome rearrangement phylogeny. *Journal of Molecular Evolution*, **63**(4), 473–483.

Wang, N., Braun, E.L., and Kimball, R.T. 2012. Testing hypotheses about the sister group of the Passeriformes using an independent 30-locus data set. *Molecular Biology and Evolution*, **29**(2), 737–750.

Wang, N., Kimball, R.T., Braun, E.L., Liang, B., and Zhang, Z. 2016. Ancestral range reconstruction of Galliformes: the effects of topology and taxon sampling. *Journal of Biogeography*.

Wareham, H.T. 1995. A simplified proof of the NP- and MAX SNP-hardness of multiple sequence tree alignment. *Journal of Computational Biology*, **2**(4), 509–514.

Warnow, T. 1993. Constructing phylogenetic trees efficiently using compatibility criteria. *New Zealand Journal of Botany*, **31**(3), 239–248.

Warnow, T. 2015. Concatenation analyses in the presence of incomplete lineage sorting. *PLoS Currents Tree of Life*. Edition 1. doi: 10.1371/currents.tol.8d41ac0f13d1abedf4c4a59f5d17b1f7.

Warnow, T., Moret, B.M.E., and St. John, K. 2001. Absolute convergence: true trees from short sequences. Pages 186–195 of: *Proceedings of ACM-SIAM Symposium on Discrete Algorithms (SODA 01)*. Philadelphia, PA: Society for Industrial and Applied Mathematics (SIAM).

Watterson, G., Ewens, W., Hall, T., and Morgan, A. 1982. The chromosome inversion problem. *Journal of Theoretical Biology*, **99**(1), 1–7.

Wehe, A., Bansal, M.S., Burleigh, J.G., and Eulenstein, O. 2008. DupTree: a program for large-scale phylogenetic analyses using gene tree parsimony. *Bioinformatics*, **24**(13), 1540–1541.

Wen, D., Yu, Y., and Nakhleh, L. 2016. Bayesian inference of reticulate phylogenies under the multispecies network coalescent. *PLoS Genetics*, **12**(5), e1006006.

Wheeler, T., and Kececioglu, J. 2007. Multiple alignment by aligning alignments. Pages 559–568 of: *Proceedings of the 15th ISCB Conference on Intelligent Systems for Molecular Biology*.

Wheeler, W.C. 2015. Phylogenetic network analysis as a parsimony optimization problem. *BMC Bioinformatics*, **16**, 296.

Wheeler, W.C., Aagesen, L., Arango, C.P., Faivovich, J., Grant, T., D'Haese, C., Janies, D., Smith, W.L., Varón, A., and Giribet, G. 2006. *Dynamic Homology and Phylogenetic Systematics: A Unified Approach Using POY*. New York: American Museum of Natural History.

Whelan, S., and Goldman, N. 2001. A general empirical model of protein evolution derived from multiple protein families using a maximum-likelihood approach. *Molecular Biology and Evolution*, **18**(5), 691–699.

Wickett, N.J., Mirarab, S., Nguyen, N., Warnow, T., Carpenter, E., Matasci, N., Ayyampalayam, S., Barker, M.S., Burleigh, J.G., Gitzendanner, M.A., et al. 2014. Phylotranscriptomic analysis of the origin and early diversification of land plants. *Proceedings of the National Academy of Sciences (USA)*, **111**(45), E4859–E4868.

Wilkinson, M. 1994. Common cladistic information and its consensus representations: reduced Adams and reduced cladistic consensus trees and profiles. *Systematic Biology*, **43**, 343–368.

Wilkinson, M. 1995. More on reduced consensus methods. *Systematic Biology*, **44**(3), 435–439.

Wilkinson, M., and Cotton, J.A. 2006. Supertree methods for building the Tree of Life: divide-and-conquer approaches to large phylogenetic problems. *Reconstructing the Tree of Life: Taxonomy and Systematics of Species Rich Taxa*, **61**.

Wilkinson, M., Cotton, J.A., Creevey, C., Eulenstien, O., Harris, S.R., Lapointe, F.-J., Levasseur, C., McInerney, J.O., Pisani, D., and Thorley, J.L. 2005. The shape of supertrees to come: tree shape related properties of fourteen supertree methods. *Systematic Biology*, **54**(3), 419–431.

Willson, S.J. 2004. Constructing rooted supertrees using distances. *Bulletin of Mathematical Biology*, **66**(6), 1755–1783.

Willson, S.J. 2012. Tree-average distances on certain phylogenetic networks have their weights uniquely determined. *Algorithms for Molecular Biology*, **7**(1), 1.

Willson, S.J. 2013. Reconstruction of certain phylogenetic networks from their tree-average distances. *Bulletin of Mathematical Biology*, **75**(10), 1840–1878.

Wright, S. 1931. Evolution in Mendelian populations. *Genetics*, **16**, 97–159.

Wu, Y. 2012. Coalescent-based species tree inference from gene tree topologies under incomplete lineage sorting by maximum likelihood. *Evolution*, **66**, 763–775.

Wu, Y. 2013. An algorithm for constructing parsimonious hybridization networks with multiple phylogenetic trees. In: *Annual International Conference on Computational Molecular Biology (RECOMB 2013)*.

Xin, L., Ma, B., and Zhang, K. 2007. A new quartet approach for reconstructing phylogenetic trees: Quartet Joining method. Pages 40–50 of: *Proceedings, Computing and Combinatorics (COCOON) 2007*. Berlin: Springer.

Yamada, K., Tomii, K., and Katoh, K. 2016. Application of the MAFFT sequence alignment program to large data: reexamination of the usefulness of chained guide trees. *Bioinformatics*, **32**(21), 3246–3251.

Yancopoulos, S., Attie, O., and Friedberg, R. 2005. Efficient sorting of genomic permutations by translocation, inversion and block interchange. *Bioinformatics*, **21**(16), 3340–3346.

Yang, Z. 1995. A space-time process model for the evolution of DNA sequences. *Genetics*, **139**, 993–1005.

Yang, Z. 2009. *Computational Molecular Evolution*. Oxford: Oxford University Press.

Yang, Z. 2014. *Molecular Evolution: A Statistical Approach.* Oxford: Oxford University Press.

Yang, Z., and Goldman, N. 1997. Are the big indeed easy? *Trends in Ecology & Evolution*, **12**(9), 357.

Yang, Z., Nielsen, R., Goldman, N., and Pedersen, A.M.K. 2000. Codon-substitution models for heterogeneous selection pressure at amino acid sites. *Genetics*, **155**(1), 431–449.

Yona, G., and Levitt, M. 2002. Within the twilight zone: a sensitive profile–profile comparison tool based on information theory. *Journal of Molecular Biology*, **315**, 1257–1275.

Yu, Y., and Nakhleh, L. 2015a. A distance-based method for inferring phylogenetic networks. Pages 378–389 of: *Proceedings International Symposium on Bioinformatics Research and Applications (ISBRA)*. Berlin: Springer.

Yu, Y., and Nakhleh, L. 2015b. A maximum pseudo-likelihood approach to phylogenetic networks. *BMC Genomics*, **16**(10), 1–10.

Yu, Y., Warnow, T., and Nakhleh, L. 2011a. Algorithms for MDC-based multi-locus phylogeny inference: beyond rooted binary gene trees on single alleles. *Journal of Computational Biology*, **18**, 1543–1559.

Yu, Y., Than, C., Degnan, J.H., and Nakhleh, L. 2011b. Coalescent histories on phylogenetic networks and detection of hybridization despite incomplete lineage sorting. *Systematic Biology*, **60**(2), 138–149.

Yu, Y., Degnan, J.H., and Nakhleh, L. 2012. The probability of a gene tree topology within a phylogenetic network with applications to hybridization detection. *PLoS Genetics*, **8**(4), e1002660.

Yu, Y., Ristic, N., and Nakhleh, L. 2013. Fast algorithms and heuristics for phylogenomics under ILS and hybridization. *BMC Bioinformatics*, **14**(Suppl 15), S6.

Yu, Y., Dong, J., Liu, K., and Nakhleh, L. 2014. Maximum likelihood inference of reticulate evolutionary histories. *Proceedings of the National Academy of Sciences (USA)*, **111**(46), 16448–16453.

Yule, G.U. 1924. A mathematical theory of evolution, based on the conclusions of Dr. J.C. Willis. *Philosophical Transactions of the Royal Society London, Series B*, **213**, 21–87.

Zharkikh, A., and Li, W.-H. 1993. Inconsistency of the maximum-parsimony method: the case of five taxa with a molecular clock. *Systematic Biology*, **42**, 113–125.

Zhou, H., and Zhou, Y. 2005. SPEM: improving multiple sequence alignment with sequence profiles and predicted secondary structures. *Bioinformatics*, **21**(18), 3615–3261.

Zhou, Y., Rodrigue, N., Lartillot, N., and Philippe, H. 2007. Evaluation of the models handling heterotachy in phylogenetic inference. *BMC Evolutionary Biology*, **7**, 206.

Zhu, J., Yu, Y., and Nakhleh, L. 2016. In the light of deep coalescence: revisiting trees within networks. arXiv preprint arXiv:1606.07350.

Zimmermann, T., Mirarab, S., and Warnow, T. 2014. BBCA: improving the scalability of *BEAST using random binning. *BMC Genomics*, **15**(Suppl 6), S11.

Zuckerman, D. 2006. Linear degree extractors and the inapproximability of max clique and chromatic number. Pages 681–690 of: *Proceedings 38th ACM Symposium on the Theory of Computing (STOC)*.

Zwickl, D.J., and Hillis, D.M. 2002. Increased taxon sampling greatly reduces phylogenetic error. *Systematic Biology*, **51**(4), 588–598.

Index

absolute fast converging methods, 101, 102, 167–170, 296
additive matrix, 10, 11, 13–15, 84–89, 93–101, 103, 152–154, 248, 282–283
   topology invariant neighborhood, 93
agreement subtrees, 116, 117
   maximum agreement subtree, 116
   maximum compatibility subtree, 116
Aho, Sagiv, Szymanski, and Ullman algorithm, 52, 53, 57, 140, 242, 243
algorithms, 304, 307–312, 318–320
   "Big-O" running time analysis, 304, 307–311
   design techniques, 24, 26, 285, 317–320
   divide-and-conquer methods, 24, 121, 254, 274, 279, 284–286, 289, 291, 296, 320
   dynamic programming, 318–320
   exhaustive search, 311, 312
   greedy algorithms, 311
   iteration, 24, 84, 91, 211, 214, 218–221, 254, 285, 286, 288–293
   pseudocode, 308, 309, 320
   recursive algorithms, 8, 12–14, 215, 254, 287, 319, 320
alignment-free phylogeny estimation, 102, 103, 289

Bayesian phylogenetics, 18, 20, 141, 159–160, 166–167, 172–173, 217, 224, 241, 247, 252, 258, 271, 277–278
   branch support, 159, 160, 166
   computational challenges, 17, 159, 160, 277–279
   detailed balance property, 159
   maximum *a posteriori* (MAP) tree, 159, 160, 166, 278
   MCMC techniques, 159, 166, 217, 277, 278
   MrBayes, 247
   posterior probability distribution, 160
   statistical consistency, 160
binary relation, 305
bipartition compatibility, 39–41
bipartition encoding of a tree, 21, 114
BLAST, 290

branch support calculation, 165, 299, 300
   Bayesian support values, 159, 166
   bootstrapping, 165, 166, 299, 300

centroid edge, xvii, 214, 218, 221, 233, 287–289, 298
character compatibility, 63, 69, 70
character data, 5, 61
clade compatibility, 33–35
cladogenesis models, 172, 173
   birth–death model, 18, 172
   Yule model, 172
clustering sets of trees, 117
coalescent-based species tree estimation, 102, 234–254, 268–270, 292–293
   *BEAST, 252–254
   anomaly zone, 239–241, 244
   ASTRAL, 246–248, 253, 256, 259, 270
   ASTRID, 102, 103, 247, 248, 253
   BBCA, 253, 254
   BUCKy, 247
   concatenation analysis, 238, 239
   fixed-length statistical consistency, 252, 253
   impact of gene tree estimation error, 248–250
   MDC, 256
   METAL, 250, 253
   MP-EST, 24, 243, 249, 253, 254, 292, 293
   NJst, 102, 103, 247, 248, 259
   quartet-tree methods, 129, 244–247, 250, 251
   site-based methods, 241, 250, 252
   SMRT-ML, 250, 253
   SNAPP, 250
   SRSTE, 243, 244
   summary methods, 129, 241–250
   SuperTriplets, 243
   SUSTE, 244, 245
   SVDquartets, 250–252
combinatorial counting, 307
consensus trees, 41, 44, 109–118, 159, 300
   asymmetric median tree, 110, 113–115
   characteristic tree, 115–117
   compatibility tree, 110, 112, 113

extended majority consensus, 112
greedy consensus tree, 110–112, 115, 278
local consensus tree, 118
majority consensus tree, 110–115, 122, 159
strict consensus tree, 41, 44, 110–112, 115, 122

DACTAL, 24, 132, 254, 286, 288–292
direct optimization, 216, 217
disk-covering methods (DCMs), 24, 254, 279–288, 290, 293, 296, 297
dissimilarity matrix, 13, 83
distance matrix, 83
distance-based tree estimation, 7–17, 21–23, 83–103, 140–141, 152–154, 167–170, 247–248, 278, 282–285, 296
    Agarwala et al. algorithm, 94, 96–100, 153, 284
    balanced minimum evolution, 95, 96, 100
    BioNJ*, 248
    branch length estimation, 94, 95
    Buneman Tree, 90, 96–100, 284
    computing distances under statistical models, 14, 83, 102, 103, 152–154, 161
    error tolerance, 89
    FastME, 17, 96, 97, 100–103, 140, 153, 248
    Four Point Condition, 11, 86, 87
    Four Point Method, 11–15, 87–89, 168–170
    impact of missing data, 103
    minimum evolution, 95
    Naive Quartet Method, 14, 15, 17, 18, 20–22, 24, 89, 90, 97–100, 165, 167, 168
    neighbor joining, 17, 18, 20–24, 90–92, 97–103, 165, 170, 248, 275, 283–285, 296
    optimization problems, 94, 95, 100
    safety radius, 96–101, 153, 154, 248
    statistical consistency, 92–94, 96–101, 153, 154
    UPGMA, 9, 16–18, 20–22, 24, 84–86, 91, 210, 215
    using DCMs, 282, 283, 285
Dobzhansky, xiii
dynamic programming, 63–68, 156, 157, 185, 187–192, 200–203, 256, 318–320

evolutionary diameter, 22

false discovery rate, 321
Felsenstein Zone tree, 161–165, 171

gene duplication and loss, 229, 230, 234, 254–258
    MixTreEm, 258
    Phyldog, 258
gene tree parsimony, 256, 258
genome rearrangements, 72, 102, 229, 230, 270–272, 296
genome-scale evolution, xv, 4, 23, 24, 102, 270–272
graph theory, 304, 305, 309, 312–317
    adjacency matrix and adjacency list, 309, 315
    directed acyclic graph (DAG), 196, 224, 225, 305
    Eulerian graph, 305, 313
    maximum clique, 312
    minimum vertex coloring, 249, 312–317

Hamming distance, 13, 17
Hasse Diagram, 33–35, 39, 40, 70, 305, 306
heterotachy, 170, 171
homology, xvii, 178–183, 194, 204, 205, 212–216, 224, 229, 230
homoplasy, xvii, 62
horizontal gene transfer, xv, 4, 23, 24, 61, 109, 234, 235, 259, 300
hybridization, 4, 23, 24, 61, 234, 300

incomplete lineage sorting, 102, 235–238, 241–254, 270
ingroup taxa, 44
insertions and deletions (indels), xvii, 4, 180

long branch attraction, 162, 164

maximum compatibility tree estimation, 62, 63, 69, 70, 72–77, 161, 163
    compatibility informative characters, 76, 77, 164
    positively misleading, 161, 163, 164
maximum likelihood tree estimation, 17, 18, 102, 128, 139–141, 147, 159–164, 171–172, 216, 218, 219, 238, 239, 249, 252, 253, 258, 275, 276, 278, 285, 296
    FastTree-2, 160, 279, 291
    heuristics, 158, 275, 276
    IQTree, 279
    nhPhyml, 279
    NP-hard, 17, 158
    PhyML, 279
    PoMo, 279
    RAxML, 139, 279, 291
    use within supertree methods, 128
    using DCMs, 296
maximum parsimony tree estimation, 17, 22, 23, 62–69, 73, 75–77, 127, 161–164, 171, 192, 275, 276, 278
    Fitch algorithm, 63–66
    heuristics, 17, 69, 275, 276
    missing data treatment, 127
    NP-hard, 17, 68
    parsimony informative characters, 76, 77, 137, 161
    positively misleading, 17, 21, 161, 163, 164, 171
    Sankoff algorithm, 66, 68
    small parsimony and large parsimony problems, 63, 68, 69
    statistical inconsistency, 17, 164
    using DCMs, 296
minimum spanning tree, 193
missing data, 71, 75, 103, 127, 128, 178, 248, 257, 334
multi-species coalescent model, 234–240, 243–246

multiple sequence alignment, xiii, xv, 4, 19, 24, 102, 178–233, 299
  aligning alignments, 207–209, 219, 226
  alignment error measurement, 180, 182, 183
  BAli-Phy, 217, 220, 224, 226, 229
  benchmarks, 178, 181
  Clustal, 210, 217
  co-estimation of alignments and trees, 211, 215–223, 291
  consensus alignment, 224–226
  consistency, 212–214, 224
  divide-and-conquer methods, 214–216
  FastSP, 183
  impact of guide tree, 212
  impact on tree estimation, 181, 300
  MAFFT, 211, 213, 215–217, 219, 220
  MAPGAPS, 214, 215
  mega-phylogeny, 214, 215
  Muscle, 211, 217, 219, 220
  OPAL, 219, 220
  optimization problems, 190, 191
  PAGAN, 204, 205, 211, 217, 228, 229
  PASTA, 24, 212, 214, 215, 218–223, 226, 229, 291
  phylogeny-aware, 211, 217, 228, 229
  Prank, 211, 217, 228, 229
  ProbCons, 211, 213, 217
  progressive alignment, 209–212, 226, 229
  PROMALS, 205, 213–215
  reference-based alignment, 204, 205
  SATé, 24, 212, 214, 215, 220, 229, 291, 292
  SATCHMO-JS, 214, 215, 218
  seed alignment, 206
  sequence evolution models, 18, 25, 103, 216, 217
  StatAlign, 217, 224–226
  statistical models, 194, 196–198, 200–206, 220, 221, 223, 226, 227
  Sum-of-Pairs Alignment, 190, 191
  T-Coffee, 217
  template-based alignment, 205, 206
  tree alignment, 190–194, 215, 216
  UPP, 214, 220, 228, 229

NNI, SPR, and TBR moves, 117, 275–277, 296
NP-hard problems, xvii, 17, 68, 69, 72, 74, 122, 158, 274–277, 283, 295, 299, 311–318, 320
  heuristic search strategies, 274–277, 299
  Karp reduction, 315–317
  NP-complete problems, xvii, 314, 316
  optimization, decision, and construction problems, 311–313

orthology, 224, 228, 255, 272
outgroup taxa, 44

pairwise sequence alignment, 180, 185–190, 227
  edit distances, 184
  local alignment, 227
  Needleman–Wunsch, 185, 187–190, 207, 227
  Smith–Waterman, 227
perfect phylogeny, 62, 63, 70–72, 74, 77
Pfam, 204
phylogenetic network, 23, 24, 259–267, 269, 270, 300
  data-display network, 263–267, 270
  evolutionary network, 235, 261–264, 267
phylogenetic placement, 206, 297
phylogenomics, 129, 234–246, 248–261, 263–265, 267, 270–272
polytomy, xvii, 116
post-tree analysis, 228
profile hidden Markov models, 194, 196–204, 206, 207, 220, 226
  Baum–Welch method, 204
  building profile HMMs, 203, 204
  database searches, 204
  ensembles of hidden Markov models, 220, 221, 223, 224
  Forward algorithm, 203
  HMMER, 205
  Viterbi algorithm, 201–203, 207
proof techniques, 307, 308
  proof by contradiction, 308
  proof by induction, 307, 308

quartet trees, 12, 13, 46, 53–58, 87–91, 118, 128–132, 165
  All Quartets Method, 12, 13, 53–57, 87, 89, 244, 245, 251
  dyadic closure method, 54–56, 153, 168–170
  Q(T), 12, 13
  quartet amalgamation methods, 53, 89, 90, 128, 129, 132, 139, 251, 274, 276
  quartet-based tree construction methods, 102, 128
  quartet compatibility, 54, 57
  Quartets MaxCut, 128, 135, 139
  short quartets, 54–56, 284, 286

Robinson–Foulds tree error, 19, 21, 22, 42, 46, 283
rogue taxon, 43, 44, 46
rooted tree compatibility, 52, 57

sample complexity, 145, 167
sensitivity, specificity, precision, and recall, 321, 322
sequence evolution models, xiv, xv, 4–8, 17, 19, 21, 24, 26, 29, 102, 140, 145–173, 180, 278, 279
  amino acid models, 150
  Cavender–Farris–Neyman (CFN), xiv, 4–11, 13–17, 91–94, 96–98, 102, 128, 145, 148, 153, 156, 158, 161–163
  codon models, 145
  computing the probability of a sequence dataset, 154, 156
  Felsenstein's Pruning Algorithm, 156–159

General Markov model, 149, 153, 155, 158, 161
Generalised Time Reversible (GTR) model, 19, 148, 149, 156, 158, 160
heterotachy, 170, 171
i.i.d. assumption, 16, 147, 149
Jukes–Cantor (JC69) model, 146–149, 153, 156–158, 160, 164, 167–170
Jukes-Cantor (JC69) model, 19
long branch attraction, 164
Markov property, 147
model selection, 151, 152, 296
No Common Mechanism model, 149, 158, 170, 171
rates-across-sites, 16, 149, 150, 172
similarity to coin tosses, 6
stationary assumption, 147
strict molecular clock, 163
Thorne–Kishino–Felsenstein 1991 (TKF91), 103
time-reversible, 147, 154, 158
sequence profiles, 194–198
sequencing technologies
 impact on phylogeny estimation, xiii, 268, 269
simulation study, 18–21, 23, 26, 41, 90, 146, 164, 165, 167, 173, 178, 180, 181, 238, 258, 267, 283, 328, 329
statistical binning, 249, 250
statistical consistency, xiv, 6, 13–18, 21, 24, 145–147, 153, 164, 167, 171, 278
statistical identifiability, 145–147
statistical significance, 328
strict molecular clock, 7–10, 15, 44, 84, 251–253
supertree methods, xv, 52, 56, 57, 109, 121–142, 245, 258, 279, 281, 285, 286, 288–290, 292
 Matrix Representation with Parsimony (MRP)
  missing data treatment, 127
 Asymmetric Median Supertree, 123, 124, 126
 compatibility supertree, 52, 56, 57, 122–126, 135
 distance-based, 140
 FastRFS, 126, 141, 256
 guenomu, 258, 270
 majority rule supertrees, 122
 Matrix Representation with Likelihood (MRL), 122, 128, 135
  missing data treatment, 128
 Matrix Representation with Parsimony (MRP), 122, 126–128, 135–139, 291
  MRP matrix, 126, 127
 maximum likelihood supertree, 141
 Maximum Quartet Support Supertree, 122, 128, 129, 245
 MinCut Supertree, 122, 140
 Quartet Median Tree, 122, 128, 129, 245, 246, 254
 quartet-based methods, 139
 Robinson–Foulds Supertree, 122, 124–126, 141
 Split-Constrained Quartet Support Supertree, 129–132, 247
  statistical aspects of, 132, 141

Strict Consensus Merger (SCM), 132–139, 285, 288–290
SuperFine, 132, 135–139, 289–291
SuperTriplets, 243
use within DCMs, 279, 281, 289–293
using rooted source trees, 139, 140

taxon sampling, 164, 165
transitive closure, 181, 204, 215, 306
tree compatibility, 57, 112
tree error rates, 18–22, 41, 42, 46
trees, 3, 29–46, 57
 binary tree, 4, 21
 binary vs. multifurcating, 29, 31, 38, 39, 42
 bipartitions of unrooted trees, 19, 36–38, 42
 caterpillar tree, 45, 46, 56, 164
 clade representations of rooted trees, 31, 32
 clades of rooted trees, 40
 comparing two unrooted trees using bipartitions, 38
 completely balanced tree, 45
 constructing a rooted tree from its clades, 33
 constructing an unrooted tree from its bipartition set, 39
 determining if a set of bipartitions is compatible, 40
 different graphical ways of representing trees, 30
 edges and branches, 5, 29
 fully resolved trees, 38
 homeomorphic subtree, 12, 45, 46
 leaves, 29
 MRCA, 29, 236, 237
 Newick notation, 30–32, 36, 37
 number of trees on *n* leaves, 43
 polytomy, 29, 57, 59, 87, 100, 133, 135
 rooted vs. unrooted, 29, 44, 45
 star tree, 42, 44, 87–90, 92, 100, 113, 115
 tree refinement, 38, 111, 112
 vertices, 29
treespace, 43, 117, 270, 296
triangle inequality, 14, 83
triangulated graphs, 279–283, 285–288, 293–296
 decompositions of, 280, 281, 287, 288, 294, 296
 definition, 279, 293
 perfect elimination ordering, 293–296
 short subtree graph, 282, 285, 286, 296
 threshold graph, 282, 283, 296
 use within DCMs, 280, 285, 289
triplet trees, 52, 53, 57, 118, 139, 140
 maximum triplet support problem, 139
 triplet tree compatibility problem, 57
two-phase phylogeny estimation, 178, 179, 216, 291

ultrametric matrix, 8, 9, 84, 85, 140, 141
unrooted tree compatibility, 51, 57, 125, 135

whole genome alignment, 229, 230
writing papers about computational methods, 327–330